$\mathcal{D}$

Model Selection and Multimodel Inference
Second Edition

Springer

New York
Berlin
Heidelberg
Barcelona
Hong Kong
London
Milan
Paris
Singapore
Tokyo

Kenneth P. Burnham   David R. Anderson

Model Selection and Multimodel Inference

A Practical Information-Theoretic Approach

Second Edition

With 31 Illustrations

Springer

Kenneth P. Burnham
David R. Anderson
Colorado Cooperative Fish
  and Wildlife Research Unit
Colorado State University
Fort Collins, CO 80523-1484
USA

Cover Illustration: The cover was assembled from photos of the yellow-bellied toad (*Bombina varie-gata*) taken by Jonas Barandum as part of his Ph.D. program at the University of Zurich. These toads have individually identifiable patterns on their abdomen from a few weeks following metamorphosis that remain unchanged until death. Two pairs are duplicates—but which two?
Cover photographs by Dr. Jonas Barandum, St. Gallen, Switzerland. Cover design by Kenton Allred.

Library of Congress Cataloging-in-Publication Data
Burnham, Kenneth P.
    Model selection and multimodel inference : a practical information-theoretic approach
    / Kenneth P. Burnham, David R. Anderson.—2nd ed.
        p.   cm.
    Rev. ed. of: Model selection and inference. © 1998.
    Includes bibliographical references (p.).
    ISBN 0-387-95364-7 (alk. paper)
    1. Biology—Mathematical models.   2. Mathematical statistics.   I. Burnham, Kenneth P.
    Model selection and inference.   II. Title.
QH323.5   B87   2002
570′.1′51—dc21   2001057677

ISBN 0-387-95364-7   Printed on acid-free paper.

Printed in the United States of America.

9 8 7 6 5 4 3 2   SPIN 10853081

www.springer-ny.com

Springer-Verlag   New York Berlin Heidelberg
A member of BertelsmannSpringer Science+Business Media GmbH

To my mother and father, Lucille R. (deceased) and J. Calvin Burnham (deceased), and my son and daughter, Shawn P. and Sally A. Burnham

To my parents, Charles R. (deceased) and Leta M. Anderson; my wife, Dalene F. Anderson; and my daughters, Tamara E. and Adrienne M. Anderson

Preface

We wrote this book to introduce graduate students and research workers in various scientific disciplines to the use of information-theoretic approaches in the analysis of empirical data. These methods allow the data-based selection of a "best" model and a ranking and weighting of the remaining models in a pre-defined set. Traditional statistical inference can then be based on this selected best model. However, we now emphasize that information-theoretic approaches allow formal inference to be based on more than one model (multimodel inference). Such procedures lead to more robust inferences in many cases, and we advocate these approaches throughout the book.

The second edition was prepared with three goals in mind. First, we have tried to improve the presentation of the material. Boxes now highlight essential expressions and points. Some reorganization has been done to improve the flow of concepts, and a new chapter has been added. Chapters 2 and 4 have been streamlined in view of the detailed theory provided in Chapter 7. Second, concepts related to making formal inferences from more than one model (multimodel inference) have been emphasized throughout the book, but particularly in Chapters 4, 5, and 6. Third, new technical material has been added to Chapters 5 and 6. Well over 100 new references to the technical literature are given. These changes result primarily from our experiences while giving several seminars, workshops, and graduate courses on material in the first edition. In addition, we have done substantially more thinking about the issue and reading the literature since writing the first edition, and these activities have led to further insights.

Information theory includes the celebrated Kullback–Leibler "distance" between two models (actually, probability distributions), and this represents a fundamental quantity in science. In 1973, Hirotugu Akaike derived an estimator of the (relative) expectation of Kullback–Leibler distance based on Fisher's maximized log-likelihood. His measure, now called *Akaike's information criterion* (AIC), provided a new paradigm for model selection in the analysis of empirical data. His approach, with a fundamental link to information theory, is relatively simple and easy to use in practice, but little taught in statistics classes and far less understood in the applied sciences than should be the case.

We do not accept the notion that there is a simple "true model" in the biological sciences. Instead, we view modeling as an exercise in the approximation of the explainable information in the empirical data, in the context of the data being a sample from some well-defined population or process. Rexstad (2001) views modeling as a fabric in the tapestry of science. Selection of a best approximating model represents the inference from the data and tells us what "effects" (represented by parameters) can be supported by the data. We focus on Akaike's information criterion (and various extensions) for selection of a parsimonious model as a basis for statistical inference. Model selection based on information theory represents a quite different approach in the statistical sciences, and the resulting selected model may differ substantially from model selection based on some form of statistical null hypothesis testing.

We recommend the information-theoretic approach for the analysis of data from observational studies. In this broad class of studies, we find that all the various hypothesis-testing approaches have no theoretical justification and may often perform poorly. For classic experiments (control–treatment, with randomization and replication) we generally support the traditional approaches (e.g., analysis of variance); there is a very large literature on this classic subject. However, for complex experiments we suggest consideration of fitting explanatory models, hence on estimation of the size and precision of the treatment effects and on parsimony, with far less emphasis on "tests" of null hypotheses, leading to the arbitrary classification "significant" versus "not significant." Instead, a strength of evidence approach is advocated.

We do not claim that the information-theoretic methods are always the very best for a particular situation. They do represent a unified and rigorous theory, an extension of likelihood theory, an important application of information theory, and they are objective and practical to employ across a very wide class of empirical problems. Inference from multiple models, or the selection of a single "best" model, by methods based on the Kullback–Leibler distance are almost certainly better than other methods commonly in use now (e.g., null hypothesis testing of various sorts, the use of R^2, or merely the use of just one available model). In particular, subjective data dredging leads to overfitted models and the attendant problems in inference, and is to be strongly discouraged, at least in more confirmatory studies.

Parameter estimation has been viewed as an optimization problem for at least eight decades (e.g., maximize the log-likelihood or minimize the residual sum of squared deviations). Akaike viewed his AIC and model selection as "... a natural extension of the classical maximum likelihood principle." This extension brings model selection and parameter estimation under a common framework—optimization. However, the paradigm described in this book goes beyond merely the computation and interpretation of AIC to select a parsimonious model for inference from empirical data; it refocuses increased attention on a variety of considerations and modeling prior to the actual analysis of data. Model selection, under the information-theoretic approach presented here, attempts to identify the (likely) best model, orders the models from best to worst, and produces a weight of evidence that each model is really the best as an inference.

Several methods are given that allow model selection uncertainty to be incorporated into estimates of precision (i.e., multimodel inference). Our intention is to present and illustrate a consistent methodology that treats model formulation, model selection, estimation of model parameters and their uncertainty in a unified manner, under a compelling common framework. We review and explain other information criteria (e.g., AIC_c, $QAIC_c$, and TIC) and present several examples to illustrate various technical issues, including some comparisons with BIC, a type of dimension consistent criterion. In addition, we provide many references to the technical literature for those wishing to read further on these topics.

This is an applied book written primarily for biologists and statisticians using models for making inferences from empirical data. This is primarily a science book; we say relatively little about decision making in management or management science. Research biologists working either in the field or in the laboratory will find simple methods that are likely to be useful in their investigations. Researchers in other life sciences, econometrics, the social sciences, and medicine might also find the material useful but will have to deal with examples that have been taken largely from ecological studies of free-ranging vertebrates, as these are our interests. Applied statisticians might consider the information-theoretic methods presented here quite useful and a superior alternative to the null hypothesis testing approach that has become so tortuous and uninformative. We hope material such as this will find its way into classrooms where applied data analysis and associated science philosophy are taught. This book might be useful as a text for a course for students with substantial experience and education in statistics and applied data analysis. A second primary audience includes honors or graduate students in the biological, medical, or statistical sciences. Those interested in the empirical sciences will find this material useful because it offers an effective alternative to (1) the widely taught, yet often both complex and uninformative, null hypothesis testing approaches and (2) the far less taught, but potentially very useful, Bayesian approaches.

Readers should ideally have some maturity in the quantitative sciences and experience in data analysis. Several courses in contemporary statistical theory and methods as well as some philosophy of science would be particularly useful in understanding the material. Some exposure to likelihood theory is nearly essential, but those with experience only in least squares regression modeling will gain some useful insights. Biologists working in a team situation with someone in the quantitative sciences might also find the material to be useful. The book is meant to be relatively easy to read and understand, but the conceptual issues may preclude beginners. Chapters 1–4 are recommended for all readers because they provide the essential material, including concepts of multimodel inference. Chapters 5 and 6 present more difficult material and some new research results. Few readers will be able to absorb the concepts presented here after just one reading of the material; some rereading and additional consideration will often be necessary to understand the deeper points. Underlying theory is presented in Chapter 7, and this material is much deeper and more mathematical. A high-level summary of the main points of the book is provided in Chapter 8.

We intend to remain active in this subject area after this second edition has been published, and we invite comments from colleagues as an ideal way to learn more and understand differing points of view. We hope that the text does not appear too dogmatic or idealized. We have tried to synthesize concepts that we believe are important and incorporate these as recommendations or advice in several of the chapters. This book is an effort to explore the K-L–based multimodel inference in some depth. We realize that there are other approaches, and that some people may still wish to test null hypotheses as the basis for building models of empirical data, and that others may have a more lenient attitude toward data dredging than we advocate here. We do not want to deny other model selection methods, such as cross-validation, nor deny the value of Bayesian methods. Indeed, we just learned (March, 2002) that AIC can be derived as a Bayesian result and have added a note on this issue while reviewing the final page proofs (see Section 6.4.5). However, in the context of objective science, we are compelled by the a priori approach of building candidate models to represent research hypotheses, the use of information-theoretic criteria as a basis for selecting a best approximating model; model averaging, or other multimodel inference methods, when truth is surely very complex; the use of likelihood theory for deriving parameter estimators; and incorporating model selection uncertainty into statistical inferences. In particular, we recommend moving beyond mere selection of a single best model by using concepts and methods of multimodel inference.

Several people have helped us as we prepared the two editions of this book. In particular, we acknowledge C. Chatfield, C. Hurvich, B. Morgan, D. Otis, J. Rotella, R. Shibata, and K. Wilson for comments on earlier drafts of the original manuscript. We are grateful to three anonymous reviewers for comments that allowed us to improve the first edition. D. Otis and W. Thompson served as the reviewers for the second edition and offered many suggestions that were helpful; we greatly appreciate their excellent suggestions. Early discussions with S. Buckland, R. Davis, R. Shibata, and G. White were very useful. S. Beck, K. Bestgen, D. Beyers, L. Ellison, A. Franklin, W. Gasaway, B. Lubow, C. McCarty, M. Miller, and T. Shenk provided comments and insights as part of a graduate course on model selection methods that they took from the authors. C. Flather allowed us to use his data on species accumulation curves as our first example, and we thank C. Braun and the Colorado Division of Wildlife for the data on sage grouse; these data were analyzed by M. Zablan under the supervision of G. White. C. Southwell allowed us to use his kangaroo data from Wallaby Creek. P. Lukacs conducted the bootstrap analysis and some of the Monte Carlo studies of the body fat data in Chapter 5. J. Kullback allowed us to use a photo of his father, and H. Akaike, R. Leibler, R. Shibata, and K. Takeuchi kindly sent us photos and biographical material that appear in the book. Chelsea Publishing Company allowed our use of the photo of L. Boltzmann from the book *Wissenschaftliche Abhandlungen von Ludwig Boltzmann*, and the International Biometric Society authorized our use of a photo of R. Fisher (from *Biometrics* 1964, taken in 1946 by A. Norton). J. Barandun provided the toad photos for the cover, K. Allred provided the cover design, and B. Schmidt helped in coordination. C. Dion, R. Fulton, S. Kane, B. Klein, A. Lyman, and T. Sundlov helped obtain library materials. J. Kimmel and L. Farkas helped in countless ways as we prepared both editions of this book.

We are happy to acknowledge the long-term cooperators of the Colorado Cooperative Fish and Wildlife Research Unit: the Colorado Division of Wildlife, Colorado State University, the Biological Resources Division of the U.S. Geological Survey, and the Wildlife Management Institute. Graduate students and faculty within the Department of Fisheries and Wildlife Biology at Colorado State University provided a forum for our interests in the analysis of empirical data. We extend our appreciation to several federal agencies within the Department of the Interior, particularly the U.S. Geological Survey, for their support of our long-term research interests.

Fort Collins, Colorado   Kenneth P. Burnham
                                                                    David R. Anderson
                                                                    January 2002

Contents

About the Authors

Drs. Kenneth P. Burnham and David R. Anderson have worked closely together for the past 28 years and have jointly published 9 books and research monographs and 66 journal papers on a variety of scientific issues. Currently, they are both in the Colorado Cooperative Fish and Wildlife Research Unit at Colorado State University, where they conduct research, teach graduate courses, and mentor graduate students.

Ken Burnham has a B.S. in biology and M.S. and Ph.D. degrees in statistics. For 29 years post-Ph.D. he has worked as a statistician, applying and developing statistical theory in several areas of life sciences, especially ecology and wildlife, most often in collaboration with subject-area specialists. Ken has worked (and lived) in Oregon, Alaska, Maryland (Patuxent Wildlife Research Center), North Carolina (U.S. Department of Agriculture at North Carolina State University, Statistics Department), and Colorado (currently USGS-BRD). He is the recipient of numerous professional awards including Distinguished Achievement Medal from the American Statistical Association, and Distinguished Statistical Ecologist Award from INTECOL (International Congress of Ecology). Ken is a Fellow of the American Statistical Association.

David Anderson received B.S. and M.S. degrees in wildlife biology and a Ph.D. in theoretical ecology. He is currently a Senior Scientist with the Biological Resources Division within the U.S. Geological Survey and a professor in the Department of Fishery and Wildlife Biology. He spent 9 years at the Patuxent Wildlife Research Center in Maryland and 9 years as leader of the Utah Cooperative Wildlife Research Unit and professor in the Wildlife Science Department at Utah State University. He has been at Colorado State University since 1984. He is the recipient of numerous professional awards for scientific and academic contributions, including the Meritorious Service Award from the U.S. Department of the Interior.

Glossary

Notation and abbreviations generally used are given below. Special notation for specific examples can be found in those sections.

AIC   Akaike's information criterion.

AIC_{min} The estimate of relative, expected K-L information for the best model in the set, given the data. For example, given the models $g_1, g_2, \ldots, g_R$ and the data x, if the information criterion is minimized for model g_6, then $min = 6$, signifying that AIC_6 is the minimum over $AIC_1, \ldots, AIC_R$. The minimum AIC is a random variable over samples. This notation, indicating the index number in $\{1, 2, \ldots, R\}$ that minimizes expected K-L information, also applies to AIC_c, $QAIC_c$, and TIC.

AIC_{best} In any set of models, one will be the best expected K-L model, hence the actual best AIC model. The model for which $E_f(AIC)$ is minimized is denoted by the index *best*, whereas *min* is a random variable (like $\hat{\theta}$), *best* is fixed (like θ). This value can be determined using Monte Carlo methods. This "best" model is the same model over all possible samples (of which we have only a single sample). This notation also applies to AIC_c, $QAIC_c$, and TIC.

AIC_c A second-order AIC, necessary for small samples.

Akaike weights The relative likelihood of the model, given the data. These are normalized to sum to 1, are denoted by w_i, and interpreted as probabilities.

best   An index to denote the theoretically best fitted model; this model is best in the sense of expected K-L information, given the data. Such a best model can be found from Monte Carlo methods and represents a statistical expectation. For example, consider the set $E(AIC_i)$, where $i = 1, 2., \ldots, R$. Then, the model where $E(AIC_i)$ is minimized is denoted by AIC_{best}. AIC, AIC_c, $QAIC_c$, or TIC could be used in this context.

Bias   (of an estimator) Bias $= E(\hat{\theta}) - \theta$.

BIC   Bayesian information criterion (Akaike 1978a,b; Schwarz 1978), also termed SIC in some literature.

c   A simple variance inflation factor used in quasi-likelihood methods where there is overdispersion of count data (e.g., extra binomial variation).

Δ_i AIC differences, relative to the smallest AIC value in the set of R models. Hence, AIC values are rescaled by a simple additive constant such that the model with the minimum AIC value has $\Delta_i = 0$. Formally, $\Delta_i = AIC_i - AIC_{min}$. These values are estimates of the expected K-L information (or distance) between the selected (best) model and the ith model. These differences apply to AIC, AIC_c, $QAIC_c$, or TIC.

Δ_p A "pivotal" value, analogous to $(\theta - \hat{\theta})/\hat{se}(\hat{\theta})$; $\Delta_p = AIC_{best} - AIC_{min}$.

df   Degrees of freedom as associated with hypothesis testing. The df is the difference between the number of parameters in the null and alternative hypotheses in standard likelihood ratio tests.

$E(\hat{\theta})$ The statistical expectation of the estimator $\hat{\theta}$.

Estimate   The computed value of an estimator, given a particular set of sample data (e.g., $\hat{\theta} = 9.8$).

Estimator   A function of the sample data that is used to estimate some parameter. An estimator is a random variable and is denoted by a "hat" (e.g., $\hat{\theta}$).

Evidence ratio   The relative likelihood of model i versus model j (e.g., $\mathcal{L}(g_i|data)/\mathcal{L}(g_j|data)$, which is identical to w_i/w_j).

$f(x)$ Used to denote "truth" or "full reality," the process that produces multivariate data x. This conceptual probability distribution is often considered to be a mapping from an infinite-dimensional space.

$g_i(x)$ Used to denote the set of candidate models that are hypothesized to provide an adequate approximation for the distribution of empirical data. The expression $g_i(x \mid \theta)$ is used when it is necessary to clarify that the function involves parameters θ.

Often, the parameters have been estimated; thus the estimated approximating model is denoted by $g_i(x \mid \hat{\theta})$. Often, the set of R candidate models is represented as simply $g_1, g_2, \ldots, g_R$. Also, $\hat{g}_i = g_i(x \mid \hat{\theta})$.

Global model	A highly parameterized model containing the variables and associated parameters thought to be important as judged from an a priori consideration of the problem at hand. When there is a global model, all other models in the set are special cases of this global model.
K	The number of estimable parameters in an approximating model.
K-L	Kullback–Leibler distance (or discrepancy, information, number).
LRT	Likelihood ratio test.
LS	Least squares method of estimation.
$\mathcal{L}(\theta \mid x, g)$	Likelihood function of the model parameters, given the data x and the model g.
$\mathcal{L}(g_i \mid x)$	The discrete likelihood of model g_i, given the data x.
$\log(\cdot)$	The natural logarithm ($\log_e$).
$\mathrm{logit}(\theta)$	The logit transform: $\mathrm{logit}(\theta) = \log(\theta/(1 - \theta))$, where $0 < \theta < 1$.
g_i	Shorthand notation for the candidate models considered.
min	An index to denote the fitted model that minimizes the information criterion, given the data. Then, model g_{min} is the model selected, based on minimizing the appropriate criterion, given the data. AIC, AIC_c, QAIC_c, or TIC could be used in this context.
ML	Maximum likelihood method of estimation.
MLE	Maximum likelihood estimate (or estimator).
n	Sample size. In some applications there may be more than one relevant sample size (e.g., in random effects models).
Parsimony	The concept that a model should be as simple as possible concerning the included variables, model structure, and number of parameters. Parsimony is a desired characteristic of a model used for inference, and it is usually defined by a suitable tradeoff between squared bias and variance of parameter estimators. Parsimony lies between the evils of under- and over-fitting.
Precision	A property of an estimator related to the amount of variation among estimates from repeated samples.

$\propto$	A symbol meaning "proportional to."
QAIC or QAIC$_c$	Versions of AIC or AIC$_c$ for overdispersed count data where quasi-likelihood adjustments are required, hence $\hat{c}$ used.
π_i	Model selection probabilities (or relative frequencies), often from Monte Carlo studies or the bootstrap.
R	The number of candidate models in the set; $i = 1, 2, \ldots, R$. One of these models is the estimated best model (i.e., in the sense of a specific model $g(x|\hat{\theta})$, where the model parameters have been estimated) for the data at hand (g_{min}). One model (possibly the same model) is the theoretically best model (g_{best}) to use as a basis for inference from the data.
τ_i	Prior probability of model i. Also used to cope with model redundancy (Section 4.6).
θ	Used to denote a generic parameter vector (such as a set of conditional survival probabilities S_i).
$\hat{\theta}$	An estimator of the generic parameter θ.
θ_0	The optimal parameter value in a given model g, given a fixed sample size, but ignoring estimation issues (see Section 7.1). This is the value that minimizes K-L information, given the model structure.
TIC	Takeuchi's information criterion.
w_i	Akaike weights. Used with any of the information criteria that are estimates of expected Kullback–Leibler information (AIC, AIC$_c$, QAIC, TIC). The w_i sum to 1 and may be interpreted as the probability that model i is the actual expected K-L best model for the sampling situation considered.
$w_+(j)$	Sum of Akaike weights over all models that include the explanatory variable j. These sums are useful in variable-selection problems where one wants a measure of relative importance of the explanatory variables and in computing estimates that are robust to model selection bias.
χ^2	A test statistic distributed as chi-squared with specified degrees of freedom df. Used here primarily in relation to a goodness-of-fit test of the global model in analyzing count data.
$\approx$	Approximately equal to.
$\sim$	Distributed as.

1
Introduction

1.1 Objectives of the Book

This book is about making valid inferences from scientific data when a meaningful analysis depends on a model of the information in the data. Our general objective is to provide scientists, including statisticians, with a readable text giving practical advice for the analysis of empirical data under an information-theoretic paradigm. We first assume that an exciting scientific question has been carefully posed and relevant data have been collected, following a sound experimental design or probabilistic sampling program. Alternative hypotheses, and models to represent them, should be carefully considered in the design stage of the investigation. Often, little can be salvaged if data collection has been seriously flawed or if the question was poorly posed (Hand 1994). We realize, of course, that these issues are never as ideal as one would like. However, proper attention must be placed on the collection of data (Chatfield 1991, 1995a Anderson 2001). We stress inferences concerning the structure and function of biological systems, relevant parameters, valid measures of precision, and formal prediction.

There are many studies where we seek an understanding of relationships, especially causal ones. There are many studies to understand our world; models are important because of the parameters in them and relationships expressed between and among variables. These parameters have relevant, useful interpretations, even when they relate to quantities that are not directly observable (e.g., survival probabilities, animal density in an area, gene frequencies, and interaction terms). Science would be very limited without such unobservables as constructs in models. We make statistical inferences from the data, to a real or conceptual population or process, based on models involving such parameters. Observables and prediction are often critical, but science is broader than these issues.

The first objective of this book is to outline a consistent *strategy* for issues surrounding the analysis of empirical data. Induction is used to make statistical inference about a defined population or process, given an empirical sample or experimental data set. "Data analysis" leading to valid inference is the integrated process of careful a priori model formulation, model selection, parameter estimation, and measurement of precision (including a variance component due to model selection uncertainty). We do not believe that model selection should be treated as an activity that precedes the analysis; rather, model selection is a critical and integral aspect of scientific data analysis that leads to valid inference.

A philosophy of thoughtful, science-based, a priori modeling is advocated. Often, one first develops a global model (or set of models) and then derives several other plausible candidate (sub)models postulated to represent good approximations to information in the data at hand. This forms the *set of candidate models*. Science and biology play a lead role in this a priori model building and careful consideration of the problem. A simple example of models to represent alternative scientific hypotheses might be helpful at this early point. Consider the importance of an interaction between age (a) and winter severity (w) in a particular animal population. A model including such an interaction would have the main effects plus the interaction; $a + w + a * w$, while the model $a + w$ lacks the interaction term. Information-theoretic methods allow several lines of quantitative evidence concerning the importance of this hypothesized interaction.

The modeling and careful thinking about the problem are critical elements that have often received relatively little attention in statistics classes (especially for nonmajors), partly because such classes rarely consider an overall strategy or philosophy of data analysis. A proper a priori model-building strategy tends to avoid "data dredging," which leads to overfitted models, that is, to the "discovery" of effects that are actually spurious (Anderson 2001a). Instead, there has often been a rush to "get to the data analysis" and begin to rummage through the data and compute various estimates of interest or conduct null hypothesis tests. We realize that these other philosophies may have their place, especially in more exploratory investigations.

The second objective is to explain and illustrate methods developed recently at the interface of information theory and mathematical statistics for selection of an estimated "best approximating model" from the a priori set of candidate models. In particular, we review and explain the use of Akaike's information criterion (AIC) in the selection of a model (or small set of good models) for statistical inference. AIC provides a simple, effective, and objective means for the selection of an estimated "best approximating model" for data analysis and inference. Model selection includes "variable selection" as frequently practiced in regression analysis. Model selection based on information theory is a relatively new paradigm in the biological and statistical sciences and is quite different from the usual methods based on null hypothesis testing. Model selection based on information theory is not the only reasonable approach, but it is what we are focusing on here because of its philosophical and computational advantages.

The practical use of information criteria, such as Akaike's, for model selection is relatively recent (the major exception being in time series analysis, where AIC has been used routinely for the past two decades). The marriage of information theory and mathematical statistics started with Kullback's (1959) book. Akaike considered AIC to be an extension of R. A. Fisher's likelihood theory. These are all complex issues, and the literature is often highly technical and scattered widely throughout books and research journals. Here we attempt to bring this relatively new material into a readable text for people in (primarily) the biological and statistical sciences. We provide a series of examples, many of which are biological, to illustrate various aspects of the theory and application.

In contrast, hypothesis testing as a means of selecting a model has had a much longer exposure in science. Many seem to feel more comfortable with the hypothesis testing paradigm in model selection, and some even consider the results of a test as *the* standard by which other approaches should be judged (we believe that they are wrong to do so). Bayesian methods in model selection and inference have been the focus of much recent research. However, the technical level of this material often makes these approaches unavailable to many in the biological sciences. A variety of cross-validation and bootstrap-based methods have been proposed for model selection, and these, too, seem like very reasonable approaches. The computational demands of many of the Bayesian and cross-validation methods for model selection are often quite high (often 1–3 orders of magnitude higher than information-theoretic approaches), especially if there are more than a dozen or so high-dimensional candidate models.

The theory presented here allows estimates of "model selection uncertainty," inference problems that arise in using the same data for both model selection and the associated parameter estimation and inference. If model selection uncertainty is ignored, precision is often overestimated, achieved confidence interval coverage is below the nominal level, and predictions are less accurate than expected. Another problem is the inclusion of spurious variables, or factors, with no assessment of the reliability of their selection. Some general methods for dealing with model- and variable-selection uncertainty are suggested and examples provided. Incorporating model selection uncertainty into estimators of precision is an active area of research, and we expect to see additional approaches developed in the coming years.

The third objective is to present a number of approaches to making formal inference from more than one model in the set. That is, rather than making inferences from only the model estimated to be the best, robust inferences can be made from several, even all, models being considered. These procedures are termed *multimodel inference* (MMI). Model averaging has been an active research area for Bayesians for the past several years (Hoeting et al. 1999). Model averaging can be easily done under an information-theoretic approach. Model averaging has several practical and theoretical advantages, particularly in prediction or in cases where a parameter of interest occurs in all the models. Confidence sets on models is another useful approach, particularly when models in the set represent a logical ordering (e.g., a set of models representing chronic treatment effects over 1, 2, . . . , t time periods). Finally, the relative importance of explantory variables in a general regression setting can be easily assessed by summing certain quantities across models. MMI is also potentially useful in certain conflict resolution issues (Anderson et al. 2001c).

Current practice often would judge a variable as important or unimportant, based on whether that variable was in or out of the selected model (e.g., stepwise regression, based on hypothesis testing). Such procedures provide a misleading dichotomy (see Breiman 2001) and are not in the spirit of a weight of evidence. MMI allows us to discard simplistic dichotomies and focus on quantitatively ranking models and variables as to their relative value and importance.

Modeling is an art as well as a science and is directed toward finding a good approximating model of the information in empirical data as the basis for statistical inference from those data. In particular, the number of parameters estimated from data should be substantially less than the sample size, or inference is likely to remain somewhat preliminary (e.g., Miller (1990: x)) mentions a regression problem with 757 variables and a sample size of 42 (it is absurd to think that valid inference is likely to come from the analysis of these data). In cases where there are relatively few data per estimated parameter, a small-sample version of AIC is available (termed AIC_c) and should be used routinely rather than AIC. There are cases where quasi-likelihood methods are appropriate when count data are overdispersed; this theory leads to modified criteria such as QAIC and $QAIC_c$, and these extensions are covered in the following material.

Simple models with only 1-2 parameters are not the central focus of this book; rather, we focus on models of more complex systems. Parameter estimation has been firmly considered to be an optimization problem for many decades, and AIC formulates the problem of model selection as an optimization problem across a set of candidate models. Minimizing AIC is a simple operation with results that are easy to interpret. Models can be clearly ranked and scaled, allowing full consideration of other good models, in addition to the estimated "best approximating model." Evidence ratios allow a formal strength of evidence for alternative hypotheses. Competing models, those with AIC values close to the minimum, are also useful in the estimation of model selection uncertainty. Inference should often be based on more than a single model, unless the data clearly support only a single model fit to the data. Thus, some approaches are provided to allow inference from several or all of the models, including model averaging.

This is primarily an applied book. A person with a good background in mathematics and theoretical statistics would benefit from studying Chapter 7. McQuarrie and Tsai (1998) present both theoretical and applied aspects of model selection in regression and time series analysis, including extensive results of large-scale Monte Carlo simulation studies.

1.2 Background Material

Data and stochastic models of data are used in the empirical sciences to make inferences concerning both processes and parameters of interest (see Box et al. 1981, Lunneborg 1994, and Shenk and Franklin 2001 for a review of principles). Statistical scientists have worked with researchers in the biological sciences for many years to improve methods and understanding of biological processes. This book provides practical, omnibus methods to achieve valid inference from models that are good approximations to biological processes and data. We focus on statistical evidence and try to avoid arbitrary dichotomies such as "significant or not significant." A broad definition of data is employed here. A single, simple data set might be the subject of analysis, but more often, data collected from several field sites or laboratories are the subject of a more comprehensive analysis. The data might commonly be extensive and partitioned by age, sex, species, treatment group, or within several habitat types or geographic areas. In linear and nonlinear regression models there may be many explanatory variables. There are often factors (variables) with small, moderate, and large effects in these information-rich data sets (the concept of tapering effect sizes). Parameters in the model represent the effects of these factors. We focus on modeling philosophy, model selection, estimation of model parameters, and valid measures of precision under the relatively new paradigm of information-theoretic methods. Valid inference rests upon these four issues, in addition to the critical considerations relating to problem formulation, study design, and protocol for data collection.

1.2.1   Inference from Data, Given a Model

R. A. Fisher (1922) discussed three aspects of the general problem of valid inference: (1) model specification, (2) estimation of model parameters, and (3) estimation of precision. Here, we prefer to partition model specification into two components: formulation of a set of candidate models and selection of a model (or small number of models) to be used in making inferences. For much of the twentieth century, methods have been available to objectively and efficiently estimate model parameters and their precision (i.e., the sampling covariance matrix). Fisher's *likelihood theory* has been the primary omnibus approach to these issues, but it *assumes* that the model structure is known (and correct, i.e., a true model) and that only the parameters in that structural model are to be estimated. Simple examples include a linear model such as $y = \alpha + \beta x + \epsilon$ where the residuals (ϵ) are assumed to be normally distributed, or a log-linear model for the analysis of count data displayed in a contingency table. The parameters in these models can be estimated using *maximum likelihood* (ML) methods. That is, if one assumes or somehow chooses a particular model, methods exist that are objective and asymptotically optimal for estimating model parameters and the sampling covariance structure, conditional on that model. A more challenging example might be to assume that data are appropriately modeled by a 3-parameter gamma distribution; one can routinely use the method of maximum likelihood to estimate these model parameters and the model-based 3×3 sampling covariance matrix. Given an appropriate model, and if the sample size is "large," then maximum likelihood provides estimators of parameters that are consistent (i.e., asymptotically unbiased with variance tending to zero), fully efficient (i.e., minimum variance among consistent estimators), and normally distributed. With small samples, but still assuming an appropriate model, ML estimators often have small-sample bias, where bias $\equiv E(\hat{\theta}) - \theta$. Such bias is usually a trivial consideration, as it is often substantially less than the se($\hat{\theta}$), and bias-adjusted estimators can often be found if this is deemed necessary. The sampling distributions of ML estimators are often skewed with small samples, but profile likelihood intervals or log-based intervals or bootstrap procedures can be used to achieve asymmetric confidence intervals with good coverage properties. **In general, the maximum likelihood method provides an objective, omnibus theory for estimation of model parameters and the sampling covariance matrix,** *given an appropriate model*.

1.2.2   Likelihood and Least Squares Theory

Biologists have typically been exposed to least squares (LS) theory in their classes in applied statistics. LS methods for linear models are relatively simple to compute, and therefore they enjoyed an early history of application (Weisburg 1985). In contrast, Fisher's likelihood methods often require iterative numerical methods and were thus not popular prior to the widespread availability of personal computers and the development of easy-to-use software. LS theory has many similarities with likelihood theory, and it yields identical estimators of the structural parameters (but not σ^2) for linear and nonlinear models when the residuals are assumed to be independent and normally distributed. It is now easy to allow alternative error structures (i.e., nonnormal residuals such as Poisson, gamma or log-normal) for regression and other similar problems in either a likelihood or quasi-likelihood framework (e.g., McCullagh and Nelder 1989, Heyde 1997), but more difficult in an LS framework.

The concepts underlying both estimation methods are relatively simple to understand (Silvey 1975). Consider the simple linear regression, where a response variable (y) is modeled as a linear function of an explanatory variable

(x) as $y_i = \beta_0 + \beta_1 \cdot x_i + \epsilon_i$. The ϵ_i are error terms (residuals) which are often modeled as independent normal random variables with mean 0 and constant variance σ^2. Under LS the estimates of β_0 and β_1 are those that minimize $\sum(\epsilon_i)^2$ — hence the name *least squares*. The parameter estimates $\hat{\beta}_0$ and $\hat{\beta}_1$ minimize the average squared error terms (ϵ_i) and define a regression line that is the "best fit." Hundreds of statistics books cover the theory and application for least squares estimation in linear and nonlinear models, particularly when the ϵ_i are assumed to be independent, normally distributed random variables.

Likelihood methods are much more general, far less taught in applied statistics courses, and slightly more difficult to understand at first. The material in much of this book relies on an understanding of likelihood theory, so some brief introduction is given here. While likelihood theory is a paradigm underlying both frequentist and Bayesian statistics, there are no more than a handful of applied books solely on this important subject (good examples include McCullagh and Nelder 1989, Edwards 1992, Azzalini 1996, Morgan 2000, and Severini 2000).

The theory underlying likelihood begins with a probability model, given the parameters (θ). Specifically, model g describes the probability distribution of the data, given the model parameters and a specific model form; denoted by $g(x|\theta, model)$. A simple example is the binomial probability function where θ is the probability of a "success"; let this be the parameter $p = 0.4$. The data could be the observation of $y = 15$ successes out of $n = 40$ independent trials. Then, the discrete probability of getting 15 successes out of 40 trials, given the parameter $(p = 0.4)$ and the binomial model, is

$$g(y, n|p, binomial) = \binom{n}{y} p^y (1 - p)^{n-y},$$

$$g(15, 40|p = 0.4, binomial) = \frac{40!}{15!25!}(0.4)^{15}(1 - 0.4)^{25} = 0.123.$$

The key point is that for this calculation, the model (here a binomial model) and its parameters (here $p = 0.4$) are known in advance (i.e., they are *given*). In very simple problems such as this, an excellent model is available and can be considered given (such is rarely the case in the real world, where one is not sure what model might be used). Then one observes the data ($y = 15$ and $n = 40$) and can compute the probability of the data, given the model and its parameters.

In much of science, neither the model parameters nor the model is known. However, data can be collected in a way that allows the parameters to be estimated if a good model can be found or assumed. The likelihood function is the basis for such parameter estimation and is a function of the parameter p, given the data and the binomial model:

$$\mathcal{L}(p|y, n, binomial) = \binom{n}{y} p^y (1 - p)^{n-y}$$

or

$$\mathcal{L}(p|15, 40, binomial) = \frac{40!}{15!25!}(p)^{15}(1 - p)^{25}.$$

Clearly, the likelihood is a function of (only) the unknown parameter (p in this example); everything else is known or assumed. The probability model and the associated likelihood function differ only in terms of what is known or given. In the probability model, the parameters, the model, and the sample size are known, and interest lies in the probability of observing a particular event (the data, y given n in this simple example). In the likelihood function, the data are given (observed) and the model is assumed (but given), and interest lies in estimating the unknown parameters; thus, the likelihood is a function of only the parameters. The probability model of the data and the likelihood function of the parameters are closely related; they merely reverse the roles of the data and the parameters, given a model. The binomial coefficient $\binom{n}{y}$ does not contain the unknown parameter p and is often omitted (it does not contain any information about the unknown parameters and is often difficult to compute if $n > 50$).

The notation for the likelihood function is very helpful in its understanding; consider the general expression $\mathcal{L}(\theta|data, model)$. If we follow the usual convention of letting x represent the empirical data and g a *given* approximating model, then $\mathcal{L}(\theta|x, g)$ is read as "the likelihood of a particular numerical value of the unknown parameter θ (θ is usually a vector), given the data x and a particular model g."

A well-known example will help illustrate the concept. Consider flipping n pennies and observing y "heads." Assuming that the flips are independent and that each penny has an equal probability of a head, the binomial model is an obvious model choice in this simple setting. The likelihood function is $\mathcal{L}(p|y, n,$ binomial), where p is the (unknown) probability of a head. Thus, given the data (y and n) and the binomial model, one can compute the *likelihood* that p is 0.15 or 0.73 or any other value between 0 and 1. The likelihood (a relative, not absolute, value) is a function of the unknown parameter p. Given this formalism, one might compute the likelihood of many values of the unknown parameter p and pick the most likely one as the *best* estimate of p, *given* the data and the model. It seems compelling to pick the value of p that is "most likely." This is Fisher's concept of *maximum likelihood estimation*; he published this when he was 22 years old as a third-year undergraduate at Cambridge University! He reasoned that the best estimate of an unknown parameter (given data and a model) was that which was the most likely; thus the name *maximum likelihood*, ML. The ML estimate (MLE) for the binomial model happens to have a closed-form expression that is well known: $\hat{p} = y/n = 7/11 = 0.6363$. That is, the numerical value of y/n exactly maximizes the likelihood function. In most real-world cases a simple, closed form estimator either does not exist or cannot be found without substantial difficulty.

Likelihood theory includes asymptotically optimal methods for estimation of unknown parameters and their variance–covariance matrix, derivation of hypothesis tests, the basis for profile likelihood intervals, and other important quantities (such as model selection criteria). More generally, likelihood theory includes the broad concept of *support* (Edwards 1992). Likelihood is also the essential basis for Bayesian approaches to statistical inference. In fact, likelihood is the backbone of statistical theory, whereas least squares can be viewed as a limited special case and, while very useful in several important applications, is not foundational in modern statistics.

For many purposes the natural logarithm of the likelihood function is essential; written as $\log(\mathcal{L}(\theta|data, model))$, or $\log(\mathcal{L}(\theta|x, model))$, or if the context is clear, just $\log(\mathcal{L}(\theta))$ or even just $\log(\mathcal{L})$. Often, one sees notation such as $\log(\mathcal{L}(\theta|x))$, without it being clear that a particular model is assumed. An advanced feature of $\log(\mathcal{L})$ is that it, by itself, is a type of *information* concerning θ and the model (Edwards 1992:22–23). The log-likelihood for the binomial model where 11 pennies are flipped and 7 heads are observed is

$$\log(\mathcal{L}(p|y, n, binomial)) = \log \binom{n}{y} + y \cdot \log(p) - (n - y) \cdot \log(1 - p),$$

$$= \log \binom{11}{7} + 7 \cdot \log(p) - (11 - 7) \cdot \log(1 - p)$$

$$= 5.79909 + 7 \cdot \log(p) - (4) \cdot \log(1 - p).$$

A property of logarithms for values between 0 and 1 is that they lie in the negative quadrant; thus, values of discrete log-likelihood functions are negative (unless some additive constants have been omitted). Figure 1.1 shows a plot of the likelihood (*a*) and log-likelihood (*b*) functions where 11 pennies were flipped, 7 heads were observed, and the binomial model was assumed. The value of $p = 0.636$ maximizes both the likelihood and the log-likelihood function; this value is denoted by $\hat{p}$ and is the maximum likelihood estimate (MLE). Relatively little information is contained in such a small sample size ($n = 11$) and this is reflected in the broad shape of the plots. Had the sample size been 5 times larger, with $n = 55$ and 35 heads observed, the likelihood and log-likelihood functions would be more peaked (Figure 1.1c and d). In fact, the sampling variance is derived from the shape of the log-likelihood function around its maximum point. In the usual case where θ is a vector, a variance–covariance matrix can be estimated based on partial derivatives of the log-likelihood function. These procedures will not be developed here.

The value of the log-likelihood function at its maximum point is a very important quantity, and it is this point that defines the *maximum likelihood estimate*. In the example with 11 flips and 7 heads, the value of the maximized log-likelihood is -1.411 (Figure 1.1b). This result is computed by taking the

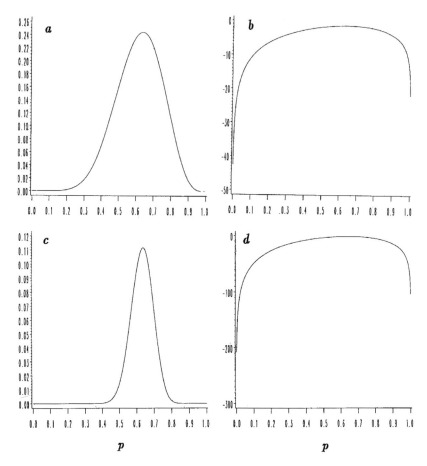

FIGURE 1.1. Plots of the binomial likelihood (a) and log-likelihood (b) function, given $n = 11$ penny flips and the observation that $y = 7$ of these were heads. Also shown are plots of the binomial likelihood (c) and log-likelihood (d) function, given a sample size 5 times larger; $n = 55$ penny flips and the observation that $y = 35$ of these were heads. Note the differing scales on the Y axis.

log-likelihood function

$$\log(\mathcal{L}(p|y, n, \textit{binomial})) = \log \binom{n}{y} + y \cdot \log(p) - (n - y) \cdot \log(1 - p)$$

and substituting the MLE ($\hat{p} = 0.6363$) and the data (y and n),

$$-1.411 = 5.79909 + 7 \cdot \log(0.6363) - (4) \cdot \log(1 - 0.6363).$$

Thus, when one sees reference to a maximized $\log(\mathcal{L}(\theta))$ this merely represents a numerical value (e.g., -1.411).

Many do not realize that the common procedure for setting a 95% confidence interval (i.e., $\hat{\theta} \pm 1.96 \cdot \hat{se}(\hat{\theta})$) is merely an approximation. The estimator $\hat{\theta}$ is only asymptotically normal, and if the sample size is too small, the sampling distribution will often be nonnormal and the approximation will be poor (i.e., achieved confidence interval coverage can be much less than the nominal value, say, 95%). For example, if the binomial parameter is near 0 or 1, the distribution of the estimator $\hat{\theta}$ will be nonnormal (asymmetric) unless the sample size is very large. In general, rather than use the simple approximation, one can set a 95% interval using the log-likelihood function; this procedure, in general, is called a *profile likelihood interval*. This is not a simple procedure; thus the approximation has seen heavy use in applied data analysis. We cannot provide the full theory for profile likelihood intervals here, but will give an example for the binomial case where $n = 11$, $y = 7$, $\hat{p} = 0.6363$, and the maximized log-likelihood value is -1.411. Here, we start with 3.84, which is the 0.05 point of the chi-squared distribution with 1 degree of freedom. One-half of this value is 1.92, and this value is subtracted from the maximum point of the log-likelihood function: $-1.411 - 1.92 = -3.331$. Now, numerically, one must find the 2 values of p that are associated with the values of the log-likelihood function at -3.331. These 2 values are the endpoints of an exact 95% likelihood confidence interval. In this example, the 95% likelihood interval is (0.346, 0.870).

Biologists familiar with LS but lacking insight into likelihood methods might benefit from an example. Consider a multiple linear regression model where a dependent variable y is hypothesized to be a function of p explanatory (predictor) variables x_j ($j = 1, 2, \ldots, r$). Here the residuals ϵ_i of the n observations are assumed to be independent, normally distributed with a constant variance σ^2, and the model structure is expressed as

$$y_i = \beta_0 + \beta_1 x_1 + \beta_2 x_2 + \cdots + \beta_r x_r + \epsilon_i, \qquad i = 1, \ldots, n.$$

Hence

$$E(y_i) = \beta_0 + \beta_1 x_1 + \beta_2 x_2 + \cdots + \beta_r x_r, \qquad i = 1, \ldots, n,$$

and $E(y_i)$ is a linear function of $r + 1$ parameters. The conceptual residuals,

$$\epsilon_i = y_i - (\beta_0 + \beta_1 x_1 + \beta_2 x_2 + \cdots + \beta_r x_r) = y_i - E(y_i),$$

have the joint probability distribution $g(\underline{\epsilon}|\underline{\theta})$, where $\underline{\theta}$ is a vector of $K = r + 2$ parameters ($\beta_0, \beta_1, \ldots, \beta_r$, and σ). Here, corresponding to observation i one has the model

$$g(\epsilon_i|\underline{\theta}) = \frac{1}{\sqrt{2\pi}\,\sigma}\, e^{-\frac{1}{2}\left[\frac{\epsilon_i}{\sigma}\right]^2}.$$

The likelihood is simply the product of these over the n observations, interpreted as a function of the unknown parameters, given the data, the linear model structure, and the normality assumption:

$$\mathcal{L}(\underline{\theta}|\underline{x}) = \prod_{i=1}^{n} \frac{1}{\sqrt{2\pi}\,\sigma}\, e^{-\frac{1}{2}\left[\frac{\epsilon_i}{\sigma}\right]^2} = \left(\frac{1}{\sqrt{2\pi}\,\sigma}\right)^n e^{-\frac{1}{2}\sum_{i=1}^{n}\left[\frac{\epsilon_i}{\sigma}\right]^2}.$$

Here we use "x" in $\mathcal{L}(\theta|x)$ to denote the full data. When the ϵ_i are normally distributed with constant variance σ^2, the maximum likelihood estimator (MLE) of β is identical to the usual LS regression estimators (however, the estimator of σ^2 differs slightly). This formalism shows, *given the model*, the link between the data, the model, and the parameters to be objectively estimated, using either LS or ML.

In all fitted linear models the residual sum of squares (RSS) is

$$\text{RSS} = \sum_{i=1}^{n} \hat{\epsilon}_i^2,$$

where

$$\hat{\epsilon}_i = y_i - (\hat{\beta}_0 + \hat{\beta}_1 x_1 + \hat{\beta}_2 x_2 + \cdots + \hat{\beta}_r x_r),$$
$$= y_i - \hat{E}(y_i)$$

The ML estimator is $\hat{\sigma}^2 = \text{RSS}/n$, while the estimator universally used in the LS case is $\hat{\sigma}^2 = \text{RSS}/(n - (r + 1))$. This shows that ML and LS estimators of σ^2 differ by a factor of $n/(n - (r + 1))$; often a trivial difference unless the sample size is small. The maximized likelihood is

$$\mathcal{L}(\hat{\theta}|x) = \left[\frac{1}{\sqrt{2\pi}\hat{\sigma}}\right]^n e^{-\frac{1}{2}n},$$

or

$$\log(\mathcal{L}(\hat{\theta})) = -\frac{1}{2}n \log(\hat{\sigma}^2) - \frac{n}{2}\log(2\pi) - \frac{n}{2}.$$

The additive constants can often be discarded from the log-likelihood because they are constants that do not influence likelihood-based inference. Thus for all standard linear models, we can take

$$\log(\mathcal{L}(\hat{\theta})) \approx -\frac{1}{2}n \log(\hat{\sigma})^2.$$

This result is important in model selection theory because it allows a simple mapping from LS analysis results (e.g., the RSS or the MLE of σ^2) into the maximized value of the log-likelihood function for comparisons over such linear models with normal residuals. Note that the log-likelihood is defined up to an arbitrary additive constant in this usual case. If the model set includes linear and nonlinear models or if the residual distributions differ (e.g., normal, gamma, and log-normal), then all the terms in the log-likelihood must be retained, without omitting any constants. Most uses of the log-likelihood are relative to its maximum, or to other likelihoods at their maxima, or to the curvature of the log-likelihood function at the maximum.

The number of parameters $K = r + 2$ in these linear models must include the intercept (say, β_0), the r regression coefficients ($\beta_1, \ldots, \beta_r$), and the residual variance (σ^2). Often, one (erroneously) considers only the number of parameters being estimated as the intercept and the slope parameters (ignoring σ^2);

Sir Ronald Aylmer Fisher was born in 1890 in East Finchley, London, and died in Adelaide, Australia, in 1962. This photo was taken when he was approximately 66 years of age. Fisher was one of the foremost scientists of his time, making incredible contributions in theoretical and applied statistics and genetics. Details of his life and many scientific accomplishments are found in Box (1978). He published 7 books (one of these had 14 editions and was printed in 7 languages) and nearly 300 journal papers. Most relevant to the subject of this book is Fisher's likelihood theory and parameter estimation using his method of maximum likelihood.

however, in the context of model selection, the number of parameters must include σ^2 and thus $K = r + 2$. If the method of LS is used to obtain parameter estimators, one must use the regression-based estimate of σ^2 times $(n - (r + 1))/n = (n - K + 1)/n$ to obtain the ML estimator of σ^2. In LS estimation, we minimize RSS $= n\hat{\sigma}^2$, which for all parameters other than σ^2 itself is equivalent to maximizing $-\frac{1}{2} \cdot n \log(\hat{\sigma}^2)$.

There is a close relationship between LS and ML methods for linear and nonlinear models, where the ϵ_i are assumed to be normally distributed. For example, the LS estimates of the structural model parameters (but not σ^2) are equivalent to the MLEs. Likelihood (and related Bayesian) methods allow easy extensions to the many other classes of models and, with the exploding power of computing equipment, likelihood methods are finding increasing use by both statisticians and researchers in other scientific disciplines (see Garthwaite et al. 1995 for background).

1.2.3   The Critical Issue: "What Is the Best Model to Use?"

While hundreds of books and countless journal papers deal with estimation of model parameters and their associated precision, relatively little has appeared concerning model specification (what set of candidate models to consider) and model selection (what model(s) to use for inference) (see Peirce 1955). In fact, Fisher believed at one time that model specification was outside the field of mathematical statistics, and this attitude prevailed within the statistical community until at least the early 1970s. *"What is the best model to use?"* is *the* critical question in making valid inference from data in the biological sciences.

The likelihood function $\mathcal{L}(\theta|x, model)$ makes it clear that for inference about θ, data and the model are taken as *given*. Before one can compute the likelihood that $\theta = 5.3$, one must have data and a particular statistical model. While an investigator will have empirical data for analysis, it is unusual that the model is known or given. Rather, a number of alternative model forms must be somehow considered as well as the specific explanatory variables to be used in modeling a response variable. This issue includes the *variable selection problem* in multiple regression analysis. If one has data and a model, LS or ML theory can be used to estimate the unknown parameters (θ) and other quantities useful in making statistical inferences. However, which model is the best to use for making inferences? What is the basis for saying a model is "best"?

Model selection relates to fitted models: given the data and the form of the model, then the MLEs of the model parameters have been found ("fitted"). Inference relates to theoretical models. It is necessary to consider four cases;

(1) models as structure only (θ value irrelevant),
(2) models as structure, plus specific θ_o (this is the theoretical best value),
(3) models as structure, plus MLE $\hat{\theta}$, fitted to data,
(4) models as structure by fitting, downplaying θ.

If a poor or inappropriate model (3, above) is used, then inference based on the data and this model will often be poor. Thus, it is clearly important to select (i.e., infer) an appropriate model (1, above) for the analysis of a specific data set; however, this is not the same as trying to find the "true model." Model selection methods with a deep level of theoretical support are required and, particularly, methods that are easy to use and widely applicable in practice. Part of "applicability" means that the methods have good operating characteristics for realistic sample sizes. As Potscher (1991) noted, asymptotic properties are of little value unless they hold for realized sample sizes.

A simple example will motivate some of the concepts presented. Flather (1992 and 1996) studied patterns of avian species-accumulation rates among forested landscapes in the eastern United States using index data from the Breeding Bird Survey (Bystrak 1981). He derived an a priori set of 9 candidate models from two sources: (1) the literature on species area curves (most often the power or exponential models were suggested) and (2) a broader search of the literature for functions that increased monotonically to an asymptote (Table 1.1). Which model should be used for the analysis of these ecological data? Clearly, none of these 9 models are likely to be the "truth" that generated

TABLE 1.1. Summary of a priori models of avian species-accumulation curves from Breeding Bird Survey index data for Indiana and Ohio (from Flather 1992:51 and 1996). The response variable (y) is the number of accumulated species, and the explanatory variable (x) is the accumulated number of samples. Nine models and their number of parameters are shown to motivate the question, "Which fitted model should be used for making inference from these data?"

Model structure	Number of parameters $(K)^a$
$E(y) = ax^b$	3
$E(y) = a + b \log(x)$	3
$E(y) = a(x/(b+x))$	3
$E(y) = a(1 - e^{-bx})$	3
$E(y) = a - bc^x$	4
$E(y) = (a + bx)/(1 + cx)$	4
$E(y) = a(1 - e^{-bx})^c$	4
$E(y) = a \left(1 - [1 + (x/c)^d]^{-b}\right)$	5
$E(y) = a[1 - e^{-(b(x-c))^d}]$	5

^{a}There are $K-1$ structural parameters and one residual variance parameter, σ^2. Assumed: $y = E(y)+\epsilon$, $E(\epsilon) = 0$, $V(\epsilon) = \sigma^2$.

the index data from the Breeding Bird Survey over the years of study. Instead, Flather wanted an approximating model that fit the data well and could be used in making inferences about bird communities on the scale of large landscapes. In this first example, the number of parameters in the candidate models ranges only from 3 to 5. Which approximating model is "best" for making inferences from these data is answered philosophically by the principle of parsimony (Section 1.4) and operationally by several information-theoretic criteria in Chapter 2. Methods for estimating model selection uncertainty and incorporating this into inferences are given in Chapter 2 and illustrated in Chapters 4 and 5.

Note, in each case, that the response variable y is being modeled, rather than mixing models of y with $\log(y)$, or other transformations of the response variable (Table 1.1). These models are in the sense of 1 above, as the structure is given but the parameter values are unspecified. Given appropriate data, ML can be used to obtain $\hat{\theta}$ in the sense of 3 above. In some of the physical sciences the model parameters are derived from theory, without the need for problem-specific empirical data. Such cases seem to be the exception in the biological sciences, where model parameters must usually be estimated from the data using least squares or likelihood theory.

1.2.4 Science Inputs: Formulation of the Set of Candidate Models

Model specification or formulation, in its widest sense, is conceptually more difficult than estimating the model parameters and their precision. Model formulation is the point where the scientific and biological information formally enter the investigation. Building the set of candidate models is partially a subjective art; that is why scientists must be trained, educated, and experienced in their discipline. The published literature and experience in the biological sciences can be used to help formulate a set of a priori candidate models. The most original, innovative part of scientific work is the phase leading to the proper question. Good approximating models, each representing a scientific hypothesis, in conjunction with a good set of relevant data can provide insight into the underlying biological process and structure.

Lehmann (1990) asks, "where do models come from," and cites some biological examples (also see Ludwig 1989, Walters 1996, Lindsey 1995). Models arise from questions about biology and the manner in which biological systems function. Relevant theoretical and practical questions arise from a wide variety of sources (see Box et al. 1978, O'Connor and Spotila 1992). Traditionally, these questions come from the scientific literature, results of manipulative experiments, personal experience, or contemporary debate within the scientific community. More practical questions stem from resource management controversies, biomonitoring programs, quasi-experiments, and even judicial hearings.

Chatfield (1995b) suggests that there is a need for more *careful thinking* (than is usually evident) and a *better balance* between the problem (biological question), analysis theory, and data. This suggestion has been made in the literature for decades. One must conclude that it has not been taught sufficiently in applied science or statistics courses. **Our science culture does not regularly do enough to expect and enforce critical thinking.** Too often, the emphasis is focused on the analysis theory and data analysis, with too little thought about the reason for the study in the first place (see Hayne 1978 for convincing examples).

Tukey (1980) argues for the need for deep thinking and early exploratory data analysis, and that the results of these activities lead to good scientific questions and confirmatory data analysis. In the exploratory phases, he suggests the importance of a flexible attitude and plotting of the data. He does not advocate the computation of test statistics, P-values, and so forth during exploratory data analysis. Tukey concludes that to implement the confirmatory paradigm properly we need to do a lot of exploratory work.

The philosophy and theory presented here must rest on well-designed studies and careful planning and execution of field or laboratory protocol. Many good books exist giving information on these important issues (Burnham et al. 1987, Cook and Campbell 1979, Mead 1988, Hairston 1989, Desu and Roghavarao 1991, Eberhardt and Thomas 1991, Manly 1992, Skalski and Robson 1992, Thompson 1992, Scheiner and Gurevitch 1993, Cox and Reid 2000, and Guisan and Zimmermann 2000). Chatfield (1991) reviews statistical pitfalls and ways that these might be avoided. Research workers are urged to pay close attention to these critical issues. Methods given here should not be thought to salvage poorly designed work. In the following material we will assume that the data are "sound" and that inference to some larger population is reasonably justified by the manner in which the data were collected.

Development of the a priori set of candidate models often should include a global model: a model that has many parameters, includes all potentially relevant effects, and reflects causal mechanisms thought likely, based on *the science of the situation*. The global model should also reflect the study design and attributes of the system studied. Specification of the global model should not be based on a probing examination of the data to be analyzed. At some early point, one should investigate the fit of the global model to the data (e.g., examine residuals and measures of fit such as R^2, deviance, or formal χ^2 goodness-of-fit tests) and proceed with analysis only if it is judged that the global model provides an acceptable fit to the data. Models with fewer parameters can then be derived as special cases of the global model. This set of reduced models represents plausible alternatives based on what is known or hypothesized about the process under study. Generally, alternative models will involve differing numbers of parameters; the number of parameters will often differ by at least an order of magnitude across the set of candidate models. Chatfield (1995b) writes concerning the importance of subject-matter considerations such as accepted theory, expert background knowledge, and prior information in addition to known constraints on both the model parameters and the variables in the models. All these factors should be brought to bear on the makeup of the set of candidate models, prior to actual data analysis.

The more parameters used, the better the fit of the model to the data that is achieved. Large and extensive data sets are likely to support more complexity, and this should be considered in the development of the set of candidate models. **If a particular model (parametrization) does not make biological sense, this is reason to exclude it from the set of candidate models, particularly in the case where causation is of interest**. In developing the set of candidate models, one must recognize a certain balance between keeping the set small and focused on plausible hypotheses, while making it big enough to guard against omitting a very good a priori model. While this balance should be considered, we advise the inclusion of all models that seem to have a reasonable justification, prior to data analysis. While one must worry about errors due to both underfitting and overfitting, it seems that modest overfitting is less damaging than underfitting (Shibata 1989). We recommend and encourage a considerable amount of careful, a priori thinking in arriving at a set of candidate models (see Peirce 1955, Burnham and Anderson 1992, Chatfield 1995b).

Freedman (1983) noted that when there are many, say 50, explanatory variables $(x_1, x_2, \ldots, x_{50})$ used to predict a response variable (y), variable-selection methods will provide regression equations with high R^2 values, "significant" F values, and many "significant" regression coefficients, as shown by large t values, *even if the explanatory variables are independent of y*. This undesirable situation occurs most frequently when the number of variables is of the same order as the number of observations. This finding, known as Freedman's paradox, was illustrated by Freedman using hypothesis testing as a means to select a model of y as a function of the x's, but the same type of problematic result can be found in using other model selection methods. Miller (1990) notes that estimated regression coefficients are biased away from zero in such cases; this is a type of model selection bias. The partial resolution of this paradox is in the a priori modeling considerations, keeping the number of candidate models small, achieving a large sample size relative to the number of parameters to be estimated, and basing inference on more than one model.

It is not uncommon to see biologists collect data on 50–130 "ecological" variables in the blind hope that some analysis method and computer system will "find the variables that are significant" and sort out the "interesting" results (Olden and Jackson 2000). This shotgun strategy will likely uncover mainly spurious correlations (Anderson et al. 2001b), and it is prevalent in the naive use of many of the traditional multivariate analysis methods (e.g., principal components, stepwise discriminant function analysis, canonical correlation methods, and factor analysis) found in the biological literature. We believe that mostly spurious results will be found using this unthinking approach (also see Flack and Chang 1987 and Miller 1990), and we encourage investigators to give very serious consideration to a well-founded set of candidate models and predictor variables (as a reduced set of possible prediction) as a means of minimizing the inclusion of spurious variables and relationships. Ecologists are not alone in collecting a small amount of data on a very large number of variables. A. J. Miller (personal communication) indicates that he has seen data sets in other fields with as many as 1,500 variables where the number of cases is less than 40 (a purely statistical search for meaningful relationships in such data is doomed to failure).

After a carefully defined set of candidate models has been developed, one is left with the evidence contained in the data; the task of the analyst is to interpret this evidence from analyzing the data. Questions such as, "What effects are supported by the data?" can be answered objectively. This modeling approach allows a clear place for experience (i.e., prior knowledge and beliefs), the results of past studies, the biological literature, and current hypotheses to enter the modeling process formally. Then, one turns to the data to see "what is important" within a sense of parsimony. In some cases, careful consideration of the number and nature of the predictor variables to be used in the analysis will suffice in defining the candidate models. This process may result in an initial set of, say, 15–40 predictor variables and a consolidation to a much smaller set to use in the set of candidate models. Using AIC and other similar methods one can only hope to select the best model from this set; if good models are not in the set of candidates, they cannot be discovered by model selection (i.e., data analysis) algorithms.

We lament the practice of generating models (i.e., "modeling") that is done in the total absence of real data, and yet "inferences" are made about the status, structure, and functioning of the real world based on studying these models. We do not object to the often challenging and stimulating intellectual exercise of model construction as a means to integrate and explore our myriad ideas about various subjects. For example, Berryman et al. (1995) provide a nice list of 26 candidate models for predator–prey relationships and are interested in their "credibility" and "parsimony." However, as is often the case, there are no empirical data available on a variety of taxa to pursue these issues in a rigorous manner (also see Turchin and Batzli (2001), who suggest 8 models, each a system of 2–3 differential equations, for vegetation–herbivore population interactions). Such exercises help us sort out ideas that in fact conflict when their logical consequences are explored. Modeling exercises can strengthen our logical and quantitative abilities. Modeling exercises can give us insights into how the world *might* function, and hence modeling efforts can lead to alternative hypotheses to be explored with real data. Our objection is only to the confusing of presumed insights from such models with inferences about the real world (see Peters 1991, Weiner 1995). An inference from a model to some aspect of the real world is justified only after the model has been shown to adequately fit relevant empirical data (this will certainly be the case when the model in its totality has been fit to and tested against reliable data). Gause (1934) had similar beliefs when he stated, "Mathematical investigations independent of experiments are of but small importance . . . ."

The underlying philosophy of analysis is important here. We advocate a conservative approach to the overall issue of *strategy* in the analysis of data in the biological sciences with an emphasis on a priori considerations and models to be considered. *Careful, a priori consideration of alternative models will often require a major change in emphasis among many people.* This is often an unfamiliar concept to both biologists and statisticians, where there has been a tendency to use either a traditional model or a model with associated computer software, making its use easy (Lunneborg 1994). This a priori strategy is in contrast to strategies advocated by others who view modeling and data analysis as a highly iterative and interactive exercise. Such a strategy, to us, represents deliberate data dredging and should be reserved for early exploratory phases of initial investigation. Such an exploratory avenue is not the subject of this book.

Here, we advocate the deliberate exercise of carefully developing a set of, say, 4–20 alternative models as potential approximations to the population-level information in the data available and the scientific question being addressed (Lytle 2002 provides an advanced example). Some practical problems might have as many as 70–100 or more models that one might want to consider. The number of candidate models is often larger with large data sets. We find that people tend to include many models that are far more general than the data could reasonably support (e.g., models with several interaction parameters). There need to be some well-supported guidelines on this issue to help analysts better define the models to be considered. This set of models, developed without first deeply examining the data, constitutes the "*set of candidate models.*" The science of the issue enters the analysis through the a priori set of candidate models.

1.2.5   Models Versus Full Reality

Fundamental to our paradigm is that none of the models considered as the basis for data analysis are the "true model" that generates the biological data we observe (see, for example, Bancroft and Han 1977). We believe that "truth" (full reality) in the biological sciences has essentially infinite dimension, and hence full reality cannot be revealed with only finite samples of data and a "model" of those data. It is generally a mistake to believe that there is a simple "true model" in the biological sciences and that during data analysis this model can be uncovered and its parameters estimated. Instead, biological systems are complex, with many small effects, interactions, individual heterogeneity, and individual and environmental covariates (most being unknown to us); we can only hope to identify a model that provides a good *approximation* to the data available. The words "true model" represent an oxymoron, except in the case of Monte Carlo studies, whereby a model is used to generate "data" using pseudorandom numbers (we will use the term "generating model" for such computer-based studies). The concept of a "true model" in biology seems of little utility and may even be a source of confusion about the nature of approximating models (e.g., see material on BIC and related criteria in Chapter 6).

A model is a simplification or approximation of reality and hence will not reflect all of reality. Taub (1993) suggests that unproductive debate concerning true models can be avoided by simply recognizing that a model is not truth by definition. Box (1976) noted that "all models are wrong, but some are useful." While a model can never be "truth," a model might be ranked from very useful, to useful, to somewhat useful to, finally, essentially useless. Model selection methods try to rank models in the candidate set relative to each other; whether any of the models is actually "good" depends primarily on the quality of the data and the science and a priori thinking that went into the modeling. Full truth (reality) is elusive (see deLeeuw 1988). Proper modeling and data analysis tell what inferences the data support, not what full reality might be (White et al. 1982:14–15, Lindley 1986). Models, used cautiously, tell us "what effects are supported by the (finite) data available." Increased sample size (information) allows us to chase full reality, but never quite catch it.

The concept of truth and the false concept of a true model are deep and surprisingly important. Often, in the literature, one sees the words *correct* model or simply *the* model as if to be vague as to the exact meaning intended. Bayesians seem to say little about the subject, even as to the exact meaning of the prior probabilities on models. Consider the simple model of population size (n) at time t,

$$n_{t+1} = n_t \cdot s_t,$$

where s is the survival probability during the interval from t to $t + 1$. This is a correct model in the sense that it is algebraically and deterministically correct; however, it is not an exact representation or model of truth. This model is not explanatory; it is definitional (it is a tautology, because it implies that

$s_t = n_{t+1}/n_t$). For example, from the theory of natural selection, the survival probability differs among the n animals. Perhaps the model above could be improved if average population survival probability was a random variable from a beta distribution; still, this is far from a model of full reality or truth, even in this very simple setting. Individual variation in survival could be caused by biotic and abiotic variables in the environment. Thus, a more exact model of full reality would have, at the very least, the survival of each individual as a nonlinear function of a large number of environmental variables and their interaction terms. Even in this simple case, it is surely clear that one cannot expect any mathematical model to represent full reality; there are no true models in the biological sciences. We will take a set of approximating models g_i, without pretending that one represents full reality and is therefore "true."

In using some model selection methods it is assumed that the set of candidate models contains the "true model" that generated the data. We will not make this assumption, unless we use a data set generated by Monte Carlo methods as a tutorial example (e.g., Section 3.4), and then we will make this artificial condition clear. In the analysis of real data, it seems unwarranted to pretend that the "true model" is included in the set of candidate models, or even that the true model exists at all. Even if a "true model" did exist and if it could be found using some method, it would not be good as a fitted model for general inference (i.e., understanding or prediction) about some biological system, because its numerous parameters would have to be estimated from the finite data, and the precision of these estimated parameters would be quite low.

Often the investigator wants to simplify some representation of reality in order to achieve an understanding of the dominant aspects of the system under study. If we were given a nonlinear formula with 200 parameter values, we could make correct predictions, but it would be difficult to *understand* the main dynamics of the system without some further simplification or analysis. Thus, one should tolerate some inexactness (an inflated error term) to facilitate a simpler and more useful understanding of the phenomenon.

In particular, we believe that there are tapering effect sizes in many biological systems; that is, there are often several large, important effects, followed by many smaller effects, and, finally, followed by a myriad of yet smaller effects. These effects may be sequentially unveiled as sample size increases. The main, dominant, effects might be relatively easy to identify and support, even using fairly poor analysis methods, while the second-order effects (e.g., a chronic treatment effect or an interaction term) might be more difficult to detect. The still smaller effects can be detected only with very large sample sizes (cf. Kareiva 1994 and related papers), while the smallest effects have little chance of being detected, even with very large samples. Rare events that have large effects may be very important but quite difficult to study. Approximating models must be related to the amount of data and information available; small data sets will appropriately support only simple models with few parameters, while more comprehensive data sets will support, if necessary, more complex models.

This tapering in "effect size" and high dimensionality in biological systems might be quite different from some physical systems where a small-dimensioned model with relatively few parameters might accurately represent full truth or reality. Biologists should not believe that a simple "true model" exists that generates the data observed, although some biological questions might be of relatively low dimension and could be well approximated using a fairly simple model. The issue of a range of tapering effects has been realized in epidemiology, where Michael Thun notes, "... you can tell a little thing from a big thing. What's very hard to do is to tell a little thing from nothing at all" (Taubes 1995). *Full reality will always remain elusive in the biological sciences.*

At a more advanced conceptual level, these is a concept that "information" about the population (or process or system) under study exists in the data and the goal is to express this information in a more compact, understandable form using a "model." Conceptually, this is a change in coding system, similar to using a different "alphabet." The data have only a finite, fixed amount of information. The *goal* of model selection is to achieve a perfect one-to-one translation so that no information is lost; in fact, we cannot achieve this ideal. The data can be ideally partitioned into *information* and *noise*. The noise part of the data is not information. However, noise could contain information that we cannot decode. Conceptually, the role of a good model is to filter the data so as to separate information from noise.

Our main emphasis in modeling empirical data is to understand the biological structure, process, or system. Sometimes prediction will be of interest; here, however, one would hopefully have an understanding of the structure of the system as a basis for making trustworthy predictions. We recommend developing a set of candidate models prior to intensive data analysis, selecting one that is "best," and estimating the parameters of that model and their precision (using maximum likelihood or least squares methods). This unified strategy is a basis for valid inferences, and there are several more advanced methods to allow additional inferences and insights. In particular, models exist to allow formal inference from more than one model, and this has a number of advantages (Hoeting et al. 1999). Statistical science is not so much a branch of mathematics, but rather it is concerned with the development of a practical theory of information using what is known or postulated about the science of the matter. In our investigations into these issues we were often surprised by how much uncertainty there is in selecting a good approximating model; the variability in terms of what model is selected or considered best from independent data sets, for example, is often large.

1.2.6   An Ideal Approximating Model

We consider some properties of an ideal model for valid inference in the analysis of data. It is important that the best model is selected from a set of models that were defined prior to data analysis and based on the science of the issue at hand. Ideally, the process by which a "best" model is selected would be objective and repeatable; these are fundamental tenets of science. The ideal model would be appropriately simple, based on concepts of parsimony. Furthermore, precise, unbiased estimators of parameters would be ideal, as would accurate estimators of precision. The best model would ideally yield achieved confidence interval coverage close to the nominal level (often 0.95) and have confidence intervals of minimum width. Achieved confidence interval coverage is a convenient index to whether parameter estimators and measures of precision are adequate. Finally, one would like as good an approximation of the structure of the system as the information permits. Thus, in many cases adjusted R^2 can be computed and σ^2 estimated as a measure of variation explained or residual variation, respectively. Ideally, the parameters in the best model would have biological interpretations. If prediction was the goal, then having the above issues in place might warrant some tentative trust in model predictions. There are many cases where two or more models are essentially tied for "best," and this should be fully recognized in further analysis and inference, especially when they produce different predictions. In other cases there might be 4–10 models that have at least some support, and these, too, deserve scrutiny in reaching conclusions from the data, based on inferences from more than a single model.

1.3 Model Fundamentals and Notation

This section provides a conceptualization of some important classes of models as they are used in this book. Some of these classes are particularly important in model selection. A general notation is introduced that is intended to be helpful to readers.

1.3.1 Truth or Full Reality f

While there are no models that exactly represent full reality (cf. Section 1.2.5), full truth can be denoted as f. The concept of f is abstract. It is this truth to which we want to make inferences, based on data and approximating models. We use the notation $f(x)$ to denote that integration is over the variable x, but we do not want to convey the notion that f is a function of the data x. Data arise from full reality and can be used to make formal inferences back to this truth, if data collection has been carefully planned and proper sampling or experimental design has been achieved.

1.3.2 Approximating Models $g_i(x|\theta)$

We use the notation $g_i(x|\theta)$ or often, if the context is clear, g_i to denote the ith approximating model. We use θ to represent generally a parameter or vector of parameters. Thus, θ is generic and might represent parameters in a regression model $(\beta_0, \beta_1, \beta_2)$ or the probability of a head in penny flipping trials (p). The models g_i are discrete or continuous probability distributions, and our focus will be on their associated likelihoods, $\mathcal{L}(\theta|data, model)$ or log-likelihoods $\log(\mathcal{L}(\theta|data, model))$. Notation for the log-likelihood will sometimes be shortened to $\log(\mathcal{L}(\theta|x, g))$ or even $\log(\mathcal{L})$. Ideally, the set of R models will have been defined prior to data analysis. These models specify only the form of the model, leaving the unknown parameters (θ) unspecified.

A simple example will aid in the understanding of this section. Consider a study of mortality (μ_c) as a function of concentration (c) of some chemical compound. The size (s) of the animal (binary as small or large) and a group covariate $(z$, such as gender) are also recorded, because they are hypothesized to be important in better understanding the concentration–mortality function. Investigators might consider mortality probability during some fixed time interval to be a logistic function of concentration, where, for example, $c = 0, 1, 2, 4, 8$, and 16. The full structure of the logistic model when all 3 variables are included in the model can be written as,

$$\mu_c = \frac{1}{1 + \exp\{-(\beta_0 + \beta_1 c + \beta_2 s + \beta_3 z)\}}.$$

Use of the logistic link function allows the expression to be written as a linear model structure,

$$\text{logit}(\mu_c) = \log_e\left(\frac{\mu_c}{1 - \mu_c}\right) = \beta_0 + \beta_1 c + \beta_2 s + \beta_3 z.$$

Here the data (y) are binary for mortality (dead or alive), size (small or large), and gender (male and female), while concentration is recorded at 6 fixed levels. The response variable $y = 1$ if the animal died and 0 if it lived, given a particular concentration. Then,

$$\text{Prob}\{y = 1|c, s, z\} = \mu_c$$

for n individuals at concentration c, size s, and gender z. Then, the likelihood is proportional to

$$\mathcal{L}(\mu_c|data, model) = \prod_{i=1}^{n} (\mu_c(i))^{y_i} (1 - \mu_c(i))^{1-y_i}.$$

Thus, a set of approximating structural models might be defined, based on the science of the issue. The stochastic part of the model is assumed to be Bernoulli. The models are alternatives, defined prior to data analysis, and the interest is in the strength of evidence for each of the alternative hypotheses, represented by models. Five $(R = 5)$ structural models will be used for illustration:

$$g_1(x): \quad \text{logit}(\mu_c) = \beta_0 + \beta_1 c + \beta_2 s + \beta_3 z,$$
$$g_2(x): \quad \text{logit}(\mu_c) = \beta_0 + \beta_1 c + \beta_2 s,$$
$$g_3(x): \quad \text{logit}(\mu_c) = \beta_0 + \beta_1 c \qquad + \beta_3 z,$$

$$g_4(x): \quad \text{logit}(\mu_c) = \beta_0 + \beta_1 c,$$
$$g_5(x): \quad \text{logit}(\mu_c) = \beta_0.$$

These models specify the structural form (including how the parameters and covariates enter), but not the parameter values (the β_i); each assumes that the y are independent Bernoulli random variables. The first model serves as a global model. The second model represents the hypothesis that the group covariate (z) is unimportant, while the third model is like the first, except that the size is hypothesized to be unimportant. The fifth model implies that mortality is constant and not a function of concentration. Often, enough is known about the compound that model g_5 is not worth exploration. Of course, the log-log or complementary log-log, or probit function could have been used to model the hypothesized relationships in this example, rather than the logistic.

1.3.3 The Kullback–Leibler Best Model $g_i(x|\theta_0)$

For given full reality (f), data (x), sample size (n), and model set (R) there is a best model in the sense of Kullback-Leibler information (introduced in Chapter 2). That is, given the possible data, the form of each model, and the possible parameter values, K-L information can be computed for each model in the set and the model best approximating full reality determined.

The parameters that produce this conceptually best single model, in the class $g(x|\theta)$, are denoted by θ_0, Of course, this model is generally unknown to us but can be estimated; such estimation involves computing the MLEs of the parameters in each model ($\hat{\theta}$) and then *estimating* K-L information as a basis for model selection and inference. The MLEs converge asymptotically to θ_0 and the concept of bias is with respect to θ_0, rather than our conceptual "true parameters" associated with full reality f.

1.3.4 Estimated Models $g_i(x|\hat{\theta})$

Estimated models have specific parameter values from ML or LS estimation, based on the given data and model. If another, replicate data set were available and based on the same sample size, the parameter estimates would differ somewhat; the amount of difference expected is related to measures of precision (e.g., standard errors and confidence intervals). It is important to keep separate the model form $g_i(x|\theta)$ from specific estimates of this model, based on data and the process of parameter estimation, $g_i(x|\hat{\theta})$.

In the models of mortality as a function of concentration and other variables (above), there are associated likelihoods and log-likelihoods. Likelihood theory can be used to obtain the MLEs $\hat{\beta}_0$ and $\hat{\beta}_1$ for model g_4, for example. The likelihood function is

$$\mathcal{L}(\beta_0, \beta_1 | data, \ model) = \prod_{i=1}^{n} (\mu_c(i))^{y_i} \, (1 - \mu_c(i))^{1-y_i},$$

where

$$\mu_c = \frac{1}{1 + \exp\{-(\beta_0 + \beta_1 c)\}}.$$

Thus, the only parameters in the likelihood are β_0 and β_1 and given the data, one can obtain the MLEs. The value of the maximized log-likelihood and the estimated variance–covariance matrix can also be computed. In a sense, when we have only the model form $g(x|\theta)$ we have an infinite number of models, where all such models have the same form but different values of θ. Yet, in all of these models there is a unique K-L best model. Conceptually, we know how to find this model, given f.

1.3.5   Generating Models

Monte Carlo simulation is a very useful and general approach in theoretical and applied statistics (Manly 1991). These procedures require that a model be specified as the basis for generating Monte Carlo data. Such a model is not full reality, and thus we call it a *generating model*. It is "truth" only in the sense of computerized truth. One should not confuse a generating model or results based on Monte Carlo data with full reality f.

1.3.6   Global Model

Ideally, the global model has in it all the factors or variables thought to be important. Other models are often special cases of this global model. There is not always a global model. If sample size is small, it may be impossible to fit the global model. Goodness-of-fit tests and estimates of an overdispersion parameter for count data should be based (only) on the global model. The concept of overdispersion is relatively model-independent; however, some model must be used to compute or model any overdispersion thought to exist in count data. Thus, the most highly parametrized model will serve best as the basis for assessing overall fit and estimating a parameter associated with overdispersion. In the models of mortality (above), model g_1 would serve as the global model.

The advantage of this approach is that if the global model fits the data adequately, then a selected model that is more parsimonious will also fit the data (this is an empirical result, not a theorem). Parsimonious model selection should not lead to a model that does not fit the data (this property seems to hold for the selection methods we advocate here). Thus, goodness-of-fit assessment and the estimation of overdispersion parameters should be addressed using the global model (this could also be computed for the selected model).

In summary, we will use the word "model" to mean different things; hopefully, the context will be clear. Certainly it is important to distinguish clearly between f and g. The general structural form is denoted by $g(x|\theta)$, without specifying the numerical value of the parameter θ (e.g., models given in Table

1.1). If one considers estimation of θ, then there are an infinite number of possible values of θ. Therefore, there is an entire class of models $g(x|\theta)$, defined by the space over which θ varies. Frequently, we will refer to the model where MLEs (the most likely, given the data and the model) have been found. In other cases we will mean the best model, $g(x|\theta_0)$, which is one specific model (the K-L best relative to f).

1.3.7   Overview of Stochastic Models in the Biological Sciences

Models are useful in the biological sciences for understanding the structure of systems, estimating parameters of interest and their associated variance–covariance matrix, predicting outcomes and responses, and testing scientific hypotheses. Such models might be used for "relational" or "explanatory" purposes or might be used for prediction. In the following material we will review the main types of models used in the biological sciences. Although the list is not meant to be exhaustive, it will allow the reader an impression of the wide class of models of empirical data that we will treat under an information-theoretic framework.

Simple linear and multiple linear regression models (Seber 1977, Draper and Smith 1981, Brown 1993) have seen heavy use in the biological sciences over the past four decades. These models commonly employ one to perhaps 8–12 parameters, and the statistical theory is fully developed (either based on least squares or likelihood theory). Similarly, analysis of variance and covariance models have been widely used, and the theory underlying these methods is closely related to regression models and is fully developed (both are examples of general linear models). Theory and software for this wide class of methods are readily available.

Nonlinear regression models (Gallant 1987, Seber and Wild 1989, Carroll et al. 1995) have also seen abundant use in the biological sciences (logistic regression is a common example). Here, the underlying theory is often likelihood based, and some classes of nonlinear models require very specialized software. In general, nonlinear estimation is a more advanced problem and is somewhat less well understood by many practicing researchers.

Other types of models used in the biological sciences include generalized linear (McCullagh and Nelder 1989, Morgan 1992, 2000) and generalized additive (Hastie and Tibshirani 1990) models (these can be types of nonlinear regression models). These modeling techniques have seen increasing use in the past decade. Multivariate modeling approaches such as multivariate ANOVA and regression, canonical correlation, factor analysis, principal components analysis, and discriminate function analysis have had a checkered history in the biological and social sciences, but still see substantial use (see review by James and McCulloch 1990). Log-linear and logistic models (Agresti 1990) have become widely used for count data. Time series models (Brockwell and Davis 1987, 1991) are used in many biological disciplines. Various models of an organism's growth (Brisbin et al. 1987, Gochfeld 1987) have been proposed and used in biology. Caswell (2001) provides a large number of matrix population models that have seen wide use in the biological sciences.

Compartmental models are a type of state transition in continuous time and continuous response and are usually based on systems of differential or partial differential equations (Brown and Rothery 1993, Matis and Kiffe 2000). There are discrete state transition models using the theory of Markov chains (Howard 1971); these have found use in a wide variety of fields including epidemiological models of disease transmission. More advanced methods with potentially wide application include the class of models called "random effects" (Kreft and deLeeuw 1998).

Models to predict population viability (Boyce 1992), often based on some type of Leslie matrix, are much used in conservation biology, but rarely are alternative model forms given serious evaluation. A common problem here is that these models are rarely based on empirical data; the form of the model and its parameter values are often merely only "very rough guesses" necessitated by the lack of empirical data (White 2000).

Biologists in several disciplines employ differential equation models in their research (see Pascual and Kareiva 1996 for a reanalysis of Gause's competition data and Roughgarden 1979 for examples in population genetics and evolutionary ecology). Many important applications involve exploited fish populations (Myers et al. 1995). Computer software exists to allow model parameters to be estimated using least squares or maximum likelihood methods (e.g., SAS and Splus). These are powerful tools in the analysis of empirical data, but also beg the issue of "what model to use."

Open and closed capture–recapture (Lebreton et al. 1992) and band recovery (Brownie et al. 1985) models represent a class of models based on product multinomial distributions (see issues 5 and 6 of volume 22 of the *Journal of Applied Statistics*, 1995). Distance sampling theory (Buckland et al. 1993, 2001) relies on models of the detection function and often employs semiparametric models. Parameters in these models are nearly always estimated using maximum likelihood.

Spatial models (Cressie 1991 and Renshaw 1991) are now widely used in the biological sciences, allowing the biologist to take advantage of spatial data sets (e.g., geographic information systems). Stein and Corsten (1991) have shown how Kriging (perhaps the most widely used spatial technique) can be expressed as a least squares problem, and the development of Markov chain Monte Carlo methods such as the Gibbs sampler (Robert and Casella 1999, Chen et al. 2000) allow other forms of spatial models to be fitted by least squares or maximum likelihood (Augustin et al. 1996). Further unifying work for methods widely used on biological data has been carried out by Stone and Brooks (1990). Geographic information systems potentially provide large numbers of covariates for biological models, so that model selection issues are particularly important.

Spatiotemporal models are potentially invaluable to the biologist, though most researchers model changes over space or time, and not both simultaneously. The advent of Markov chain Monte Carlo methods (Gilks et al. 1996, Gamerman 1997) may soon give rise to a general but practical framework for spatiotemporal modeling; model selection will be an important component of such a framework. A step towards this general framework was made by Buckland and Elston (1993), who modeled changes in the spatial distribution of wildlife.

There are many other examples where modeling of data plays a fundamental role in the biological sciences. Henceforth, we will exclude only modeling that cannot be put into a likelihood or quasi-likelihood (Wedderburn 1974) framework and models that do not explicitly relate to empirical data. All least squares formulations are merely special cases that have an equivalent likelihood formulation in usual practice. There are general information-theoretic approaches for models well outside the likelihood framework (Qin and Lawless 1994, Ishiguo et al. 1997, Hurvich and Simonoff 1998, and Pan 2001a and b). There are now model selection methods for nonparametric regression, splines, kernel methods, martingales, and generalized estimation equations. Thus, methods exist for nearly all classes of models we might expect to see in the theoretical or applied biological sciences.

1.4   Inference and the Principle of Parsimony

1.4.1   Avoid Overfitting to Achieve a Good Model Fit

Consideroverfitting two analysts studying a small set of biological data using a multiple linear regression model. The first exclaims that a particular model provides an excellent fit to the data. The second notices that 22 parameters were used in the regression and states, "Yes, but you have used enough parameters to fit an elephant!" This seeming conflict between increasing model fit and increasing numbers of parameters to be estimated from the data led Wel (1975) to answer the question, "How many parameters *does* it take to fit an elephant?" Wel finds that about 30 parameters would do reasonably well (Figure 1.2); of course, had he fit 36 parameters to his data, he could have achieved a perfect fit.

Wel's finding is both insightful and humorous, but it deserves further interpretation for our purposes here. His "standard" is itself only a crude drawing—it even lacks ears, a prominent elephantine feature; hardly truth. A better target would have been a large, digitized, high-resolution photograph; however, this, too, would have been only a model (and not truth). Perhaps a real elephant should have been used as truth, but this begs the question, "Which elephant should we use?" This simple example will encourage thinking about full reality, "true models," and approximating models and motivate the *principle of parsimony* in the following section. **William of Occam suggested in the fourteenth century that one "shave away all that is unnecessary"—a dictum often referred to as *Occam's razor*. Occam's razor has had a long history**

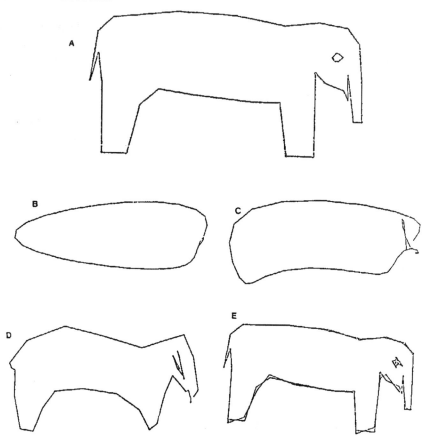

FIGURE 1.2. "How many parameters does does it take to fit an elephant?" was answered by Wel (1975). He started with an idealized drawing (A) defined by 36 points and used least squares Fourier sine series fits of the form $x(t) = \alpha_0 + \sum \alpha_i \sin(it\pi/36)$ and $y(t) = \beta_0 + \sum \beta_i \sin(it\pi/36)$ for $i = 1, \ldots, N$. He examined fits for $K = 5, 10, 20$, and 30 (shown in B–E) and stopped with the fit of a 30 term model. He concluded that the 30-term model "may not satisfy the third-grade art teacher, but would carry most chemical engineers into preliminary design."

in both science and technology, and it is embodied in the principle of parsimony. Albert Einstein is supposed to have said, "Everything should be made as simple as possible, but no simpler."

Success in the analysis of real data and the resulting inference often depends importantly on the choice of a best approximating model. Data analysis in the biological sciences should be based on a parsimonious model that provides an accurate approximation to the structural information in the data at hand; this should not be viewed as searching for the "true model." Modeling and model selection are essentially concerned with the "art of approximation" (Akaike 1974).

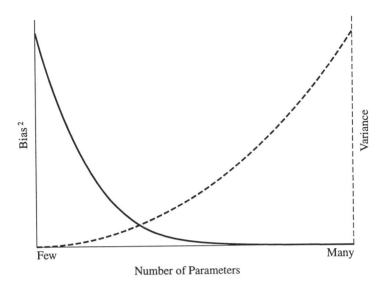

FIGURE 1.3. The *principle of parsimony*: the conceptual tradeoff between squared bias (solid line) and variance vs. the number of estimable parameters in the model (K). All model selection methods implicitly employ some notion of this tradeoff. The best approximating model need not occur exactly where the two curves intersect. Full truth or reality is not attainable with finite samples and usually lies well to the right of the region in which the best approximating model lies (the tradeoff region). Bias decreases and variance (uncertainty) increases as the number of parameters in a model increases.

1.4.2   The Principle of Parsimony

If the fit is improved by a model with more parameters, then where should one stop? Box and Jenkins (1970:17) suggested that the *principle of parsimony* should lead to a model with "... the smallest possible number of parameters for adequate representation of the data." Statisticians view the principle of parsimony as a bias versus variance tradeoff. In general, bias decreases and variance increases as the dimension of the model (K) increases (Figure 1.3). Often, we may use the number of parameters in a model as a measure of the degree of structure inferred from the data. The fit of any model can be improved by increasing the number of parameters (e.g., the elephant-fitting problem); however, a tradeoff with the increasing variance must be considered in selecting a model for inference. Parsimonious models achieve a proper tradeoff between bias and variance. All model selection methods are based to some extent on the principle of parsimony (Breiman 1992, Zhang 1994).

   In understanding the utility of an approximate model for a given data set, it is convenient to consider two undesirable possibilities: underfitted and over-fitted models. Here, we must avoid judging a selected model in terms of some supposed "true model," as occurs when data are simulated from a known, often very simple, model using Monte Carlo methods. In this case, if the generating model had 10 parameters, it is often said that an approximating model with only 7 parameters is underfitted (compared with the generating model with 10 parameters). This interpretation is often of little value, because it largely ignores the principle of parsimony and its implications and hinges on the misconception that such a simple true model exists in biological problems. If we believe that truth is essentially infinite-dimensional, then overfitting is not even defined in terms of the number of parameters in the fitted model. We will avoid this use of the terms "underfitted" and "overfitted" that suppose the existence of a low-dimensional "true model" as a "standard."

Instead, we reserve the terms underfitted and overfitted for use in relation to a "best approximating model" (Section 1.2.6). Here, an underfitted model would ignore some important replicable (i.e., conceptually replicable in most other samples) structure in the data and thus fail to identify effects that were actually supported by the data. In this case, bias in the parameter estimators is often substantial, and the sampling variance is underestimated, both factors resulting in poor confidence interval coverage. Underfitted models tend to miss important treatment effects in experimental settings. Overfitted models, as judged against a best approximating model, are often free of bias in the parameter estimators, but have estimated (and actual) sampling variances that are needlessly large (the precision of the estimators is poor, relative to what could have been accomplished with a more parsimonious model). Spurious treatment effects tend to be identified, and spurious variables are included with overfitted models. Shibata (1989) argues that underfitted models are a more serious issue in data analysis and inference than overfitted models. This assessment breaks down in many exploratory studies where sample size might be only 35–80 and there are 20–80 explanatory variables. In these cases, one may expect substantial overfitting and many effects that are actually spurious (Freedman 1983, Anderson et al. 2001b).

The concept of parsimony and a bias versus variance tradeoff is very important. Thus we will provide some additional insights (also see Forster 1995, Forster and Sober 1994, and Jaffe and Spirer 1987). The goal of data collection and analysis is to make inferences from the sample that properly apply to the population. The inferences relate to the *information* about structure of the system under study as inferred from the models considered and the parameters estimated in each model. A paramount consideration is the repeatability, with good precision, of any inference reached. When we imagine many replicate samples, there will be some recognizable features common to almost all of the samples. Such features are the sort of inference about which we seek to make strong inferences (from our single sample). Other features might appear in, say, 60% of the samples yet still reflect something real about the population or process under study, and we would hope to make weaker inferences concerning these. Yet additional features appear in only a few samples, and these might be best included in the error term (σ^2) in modeling. If one were to make an inference about these features quite unique to just the single data set at hand, as if they applied to all (or most all) samples (hence to the population), then we would say that the sample is overfitted by the model (we have overfitted the *data*). Conversely, failure to identify the features present that are strongly replicable over samples is underfitting. The data are not being approximated; rather we approximate the structural information in the data that is replicable over such samples (see Chatfield 1996, Collopy et al. 1994). Quantifying that structure with a model form and parameter estimates is subject to some "sampling variation" that must also be estimated (inferred) from the data.

True replication is very advantageous, but this tends to be possible only in the case of strict experiments where replication and randomization are a foundation. Such experimental replication allows a valid estimate of residual variation (σ^2). An understanding of these issues makes one realize what is lost when observational studies seem possible and practical, and strict experiments seem less feasible.

A best approximating model is achieved by properly balancing the errors of underfitting and overfitting. Stone and Brooks (1990) comment on the "... straddling pitfalls of underfitting and overfitting." The proper balance is achieved when bias and variance are controlled to achieve confidence interval coverage at approximately the nominal level and where interval width is at a minimum. Proper model selection rejects a model that is far from reality and attempts to identify a model in which the error of approximation and the error due to random fluctuations are well balanced (Shibata 1983, 1989). Some model selection methods are "parsimonious" (e.g., BIC, Schwarz 1978) but tend, in realistic situations, to select models that are too simple (i.e., underfitted); thus, bias is large, precision is overestimated, and achieved confidence interval coverage is well below the nominal level. Such instances are not satisfactory for inference. One has only a highly precise, quite biased result.

Sakamoto et al. (1986) simulated data to illustrate the concept of parsimony and the errors of underfitting and overfitting models (Figure 1.4). Ten data sets (each with $n = 21$) were generated from the simple model

$$y = e^{(x-0.3)^2} - 1 + \epsilon,$$

where x varied from 0 to 1 in equally spaced steps of 0.05, and $\epsilon \sim N(0, 0.01)$. Thus, in this case, they considered the generating model to have $K = 3$ parameters: 0.3, -1, and 0.01. They considered the set of candidate models (i.e., the approximating models) to be simple polynomials of order 0 to 5, as in the table below.

Order	K	Approximating Model
0	2	$E(y) = \beta_0$
1	3	$E(y) = \beta_0 + \beta_1(x)$
2	4	$E(y) = \beta_0 + \beta_1(x) + \beta_2(x^2)$
3	5	$E(y) = \beta_0 + \beta_1(x) + \beta_2(x^2) + \beta_3(x^3)$
4	6	$E(y) = \beta_0 + \beta_1(x) + \beta_2(x^2) + \beta_3(x^3) + \beta_4(x^4)$
5	7	$E(y) = \beta_0 + \beta_1(x) + \beta_2(x^2) + \beta_3(x^3) + \beta_4(x^4) + \beta_5(x^5)$.

Thus, each of these 6 models was fit to each of the 10 simulated data sets.

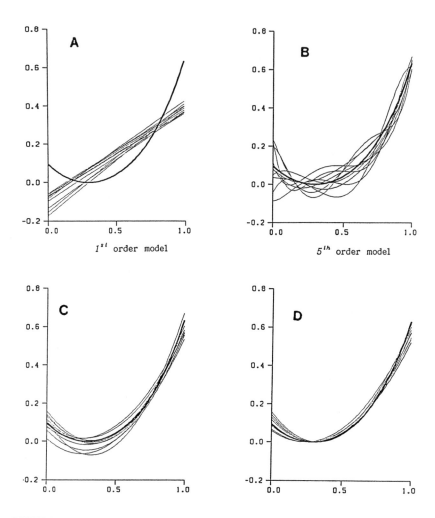

FIGURE 1.4. Ten Monte Carlo repetitions of data sets ($n = 21$) generated from the model $y = e^{(x-0.3)^2} - 1 + \epsilon; 0 \leq x \leq 1, \epsilon \sim N(0, .01)$ (from Sakamoto et al. 1986:164–179). A 1st-order polynomial (A) clearly misidentifies the basic nonlinear structure, and is underfitted and unsatisfactory. A 5th-order polynomial (B) has too many parameters, an unnecessarily large variance, and will have poor predictive qualities because it is unstable (overfitted). Neither A nor B is properly parsimonious, nor do they represent a best approximating model. A 2nd-order polynomial seems quite good as an approximating model (C). If it is known that the function is nonnegative and has its minimum at $x = 0.3$, then the approximating model that enforces these conditions is improved further (D). In more realistic situations, one lacks the benefit of simple plots and 10 independent data sets, such as those shown in A–D. See Section 3.7 for a full analysis of these data.

Strong model bias occurs when an underfitting (e.g., the mean-only model with $K = 2$ or the linear, 1st order, $K = 3$) model is employed (Figure 1.4A). Here bias is obvious, the nonlinear structure of the generating model is poorly approximated, and confidence interval coverage and predictions from the model will be quite poor. Of course, there is *some* model bias for each of the 5 models because they are only simple polynomial approximations. Overfitting is illustrated in Figure 1.4B, where a 5th-order polynomial ($K = 7$) is used as an approximating model. Here, there is little evidence of bias (an average quantity), precision is obviously poor, and it is difficult to identify the simple structure of the model. Prediction will be quite imprecise from this model, and it has features that do not occur in the generating model, particularly if one extrapolates beyond the range of the data (always a risky practice). Both underfitting and overfitting are undesirable in judging approximating models for data analysis.

If a second-order polynomial ($K = 4$) is used as the approximating model, the fits seem quite reasonable (Figure 1.4C), and one might expect valid inference from this model. Finally, *if* it were known a priori from the science of the situation that the function was nonnegative and had a minimum of zero at $x = 0.3$, then an improved quadratic approximating model could use this information very effectively (Figure 1.4D). The form of this model is

$$E(y) = \beta_0(x + \beta_1)^2$$

with $K = 3$ (i.e., β_0, β_1, and σ^2), whereas the second-order polynomial has 4 parameters. This example illustrates that valid statistical inference is only partially dependent on the analysis process; the science of the situation must play an important role through modeling. This particular example provides a visual image of underfitting and overfitting in a simple case where the generating model and various approximating models can be easily graphed in two dimensions. Parsimony issues with real data in the biological sciences nearly always defy such a simple graphical approach because truth is not known; one rarely has 10 independent data sets on exactly the same process, and plots in high dimensions are problematic to produce and interpret. Note, also, that the generating model contained no tapering effects. However, the approximating models do have tapering effects. Therefore, objective and effective methods are needed that do not rely on simple graphics and can cope with the real-world complexities and high dimensionality.

1.4.3   Model Selection Methods

Model selection has most often been viewed, and hence taught, in a context of null hypothesis testing. Sequential testing has most often been employed, either stepup (forward) or stepdown (backward) methods. Stepwise procedures allow for variables to be added or deleted at each step. These testing-based methods remain popular in many computer software packages in spite of their poor operating characteristics. Testing schemes are based on subjective α levels;

commonly 0.05 or 0.01; however, Rawlings (1988) recommends 0.15 in the context of stepwise regression. The multiple testing problem is serious if many tests are to be made (see Westfall and Young 1993), and the tests are not independent. Tests between models that are not nested are problematic. A model is nested if it is a special case of another model; for example, a third-degree polynomial is nested within a fourth-degree polynomial. Generally, hypothesis testing is a very poor basis for model selection (Akaike 1974 and Sclove 1994b). McQuarrie and Tsai (1998) do not even treat this subject except for a short appendix on stepwise regression—the final three pages in their book.

Cross-validation has been suggested and well studied as a basis for model selection (Mosteller and Tukey 1968, Stone 1974, 1977; Geisser 1975). Here, the data are divided into two partitions. The first partition is used for model fitting; and the second is used for model validation (sometimes the second partition has only one observation). Then a new partition is selected, and this whole process is repeated hundreds or thousands of times. Some criterion is then chosen, such as minimum squared prediction error, as a basis for model selection. There are several variations on this theme, and it is a useful methodology (Craven and Wahba 1979, Burman 1989, Shao 1993, Zhang 1993a, and Hjorth 1994). These methods are quite computer intensive and tend to be impractical if more than about 15–20 models must be evaluated or if sample size is large. Still, cross-validation offers an interesting alternative for model selection.

Some analysts favor using a very general model in all cases (e.g., an over-fitted model). We believe that this is generally poor practice (Figure 1.3B). Others have a "favorite" model that they believe is good, and they use it in nearly all situations. For example, some researchers always use the hazard rate model (Buckland et al. 1993) with 2 parameters ($K = 2$) as an approximating model to the detection function in line transect sampling. This might be somewhat reasonable for situations where a simple model suffices (e.g., $K = 2$ to 3), but will be poor practice in more challenging modeling contexts where $10 \leq K \leq 30$ or more is required. These *ad hoc* rules ignore the principle of parsimony and data-based model selection, in which the data help select the model to be used for inference.

If goodness-of-fit tests can be computed for all alternative models even if some are not nested within others, then one could use the model with the fewest parameters that "fits" (i.e., $P > 0.05$ or 0.10). However, increasingly better fits can often be achieved by using models with more and more parameters (e.g., the elephant-fitting problem), and this can make the arbitrary choice of α very critical. A large α-level leads to overfitted models and their resulting problems. In addition, other problems may be encountered such as over- or underdispersion and low power if one must pool small expectations to ensure that the test statistic is chi-square distributed. Perhaps, most importantly, there is no theory to suggest that this approach will lead to selected models with good inferential properties (i.e., an adequate bias vs. variance tradeoff or good achieved confidence interval coverage and width).

The adjusted coefficient of multiple determination has been used in model selection in an LS setting (the adjusted coefficient $= 1 - (1 - R^2)\left(\frac{n-1}{n-p}\right)$, where R^2 is the usual coefficient of multiple determination; Draper and Smith 1981:91–92). Under this method, one selects the model in which this adjusted statistic is largest. McQuarrie and Tsai (1998) found this approach to be very poor (also see Rencher and Pun (1980). While adjusted R^2 is useful as a descriptive statistic, it is not useful in model selection. Mallows's C_p statistic (Mallows 1973, 1995) is also used in LS regression with normal residuals and a constant variance and in this special case provides a ranking of the candidate models that is the same as the rankings under AIC (the numerical values, C_p vs. AIC, will differ, see Atilgan 1996). The selection of models using the adjusted R^2 statistic and Mallows's C_p are related for simple LS problems (see Seber 1977:362–369). Hurvich and Tsai (1989) and McQuarrie and Tsai (1998) provide some comparisons of AIC_c vs. several competitors for linear regression problems.

Bayesian researchers have taken somewhat different approaches and assumptions, and have proposed several alternative methods for model selection. Methods such as CAIC, BIC (SIC), WIC, and HQ are mentioned in Section 2.8, as well as full Bayesian model selection (see especially Hoeting et al. 1999). These other Bayesian approaches to model selection and inference are at the current state of the art in statistics but may seem very difficult to understand and implement and are very computer intensive (e.g., Laud and Ibrahim 1995 and Carlin and Chib 1995). Draper (1995) provides a recent review of these advanced methods (also see Potscher 1991). Spiegelhalter et al. (2002) have developed a deviance information criterion (DIC) from a Bayesian perspective that is analogous to AIC. This seems to represent a blending of frequentist and Bayesian thinking, resulting in an AIC-like criterion.

The general approach that we advocate here is one derived by Akaike (1973, 1974, 1977, 1978a and b, and 1981a and b), based on information theory, and it is discussed at length in this book. Akaike's information-theoretic approach has led to a number of alternative methods having desirable properties for the selection of best approximating models in practice (e.g., AIC, AIC_c, $QAIC_c$, and TIC—Chapters 2 and 7). Our general advocacy concerning AIC and the associated criteria is somewhat stronger than that of Linhart and Zucchini (1986) but similar in that they also recommend objective procedures based on some well-defined criterion with a strong, fundamental basis.

1.5   Data Dredging, Overanalysis of Data, and Spurious Effects

The process of analyzing data with few or no a priori questions, by subjectively and iteratively searching the data for patterns and "significance," is often called by the derogatory term "data dredging." Other terms include "post hoc data analysis" or "data snooping," or "data mining," but see Hand (1998) and Hand et al. (2000) for a different meaning of data mining with respect to very large data sets. Often the problem arises when data on many variables have been taken with little or no a priori motive or without benefit of supporting science. No specific objectives or alternatives were in place prior to the analysis; thus the data are submitted for analysis in the hope that the computer and a plethora of null hypothesis test results will provide information on "what is significant." A model is fit, and variables not in that model are added to create a new model, letting the data and intermediate results suggest still further models and variables to be investigated. Patterns seen in the early part of the analysis are "chased" as new variables, cross products, or powers of variables are added to the model and alternative transformations tried. These new models are clearly based on the intermediate results from earlier waves of analyses. The final model is the result of effective dredging, and often nearly everything remaining is "significant." Under this view, Hosmer and Lemeshow (1989:169) comment that "Model fitting is an iterative procedure. We rarely obtain the final model on the first pass through the data." However, we believe that such a final model is probably overfitted and unstable (i.e., likely to vary considerably if other sample data were available on the same process) with actual predictive performance (i.e., on new data) often well below what might be expected from the statistics provided by the terminal analysis (e.g., Chatfield 1996, Wang 1993). The inferential properties of a priori versus post hoc data analysis are very different. For example, (traditionally) no valid estimates of precision can be made from the model following data dredging (but see Ye 1998).

1.5.1   Overanalysis of Data

If data dredging is done, the resulting model is very much tailored (i.e., overfitted) to the data in a post hoc fashion, and the estimates of precision are likely to be overestimated. Such tailoring overdescribes the data and diminishes the validity of inferences made about the information in the data to the population of interest. Many naive applications of classical multivariate analyses are merely "fishing trips" hoping to find "significant" linear relationships among the many variables subjected to analysis (Rexstad et al. 1988, 1990, Cox and Reid 2000).

Computer routines (e.g., SAS INSIGHT) and associated manuals make data dredging both easy and "effective." Some statistical literature deals with the so-called *iterative process of model building* (e.g., Henderson and Velleman 1981). One looks for patterns in the residuals, employs various tests for selecting variables in their decreasing order of "importance," and tries all possible models. Stepwise regression and discriminant functions, for example, are used to search for "significant" variables; such methods are especially problematic if many variables (Freedman's paradox) are available for analysis (sometimes data are available on over 100 variables, and the sample size may often be less than the number of variables). These problems of overfitting can escalate when flexible generalized linear or generalized additive models are employed.

White (2000:1097) notes, "It is widely acknowledged by empirical researchers that data snooping [dredging] is a dangerous practice to be avoided, but in fact it is endemic." Examples of data dredging include the examination of crossplots or a correlation matrix of the explanatory variables versus the response variable. These data-dependent activities can suggest apparent linear or nonlinear relationships and interactions *in the sample* and therefore lead the investigator to consider additional models. These activities should be avoided, because they probably lead to overfitted models with spurious parameter estimates and inclusion of unimportant variables as regards the *population* (Anderson et al. 2001b). The sample may be well fit, but the goal is to make a valid inference from the sample to the population. This type of data-dependent, exploratory data analysis has a place in the earliest stages of investigating a biological relationship but should probably remain unpublished. However, such cases are not the subject of this book, and we can only recommend that the results of such procedures be treated as possible hypotheses (Lindsey 1999c, Longford and Nelder 1999). New data should be collected to address these hypotheses effectively and then submitted for a comprehensive and largely a priori strategy of analysis such as we advocate here.

Two types of data dredging might be distinguished. The first is that described above; a highly interactive, data dependent, iterative post hoc approach. The second is also common and also leads to likely overfitting and the finding of effects that are actually spurious. In this type, the investigator also has little a priori information; thus "all possible models" are considered as candidates (e.g., SAS PROC REG allows this as an option). Note that the "all possible models" approach usually does not include interaction terms (e.g., $x_2 * x_5$) or various transformations such as $(x_1)^2$ or $1/x_3$ or $\log(x_2)$. In even moderate-sized problems, the number of candidate models in this approach can be very large (e.g., 20 variables > a million models, 30 variables > a billion models). At least this second type is not explicitly data dependent, but it is implicitly data dependent and leads to the same "sins." Also, it is usually a one-pass strategy, rather than taking the results of one set of analyses and inputting some of these into the consideration of new models. Still, in some applications, computer software often can systematically search all such models nearly automatically, and thus the strategy of trying all possible models (or at least a very large number of models) continues, unfortunately, to be popular. We believe that many situations could be substantially improved if the researcher tried harder to focus on the science of the situation before proceeding with such an unthoughtful approach.

Standard inferential tests and estimates of precision (e.g., ML or LS estimators of the sampling covariance matrix, given a model) are invalid when a final model results from the first type of data dredging. Resulting "P-values" are misleading, and there is no valid basis to claim "significance." Even conceptually there is no way to estimate precision because of the subjectivity involved in iterative data dredging and the high probability of overfitting. In the second type of data dredging one might consider Bonferroni adjustments of the α-levels or P-values. However, if there were 1,000 models, then the α-level would be 0.00005, instead of the usual 0.05! Problems with data dredging are often linked with the problems with hypothesis testing (Johnson 1999, Anderson et al. 2000). This approach is hardly satisfactory; thus analysts have ignored the issue and merely pretended that data dredging is without peril and that the usual inferential methods somehow still apply. **Journal editors and referees rarely seem to show concern for the validity of results and conclusions where substantial data dredging has occurred. Thus, the entire methodology based on data dredging has been allowed to be perpetuated in an unthinking manner.**

We certainly encourage people to understand their data and attempt to answer the scientific questions of interest. We advocate some examination of the data prior to the formal analysis to detect obvious outliers and outright errors (e.g., determine a preliminary truncation point or the need for grouping in the analysis of distance sampling data). One might examine the residuals from a carefully chosen global model to determine likely error distributions in the candidate models (e.g., normal, lognormal, Poisson). However, if a particular pattern is noticed while examining the residuals and this leads to including another variable, then we might suggest caution concerning data dredging. Often, there can be a fine line between a largely a priori approach and some degree of data dredging.

Thus, this book will address primarily cases where there is substantial a priori knowledge concerning the issue at hand and where a relatively small set of good candidate models can be specified in advance of actual data analysis. Of course, there is some latitude where some (few) additional models might be investigated as the analysis proceeds; however, results from these explorations should be kept clearly separate from the purely a priori science. We believe that objective science is best served using a priori considerations with very limited peeking at plots of the data, parameter estimates from particular models, correlation matrices, or test statistics as the analysis proceeds. We do not condone data dredging in confirmatory analyses, but allow substantial latitude in more preliminary explorations. If some limited data dredging is done after a careful analysis based on prior considerations, then we believe that these two types of results should be carefully explained in resulting publications (Tukey 1980). For this philosophy to succeed, there should be more careful a priori consideration of alternative candidate models than has been the case in the past.

1.5.2   Some Trends

At the present time, nearly every analysis is done using a computer; thus biologists and researchers in other disciplines are increasingly using likelihood methods for more generalized analyses. Standard computer software packages

---

Data Dredging

Data dredging (also called data snooping, data mining, post hoc data analysis) should generally be avoided, except in (1) the early stages of exploratory work or (2) *after* a more confirmatory analysis has been done. In this latter case, the investigator should fully admit to the process that led to the post hoc results and should treat them much more cautiously than those found under the initial, a priori, approach. When done carefully, we encourage people to explore their data beyond the important a priori phase.

We recommend a substantial, deliberate effort to get the a priori thinking and models in place and try to obtain more confirmatory results; *then* explore the post hoc issues that often arise after one has seen the more confirmatory results.

Data dredging activities form a continuum, ranging from fairly trivial (venial) to the grievous (mortal). There is often a fine line between dredging and not; our advice is to stay well toward the a priori end of the continuum and thus achieve a more confirmatory result.

One can always do post hoc analyses after the a priori analysis; but one can never go from post hoc to a priori. Why not keep one's options open in this regard?

Grievous data dredging is endemic in the applied literature and still frequently taught or implied in statistics courses without the needed caveats concerning the attendant inferential problems.

Running all possible models is a thoughtless approach and runs the high risk of finding effects that are, in fact, spurious if only a single model is chosen for inference. If prediction is the objective, model averaging is useful, and estimates of precision should include model selection uncertainty. Even in this case, surely one can often rule out many models on a priori grounds.

---

allow likelihood methods to be used where LS methods have been used in the past. LS methods will see decreasing use, and likelihood methods will see increasing use as we proceed into the twenty-first century. Likelihood methods allow a much more general framework for addressing statistical issues (e.g., a choice of link functions and error distributions as in log linear and logistic regression models). Another advantage in a likelihood approach is that confidence intervals with good properties can be set using profile likelihood intervals. Edwards (1976), Berger and Wolpert (1984), Azzalini (1996), Royall (1997), and Morgan (2000) provide additional insights into likelihood methods, while Box (1978) provides the historical setting relating to Fisher's general methods.

During the past twenty years, modern statistical science has been moving away from traditional formal methodologies based on statistical hypothesis testing (Clayton et al. 1986, Jones and Matloff 1986, Yoccoz 1991, Bozdogan 1994, Johnson 1995, Stewart-Oaten 1995, Nester 1996, Johnson 1999, Anderson et al. 2000). The historic emphasis on hypothesis testing will continue to diminish in the years ahead (e.g., see Quinn and Dunham 1983, Bozdogan 1994), with increasing emphasis on estimation of effects or effect sizes and associated confidence intervals (Graybill and Iyer 1994:35, Cox and Reid 2000).

Most researchers recognize that we do not conduct experiments merely to reject null hypotheses or claim statistical significance; we want deeper insights than this. We typically want to compare meaningful (i.e., plausible) alternatives, or seek information about effects and their size and precision, or are interested in causation. **There has been too much formalism, tradition, and confusion that leads people to think that statistics and statistical science is mostly about testing uninteresting or trivial null hypotheses, whereas science is much more than this. We must move beyond the traditional testing-based thinking because it is so uninformative.**

In particular, hypothesis testing for model selection is often poor (Akaike 1981a) and will surely diminish in the years ahead. There is no statistical theory that supports the notion that hypothesis testing with a fixed α level is a basis for model selection. There are not even general formal rules (or even guidelines) that rigorously define how the various P-values might be used to arrive at a final model. How does one interpret dozens of P-values, from tests with differing power, to arrive at a good model? Only *ad hoc* rules exist in this case and generally fail to result in a final parsimonious model with good inferential properties. The multiple testing issue is problematic as is the fact that likelihood ratio tests exist only for nested models. Tests of hypotheses within a data set are not independent, making inferences difficult. The order of testing is arbitrary, and differing test order will often lead to different final models. Model selection is dependent on the arbitrary choice of α, but α should depend on both n and K to be useful in model selection; however, theory for this is lacking. Testing theory is problematic when nuisance parameters occur in the models being considered. Finally, there is the fact that the so-called null is probably false on simple a priori grounds (e.g., H_0: the treatment had *no* effect, so the parameter θ is constant across treatment groups or years, $\theta_1 = \theta_2 = \cdots = \theta_k$). Rejection of such null hypotheses does not mean that the effect or parameter should be included in the approximating model! The entire testing approach is both common and somewhat absurd. All of these problems have been well known in the literature for many years; they have merely been ignored in the practical analysis of empirical data. Nester (1996) provides an interesting summary of quotations regarding hypothesis testing.

Unfortunately, it has become common to compute estimated test power after a hypothesis test has been conducted and found to be nonsignificant. Such post hoc power is not valid (Goodman and Berlin 1994, Gerard et al. 1998, Hoenig and Heisey 2001). While a priori power and sample size considerations are important in planning an experiment or observational study, estimates of post hoc power are not valid and should not be reported (Anderson et al. 2001d).

Computational restrictions prevented biologists from evaluating alternative models until the past two decades or so. Thus, people tended to use an available model, often without careful consideration of alternatives. Present computer hardware and software make it possible to consider a number of alternative models as an integral component of data analysis. Computing power has permitted more computer-intensive methods such as the various cross-validation and bootstrapping approaches and other resampling schemes (Mooney and Duval 1993, Efron and Tibshirani 1993), and such techniques will see ever increasing use in the future.

The size or dimension (K) of some biological models can be quite high, and this has tended to increase over the past two decades. Open capture–recapture and band recovery models commonly have 20–40 estimable parameters for a single data set and might have well over 200 parameters for the joint analysis of several data sets (see Burnham et al. 1987, Preface, for a striking example of these trends). Analysis methods for structural equations commonly involve 10–30 parameters (Bollen and Long 1993). These are applications where objective model specification and selection is essential to answer the question, *"What inferences do the data support about the population?"*

1.6   Model Selection Bias

The literature on model selection methods has increased substantially in the past 15–25 years; much of this has been the result of Akaike's influential papers in the mid-1970s. However, relatively little appears in the literature concerning the properties of the parameter estimators, given that a data-dependent model selection procedure has been used (see Rencher and Pun 1980, Hurvich and Tsai 1990, Miller 1990, Goutis and Casella 1995, Ye 1998). Here, data are used to both select a parsimonious model and estimate the model parameters and their precision (i.e., the conditional sampling covariance matrix, given the selected model). These issues prompt a concern for both model selection bias and model selection uncertainty (Section 1.7).

Bias in estimates of model parameters often arises when data-based selection has been done. Miller (1990) provides a technical discussion of model selection bias in the context of linear regression. He notes his experience in the stepwise analysis of meteorological data with large sample sizes and 150 candidate models. When selecting only about 5 variables from the 150 he observed, he found t statistics as large as 6, suggesting that a particular variable was very highly significant, and yet even the sign of the corresponding regression coefficient could be incorrect. Miller warns that P-values from subset selection software are totally without foundation, and large biases in regression coefficients are often caused by data-based model selection.

Consider a linear model where there is a response variable (y) and 4 explanatory variables x_j, where $j = 1, \ldots, 4$. Order is not important in this example, so for convenience let x_1 be, in fact, very important, x_2 important, x_3 somewhat important, while x_4 is barely important. Given a decent sample size, nearly any model selection method will indicate that x_1 and probably x_2 are important (Miller called such variables "dominant"). If one had 1,000 replicate data sets of the same size, from the same stochastic process, x_1 (particularly) and x_2 would be included in the model in nearly all cases. In these cases, an inference from a sample data set to the population would be valid. For models selected that included predictors x_1 and x_2 (essentially all 1,000 models), the estimators of the regression coefficients associated with variables x_1 and x_2 would have good statistical properties with respect to bias and precision (i.e., standard theory tends to hold for the estimators $\hat{\beta}_1$ and $\hat{\beta}_2$).

Variable x_3 is somewhat marginal in its importance; assume, for example, that $|\beta_3|/\text{se}(\beta_3) \approx 1$, and thus its importance is somewhat small. This variable might be included in the model in only 15–30% of the 1,000 data sets. In data sets where it is selected, it tends to have an estimated regression coefficient that is biased away from zero. Thus, an inference from one of the data sets concerning the population tend to exaggerate the importance of the variable x_3. An inference from a data set in one of the remaining 70–85% of the data sets would imply that x_3 was of no importance. Neither of these cases is satisfactory.

Variable x_4 is barely important at all (a tapering effect), and it might have $|\beta_4|/\text{se}(\beta_4) \approx \frac{1}{4}$. This variable might be included in only a few (e.g., 5–10%) of the 1,000 data sets and, when it is selected, there will likely be a large bias (away from 0) in the estimator of this regression parameter. Inference from a particular sample where this variable is included in the model would imply that the variable x_4 was much more important than is actually the case (of course, the investigator has no way to know that $\hat{\beta}_4$, when selected, might be in the upper 5–10% of its sampling distribution). Then, if one examines the usual t-test, where $t = \hat{\beta}_4/\widehat{\text{se}}(\hat{\beta}_4)$, the likely decision will often be that the variable x_4 is significant, and should be retained in the model. This misleading result comes from the fact that the numerator in the test is biased high, while the denominator is biased low. The analyst has no way to know that this test result is probably spurious.

When predictor variables x_3 and x_4 are included in models, the associated estimator for a σ^2 is negatively biased and precision is exaggerated. These two types of bias are called model selection bias and can often be quite serious (Miller 1990, Ye 1998). Ye (1998) warns, "...the identification of a clear structure bears little cost [i.e., including variables x_1 and x_2], whereas searching through white noise has a heavy cost [i.e., including variable x_4 in a model]." Of course, in the analysis of real data, the investigator typically does not know which (if any) variables are dominant versus those that are, in fact, of marginal importance. Model selection bias is related to the problem of overfitting, the notion of tapering effect sizes, and Freedman's (1983) paradox.

The problem of model selection bias is particularly serious when little theory is available to guide the analysis. Many exploratory studies have hundreds or even thousands of models, based on a large number of explanatory variables; very often the number of models exceeds the size of the sample. Once a final model has been (somehow) selected, the analyst is usually unaware that this model is likely overfit, with substantially biased parameter estimates (i.e., both the estimated structural regression coefficients, which are biased away from 0 and the estimated residual variation, which is biased low). They have unknowingly extracted some of the residual variation as if it represented model structure. When sample size is large, true replication exists, and there are relatively few models, these problems may be relatively unimportant. However, often one has only a small sample size, no true replication, and many models and variables; then model selection bias is usually severe (Zucchini 2000).

If, for example, x_3 is uncorrelated with x_1, x_2, and x_4, then the distribution of $\hat{\beta}_3$ is symmetric around β_3 and bias, given that x_3 is selected, is nil (i.e., if $\beta_3 = 0$, then $E(\hat{\beta}_3) = 0$). This is an interesting result, but probably uncommon in practice because predictor variables are almost always correlated. Consider the case where $\beta_3 = 0$, but x_3 is highly correlated with x_1 and $\beta_i > 0$. If the correlation between x_1 and x_3 is high (even 0.5) and positive, then when variable x_3 is selected, it is much more likely to be when $\hat{\beta}_3 > 0$. In all samples where x_3 is selected, $\hat{\beta}_3$ tends to be positive. In cases where the correlation between x_1 and x_3 is negative, then $\hat{\beta}_3$ tends to be negative. In either case, $\hat{\sigma}^2$ is biased low. By itself, x_3 would have some predictive value, but only because of its correlation with x_1, which is actually correlated with the response variable.

If sample size is small and there are many variables and hence models, then the negative bias in $\hat{\sigma}^2$ is often severe. If the predictor variables are highly intercorrelated and only one (say x_{11}) is actually correlated with the response variable, then the estimates of the regression coefficients will likely be substantially biased away from 0 in the subset of models where the associated predictor variable is selected. Leamer (1978), Copas (1983), Lehmann (1983) Gilchrist (1984), Breiman (1992), Zhang (1992a), and Chatfield (1995b, 1996) give insights into problems that arise when the same data are used both to select the model and to make inferences from that model.

1.7 Model Selection Uncertainty

Model selection uncertainty also arises when the data are used for both model selection and parameter estimation (Hjorth 1994:15–23). If a best model has been selected from a reasonable set of candidate models, bias in the model parameter estimators might be small for several of the more important variables, but might be substantial for variables associated with tapering effects. However, there is uncertainty as to the best model to use. From the example above, one must ask whether β_3 or β_4 should be in the model; this model uncertainty is a component of variance in the estimators.

Denote the sampling variance of an estimator $\hat{\theta}$, *given a model*, by $\text{var}(\hat{\theta}|model)$. More generally, the sampling variance of $\hat{\theta}$ should have two components: (1) $\text{var}(\hat{\theta}|model)$ and (2) a variance component due to not knowing the best approximating model to use (and, therefore, having to estimate this). Thus, if one uses a method such as AIC to select a parsimonious model, given the data, and estimates a conditional sampling variance, given the selected model. Then estimated precision will be too small because the variance component for model selection uncertainty is missing. Model selection uncertainty is the component of variance that reflects that model selection merely *estimates* which model is best, based on the single data set; a different model (in the fixed set of models considered) may be selected as best for a different replicate data set arising from the same experiment.

Failure to allow for model selection uncertainty often results in estimated sampling variances and covariances that are too low, and thus the achieved confidence interval coverage will be below the nominal value. Optimal methods for coping with model selection uncertainty are at the forefront of statistical research; better methods might be expected in the coming years, especially with the continued increases in computing power. Model selection uncertainty is problematic in making statistical inferences; if the goal is only data description, then perhaps selection uncertainty is a minor issue.

One must keep in mind that there is often considerable uncertainty in the selection of a particular model as the "best" approximating model. The observed data are conceptualized as random variables; their values would be different if another, independent sample were available. It is this "sampling variability" that results in uncertain statistical inference from the particular data set being analyzed. While we would like to make inferences that would be robust to other (hypothetical) data sets, our ability to do so is still quite limited, even with procedures such as AIC, with its cross-validation properties, and with independent and identically distributed sample data. Various computer-intensive resampling methods will further improve our assessment of the uncertainty of our inferences, but it remains important to understand that proper model selection is accompanied by a substantial amount of uncertainty. The bootstrap technique can effectively allow insights into model uncertainty; this and other similar issues are the subject of Chapter 5.

Perhaps we cannot totally overcome problems in estimating precision, following a data-dependent selection method such as AIC (e.g., see Dijkstra 1988, Ye 1998). This limitation certainly warrants exploration because model selection uncertainty is a quite difficult area of statistical inference. However, we must also consider the "cost" of *not* selecting a good parsimonious model for the analysis of a particular data set. That is, a model is just somehow "picked" independent of the data and used to approximate the data as a basis for inference. This procedure simply ignores both the uncertainty associated with model selection and the benefits of selection of a model that is parsimonious. This naive strategy certainly will incur substantial costs in terms of reliable inferences because model selection uncertainty is ignored (assumed to be zero). Alternatively, one might be tempted into an iterative, highly interactive strategy of data analysis (unadulterated data dredging). Again, there are substantial costs in terms of reliable inference using this approach. In particular, it seems impossible to objectively and validly estimate the precision of the estimators following data dredging.

1.8   Summary

Truth in the biological sciences and medicine is extremely complicated, and we cannot hope to find exact truth or full reality from the analysis of a finite amount of data. Thus, inference about truth must be based on a good approximating model. Likelihood and least squares methods provide a rigorous inference theory if the model structure is "given." However, in practical scientific problems, the model is *not* "given." Thus, the critical issue is, "what is the best model to use." This is the model selection problem.

The emphasis then shifts to the careful a priori definition of a set of candidate models. This is where the science of the problem enters the analysis. Ideally, there should be a good rationale for including each particular model in the set, as well as a careful justification for why other models were excluded. The degree to which these steps can be implemented suggests a more confirmatory analysis, rather than a more exploratory analysis. Critical thinking about the scientific question and modeling alternatives, prior to looking at the data, have been underemphasized in many statistics classes in the past. These are important issues, and one must be careful not to engage in data dredging, because this weakens inferences that might be made. Information-theoretic methods provide a simple way to select a best approximating model from the candidate set of models.

In general, the information-theoretic approach should not mean merely searching for a single best model as a basis for inference. Even if model selection uncertainty is included in estimates of precision, this is a poor approach in many cases. Instead, multimodel inference should be the usual approach to making valid inference. Here, models are ranked and scaled to enhance an understanding of model uncertainty over the set. These methods are easy to understand and compute. Specific methodologies for this more general approach are the subject of this book.

We cannot overstate the importance of the scientific issues, the careful formulation of multiple working hypotheses, and the building of a small set of models to clearly and uniquely represent these hypotheses. The methods to be presented in the following chapters are "easy" to understand, compute, and interpret; however, they rest on both good science and good data that relate to the issue. We try to emphasize a more confirmatory endeavor in the applied sciences, rather than exploratory work that has become so common and has often led to so little (Anderson et al. 2000).

Data analysis is taken to mean the entire integrated process of a priori model specification, model selection, and estimation of parameters and their precision. Scientific inference is based on this process. Information-theoretic methods free the analyst from the limiting concept that the proper approximating model is somehow "given."

The principle of parsimony is fundamental in the sciences. However, data-based selection of a parsimonious model is challenging. There are substantial rewards for proper model selection in terms of valid inferences; there are substantial dangers in either underfitting or overfitting. However, even if one has selected a good approximating model, there are issues of model selection bias and model selection uncertainty. Perhaps these cannot be fully overcome, but their effects can be lessened. These issues will be addressed in the material to follow.

Zhang (1994) notes that for the analyst who is less concerned with theoretical optimality it is more important to have available methods that are simple but flexible enough to be used in a variety of practical situations. The information-theoretic methods fall in this broad class and, when used properly, promote reliable inference.

2
Information and Likelihood Theory: A Basis for Model Selection and Inference

Full reality cannot be included in a model; thus we seek a good model to approximate the effects or factors supported by the empirical data. The selection of an appropriate approximating model is critical to statistical inference from many types of empirical data. This chapter introduces concepts from information theory (see Guiasu 1977), which has been a discipline only since the mid-1940s and covers a variety of theories and methods that are fundamental to many of the sciences (see Cover and Thomas 1991 for an exciting overview; Figure 2.1 is produced from their book and shows their view of the relationship of information theory to several other fields). In particular, the Kullback–Leibler "distance," or "information," between two models (Kullback and Leibler 1951) is introduced, discussed, and linked to Boltzmann's entropy in this chapter. Akaike (1973) found a simple relationship between the Kullback–Leibler distance and Fisher's maximized log-likelihood function (see deLeeuw 1992 for a brief review). This relationship leads to a simple, effective, and very general methodology for selecting a parsimonious model for the analysis of empirical data.

Akaike introduced his *"entropy maximization principle"* in a series of papers in the mid-1970s (Akaike 1973, 1974, 1977) as a theoretical basis for model selection. He followed this pivotal discovery with several related contributions beginning in the early 1980s (Akaike 1981a and b, 1985, 1992, and 1994). This chapter introduces AIC and related criteria such as AIC_c, $QAIC_C$, and TIC. No mathematical derivations of these criteria are given here because they are given in full detail in Chapter 7. We urge readers to understand the full derivation (given in Chapter 7), for without it, the simple and compelling idea

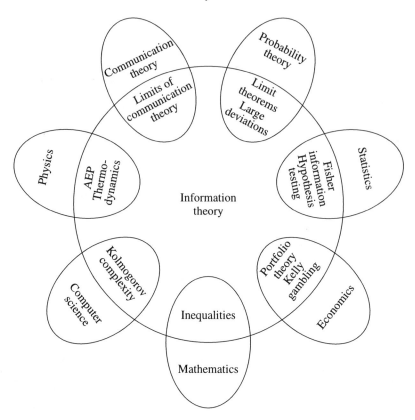

FIGURE 2.1. Information theory and its relationships to other disciplines (from Cover and Thomas 1991). Information theory began in the mid-1940s, at the close of WWII. In the context of this book, the most relevant components of information theory include Fisher information, entropy (from thermodynamics and communication theory), and Kullback–Leibler information.

underlying Kullback–Leibler information and the various information criteria cannot be fully appreciated.

2.1 Kullback–Leibler Information or Distance Between Two Models

We begin without any issues of parameter estimation and deal with very simple expressions for the models f and g, assuming that they are completely known. In initial sections of this chapter we will let both f and g be simple probability distributions, since this will allow an understanding of K-L information or distance in a simple setting. However, we will soon switch to the concept that

f is a notation for full reality or truth. We use g to denote an approximating model in terms of a probability distribution.

Kullback–Leibler Information

Kullback-Leibler information between models f and g is defined for continuous functions as the (usually multi-dimensional) integral

$$I(f,g) = \int f(x) \log \left(\frac{f(x)}{g(x|\theta)} \right) dx,$$

where log denotes the natural logarithm. The notation $I(f, g)$ denotes the **"information lost when g is used to approximate f."**

As a heuristic interpretation, $I(f,g)$ **is the distance from g to f.**

We will use both interpretations throughout this book, since both seem useful. Of course, we seek an approximating model that loses as little information as possible; this is equivalent to minimizing $I(f, g)$, over g. Full reality f is considered to be given (fixed), and only g varies over a space of models indexed by θ. Similarly, Cover and Thomas (1991) note that the K-L distance is a measure of the inefficiency of assuming that the distribution is g when the true distribution is f.

Kullback–Leibler Information

The expression for the Kullback-Leibler information or distance in the case of discrete distributions such as the Poisson, binomial, or multinomial is

$$I(f,g) = \sum_{i=1}^{k} p_i \cdot \log \left(\frac{p_i}{\pi_i} \right).$$

Here, there are k possible outcomes of the underlying random variable; the true probability of the ith outcome is given by p_i, while the $\pi_1, \ldots, \pi_k$ constitute the approximating probability distribution (i.e., the approximating model). In the discrete case, we have $0 < p_i < 1$, $0 < \pi_i < 1$, and $\sum p_i = \sum \pi_i = 1$. Hence, here f and g correspond to the p_i and π_i, respectively.

As in the continuous care the notation $I(f,g)$ denotes the **information lost when g is used to approximate f or the distance from g to f.**

In the following material we will generally think of K-L information in the continuous case and use the notation f and g for simplicity.

Well over a century ago measures were derived for assessing the "distance" between two models or probability distributions. Most relevant here is Boltzmann's (1877) concept of generalized entropy (see Section 2.12) in physics and thermodynamics (see Akaike 1985 for a brief review). Shannon (1948) employed entropy in his famous treatise on communication theory (see Atmar 2001 for an exciting review of information theory, its practicality, and relations to evolution). Kullback and Leibler (1951) derived an information measure that

Ludwig Eduard Boltzmann, 1844–1906, one of the most famous scientists of his time, made incredible contributions in theoretical physics. He received his doctorate in 1866; most of his work was done in Austria, but he spent some years in Germany. He became full professor of mathematical physics at the University of Graz, Austria, at the age of 25. His mathematical expression for entropy was of fundamental importance throughout many areas of science. The negative of Boltzmann's entropy is a measure of "information" derived over half a century later by Kullback and Leibler. J. Bronowski wrote that Boltzmann was "an irascible, extraordinary man, an early follower of Darwin, quarrelsome and delightful, and everything that a human should be." Several books chronicle the life of this great figure of science, including Cohen and Thirring (1973) and Broda (1983); his collected technical papers appear in Hasenöhrl (1909).

happened to be the negative of Boltzmann's entropy, now referred to as the Kullback–Leibler (K-L) information or distance (but see Kullback 1987, where he preferred the term *discrimination information*). The motivation for Kullback and Leibler's work was to provide a rigorous definition of "information" in relation to Fisher's "sufficient statistics." The K-L distance has also been called the K-L discrepancy, divergence, information, and number. We will treat these terms as synonyms, but tend to use *distance* or *information* in the material to follow.

The Kullback–Leibler distance can be conceptualized as a directed "distance" between two models, say f and g (Kullback 1959). Strictly speaking, this is a measure of "discrepancy"; it is not a simple distance, because the measure from f to g is not the same as the measure from g to f; it is a directed, or oriented, distance (Figure 2.2). The K-L distance is perhaps the

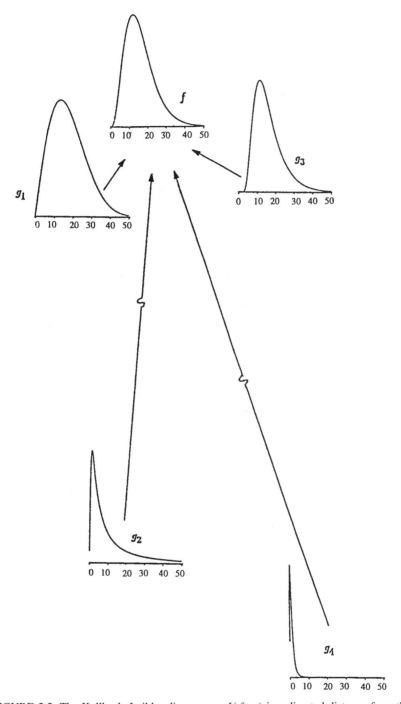

FIGURE 2.2. The Kullback–Leibler discrepancy $I(f, g_i)$ is a directed distance from the various candidate models g_i to f. Knowing the K-L distances would allow one to find which of the 4 approximating models is *closest* to model f. Here, f is gamma $(4, 4)$, and the 4 approximating models are $g_1 =$ Weibull $(2, 20)$, $g_2 =$ lognormal $(2, 2)$, $g_3 =$ inverse Gaussian $(16, 64)$, and $g_4 =$ F distribution $(4, 10)$. In each case, the model parameters are known exactly (not estimated).

most fundamental of all information measures in the sense of being derived from minimal assumptions and its additivity property. The K-L distance is an extension of Shannon's concept of information (Hobson and Cheng 1973, Soofi 1994) and is sometimes called a "relative entropy." The K-L distance between models is a *fundamental quantity* in science and information theory (see Akaike 1983) and is the logical basis for model selection in conjunction with likelihood inference.

At a heuristic level, "information" is defined as $-\log_e(f(x))$ for some continuous probability density function or $-\log_e(p_i)$ for the discrete case. Kullback–Leibler information is a type of "cross entropy," a further generalization. In either the continuous or discrete representation, the right-hand side is an expected value (i.e., $\int f(x)(\cdot)dx$ for the continuous case or $\sum_{i=1}^{k} p_i(\cdot)$ for the discrete case) of the logarithm of the ratio of the two distributions (f and g) or two discrete probabilities (p_i and π_i). In the continuous case one can think of this as an average (with respect to f) of $\log_e(f/g)$, and in the discrete case it is an average (with respect to the p_i) of the logarithm of the ratio (p_i/π_i). The foundations of these expressions are both deep and fundamental (see Boltzmann 1877, Kullback and Leibler 1951, or contemporary books on information theory).

The K-L distance ($I(f, g)$) is always positive, except when the two distributions f and g are identical (i.e., $I(f, g) = 0$ if and only if $f(x) = g(x)$ everywhere). More detail and extended notation will be introduced in Chapter 7; here we will employ a simple notation and use it to imply considerable generality in the sample data (x) and the multivariate functions f and g.

2.1.1   Examples of Kullback–Leibler Distance

An example will illustrate the K-L distances ($I(f, g_i)$). Let f be a gamma distribution with 2 parameters ($\alpha = 4, \beta = 4$). Then consider 4 approximating models g_i, each with 2 parameters (see below): Weibull, lognormal, inverse Gaussian, and the F distribution. Details on these simple probability models can be found in Johnson and Kotz (1970). The particular parameter values used for the four g_i are not material here, except to stress that they are assumed known, not estimated. "Which of these parametrized distributions is the *closest* to f?" is answered by computing the K-L distance between each g_i and f (Figure 2.2). These are as follows:

	Approximating model	$I(f, g_i)$	Rank
g_1	Weibull distribution ($\alpha = 2, \beta = 20$)	0.04620	1
g_2	lognormal distribution ($\theta = 2, \sigma^2 = 2$)	0.67235	3
g_3	inverse Gaussian ($\alpha = 16, \beta = 64$)	0.06008	2
g_4	F distribution ($\alpha = 4, \beta = 10$)	5.74555	4

Here, the Weibull distribution is closest to (loses the least information about) f, followed by the inverse Gaussian. The lognormal distribution is a poor third,

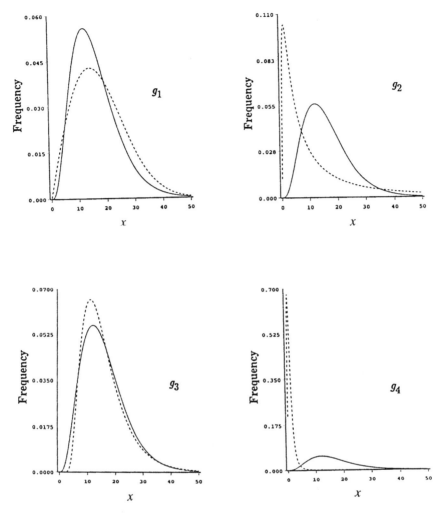

FIGURE 2.3. Plots of f (= gamma $(4, 4)$, solid line) against each of the 4 approximating models g_i (dashed lines) as a function of x. Here, g_1 = Weibull $(2, 20)$, g_2 = lognormal $(2, 2)$, g_3 = inverse Gaussian $(16, 64)$, and g_4 = F distribution $(4, 10)$. Only in the simplest cases can plots such as these be used to judge closeness between models. Model f is the same in all 4 graphs; it is merely scaled differently to allow the $g_i(x)$ to be plotted on the same graph.

while the F distribution is relatively far from the gamma distribution f (see Figure 2.3).

Further utility of the K-L distance can be illustrated by asking which of the approximating models g_i might be closest to f when the parameters of g_i are allowed to vary (i.e., what parameter values make each g_i optimally close to f?). Following a computer search of the parameter space for the Weibull, we found that the *best* Weibull had parameters $\alpha = 2.120$ and $\beta = 18.112$ and a

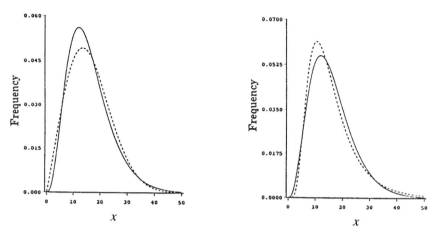

FIGURE 2.4. Plots of f (= gamma (4, 4)) against the best Weibull (left) and lognormal models. The Weibull model that was closest to f had parameters (2.120, 18.112) with K-L distance = 0.02009, while the best lognormal had parameters (2.642, 0.2838) with K-L distance = 0.02195. Compare these optimally parametrized models with those in Figure 2.3 (top).

K-L distance of 0.02009; this is somewhat closer than the original parametrization 0.04620 above. Using the same approach, the best lognormal model had parameters $\theta = 2.642$ and $\sigma^2 = 0.2838$ and a K-L distance of 0.02195, while the best inverse Gaussian model had parameters $\alpha = 16$ and $\beta = 48$ with a K-L distance of 0.03726, and the approximately best F distribution had parameters $\alpha \approx 300$, $\beta = 0.767$ and a K-L distance of approximately 1.486 (the K-L distance is not sensitive to α in this case, but is quite difficult to evaluate numerically). Thus, K-L distance indicates that the best Weibull is closer to f than is the best lognormal (Figure 2.4). Note that the formal calculation of K-L distance requires knowing the true distribution f as well as all the parameters in the models g_i (i.e., parameter estimation has not yet been addressed). Thus, K-L distance cannot be computed for real-world problems.

These values represent *directed* distances; in the first Weibull example, $I(f, g_1) = 0.04620$, while $I(g_1, f) = 0.05552$ (in fact, we would rarely be interested in $I(g_1, f)$ since this is the information lost when f is used to approximate g!). The point here is that these are directed or oriented distances and $I(f, g_1) \neq I(g_1, f)$; *nor should they be equal, because the roles of truth and model are not interchangeable.*

These are all univariate functions; thus one could merely plot them on the same scale and visually compare each g_i to f; however, this graphical method will work only in the simplest cases. In addition, if two approximating distributions are fairly close to f, it might be difficult to decide which is better by only visual inspection. Values of the K-L distance are not based on only the mean and variance of the distributions; rather, the distributions in their entirety are the subject of comparison.

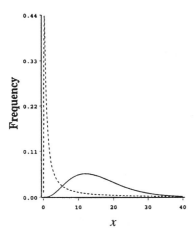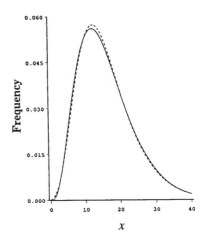

FIGURE 2.5. Plots of f (= gamma (4, 4)) against the best 2-parameter F distribution (left) and the best 3-parameter (noncentral) F distribution. The best 2-parameter model was a poor approximation to f (K-L distance = 1.486), while the best 3-parameter model is an excellent approximation (parameters 1.322, 43.308, 18.856) with K-L distance = 0.001097. Approximating models with increasing numbers of parameters typically are closer to f than approximating models with fewer parameters.

The F distribution ($\alpha = 4$, $\beta = 10$) provided a relatively poor approximation to the gamma distribution with ($\alpha = 4$, $\beta = 4$). Even the best 2-parameter F distribution remains a relatively poor approximation (K-L distance = 1.486). However, in general, adding more parameters will result in a closer approximation (e.g., the classic use of the Fourier series in the physical sciences or Wel's (1975) elephant-fitting problem). If we allow the addition of a third parameter (λ) in the F distribution (the noncentral F distribution), we find that the best model ($\alpha = 1.322$, $\beta = 43.308$, and $\lambda = 18.856$) has a K-L distance of only 0.001097; this is better than any of the other 2-parameter candidate models (Figure 2.5). Closeness of approximation can always be increased by adding more parameters to the candidate model. When we consider *estimation* of parameters and the associated uncertainty, then the principle of parsimony must be addressed (see Section 1.4), or overfitted models will be problematic.

In the remainder of the book we will want a more general, conceptual view of f, and we will use it to reflect truth or full reality. Here, reality is rarely (if ever) a model; rather, it reflects the complex biological (and measuring or sampling) process that generated the observed data x. For this reason we will not explicitly parametrize the complex function f, because it represents full reality (truth), it might not even have parameters in a sense that would be analogous to θ in a modeling framework. In fact, thinking that truth is parametrized is itself a type of (artificial) model-based conceptualization. Sometimes it is useful to think of f as full reality and let it have (conceptually) an infinite number of parameters (see Section 1.2.4). This "crutch" of infinite-dimensionality at least retains the concept of reality even though it is in some unattainable perspective. Thus, f

represents full truth, and might be conceptually based on a very large number of parameters (of a type we may have not even properly conceived) that give rise to a set of data x. Finally, we will see how this conceptualization of reality (f) collapses into a nonidentifiable constant in the context of model selection.

2.1.2   Truth, f, Drops Out as a Constant

The material above makes it obvious that both f and g (and their parameters) must be known to compute the K-L distance between these two models. However, if only relative distance is used, this requirement is diminished, since $I(f, g)$ can be written equivalently as

$$I(f, g) = \int f(x) \log(f(x)) dx - \int f(x) \log(g(x \mid \theta)) dx.$$

Note that each of the two terms on the right of the above expression is a statistical expectation with respect to f (truth). Thus, the K-L distance (above) can be expressed as a difference between two statistical expectations,

$$I(f, g) = E_f \left[\log(f(x)) \right] - E_f[\log(g(x \mid \theta))],$$

each with respect to the distribution f. This last expression provides easy insights into the derivation of AIC.

The first expectation $E_f[\log(f(x))]$ is a constant that depends only on the unknown true distribution, and it is clearly not known (i.e., we do not know f in actual data analysis). Therefore, treating this unknown term as a constant, a measure of *relative* directed distance is possible (Bozdogan 1987, Kapur and Kesavan 1992:155). Clearly, if one computed the second expectation $E_f[\log(g(x \mid \theta))]$, one could estimate $I(f, g)$ up to a constant C (namely $E_f[\log(f(x))]$),

$$I(f, g) = C - E_f[\log(g(x \mid \theta))],$$

or

$$I(f, g) - C = -E_f[\log(g(x \mid \theta))].$$

The term $\left(I(f, g) - C \right)$ is a *relative* directed distance between f and g; thus, $E_f \left[\log(g(x \mid \theta)) \right]$ becomes the quantity of interest for selecting a best model. For two models g_1 and g_2, if $I(f, g_1) < I(f, g_2)$, so g_1 is best, then $I(f, g_1) - C < I(f, g_2) - C$, and hence $-E_f[\log(g_1(x \mid \theta))] < -E_f[\log(g_2(x \mid \theta))]$. Moreover, $I(f, g_2) - I(f, g_1) \equiv -E_f[\log(g_2(x \mid \theta))] + E_f[\log(g_1(x \mid \theta))]$, so we know how much better model g_1 is than model g_2. Without knowing C we just do not know the absolute measure of how good even g_1 is, but we can identify the fact that model g_1 is better than g_2. Note that no parameter estimation is involved here, but the concepts carry over to the cases where estimation occurs. From the preceding example, where f is gamma (4, 4), then $\int f(x) \log(f(x)) dx = 3.40970$, and this term is constant across the models being compared,

$$I(f, g) - 3.40970 = -E_f[\log(g(x \mid \theta))].$$

The *relative* distances between the gamma $(4, 4)$ model and the four approximating models are shown below:

	Approximating model	Relative distance $I(f, g_i) - C$	Rank
g_1	Weibull distribution ($\alpha = 2, \beta = 20$)	3.45591	1
g_2	lognormal distribution ($\theta = 2, \sigma^2 = 2$)	4.08205	3
g_3	inverse Gaussian ($\alpha = 16, \beta = 64$)	3.46978	2
g_4	F distribution ($\alpha = 4, \beta = 10$)	9.15525	4

Note that the ranking of "closeness" of the four candidate models to f is preserved, and the relative ranking of distance between models remains unchanged, even though only relative distances are used.

Kullback-Leibler distance $I(f, g)$ is on a true ratio scale, where there is a true zero. In contrast, $- \int f(x)(\log(g(x|\theta)))dx \equiv -E_f[\log(g(x|\theta))]$ is on an interval scale and lacks a true zero. A difference of magnitude D means the same thing anywhere on the scale. Thus, $D = 10 = 12 - 2 = 1012 - 1002$; a difference of 10 means the same thing anywhere on the interval scale. Then, $10 = V_1 - V_2$, regardless of the size of V_1 and V_2.

The calculation of the two components of K-L distance (above) is in effect based on a sample size of 1. If the sample size were 100, then each component would be 100 times larger, and the difference between the two components would also be 100 times larger. For example, if $n = 100$, then $\int f(x) \log(f(x))dx = 3.40970 \times 100 = 340.970$ and $E_f[\log(g_1(x \mid \theta))]$ (the Weibull) $= 3.45591 \times 100 = 345.591$. Thus, the difference between the two components of K-L distance would be 4.620; the *relative* difference is large when sample size is large. A large sample size magnifies the separation of research hypotheses and the models used to represent them. **Adequate sample size conveys a wide variety of advantages in making valid inferences.**

Typically, as in the example above, the analyst would postulate several a priori candidate models $g_i(x \mid \theta)$ and want to select the *best* among these as a basis for data analysis and inference. Definition of "best" will involve the principle of parsimony and the related concept of a best approximating model. In data analysis, the parameters in the various candidate models are not known and must be estimated from the empirical data. This represents an important distinction from the material above, since one usually has only models with estimated parameters, denoted by $g_i(x \mid \hat{\theta})$. In this case, one needs *estimates* of the relative directed distances between the unknown f that generated the data and the various candidate models $g_i(x \mid \hat{\theta})$. Then, knowing the estimated relative distance from each $g_i(x)$ to $f(x)$, we select the candidate model that is *estimated* to be closest to truth for inference (Figure 2.2). That is, we select the model with the smallest estimated, *relative* distance. Alternatively, we select an approximating model that loses the least information about truth. The conceptual truth f becomes a constant term, and nothing need be assumed about f, since the constant is the same across the candidate models and is irrelevant for comparison. (Similarly, it is interesting to note that often the log-likelihood function also involves an additive constant that is the same across models; this term is known, but generally ignored, since it is often difficult to compute.) In practice, we can obtain only an *estimator* of the relative K-L distance from each approximating model $g_i(x \mid \hat{\theta})$ to f.

2.2   Akaike's Information Criterion: 1973

Akaike's (1973) seminal paper proposed the use of the Kullback-Leibler information or distance as a fundamental basis for model selection. However, K-L distance cannot be computed without full knowledge of both f (full reality) and the parameters (θ) in each of the candidate models $g_i(x|\theta)$. Akaike found a rigorous way to estimate K-L information, based on the empirical log-likelihood function at its maximum point.

Given a parametric structural model there is a unique value of θ that, in fact, minimizes K-L distance $I(f, g)$. This (unknown) minimizing value of the parameter depends on truth f, the model g through its structure, the parameter space, and the sample space (i.e., the structure and nature of the data that can be collected). In this sense there is a "true" value of θ underling ML estimation, let this value be θ_0. Then θ_0 is the absolute best value of θ for model g; actual K-L information loss is minimized at θ_0. If one somehow knew that model g was, in fact, the K-L best model, then the MLE $\hat{\theta}$ would estimate θ_0. This property of the model $g(x|\theta_0)$ as the minimizer of K-L, over all $\theta \in \Theta$, is an important feature involved in the derivation of AIC (Chapter 7).

In data analysis the model parameters must be estimated, and there is usually substantial uncertainty in this estimation. Models based on estimated parameters, hence on $\hat{\theta}$ not θ, represent a major distinction from the case where model parameters would be known. This distinction affects how we must use K-L distance as a basis for model selection. The difference between having θ or θ_0 (we do not) and having the estimate $\hat{\theta}$ (we do) is quite important and basi-

---

Selection Target

Akalke (1973, 1974, 1985, 1994) showed that the critical issue for getting an applied K-L model selection criterion was to estimate

$$E_y E_x[\log(g(x|\hat{\theta}(y)))],$$

where x and y are independent random samples from the same distribution and both statistical expectations are taken with respect to truth (f). This double expectation, both with respect to truth f, is the target of all model selection approaches, based on K-L information.

cally causes us to change our model selection criterion to that of minimizing *expected* estimated K-L distance rather than minimizing known K-L distance over the set of R models considered.

It is tempting to just estimate $E_y E_x[\log(g(x|\hat{\theta}(y)))]$ by the maximized $\log(\mathcal{L}(\hat{\theta})|data)$ for each model g_i. However, Akaike (1973) showed that the maximized log-likelihood is biased upward as an estimator of the model selection target (above). He also found that under certain conditions (these conditions are important, but quite technical) this bias is approximately equal to K, the number of estimable parameters in the approximating model. This is an asymptotic result of fundamental importance.

The Key Result
Thus, an approximately unbiased estimator of

$$E_y E_x[\log(g(x|\hat{\theta}(y)))]$$

for large samples and "good" models is

$$\log(\mathcal{L}(\hat{\theta}|data)) - K.$$

This result is equivalent to

$$\log(\mathcal{L}(\hat{\theta}|data)) - K = \text{constant} - \hat{E}_{\hat{\theta}}[I(f, \hat{g})],$$

where $\hat{g} = g(\cdot|\hat{\theta})$.

The bias-correction term ($K =$ the number of estimable parameters) above is a special case of a more general result derived by Takeuchi (1976) and described in the following section and in Chapter 7. **Akaike's finding of a relation between the relative expected K-L distance and the maximized log-likelihood has allowed major practical and theoretical advances in model selection and the analysis of complex data sets** (see Stone 1982, Bozdogan 1987, and deLeeuw 1992).

Akaike's Information Criterion
Akaike (1973) then defined "*an* information criterion" (AIC) by multiplying $\log(\mathcal{L}(\hat{\theta}|y)) - K$ by -2 ("taking historical reasons into account") to get

$$\text{AIC} = -2\log(\mathcal{L}(\hat{\theta}|y)) + 2K.$$

This has become known as **"Akaike's information criterion"** or AIC.

Thus, rather than having a simple measure of the directed distance between two models (i.e., the K-L distance), one has instead an *estimate* of the expected, relative distance between the fitted model and the unknown true mechanism (perhaps of infinite dimension) that actually generated the observed data.

The expression $\log(\mathcal{L}(\hat{\theta}|y))$ is the numerical value of the log-likelihood at its maximum point (see Section 1.2.2). This maximum point on the log-likelihood function corresponds to the values of the maximum likelihood estimates. The number of estimable parameters in the model is denoted by K, and it is usually clear as to what the correct count should be (see below for standard linear models). In some types of models there are some parameters that are not uniquely estimable from the data, and these should not be counted in K. Nonestimability can occur in the analysis of count data where a cell has no observations, and thus a parameter that is identifiable becomes nonestimable for that data set. Nonestimability can also arise due to inherent confounding (e.g., the parameters S_{t-1} and f_t in certain band recovery models of Brownie et al. 1985). In application, one computes AIC for each of the candidate models and selects the model with the smallest value of AIC. It is this model that is estimated to be "closest" to the unknown reality that generated the data, from among the candidate models considered. This seems a very natural, simple concept; select the fitted approximating model that is estimated, on average, to be closest to the unknown f. Basing AIC on the expectation (over $\hat{\theta}$) of $E_x[\log(g(x|\hat{\theta}(y)))]$ provides AIC with a cross-validation property for independent and identically distributed samples (see Stone 1977, Stoica et al. 1986, Tong 1994). Golub et al. (1979) show that AIC asymptotically coincides with generalized cross-validation in subset regression (also see review by Atilgan 1996).

Of course, models not in the set remain out of consideration. AIC is useful in selecting the best model in the set; however, if all the models are very poor, AIC will still select the one estimated to be best, but even that relatively best model might be poor in an absolute sense. Thus, every effort must be made to ensure that the set of models is well founded.

$I(f, g)$ can be made smaller by adding more known (not estimated) parameters in the approximating model g. Thus, for a fixed data set, the further addition of parameters in a model g_i will allow it to be closer to f. However, when these parameters must be estimated (rather than being known or "given"), further uncertainty is added to the *estimation* of the relative K-L distance. At some point, the addition of still more estimated parameters will have the opposite from desired effect (i.e., to reduce $E_{\hat{\theta}}[I(f, \hat{g})]$ as desired). At that point, the estimate of the relative K-L distance will increase because of "noise" in estimated parameters that are not really needed to achieve a good model. This phenomenon can be seen by examination of the information criterion being minimized,

$$\text{AIC} = -2\log(\mathcal{L}(\hat{\theta}|y)) + 2K,$$

where the first term on the right-hand side tends to decrease as more parameters are added to the approximating model, while the second term ($2K$) gets larger as more parameters are added to the approximating model. This is the tradeoff between bias and variance or the tradeoff between underfitting and overfitting that is fundamental to the principle of parsimony (see Section 1.4.2). Some investigators have considered K to be a measure of "complexity," but this is unnecessary, though not irrational. We consider K primarily a simple expression for the asymptotic bias in the log-likelihood as an estimator of $E_y E_x[\log(g(x|\hat{\theta}(y)))]$. Note that AIC is derived as an estimator of relative, expected K-L information; thus parsimony arises as a byproduct of this approach. Further books and papers on the derivation of AIC include Shibata (1983, 1989), Linhart and Zucchini (1986), Bozdogan (1987), and Sakamoto (1991).

Usually, AIC is positive; however, it can be shifted by any additive constant, and some shifts can result in negative values of AIC. Computing AIC from regression statistics (see Section 1.2.2) often results in negative AIC values. In our work, we have seen minimum AIC values that range from large negative numbers to as high as 340,000. **It is not the absolute size of the AIC value, it is the relative values over the set of models considered, and particularly the differences between AIC values (Section 2.5), that are important.**

The material to this point has been based on likelihood theory, which is a very general approach. In the special case of least squares (LS) estimation with normally distributed errors, and apart from an arbitrary additive constant, AIC can be expressed as a simple function of the residual sum of squares.

The Least Squares Case

If all the models in the set assume normally distributed errors with a constant variance, then AIC can be easily computed from least squares regression statistics as

$$\text{AIC} = n \log(\hat{\sigma}^2) + 2K,$$

where

$$\hat{\sigma}^2 = \frac{\sum \hat{\epsilon}_i^2}{n} \text{ (the MLE of } \sigma^2\text{),}$$

and $\hat{\epsilon}_i$ are the estimated residuals for a particular candidate model. A common mistake with LS model fitting, when computing AIC, is to take the estimate of σ^2 from the computer output, instead of computing the ML estimate, above. **Also, for LS model fitting, K is the total number of estimated regression parameters, including the intercept and σ^2.**

Thus, AIC is easy to compute from the results of LS estimation in the case of linear models and is now included in the output of many software packages for regression analysis. However, the value of K is sometimes determined incorrectly because either β_0 (the intercept) or σ^2 (or both) is mistakenly ignored in determining K.

The fact that AIC is an estimate only of relative expected K-L distance is almost unimportant. It is the fact that AIC is only an estimate of these relative distances from each model g_i to f that is less than ideal. It is important to recognize that there is usually substantial uncertainty as to the best model for a given data set. After all, these are stochastic biological processes, often with relatively high levels of uncertainty.

In as much as a statistical model can provide insight into the underlying biological process, it is important to try to determine as accurately as possible the basic underlying structure of the model that fits the data well. "Let the data speak" is of interest to both biologists and statisticians in objectively learning from empirical data. The data then help determine the proper complexity (order or dimension) of the approximating model used for inference and help determine what effects or factors are justified. In this sense, inferences for a given data set are conditional on sample size. We must admit that if much more data were available, then further effects could probably be found and supported. "Truth" is elusive; model selection tells us what inferences the data support, not what full reality might be.

Akaike (1973) multiplied the bias-corrected log-likelihood by -2 for "historical reasons" (e.g., it is well known that -2 times the logarithm of the ratio of two maximized likelihood values is asymptotically chi-squared under certain conditions and assumptions). The term -2 occurs in other statistical contexts, so it was not unreasonable that Akaike performed this simple operation to get his AIC. Two points frequently arise, and we will note these here. First, the model associated with the minimum AIC remains unchanged if the bias-corrected log-likelihood (i.e., $\log(\mathcal{L}) - K$) is multiplied by -0.17, -34, or -51.3, or any other negative number. Thus, the minimization is not changed by the multiplication of both terms by any negative constant; Akaike merely chose -2. Second, some investigators have not realized the formal link between K-L information and AIC and believed, then, that the number 2 in the second term in AIC was somehow "arbitrary" and that other numbers should also be considered. This error has led to considerable confusion in the technical literature; clearly, K is the asymptotic bias correction and is not arbitrary. Akaike chose to work with $-2\log(\mathcal{L})$, rather than $\log(\mathcal{L})$; thus the term $+2K$ is theoretically correct, for large sample size. As long as both terms (the log-likelihood and the bias correction) are multiplied by the same negative constant, the model where the criterion is minimized is unchanged and there is nothing arbitrary.

It might be argued that we should have merely defined $l = \log(\mathcal{L}(\hat{\theta} \mid data, model))$; then AIC $= -2l + 2K$, making the criterion look simpler. While this may have advantages, we believe that the full notation works for the reader and helps in understanding exactly what is meant. The full notation, or abbreviations such as $\log(\mathcal{L}(\theta \mid x, g_i))$, makes it explicit that the log-likelihood is a function of (only) the parameters (θ), while the data (x) and model (g_i, say multinomial) must be given (i.e., known). These distinctions become more important when we introduce the concept of a likelihood of a model, given the data: $\mathcal{L}(g_i \mid data)$. Both concepts are fundamental and useful in a host of ways in this book and the notation serves an important purpose here.

If the approximating models in the candidate set are poor (far from f), then Takeuchi's information criterion (TIC) is an alternative if sample size is quite large. AIC is a special case of TIC, and as such, AIC is a parsimonious approach to the estimation of relative expected K-L distance (see Section 2.3).

2.3   Takeuchi's Information Criterion: 1976

At one point in Akaike's derivation of an estimator of K-L information he made the assumption that the model set included f (full reality). This has been the subject of attention and criticism. Akaike maintained that his estimator (AIC) was asymptotically unbiased and free from any notion that full reality was a model or that such a true model was required to be in the set of candidate models. This section will indicate that such claims were justified and provides another insight into the concept of parsimony. The key to this issue is an important, little-known paper (in Japanese) by Takeuchi (1976) that appeared just 3 years after Akaike's initial breakthrough in 1973.

Takeuchi (1976) provides a very general derivation of an information criterion, without taking expectations with respect to g. His criterion is now called TIC (Takeuchi's information criterion) and was thought to be useful in cases where the candidate models were not particularly close approximations to f. TIC has a more general bias-adjustment term to allow $-2\log(\mathcal{L})$ to be adjusted to be an asymptotically unbiased estimate of relative, expected K-L information,

$$\text{TIC} = -2\log(\mathcal{L}) + 2 \cdot \text{tr}(J(\theta)I(\theta)^{-1}).$$

The $K \times K$ matrices $J(\theta)$ and $I(\theta)$ involve first and second mixed partial derivatives of the log-likelihood function, and "tr" denotes the matrix trace function. One might consider always using TIC and worry less about the adequacy of the models in the set of candidates. This consideration involves two issues that are problematic. First, one must *always* worry about the quality of the set of approximating models being considered; this is not something to shortcut. Second, using the expanded bias adjustment term in TIC involves estimation of the elements of the matrices $J(\theta)$ and $I(\theta)$ (details provided in Chapter 7). Shibata (1999) notes that estimation error of these two matrices can cause instability of the results of model selection. Consider the case where a candidate model has $K = 20$ parameters. Then the matrices $J(\theta)$ and $I(\theta)$ are of dimension 20×20, and reliable estimation of the elements of each matrix will be difficult unless sample size is very large. It turns out that $\text{tr}(J(\theta)I(\theta)^{-1})$ itself has a very simple parsimonious estimator, namely K. This is an interesting and important general result.

Thus, AIC is an approximation to TIC, where $\text{tr}(J(\theta)I(\theta)^{-1}) \approx K$. The approximation is excellent when the approximating model is "good" and becomes poor when the approximating model is a poor. However, for models that are poor, the first term, $-2\log(\mathcal{L})$, dominates the criterion because the fit is poor and this term will tend to be relatively large, compared to any much better model. Thus, with the final approximation that $\text{tr}(J(\theta)I(\theta)^{-1}) \approx K$, one can see that AIC is an asymptotically unbiased estimator of relative, expected K-L information, derived without assuming that full reality exists as a model or that such a model is in the set of candidate models. While TIC is an important contribution to the literature, it has rarely seen application. We do not recommend its use, unless sample size is very large and good estimates of the elements of the matrices $J(\theta)$ and $I(\theta)$ can be expected. Even when this can be done, we expect $\text{tr}(J(\theta)I(\theta)^{-1})$ to be very close to K.

2.4   Second-Order Information Criterion: 1978

While Akaike derived an estimator of K-L information, AIC may perform poorly if there are too many parameters in relation to the size of the sample (Sugiura 1978, Sakamoto et al. 1986). Sugiura (1978) derived a second-order variant of AIC that he called c-AIC.

A Small Sample AIC

Hurvich and Tsai (1989) further studied this small-sample (second-order) bias adjustment, which led to a criterion that is called AIC_c,

$$\text{AIC}_c = -2\log(\mathcal{L}(\hat{\theta})) + 2K\left(\frac{n}{n-K-1}\right),$$

where the penalty term is multiplied by the correction factor $n/(n-K-1)$. This can be rewritten as

$$\text{AIC}_c = -2\log(\mathcal{L}(\hat{\theta})) + 2K + \frac{2K(K+1)}{n-K-1},$$

or, equivalently,

$$\text{AIC}_c = \text{AIC} + \frac{2K(K+1)}{n-K-1},$$

where n is sample size (also see Sugiura 1978).

Unless the sample size is large with respect to the number of estimated parameters, use of AIC_c is recommended.

AIC_c merely has an additional bias-correction term. If n is large with respect to K, then the second-order correction is negligible and AIC should perform well. Findley (1985) noted that the study of bias correction is of interest in itself; the exact small-sample bias-correction term varies by type of model (e.g., normal, exponential, Poisson). Bedrick and Tsai (1994) provide a further refinement, but it is more difficult to compute (also see Hurvich and Tsai 1991 and 1995a and b, and Hurvich et al. 1990). While AIC_c was derived under Gaussian assumptions for linear models (fixed effects), Burnham et al. (1994) found this second-order approximation to the K-L distance to be useful in product multinomial models. **Generally, we advocate the use of AIC_c when the ratio n/K is small (say < 40).** In reaching a decision about the use of AIC vs. AIC_c, one must use the value of K for the highest-dimensioned (i.e., global) model in the set of candidates. If the ratio n/K is sufficiently large, then AIC and AIC_c are similar and will strongly tend to select the same model. One must use either AIC or AIC_c consistently in a given analysis, rather than mixing the two criteria. Few software packages provide AIC_c values, but these can easily be computed by hand.

2.5 Modification of Information Criterion for Overdispersed Count Data

In general, if the random variable n represents a count under some simple discrete distribution (e.g., Poisson or binomial), it has a known expectation, $\mu(\theta)$, and a known theoretical variance function, $\sigma^2(\theta)$ (θ still is unknown). In a model of overdispersed data the expectation of n is not changed, but the variance model must be generalized, for example using a multiplicative factor, e.g., $\gamma(\theta)\sigma^2(\theta)$. The form of the factor $\gamma(\theta)$ can be partly determined by theoretical considerations and can be complex (see, e.g., McCullagh and Nelder 1989). Overdispersion factors typically are small, ranging from just above 1 to perhaps 3 or 4 if the model structure is correct and overdispersion is due to small violations of assumptions such as independence and parameter homogeneity over individuals. Hence, a first approximation for dealing with overdispersion is to use a simple constant c in place of $\gamma(\theta)$, and this can be generalized to more than one c for different partitions of the data.

Count data have been known not to conform to simple variance assumptions based on binomial or multinomial distributions (e.g., Bartlett 1936, Fisher 1949, Armitage 1957, and Finney 1971). There are a number of statistical models for count data (e.g., Poisson, binomial, negative binomial, multinomial). In these, the sampling variance is theoretically determined, *by assumption* (e.g., for the Poisson model, $\text{var}(n) = E(n)$; for the binomial model, $\text{var}(\hat{p}) = p(1 - p)/n$. If the sampling variance exceeds the theoretical (model-based) variance, the situation is called "overdispersion." Our focus here is on a lack of independence in the data leading to overdispersion, or "extrabinomial variation." Eberhardt (1978) provides a clear review of these issues in the biological sciences. For example, Canada geese (*Branta* species) frequently mate for life, and the pair behaves almost as an individual, rather than as two independent "trials." The young of some species continue to live with the parents for a period of time, which can also cause a lack of independence of individual responses. Further reasons for overdispersion in biological systems include species whose members exist in schools or flocks. Members of such populations can be expected to have positive correlations among individuals within the group; such dependence causes overdispersion. A different type of overdispersion stems from parameter heterogeneity, that is, individuals having unique parameters rather than the same parameter (such as survival probability) applying to all individuals.

The estimators of model parameters often remain unbiased in the presence of overdispersion, but the model-based theoretical variances overestimate precision (McCullagh and Nelder 1989). To properly cope with overdispersion one needs to model the overdispersion and then use generalized likelihood inference methods. Quasi-likelihood (Wedderburn 1974) theory is a basis for the analysis of overdispersed data (also see Williams 1982, McCullagh and Pregibon 1985, Moore 1987, and McCullagh and Nelder 1989, Lindsey 1999a). Hurvich and Tsai (1995b) provide information on the use of AIC_c with overdispersed data.

Cox and Snell (1989) discuss modeling of count data and note that the first useful approximation is based on a single variance inflation factor (c), which can be estimated from the goodness-of-fit chi-square statistic (χ^2) of the global model and its degrees of freedom,

$$\hat{c} = \chi^2/\text{df}.$$

The variance inflation factor should be estimated from the global model. Cox and Snell (1989) assert that the simple approach of a constant variance inflation factor should often be adequate, as opposed to the much more arduous task of seeking a detailed model for the $\gamma(\theta)$. In a study of these competing approaches on five data sets, Liang and McCullagh (1993) found that modeling overdispersion was clearly better than use of a single $\hat{c}$ in only one of five cases examined.

Given $\hat{c}$, empirical estimates of sampling variances ($\text{var}_e(\hat{\theta}_i)$) and covariances ($\text{cov}_e(\hat{\theta}_i, \hat{\theta}_j)$) can be computed by multiplying the estimates of the theoretical (model-based) variances and covariances by $\hat{c}$ (a technique that has long been used; see, e.g., Finney 1971). These empirical measures of variation (i.e., $\hat{c} \cdot \widehat{\text{var}}_t(\hat{\theta}_i)$) must be treated as having the degrees of freedom used to compute $\hat{c}$ for purposes of setting confidence limits (or testing hypotheses). **The number of parameters (K) must include one for the estimation of c, the variance inflation factor, if used**. Generally, quasi-likelihood adjustments (i.e., use of $\hat{c} > 1$) are made only if some distinct lack of fit has been found (for example, if the observed significance level $P \leq 0.15$ or 0.25) and the goodness-of-fit degrees of freedom ≥ 10, as rough guidelines.

We might expect $c > 1$ with real data but would not expect c to exceed about 4 if model structure is acceptable and only overdispersion is affecting c (see Eberhardt 1978). Substantially larger values of c (say, 6–10) are usually caused partly by a model structure that is inadequate; that is, the fitted model does not account for an acceptable amount of variation in the data. Quasi-likelihood methods of variance inflation are most appropriate only after a reasonable structural adequacy of the model has been achieved. The estimate of c should be computed only for the global model; one should not make and use separate estimates of this variance inflation factor for each of the candidate models in the set. The issue of the structural adequacy of the model is at the very heart of good data analysis (i.e., the reliable identification of the structural versus residual variation in the data). Patterns in the goodness-of-fit statistics (Pearson χ^2 or G-statistics) might be an indication of structural problems with the model. Of course, the biology of the organism in question and the sampling protocol should provide clues as to the existence of overdispersion; one should not rely only on statistical considerations in this matter.

When data are overdispersed and $c > 1$, the proper likelihood is $\log(\mathcal{L})/c$ (not just $\log(\mathcal{L})$). Principles of quasi-likelihood suggest simple modifications to AIC and AIC_c; we denote these modifications by (Lebreton et al. 1992),

$$\text{QAIC} = -\left[2\log(\mathcal{L}(\hat{\theta}))/\hat{c}\right] + 2K,$$

and

$$\text{QAIC}_c = -\left[2\log(\mathcal{L}(\hat{\theta}))/\hat{c}\right] + 2K + \frac{2K(K+1)}{n-K-1},$$

$$= \text{QAIC} + \frac{2K(K+1)}{n-K-1}.$$

If an overdispersion factor is estimated, then one parameter must be added to K. Of course, when no overdispersion exists, then $c = 1$, and the formulas for QAIC and QAIC_c reduce to AIC and AIC_c, respectively. Anderson et al. (1994) found that these criteria performed well in product multinomial models of capture–recapture data in the presence of differing levels of overdispersion.

One must be careful when using some standard software packages (e.g., SAS GENMOD), since they were developed some time ago under a hypothesis testing mode (i.e., adjusting χ^2 test statistics by $\hat{c}$ to obtain F-tests). In some cases, a separate estimate of c is made for each model, and variances and covariances are multiplied by this model-specific estimate of the variance inflation factor. Some software packages compute an estimate of c for every model, thus making the correct use of model selection criteria tricky unless one is careful. Instead, we recommend that the global model be used as a basis for the estimation of a single variance inflation factor c. Then the empirical

---

Overdispersed Count Data: A Review

Try to ensure that the structural part of the data is well modeled by the global model.

If there is biological reason to suspect overdispersion, then the overdispersion parameter c can be estimated as χ^2/df, using the global model.

If overdispersion is present, the log-likelihood of the parameter θ, given the data and the model, should be computed as

$$\frac{\log(\mathcal{L}(\theta|x,g_i))}{\hat{c}}.$$

The number of parameters K is now the number of parameters θ, plus 1 to account for the estimation of the overdispersion parameter c.

The estimated overdispersion parameter should generally be $1 \leq c \leq 4$. Otherwise, some structural lack of fit is probably entering the estimate of overdispersion. If $\hat{c} < 1$, just use $c = 1$.

log-likelihood for each of the candidate models is divided by $\hat{c}$, and QAIC or QAIC$_c$ computed and used for model selection. The estimated variances and covariances should also be adjusted using $\hat{c}$ from the global model, unless there are few degrees of freedom left.

AIC for Overdispersed Count Data

Model selection should use either

$$\text{QAIC} = -[2\log(\mathcal{L}(\hat{\theta}))/\hat{c}] + 2K,$$

or

$$\text{QAIC}_c = -[2\log(\mathcal{L}(\hat{\theta}))/\hat{c}] + 2K + \frac{2K(K+1)}{n-K-1},$$

$$= \text{QAIC} + \frac{2K(K+1)}{n-K-1}$$

The variance–covariance matrix should be multiplied by the estimated overdispersion parameter $\hat{c}$ (i.e., $\hat{c}(\text{cov}(\hat{\theta}_i, \hat{\theta}_j))$).

Some commercial software computes AIC, while AIC$_c$ is rarely available, and no general software package computes QAIC or QAIC$_c$. In almost all cases, AIC, AIC$_c$, QAIC, and QAIC$_c$ can be computed easily by hand from the material that is output from standard computer packages (either likelihood or least squares estimation). In general, we recommend using this extended information-theoretic criterion for count data, and we will use QAIC$_c$ in some of the practical examples in Chapter 3. Of course, often the overdispersion parameter is near 1, negating the need for quasi-likelihood adjustments, and just as often the ratio n/K is large, negating the need for the additional bias-correction term in AIC$_c$. AIC, AIC$_c$, and QAIC$_c$ are all estimates of the relative K-L information. We often use the generic term "AIC" to mean any of these criteria.

2.6   AIC Differences, Δ_i

AIC, AIC$_c$, QAIC$_c$, and TIC are all on a relative (or interval) scale and are strongly dependent on sample size. Simple differences of AIC values allow estimates of $E_{\hat{\theta}}[\hat{I}(f, g_i)] - \min E_{\hat{\theta}}[\hat{I}(f, g_i)]$, where the expectation is over the estimated parameters and min is over the models.

The larger Δ_i is, the less plausible it is that the fitted model $g_i(x|\hat{\theta})$ is the K-L best model, given the data x. Some rough rules of thumb are available and are particularly useful for nested models:

Δ_i	Level of Empirical Support of Model i
0-2	Substantial
4-7	Considerably less
> 10	Essentially none.

AIC Differences

We recommend routinely computing (and presenting in publications) the **AIC differences**,

$$\Delta_i = AIC_i - AIC_{min},$$

over all candidate models in the set. We use the term "AIC differences" in a generic sense here to mean AIC, AIC_c, $QAIC_c$, or TIC. Such differences estimate the relative expected K-L differences between f and $g_i(x|\theta)$. These Δ_i values are easy to interpret and allow a quick comparison and ranking of candidate models and are also useful in computing Akaike weights (Section 2.9). The model estimated to be best has $\Delta_i \equiv \Delta_{min} \equiv 0$.

Models with $\Delta_i > 10$ have either essentially no support, and might be omitted from further consideration, or at least those models fail to explain some substantial explainable variation in the data. These guidelines seem useful if R is small (even as many as 100), but may break down in exploratory cases where there may be thousands of models. The guideline values may be somewhat larger for nonnested models, and more research is needed in this area (e.g., Linhart 1988). If observations are not independent, but are assumed to be independent, then these simple guidelines cannot be expected to hold. Thus, if the log-likelihood is corrected for overdispersion in count data by estimating c, then the guidelines above will be useful.

As an example, candidate models g_1, g_2, g_3, and g_4 have AIC values of 3,400, 3,560, 3,380, and 3,415, respectively. Then one would select model g_3 as the best single model as the basis for inference because g_3 has the smallest AIC value. Because these values are on a relative (interval) scale, one could subtract, say, 3,380 (the minimum of the 4 values) from each AIC value and have the following rescaled AIC values: 20, 180, 0, and 35. Of course, such rescaling does not change the ranks of the models, nor the pairwise differences in the AIC values. People are often surprised that Δ_i of only 1–10 are very important, when the associated AIC values that led to the difference are on the order of 97,000 or 243,000.

AIC Differences

It is not the absolute size of the AIC value, it is the relative values, and particularly the AIC differences (Δ_i), that are important.

An individual AIC value, by itself, is not interpretable due to the unknown constant (interval scale). AIC is only comparative, relative to other AIC values in the model set; thus such differences Δ_i are very important and useful.

We can say with considerable confidence that in real data analysis with several or more models and large sample size (say $n > 10 \times K$ for the biggest model) a model having $\Delta_i = 20$, such as model g_4, would be a very poor approximating model for the data at hand.

We can order the Δ_i from smallest to largest, and the same ordering of the models indicates how good they are as an approximation to the actual, expected K-L best model. Consider Δ_i values for 7 models as 0, 1.2, 1.9, 3.5, 4.1, 5.8, and 7.3. An important question is, how big a difference matters? This should be asked in the sense of when a model is not to be considered competitive with the selected best model as plausibly the actual K-L best model in the set of models used, for the sample size and data at hand. The question has no unambiguous answer; it is like asking how far away from an MLE $\hat{\theta}$ an alternative value of θ must be (assuming that the model is a good model) before we would say that an alternative θ is unlikely as "truth." This question ought to be answered with a confidence (or credibility) interval on θ based on $\hat{\theta}$ and its estimation uncertainty. A conventionally accepted answer here is that θ is unlikely as truth if it is further away than $\pm 2\,\widehat{se}(\hat{\theta})$ (there is a fundamental basis for using such a procedure). Relative scaling of alternative models can effectively be done using Akaike weights (Section 2.9) and evidence ratios (Section 2.10).

2.7   A Useful Analogy

In some ways, selection of a best approximating model is analogous to auto racing or other similar contests. The goal of such a race is to identify the best (fastest) car/driver combination, and the data represent results from a major race (e.g., the Indianapolis 500 in the USA, the 24 Heures du Mans in France). Only a relatively few car/driver combinations "qualify," based on prerace trials (e.g., 33 cars at Indianapolis)—this is like the set of candidate models (i.e., only certain models "qualify," based on the science of the situation). It would be chaotic if all car/driver combinations with an interest could enter the race, just as it makes little sense to include a very large number of models in the set of candidates (and risk Freedman's paradox). Cars that do not qualify do not win, even though they might indeed have been the best (fastest) had they not failed to qualify. Similarly, models, either good or bad, not in the set of candidates remain out of consideration.

At the end of the race the results provide a ranking ("placing") of each car/driver combination, from first to last. Furthermore, if a quantitative index of quality is available (e.g., elapsed time for each finisher), then a further "scaling" can be considered. Clearly, the primary interest is in "who won the race" or "which was the first"; this is like the model with the minimum AIC value. This answers the question, "Which is best in the race"; the results could differ for another (future) race or another data set, but these are, as yet, unavailable to us.

Some (secondary) interest exists in the question, "Who was in second place?" and in particular, was second place only thousandths of a second behind the winner or 5 minutes behind? The race time results provide answers to these questions, as do the Δ_i values in model selection. In the first case, the best inference might be that the first two cars are essentially tied and that neither is appreciably better than the other (still, the size of the purse certainly favors the first-place winner!), while in the second case, the inference probably favors a single car/driver combination as the clear best (with a 5-minute lead at the finish). The finishing times provide insights into the third and fourth finishers, etc. In trying to understand the performance of car/driver combinations, one has considerable information from both the rankings and their finishing times, analogous to the AIC values (both the ranks and the Δ_i values). In Sections 2.9 and 2.10 will see how the Δ_i can be used to estimate further quantities, and these will provide additional insights. Note that the absolute time of the winner is of little interest because of temperature differences, track conditions, and other variables; only the *relative* times for a given race are of critical interest. Similarly, the absolute values of AIC are also of little interest, because they reflect sample size and some constants, among other things. The value of the maximized log-likelihood (i.e., $\log(\mathcal{L}(\hat{\theta}|x))$) varies substantially from sample to sample. However, all comparisons of models are made on the same data, so this sample-to-sample variation is irrelevant. Comparing maximized log-likelihood values across data sets is like comparing race finishing times when some races are 500 miles whereas others are 400 or 600 miles.

The winner of the race is clearly the best for the particular race. If one wants to make a broader inference concerning races for an entire year, then results (i.e., ranks) from several races can be pooled or weighted. Similarly, statistical inferences beyond a single observed data set can sometimes be broadened by some type of model averaging using, for example, the nonparametric bootstrap (details in Chapters 4 and 5) and the incorporation of model selection uncertainty in estimators of precision.

The race result might not always select the best car/driver combination, because the fastest qualifying car/driver may have had bad luck (e.g., crash or engine failure) and finished well back from the leader (if at all). Similarly, in model selection one has only one realization of the stochastic process and an *estimated* relative distance as the basis for the selection of a best approximating model (a winner). If the same race is held again with the same drivers, the winner and order of finishers are likely to change somewhat. Similarly, if a new sample of data could be obtained, the model ranks would likely change somewhat.

To carry the analogy a bit further, data dredging would be equivalent to watching a race as cars dropped out and others came to the lead. Then one continually shifts the bet and predicted winner, based on the car/driver in the lead at any point in time (i.e., an unfair advantage). In this case, the final prediction would surely be improved, but the rules of play have certainly been altered! Alternatively, the definition of winning might not be established prior to the initiation of the race. Only after the race are the rules decided (e.g., based, in part, on who they think "ought" to win). Then, one might question the applicability of this specific prediction to other races. Indeed, we recommend "new rules" when data dredging has been done. That is, if a particular result was found following data dredging, then this should be fully admitted and discussed in resulting publication. We believe in fully examining the data for all the information and insights they might provide. However, the sequence leading to data dredging should be revealed, and results following should be discussed in this light.

Many realize that there is considerable variation in cars and drivers from race to race and track to track. Similarly, many are comfortable with the fact that there is often considerable sampling variation (uncertainty) associated with an estimate of a parameter from data set to data set. Similarly, if other samples (races) could be taken, the estimated best model (car/driver) might also vary from sample to sample (or race to race). Both components of sampling variation and model selection uncertainty should ideally be incorporated into measures of precision.

2.8   Likelihood of a Model, $\mathcal{L}(g_i|data)$

While the AIC differences Δ_i are useful in ranking the models, it is possible to quantify the plausibility of each model as being the actual K-L best model. This can be done by extending the concept of the likelihood of the parameters given both the data and model, i.e., $\mathcal{L}(\theta|x, g_i)$, to the concept of the likelihood of the model given the data, hence $\mathcal{L}(g_i|x)$. Such quantities are very useful in making inferences concerning the relative strength of evidence for each of the models in the set.

Likelihood of a Model, Given Data

The likelihood of model g_i, given the data, is simple to compute for each model in the set:

$$\mathcal{L}(g_i|x) \propto \exp\left(-\frac{1}{2}\Delta_i\right),$$

where "$\propto$" means "is proportional to." Such likelihoods represent the relative strength of evidence for each model.

Akaike (see, e.g., Akaike 1983b) advocates the above $\exp(-\frac{1}{2}\Delta_i)$ for the relative likelihood of the model, given the MLEs of model parameters based on the same data. Such quantities can also be expressed as

$$C\mathcal{L}(\hat{\theta}|x, g_i)e^{-K},$$

where C is an arbitrary constant.

2.9 Akaike Weights, w_i

2.9.1 Basic Formula

Model Probabilities

To better interpret the relative likelihood of a model, given the data and the set of R models, we normalize the $\mathcal{L}(g_i|x)$ to be a set of positive "Akaike weights," w_i, adding to 1:

$$w_i = \frac{\exp(-\frac{1}{2}\Delta_i)}{\sum\limits_{r=1}^{R} \exp(-\frac{1}{2}\Delta_r)}.$$

The w_i depend on the entire set; therefore, if a model is added or dropped during a post hoc analysis, the w_i must be recomputed for all the models in the newly defined set.

This idea of the likelihood of the model given the data, and hence these model weights, has been suggested for many years by Akaike (e.g., Akaike 1978b, 1979, 1980, 1981b and 1983b; also see Bozdogan 1987 and Kishino et al. 1991) and has been researched some by Buckland et al. (1997). These model weights seemed not to have a name, so we call them *Akaike weights*. This name will herein apply also when we use AIC$_c$, QAIC, QAIC$_c$, and TIC. **A given w_i is considered as the weight of evidence in favor of model i being the actual K-L best model for the situation at hand *given* that one of the R models must be the K-L best model of that set of R models.** Hence, given that there are only R models and one of them must be best in this set of models, it is convenient to normalize the relative likelihoods to sum to 1.

For the estimated K-L best model (let this be model g_{min}), $\Delta_{min} = 0$; hence, for that model $\exp(-\frac{1}{2}\Delta_{min}) \equiv 1$. The odds for the ith model actually being the K-L best model are thus $\exp(-\frac{1}{2}\Delta_i)$ to 1, or just the "ratio" $\exp(-\frac{1}{2}\Delta_i)$. It is convenient to reexpress such odds as the set of Akaike weights. The bigger a Δ_i is, the smaller the w_i, and the less plausible is model i as being the actual K-L best model for f based on the design and sample size used. The Akaike weights provide an effective way to scale and interpret the Δ_i values. These weights also have other important uses and interpretations that are given in the following chapters.

In general, likelihood provides a good measure of data-based weight of evidence about parameter values, given a model and data (see, e.g., Royall 1997). We think that this concept extends to evidence about the K-L best model, given a set of models. That is, evidence for the best model is well represented by the likelihood of a model.

2.9.2   An Extension

In the absence of any a prior information (in a Bayesian sense) about which of these models might be the K-L best model for the data at hand we are compelled by a certain aspect of information theory itself (see Jaynes 1957, Jessop 1995). Let τ_i be the prior probability that model i is the K-L best model. Lacking any prior information, we set the τ_i all equal, and hence use $\tau_i \equiv 1/R$. In fact, doing so places all R of the models on an equal footing to be selected as the K-L best model.

If there is prior information or belief, this opens the door to unequal prior probabilities. Ignoring any model redundancy (this subject is deferred to Section 4.6), τ_i is our prior state of information or belief that model g_i, fitted to the data, provides the K-L best model for the design and data at hand. This is a deceptively complex issue, as it relates both to ideas of models as best approximations to truth and to expected model fitting tradeoff of bias versus sampling variances.

To us it seems impossible to have any real prior basis for an informative differential assessment of the τ_i (other than on how the models might be structurally interrelated or partially redundant). Using the maximum entropy principle of Jaynes (1957) we should take the τ_i to represent maximal uncertainty about all unknown aspects of the probability distribution represented by the τ_i. Thus we determine the τ_i that maximize the entropy $-\sum \tau_i \log(\tau_i)$ subject to constraints that express whatever information (in the colloquial sense) we have about the distribution. In the "no information" case the only constraint we have is that $\sum \tau_i = 1$ (plus the essential $0 < \tau_i < 1$). The maximum entropy (hence maximum uncertainty) prior is then $\tau_i \equiv 1/R$. [It takes us too far a field to delve into the aspects of information theory underlying the maximum entropy principle. This principle is fundamentally tied both to Boltzmann's entropy and to information theory and can be used to justify noninformative Bayesian priors—when they exist. The interested reader is referred to Kapur and Kesavan 1992, or the less technical Jessop 1995.]

Given any set of prior probabilities (the τ_i), generalized Akaike weights are given by

$$w_i = \frac{\mathcal{L}(g_i|\underline{x})\tau_i}{\sum_{r=1}^{R} \mathcal{L}(g_r|\underline{x})\tau_r}.$$

There may be occasions to use unequal prior probabilities, hence the expression above. However, in general, by Akaike weights we mean the simple expression without the τ_i (this assumes $\tau_i = 1/R$).

The inclusion of prior probabilities (τ_i) in the w_i is not a true Bayesian approach. The full Bayesian approach to model selection requires both the prior τ_i on the model and a prior probability distribution on the parameters θ in model g_i for each model. Then the derivation of posterior results requires integration (usually achievable only by Markov chain Monte Carlo methods). Persons wishing to learn the Bayesian approach to model selection can start with the following sources: Raftery et al. (1993), Madigan and Raftery 1994, Carlin and Chib (1995), Chatfield (1995b), Draper (1995), Gelman et al. (1995), Kass and Raftery (1995), Hoeting and Ibrahim (1996), Raftery (1996a, 1996b), and Morgan (2000).

A brief comparison is given here of what we mean by the prior probabilities τ_i under this information-theoretic approach to model selection versus what seems to be meant by the prior probabilities of models in the Bayesian approach. The Bayesian approach seems generally to assume that one of the models, in the set of R models, is true. Hence, τ_i is then the prior degree of belief that model form g_i is the true model form (see, e.g., Newman 1997). Under the information-theoretic approach we do not assume that truth f is in the set of models, and $\tau_1, \ldots, \tau_R$ is a probability distribution of our prior information (or lack thereof) about which of the R models is the K-L best model for the data. Information theory itself (Kapur and Kesavan 1992) then justifies determination of the τ_i, generally as $\tau_i \equiv 1/R$. For data analysis we believe that the issue cannot be which model structure is truth, because none of the models considered is truth. Rather, the issue is, which model *when fit to the data* (i.e., when θ is *estimated*) is the best model for purposes of representing the (finite) information in the data. Letting $\tau_i = $ Prob{belief that model form g_i is the K-L best model}, then τ_i is about the "parameter" g_{best}, not about the random variable g_{min}. Here, we use only $\tau_i = 1/R$.

2.10   Evidence Ratios

Using the hypothetical example in Section 2.6, the likelihood of each model, given the data, and the Akaike weights are given below:

Model	Δ_i	$\mathcal{L}(g_i\|x)$	Akaike weight w_i
1	0	1	0.431
2	1.2	0.54881	0.237
3	1.9	0.38674	0.167
4	3.5	0.17377	0.075
5	4.1	0.12873	0.056
6	5.8	0.05502	0.024
7	7.3	0.02599	0.010.

As weight of evidence for each model we can see that the selected best model is not convincingly best; the evidence ratio for model g_1 versus model g_2 is only about 2 (i.e., $w_1/w_2 = 1.82$). This relatively weak support for the best model suggests that we should expect to see a lot of variation in the selected best model from sample to sample if we could, in this situation, draw multiple independent samples; that is, the model selection uncertainty is likely to be high. The evidence ratio for the best model versus model 6 is

---

Evidence Ratios

Evidence can be judged by the relative likelihood of model pairs as

$$\mathcal{L}(g_i|x)/\mathcal{L}(g_j|x)$$

or, equivalently, the ratio of Akaike weights w_i/w_j. Such ratios are commonly used, and we will term them **evidence ratios**. Such ratios represent the evidence about fitted models as to which is better in a K-L information sense.

In particular, there is often interest in the ratio w_1/w_j, where model 1 is the estimated best model and j indexes the rest of the models in the set. These ratios are not affected by any other model, hence do not depend on the full set of R models—just on models i and j. These evidence ratios are invariant to all other models besides i and j.

$0.431/0.024 = e^{(5.8/2)} = 18$, and we must conclude that is it unlikely that model 6 is the K-L best model; the evidence here is reasonably strong against model 6.

There is a striking nonlinearity in the evidence ratios as a function of the Δ_i values. Consider the ratio $w_1/w_j (\equiv w_{min}/w_j)$,

$$\frac{w_1}{w_j} \equiv \frac{1}{e^{-1/2\Delta_j}} \equiv e^{1/2\Delta_j}$$

in the comparison of the evidence for the best model versus the jth best model. Then, we have the following table:

Δ_j	Evidence ratio
2	2.7
4	7.4
8	54.6
10	148.4
15	1,808.0
20	22,026.5

This information helps to justify the rough rules of thumb given for judging the evidence for models being the best K-L model in the set. Jeffreys (1948) provided some likelihood-based rules similar to these over 50 years ago. See Edwards (1992) and Royall (1997) for additional perspectives on the concept of evidence in a likelihood framework.

People may, at first, be frustrated that they do not have some value or cutoff point that provides a simple dichotomy to indicate what is *important* (i.e., "significant" under the Neyman–Pearson null hypothesis testing procedure where a decision is to be reached). Even knowing that statistical significance is not particularly related to biological significance, and that the α-level is arbitrary, some investigators seem to feel comfortable being "told" what is important. This is the blind hope that the computer and its analysis software will somehow "tell" the investigator what is *important* in a yes or no sense. The approach we advocate is one of quantitative evidence; then people may interpret the quantitative evidence.

Consider a football game where the final score is 10 to 13 for teams A and B, respectively. Here, one does not ask whether the win of team B over team A was "significant." Rather, one can see that the game was close, based on the score (the evidence). Further scrutiny of the evidence could come from examining the total yards gained, the cumulative time of possession of the ball, the number of penalties, etc., for each team. Based on the totality of the evidence, one can reach a determination concerning the relative strength of the two teams. Furthermore, in this case, most rational people will reach roughly the same determination, based on the evidence. Similarly, if the score had been 40 to 3 (the evidence), it would be clear that team A hammered its hapless opponent. Even in this case there is no concept of "highly significant," much less any test of the null hypothesis based on the observed scores that the teams were of equal ability. Again, most rational people would probably agree that team A was the better team on the day of the contest, based on the evidence (40 vs. 3). Based on the evidence, people might be willing to make an inference to other games between these two teams. Of course, there are intermediate cases (10 vs. 16) where the evidence is not convincing. Perhaps the final touchdown occurred in overtime, in which case people might often interpret the evidence (10 to 16) differently. Again, a review of other game statistics might provide insights, but we should admit that not all evidence will lead to a clear determination, accepted by all. One encounters various forms of numerical evidence in everyday life and can interpret such evidence without arbitrary dichotomies.

When we learn that model g_4 has an evidence ratio of 3 in relation to model g_2, it means there is relatively little evidence in favor of model g_4. An analogy here is an auditorium containing N people (let N be large, but unspecified). Each person has a raffle ticket, except that a single person (Bob) has 3 tickets. The evidence ratio (relative likelihood) of Bob winning the raffle vs. any other individual is 3. Clearly, Bob has an edge over any other individual, but it is not strong. Of course, the probability that either Bob or any other particular individual will win is small if N is large. However, the ratio 3/1 remains the same, regardless of the value of N. In contrast, let Bob now have 100 tickets. Then his relative likelihood of winning vs. any other individual is 100, and this is relatively strong evidence. Such evidence ratios are only relative (i.e., Bob vs. another individual); nothing is to be inferred about Bob's chances (or any other individual's chances) of winning the raffle outright. Only Bob's chances relative to another individual's chances are quantified using evidence ratios. Finally, note that the probability of Bob winning, *given that either Bob or another single individual wins*, is $100/(100 + 1) = 0.99$. Evidence ratios for model pairs (e.g., model g_4 vs. model g_2) are relative values.

2.11   Important Analysis Details

Data analysis involves the proper tradeoff between bias and variance or, similarly, between underfitting and overfitting. The estimation of expected K-L information is a natural and simple way to view model selection; given a good set of candidate models, select that fitted model where information loss is minimized. Proper model selection is reflected in good achieved confidence interval coverage for the parameters in the model (or for prediction); otherwise, perhaps too much bias has been accepted in the tradeoff to gain precision, giving a false sense of high precision. This represents the worst inferential situation: a highly precise, but quite biased estimate. These ideas have had a long history in statistical thinking.

An information criterion (i.e., AIC, AIC_c, QAIC, and TIC) can be used to rank the candidate models from best to worst and scale the models using Akaike weights and evidence ratios. Often data do not support only one model as clearly best for data analysis. Instead, suppose three models are essentially tied for best, while another, larger, set of models is clearly not appropriate (either underfit or overfit). Such virtual "ties" for the best approximating model must be carefully considered and admitted. Poskitt and Tremayne (1987) discuss a "portfolio of models" that deserve final consideration. Chatfield (1995b) notes that there may be more than one model that is to be regarded as "useful."

Ambivalence
The inability to ferret out a single best model is not a defect of AIC or any other selection criterion. Rather, it is an indication that the data are simply inadequate to reach such a strong inference. That is, the data are ambivalent concerning some effect or parametrization or structure.
In such cases, all the models in the set can be used to make robust inferences: multimodel inference.

It is perfectly reasonable that several models would serve nearly equally well in approximating the information in a set of data. Inference must admit that there are sometimes competing models and the data do not support selecting only one. The issue of competing models is especially relevant in including model selection uncertainty into estimators of precision. When more than one model has substantial support, some form of multimodel inference (e.g., model averaging) should be considered (Chapter 4). The following subsections provide some important details that must be considered in a careful analysis of research data.

2.11.1   AIC Cannot Be Used to Compare Models of Different Data Sets

Models can be compared using the various information criteria, as estimates of relative, expected K-L information, only when they have been fitted to exactly the same set of data. For example, if nonlinear regression model g_1 is fitted to a data set with $n = 140$ observations, one cannot validly compare it with model g_2 when 7 outliers have been deleted, leaving only $n = 133$. Furthermore, AIC cannot be used to compare models where the data are ungrouped in one case (Model U) and grouped (e.g., grouped into histograms classes) in another (Model G).

Data Must Be Fixed

An important issue, in general, is that the data and their exact representation must be fixed and alternative models fitted to this fixed data set.

Information criteria should not be compared across different data sets, because the inference is conditional on the data in hand.

2.11.2   Order Not Important in Computing AIC Values

The order in which the information criterion is computed over the set of models is not relevant. Often, one may want to compute AIC_c, starting with the global model and proceed to simpler models with fewer parameters. Others may wish to start with the simple models and work up to the more general models with many parameters; this strategy might be best if numerical problems are encountered in fitting some high-dimensioned models. The order is irrelevant here to proper interpretation, as opposed to the various hypothesis testing approaches where the order may be both arbitrary and the results quite dependent on the choice of order (e.g., stepup (forward) vs. stepdown (backward) testing; Section 3.4.6 provides an example).

2.11.3   Transformations of the Response Variable

Model selection methods assume that some response variable (say y) is the subject of interest. Assuming that the scientific hypotheses relate to this response variable, then all the models must represent exactly this variable. Thus, the R models in the set should all have the same response variable. A common type of mistake is illustrated by the following example. An investigator is interested in modeling a response variable y and has built 4 linear regression models of y, but during the model building, he decides to include a nonlinear model. At that point he includes a model for $\log(y)$ as the fifth model. Estimates of K-L information in such cases cannot be validly compared. This is an important point, and often overlooked. In this example, one would find g_5 to be the best model followed by the other 4 models, each having large Δ_i values. Based on this result, one would erroneously conclude the importance of the nonlinearity. **Investigators should be sure that all hypotheses are modeled using the same response variable (e.g., if the whole set of models were based on $\log(y)$, no problem would be created; it is the mixing of response variables that is incorrect).**

Elaborating further, if there was interest in the normal and log-normal model forms, the models would have to be expressed, respectively, as,

$$g_1(y|\mu, \sigma) = \frac{1}{\sqrt{2\pi}\sigma} \exp\left[-\frac{1}{2}\frac{[y-\mu]^2}{\sigma^2}\right],$$

and another model,

$$g_2(y|\mu, \sigma) = \frac{1}{y\sqrt{2\pi}\sigma} \exp\left[-\frac{1}{2}\frac{[\log(y)-\mu]^2}{\sigma^2}\right].$$

Another critical matter here is that all the components of each likelihood should be retained in comparing different probability distributions. There are some comparisons of different pdfs in this spirit in Section 6.7.1. This "retain it all" requirement is not needed in cases like multiple regression with constant variance because all the comparisons are about the model structure (i.e., variables to select) with an assumption of normal errors for every model. In this case there is a global model and its associated likelihood, and the issue is how best to represent μ as a regression function.

In other cases, it is tempting to drop constants in the log-likelihood, because they do not involve the model parameters. However, alternative models may not have the same constants; this condition makes valid model comparisons impossible. The simple solution here is to retain all the terms in the log-likelihood for all the models in the set.

2.11.4   Regression Models with Differing Error Structures

This issue is related to that in Section 2.11.3. A link between the residual sum of squares (RSS) and σ^2 from regression models with normally distributed errors to the maximized log-likelihood value was provided in Section 1.2.2. This link is a special case, allowing one to work in an ordinary least squares regression framework for modeling and parameter estimation and then switch to a likelihood framework to compute $\log(\mathcal{L}(\theta|data, model))$ and various other quantities under an information-theoretic paradigm.

The mapping from $\hat{\sigma}^2$ to $\log(\mathcal{L}(\theta|data, model))$ is valid only if all the models in the set assume independent, normally distributed errors (residuals) with a constant variance. If some subset of the R models assume lognormal errors, then valid comparisons across all the models in the set are not possible. In this case, all the models, including those with differing error structures, should be put into a likelihood framework since this permits valid estimates of $\log(\mathcal{L}(\theta|data, model))$ and criteria such as AIC_c.

2.11.5   Do Not Mix Null Hypothesis Testing with Information-Theoretic Criteria

Tests of null hypotheses and information-theoretic approaches should not be used together; they are very different analysis paradigms. A very common mistake seen in the applied literature is to use AIC to rank the candidate models and then "test" to see whether the best model (the alternative hypothesis) is "significantly better" than the second-best model (the null hypothesis). This procedure is flawed, and we strongly recommend against it (Anderson et al. 2001*c*). Despite warnings about the misuse of hypothesis testing (see Anderson et al. 2000, Cox and Reid 2000), researchers are still reporting *P*-values for trivial null hypotheses, while failing to report effect size and its precision.

Some authors state that the best model (say g_3) is *significantly* better than another model (say g_6) based on a Δ value of 4–7. Alternatively, sometimes one sees that model g_6 is *rejected* relative to the best model. These statements are poor and misleading. It seems best not to associate the words *significant* or *rejected* with results under an information-theoretic paradigm. Questions concerning the strength of evidence for the models in the set are best addressed using the evidence ratio (Section 2.10), as well as an analysis of residuals, adjusted R^2, and other model diagnostics or descriptive statistics.

2.11.6   Null Hypothesis Testing Is Still Important in Strict Experiments

A priori hypothesis testing plays an important role when a formal experiment (i.e., treatment and control groups being formally contrasted in a replicated design with random assignment) has been done and specific a priori alternative hypotheses have been identified. In these cases, there is a very large body of statistical theory on testing of treatment effects in such experimental data. We certainly acknowledge the value of traditional testing approaches to the analysis of these *experimental* data. Still, the primary emphasis should be on the size of the treatment effects and their precision; too often we find a statement regarding "significance," while the treatment and control means are not even presented (Anderson et al. 2000 Cox and Reid 2000). Nearly all statisticians are calling for estimates of effect size and associated precision, rather than test statistics, *P*-values, and "significance."

Akaike (1981) suggests that the "multiple comparison" of several treatment means should be viewed as a model selection problem, rather than resorting to one of the many testing methods that have been developed (also see Berry 1988). Here, a priori considerations would be brought to bear on the issue and a set of candidate models derived, letting information criterion values aid in sorting out differences in treatment means—a refocusing on parameter estimation, instead of on testing. An alternative approach is to consider random effects modeling (Kreft and deLeeuw 1998).

In observational studies, where randomization or replication is not achievable, we believe that "data analysis" should be viewed largely as a problem in model selection and associated parameter estimation. This seems especially the case where nuisance parameters are encountered in the model, such as the recapture or resighting probabilities in capture–recapture or band–recovery studies. Here, it is not always clear what either the null or the alternative hypothesis should be in a hypothesis testing framework. In addition, often hypotheses that are tested are naive or trivial, as Johnson (1995, 1999) points out with such clarity. Should we expend resources to find out if ravens are white? Is there any reason to test formally hypotheses such as "H_0: the number of robins is the same in cities A and B"? Of course not! One should merely assume that the number is different and proceed to estimate the magnitude of the difference and its precision: an estimation problem, not a null hypothesis testing problem.

2.11.7   Information-Theoretic Criteria Are Not a "Test"

The theories underlying the information-theoretic approaches and null hypothesis testing are fundamentally quite different.

Criteria Are Not a Test

Information-theoretic criteria such as AIC, AIC_c, and $QAIC_c$ are not a "test" in any sense, and there are no associated concepts such as test power or P-values or α-levels. Statistical hypothesis testing represents a very different, and generally inferior, paradigm for the analysis of data in complex settings.

It seems best to avoid use of the word "significant" in reporting research results under an information-theoretic paradigm.

The results of model selection under the two approaches might happen to be similar with simple problems; however, in more complex situations, with many candidate models, the results of the two approaches can be quite different (see Section 3.5). **It is critical to bear in mind that there is a theoretical basis to information-theoretic approaches to model selection criteria, while the use of null hypothesis testing for model selection must be considered ad hoc** (albeit a very refined set of ad hoc procedures in some cases).

2.11.8   Exploratory Data Analysis

Hypothesis testing is commonly used in the early phases of exploratory data analysis to iteratively seek model structure and understanding. Here, one might start with 3–8 models, compute various test statistics for each, and note that several of the better models each have a gender effect. Thus, additional models are generated to include a gender effect, and more null hypothesis tests are conducted. Then the analyst notes that several of these models have a trend in time for some set of estimable parameters; thus more models with this effect are generated, and so on. While this iterative or sequential strategy violates several theoretical aspects of hypothesis testing, it is very commonly used, and the results are often published without the details of the analysis approach. We suggest that if the results are treated only as alternative hypotheses for a more confirmatory study to be conducted later, this might be an admissible practice, particularly if other information is incorporated during the design stage. Still, the sequential and arbitrary nature of such testing procedures make us wonder whether this is really a good exploratory technique because it too readily keys in on unique features of the sample data at hand (see Tukey 1980). In any event, the key here is to conduct further investigations based partially on the "hunches" from the tentative exploratory work. Conducting the further investigation has too often been ignored and the tentative "hunches" have been published as if they were a priori results. Often, the author does not admit to the post hoc activities that led to the supposed results.

We suggest that information-theoretic approaches might serve better as an exploratory tool; at least key assumptions upon which these criteria are based are not terribly violated, and there is no arbitrary α level. Exploratory data analysis using an information-theoretic criterion, instead of some form of test statistic, eliminates inferential problems in interpreting the many P-values, but one must still worry about overfitting and spurious effects (Anderson et al. 2001b). The ranking of alternative models (the Δ_i and w_i values) might be useful in the preliminary examination of data resulting from a pilot study. Based on these insights, one could design a more confirmatory study to explore the issue of interest. The results of the pilot exploration should remain unpublished. While we do not condone the use of information theoretic approaches in blatant data dredging, we suggest that it might be a more useful tool than hypothesis testing in exploratory data analysis where little a priori knowledge is available. Data dredging has enough problems and risks without using a testing-based approach that carries its own set of substantial problems and limitations.

2.12   Some History and Further Insights

Akaike (1973) considered AIC and its information theoretic foundations ". . . a natural extension of the classical maximum likelihood principle." Interestingly, Fisher (1936) anticipated such an advance over 60 years ago when he wrote,

> . . . an even wider type of inductive argument may some day be developed, which shall discuss methods of assigning from the data the functional form of the population.

This comment was quite insightful; of course, we might expect this from R. A. Fisher! Akaike was perhaps kind to consider AIC an extension of classical ML theory; he might just as well have said that classical likelihood theory was a special application of the more general information theory. In fact, Kullback believed in the importance of information theory as a unifying principle in statistics.

2.12.1 *Entropy*

Akaike's (1977) term "*entropy maximization principle*" comes from the fact that the negative of K- L information is Boltzmann's entropy (in fact, K-L information has been called negative entropy or "negentropy"). Entropy is "disorder," while max entropy is maximum disorder or minimum information. Conceptually,

$$\text{Boltzmann's entropy} = -\log\left(\frac{f(x)}{g(x)}\right).$$

Then,

$$-\text{Boltzmann's entropy} = \log\left(\frac{f(x)}{g(x)}\right),$$

and

$$\text{K-L} = E_f(-\text{Boltzmann's entropy})$$
$$= E_f\left(\log\left(\frac{f(x)}{g(x)}\right)\right),$$
$$= \int f(x)\log\left(\frac{f(x)}{g(x)}\right)\,dx.$$

Thus, minimizing the K-L distance is equivalent to maximizing the entropy; hence the name *maximum entropy principle* (see Jaynes 1957, Akaike 1983a, 1985 and Bozdogan 1987, Jessop 1995 for further historical insights). However, maximizing entropy is subject to a constraint—the model of the information in the data. A good model contains the information in the data, leaving only "noise." It is the noise (entropy or uncertainty) that is maximized under the concept of the entropy maximization principle (Section 1.2.4). Minimizing K-L information then results in an approximating model that loses a minimum amount of information in the data. Entropy maximization results in a model that maximizes the uncertainty, leaving only information (the model) "maximally" justified by the data. The concepts are equivalent, but minimizing K-L distance (or information loss) certainly seems the more direct approach.

The K-L information is *averaged* negative entropy, hence the expectation with respect to f. While the theory of entropy is a large subject by itself, readers here can think of entropy as nearly synonymous with uncertainty, or randomness or disorder in physical systems.

Boltzmann derived the fundamental theorem that

entropy is proportional to $-$ log(probability).

Entropy, information, and probability are thus linked, allowing probabilities to be multiplicative while information and entropies are additive. (This result was also derived by Shannon 1948). Fritz Hasenöhrl, a student of Boltzmann, Boltzmann's successor at Vienna University, and a famous theoretical physicist himself, noted that this result "... is one of the most profound, most beautiful theorems of theoretical physics, indeed all of science." Further information concerning Boltzmann appears in Brush (1965, 1966), while interesting insights into Akaike's career are found in Findley and Parzen (1995).

2.12.2   A Heuristic Interpretation

After Akaike's innovative derivation of AIC, people noticed a heuristic interpretation that was both interesting and sometimes misleading. The first term in AIC,

$$\text{AIC} = -2\log(\mathcal{L}(\hat{\theta}|x)) + 2K,$$

is a measure of lack of model fit, while the second term $(2K)$ can be interpreted as a "penalty" for increasing the size of the model (the penalty enforces parsimony in the number of parameters). This heuristic explanation does not do justice to the much deeper theoretical basis for AIC (i.e., the link with K-L distance and information theory). The heuristic interpretation led some statisticians to consider "alternative" penalty terms, and this has not always been productive (see Chapter 6). The so-called penalty term in AIC is not arbitrary; rather, it is the asymptotic bias-correction term. It is the result of deriving an asymptotic estimator of relative, expected K-L information. [Note, of course, that had Akaike defined $\text{AIC} = -\log(\mathcal{L}(\hat{\theta}|x)) + K$, the minimization would be unchanged; some authors use this expression, but we will use AIC as Akaike defined it.]

The heuristic view of the components of AIC clearly shows a bias vs. variance tradeoff and insight into how the principle of parsimony is met by using AIC (see Gooijer et al. 1985:316). Still, we recommend viewing AIC as an estimate of the relative expected K-L information or distance between model pairs (i.e., each g_i vs. f). Minimizing this relative, expected distance provides an estimated best approximating model for that particular data set (i.e., the *closest* approximating model to f). The relative K-L distance is the link between information theory and the log-likelihood function that is a critical element in AIC model selection.

2.12.3   More on Interpreting Information-Theoretic Criteria

Estimates of relative K-L information, the AIC differences (Δ_i), or the Akaike weights (w_i) provide a ranking of the models; thus the analyst can determine which fitted model is best, which are essentially tied for best, and which models are clearly in an inferior class (and perhaps some that are in an intermediate class). These ranks are, of course, estimates based on the data. Still, the rankings are quite useful (cf. Section 2.7 and Sakamoto et al. 1986:84) and suggest that primary inference be developed using the model for which AIC is minimized or the small number of models where there is an essential tie for the minimum

AIC (i.e., within about 1 or 2 AIC units from the minimum for nested models successively differing by one parameter). In the context of a string of nested models, when there is a single model that is clearly superior (say, the next best model is > 9–10 AIC units from the minimum) there is little model selection uncertainty and the theoretical standard errors can be used (e.g., Flather's data in Sections 1.2.3 and 2.14). When the results of model selection are less clear, then methods described in Chapter 4 can be considered. AIC allows a ranking of models and the identification of models that are nearly equally useful versus those that are clearly poor explanations for the data at hand (e.g., Table 2.2). Hypothesis testing provides no general way to rank models, even for models that are nested.

One must keep in mind that there is often considerable uncertainty in the selection of a particular model as the "best" approximating model. The observed data are conceptualized as random variables; their values would be different if another, independent set were available. It is this "sampling variability" that results in uncertain statistical inference from the particular data set being analyzed. While we would like to make inferences that would be robust to other (hypothetical) data sets, our ability to do so is still quite limited, even with procedures such as AIC, with its cross-validation properties, and with independent and identically distributed sample data. Various computer-intensive resampling methods may well further improve our assessment of the uncertainty of our inferences, but it remains important to understand that proper model selection is accompanied by a substantial amount of uncertainty. The bootstrap technique can allow insights into model uncertainty; this and other similar issues are the subject of some of the following chapters.

2.12.4   Nonnested Models

A substantial advantage in using information-theoretic criteria is that they are valid for nonnested models (e.g., Table 2.2). Of course, traditional likelihood ratio tests are defined only for nested models, and this represents another substantial limitation in the use of hypothesis testing in model selection. The ranking of models using AIC helps clarify the importance of modeling (Akaike 1973:173); for example, some models for a particular data set are simply poor and should not be used for inference.

A well-thought-out global model (where applicable) is very important, and substantial prior knowledge is required during the entire survey or experiment, including the clear statement of the question to be addressed and the collection of the data. This prior knowledge is then carefully input into the development of the set of candidate models (Section 1.2.4). Without this background science, the entire investigation should probably be considered only very preliminary.

2.12.5   Further Insights

Much of the research on model selection has been in regression and time series models, with some work being done in log-linear and classical multivariate (e.g., factor analysis) models. Bozdogan (1987) provides a review of the theory and some extensions. However, the number of published papers that critically examine the performance of AIC-selected models is quite limited. One serious problem with the statistical literature as regards the evaluation of AIC has been the use of Monte Carlo methods using only very simple generating models with a few large effects and no smaller, tapering effects. Furthermore, these Monte Carlo studies usually have a poor objective, namely, to evaluate how often a criterion selects the simple generating model. We believe that this misses the point entirely with respect to real data analysis. Such evaluations are often done even without regard for sample size (and often use AIC when AIC_c should have been used).

In Monte Carlo studies it would be useful to generate data from a much more realistic model with several big effects and a series of smaller, tapering effects (Speed and Yu 1993). Then interest is refocused onto the selection of a good approximating model and its statistical properties, rather than trying to select the simple, artificial model used to generate the data. AIC attempts to select a best approximating model for the data at hand; if (as with reality) the "true model" is at all complex, its use, with estimated parameters rather than true ones, would be poor for inference, even if it existed and its functional form (but not parameter values) were known (e.g., Sakamoto et al. 1986). This counterintuitive result occurs because the (limited) data would have to be used to estimate all the unknown parameters in the "true model," which would likely result in a substantial loss of precision (see Figure 1.3B).

AIC reformulates the problem explicitly as a problem of *approximation* of the true structure (probably infinite-dimensional, at least in the biological sciences) by a *model*. Model selection then becomes a simple function minimization, where AIC (or more properly K-L information loss) is the criterion to be minimized. AIC selection is objective and represents a very different paradigm to that of null hypothesis testing and is free from the arbitrary α levels, the multiple-testing problem, and the fact that some candidate models might not be nested. The problem of what model to use is inherently not a hypothesis testing problem (Akaike 1974). However, the fact that AIC allows a simple comparison of models does not justify the comparison of all possible models (Akaike 1985 and Section 1.3.3). If one had 10 variables, then there would be 1,024 possible models, even if interactions and squared or cubed terms are excluded. If sample size is $n \leq 1,000$, overfitting the data is almost a certainty. It is simply not sensible to consider such a large number of models, because a model that overfits the data will almost surely result, and the science of the problem has been lost. *Even in a very exploratory analysis it seems poor practice to consider all possible models; surely, some science can be brought to bear on such an unthinking approach* (otherwise, the scientist is superfluous and the work could be done by a technician).

2.13   Bootstrap Methods and Model Selection Frequencies π_i

The bootstrap is a type of Monte Carlo method used frequently is applied statistics. This computer-intensive approach is based on resampling of the observed data (Efron and Tibshirani 1993, Mooney and Duval 1993). The bootstrap was first described by Bradley Efron (1979); thousands of papers have been written on the bootstrap, with various extensions and applications in the past two decades, and it has found very wide use in applied problems. The bootstrap can be used for several purposes, particularly in the robust estimation of sampling variances or standard errors and (asymmetrical) confidence intervals. It has been used in the estimation of model selection frequencies (π_i) and in estimates of precision that include model selection uncertainty.

The bootstrap has enormous potential for the biologist with programming skills; however, its computer intensive nature will continue to hinder its use for large problems. We believe that at least 1,000 bootstrap samples are needed in many applications, and often 10,000 samples are needed for some aspects of model selection. In extreme cases, reliable results could take days of computer time to apply the bootstrap to complex data analysis cases involving large sample size and several dozen models, where the MLEs in each model must be found numerically.

The fundamental idea of the model-based sampling theory approach to statistical inference is that the data arise as a sample from some conceptual probability distribution f. Uncertainties of our inferences can be measured if we can estimate f. The bootstrap method allows the computation of measures of our inference uncertainty by having a simple empirical estimate of f and sampling from this estimated distribution. In practical application, the empirical bootstrap means using some form of resampling *with replacement* from the actual data x to generate B (e.g., $B = 1,000$ or 10,000) bootstrap samples; a bootstrap sample is denoted as x_b, where ($b = 1, 2, \ldots, B$). The sample data consist of n independent units, and it then suffices to take a simple random sample of size n, *with replacement*, from the n units of data, to get one bootstrap sample. However, the nature of the correct bootstrap data resampling can be more complex for more complex data structures.

The set of B bootstrap samples is a proxy for a set of B independent real samples from f (in reality we have only one actual sample of data). Properties expected from replicate real samples are inferred from the bootstrap samples by analyzing each bootstrap sample exactly as we first analyzed the real data sample. From the set of results of sample size B we measure our inference uncertainties from sample to (conceptual) population (Figure 2.6). For many applications it has been theoretically shown (e.g., Efron and Gong 1983, Efron and Tibshirani 1993) that the bootstrap can work well for large sample sizes (n), but it is not generally reliable for small n (say 5, 10, or perhaps even 20), regardless of how many bootstrap samples B are used. The bootstrap is not always successful in model selection (see Freedman et al. 1988).

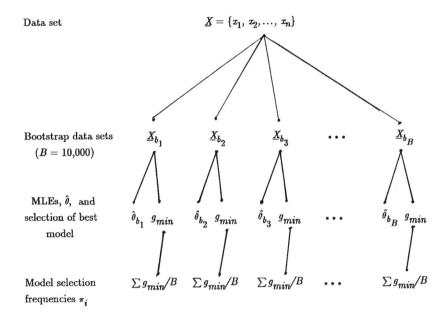

FIGURE 2.6. Diagram of the nonparametric bootstrap method as used in model selection (redrawn from Efron and Tibshirani 1993). The actual data set $\underline{X}$ is sampled with replacement, using the same sample size (n); this is done B times, to obtain B bootstrap data sets $\underline{X}_b$. Maximum likelihood theory provides estimates of the parameters ($\hat{\theta}$) for each of the models i ($i = 1, 2, \ldots, R$) and the AIC-best model (denoted by model $g_{\min}$) is found and its index stored for each of the bootstrap data sets. Finally, the model selection relative frequencies (π_i) are computed as the sums of the frequencies where model i was selected as best, divided by B. Of course, $\sum \pi_i = 1$.

2.13.1   Introduction

In many cases one can derive the sampling variance of an estimator from general likelihood theory. In other cases, an estimator may be difficult to derive or may not exist in closed form. For example, the finite rate of population change (λ) can be derived from a Leslie population projection matrix (a function of age-specific fecundity and age-specific, conditional survival probabilities). Generally, λ cannot be expressed in closed form. The bootstrap is handy for variance estimation in such nonstandard cases.

Consider a sample of weights of 27 young rats ($n = 27$); the data are (from Manly 1992),

57 60 52 49 56 46 51 63 49 57 59 54 56 59 57 52 52 61 59 53 59 51 51 56 58 46 53.

The sample mean of these data is 54.7, and the standard deviation is 4.51 with $cv = 0.0824$. For illustration, we will estimate of the standard error of the cv. Clearly, this would be nonstandard; however, it represents a way to illustrate the bootstrap.

First, we draw a random subsample of size 27 *with replacement* from the actual data. Thus, while a weight of 63 appears only once in the actual sample, perhaps it would not appear in the subsample; or it could appear more than once. Similarly, there are 3 occurrences of the weight 57 in the actual sample; perhaps the bootstrap sample would have, by chance, no values of 57. The point here is that a random sample of size 27 is taken *with replacement* from the original 27 data values. This is the first bootstrap resample ($b = 1$). From this bootstrap sample, one computes $\hat{\mu} = \bar{x}$, the $\widehat{se}(\hat{\mu}) = s/\sqrt{27}$, and the cv $= \widehat{se}(\hat{\mu})/\hat{\mu}$, and stores that value of cv in memory.

Second, the whole process is repeated B times (where we will let $B = 10,000$ samples for this example). Thus, we generate 10,000 resample data sets ($b = 1, 2, 3, \ldots, 10,000$) and from *each* of these we compute $\hat{\mu}$, $\widehat{se}(\hat{\mu})$, and the cv and store the value of the cv.

Third, we obtain the estimated standard error of the cv pertaining to the original sample by taking the standard deviation of the 10,000 cv values (corresponding to the 10,000 bootstrap samples). The process is simple; in this case, the standard error of the cv is 0.00922, or less than 1%.

Confidence intervals can be computed in the usual way, cv $\pm 2\,\widehat{se}(\text{cv})$. This gives a 95% interval of (0.0640, 0.1009) for the rat data. However, the sampling distribution may be nonnormal and a more robust interval might be required. Again, the bootstrap provides a simple approach. In this case, one sorts the $B = 10,000$ estimates of the cv in ascending order and selects the values that cut off the lower and upper 2.5 percentiles. Thus, the resulting interval might be asymmetric.

In the rat cv, the percentile bootstrap 95% confidence interval is (0.0626, 0.0984). This interval is about the same width as in the traditional approach, but shifted a bit toward 0. Incidentally, the mean of the 10,000 bootstrap samples was 0.0806 (compared to the actual sample cv of 0.0824). Even $B = 1,000$ is usually adequate for the estimation of the sampling variance or standard deviation; however, good estimates of percentile confidence intervals may require $B = 10,000$ in complicated applications.

Just as the analysis of a single data set can have many objectives, the bootstrap can be used to provide insight into a host of questions. For example, for each bootstrap sample one could compute and store the conditional variance–covariance matrix, goodness-of-fit values, the estimated variance inflation factor, the model selected, confidence interval width, and other quantities. Inference can be made concerning these quantities, based on summaries over the B bootstrap samples.

The illustration of the bootstrap on the rat data is called a nonparametric bootstrap, since no parametric distribution is assumed for the underlying process that generated the data. We assume only that the data in the original sample were "representative" and that sample size was not small. The parametric bootstrap is frequently used and allows assessment of bias and other issues. The use of the parametric bootstrap will be illustrated by the estimation of the variance inflation factor $\hat{c}$.

Consider an open population capture–recapture study in a setting where the investigators suspect a lack of independence because of the way that family groups were captured and tagged in the field. Data analysis reveals $\chi^2_{gof}/df = 3.2$. The investigators suspected some extrabinomial variation, but are surprised by the large estimate of the variance inflation factor $\hat{c}$. They suspect that the estimate is high and decide to use a parametric bootstrap to investigate their suspicion. They realize that the program RELEASE (Burnham et al. 1987) can be used to do Monte Carlo simulations and output a file with the goodness-of-fit statistics.

They input the MLEs from the real data into RELEASE as if they were parameters (ϕ_j and p_j) and use the numbers of new releases in the field data as input. Then the amount of extrabinomial variation (i.e., overdispersion, but called EBV in RELEASE) is specified. In this illustration, let EBV $\equiv 1$, meaning no overdispersion. They then run 1,000 Monte Carlo samples and obtain the information on the estimated variance inflation factor for each rep. The average of these 1,000 values gives $\hat{E}(\hat{c})$, and this can be compared to 1, the value used to generate the data. This result provides insight to the investigators on what to do about possible overdispersion in their data. More generally, the investigators could conduct several such studies for a range of EBV and see whether $E(\hat{c}|EBV) = EBV$ and assess any systematic bias in $\hat{c}$ as an estimator of EBV.

This bootstrap is parametric in that parameters were specified (in this case, from the MLEs from real data that were available) and used in a generating model to produce Monte Carlo data. The nonparametric bootstrap does not require parameters nor a model and relies on resampling the original data.

The bootstrap has been used in population biology to set confidence intervals on the median and mean life span. It is conceptually simple and has found very widespread use in applied statistics. Biologists planning a career in research or teaching should be familiar with the bootstrap. There is a very large literature on the bootstrap; see Efron and Tibshirani (1993) for an introduction to the subject and a large list of references. Some valid applications of the bootstrap are tricky (even multiple linear regression), so some care is required in more complex settings!

2.13.2   The Bootstrap in Model Selection: The Basic Idea

Consider the case where data (x) with sample size n are available and $R = 6$ models are under consideration, each representing some scientific hypothesis of interest. Let $B = 10,000$ bootstrap data sets, each of size n, and derived by resampling the data with replacement. MLEs of the parameters for each model could be computed for each bootstrap sample. Then AIC$_c$ could be computed for each of the 6 ($i = 1, 2, \ldots, 6$) models and the number of the best model (denote this by r^*, where r^* is the number of the best of the 6 models) and its associated AIC$_c$ value stored for each of the 10,000 bootstrap samples. After 10,000 such analyses, one has the bootstrap frequency of selection for each of the 6 models. These are called model selection relative frequencies π_i, the relative frequency that model i was found to be best. The relative frequency is given by $\pi_i = $ frequency/10,000 in this example. Of course, AIC or QAIC$_c$, or TIC could have been used to estimate the π_i.

Relative frequencies for model i being selected as the best model are similar to the Akalke weights, but are not identical. There is no reason, nor need, for the data-based weights of evidence (as the set of w_i) to be the same as the sampling relative frequencies at which the models are selected by an information criteria as being best. In general, likelihood provides a better measure of data-based weight of evidence about parameter values, given a model and data (see, e.g., Royall 1997), and we think that this concept (i.e., evidence for the best model is best represented by the likelihood of a model) rightly extends to evidence about a best model given an a priori set of models.

In our work we have not seen any particular advantage in the bootstrap selection frequencies over the Akaike weights. Considering the programming and computer times required for the computation of the model selection frequencies, we prefer the Akaike weights in general. We present some comparisons in Chapters 4 and 5.

We further elaborate on the interpretation of the Akaike weights as being conceptually different from the sampling-theory-based relative frequencies of model selection. It has has been noted in the literature (e.g., Akaike 1981a, 1994, Bozdogan 1987) that there is a Bayesian basis for interpreting the Akaike weight w_i as being the probability that model g_i is the expected K-L best model given the data (for convenience we usually drop this "expected" distinction and just think of the K-L best model). Once we have accepted the likelihood of model g_i given the data $\mathcal{L}(g_i|x)$, then we can compute the approximate posterior probability that model g_i is the K-L best model if we are willing to specify prior probabilities on the models (note that some Bayesians would consider this approach ad hoc since it is not the full Bayesian approach). That is, we first must specify an a priori probability distribution $\tau_1, \ldots, \tau_R$, which provides our belief that fitted model g_i will be the K-L best model for the data, given the model set. These probabilities τ_i must be specified independent of (basically, prior to) fitting any models to the data.

2.14   Return to Flather's Models

We now extend the example in Chapter 1 where 9 models for the species-accumulation curve for data from Indiana and Ohio were analyzed by Flather (1992, 1996). The simple computation of AIC was done by hand from the regression output from program NLIN in SAS (SAS Institute, Inc. 1985). In this case, apart from a constant that is the same over all models,

$$\text{AIC} = n \cdot \log(\hat{\sigma}^2) + 2K,$$

TABLE 2.1. Summary of nine a priori models of avian species-accumulation curves from the Breeding Bird Survey (from Flather 1992 and 1996). Models are shown, including the number of parameters (K), AIC values, $\Delta_i = \text{AIC}_i - \text{AIC}_{min}$ values, Akaike weights, and adjusted R^2 values for the Indian–Ohio Major Land Resource Area. AIC is computed for each model; the order is not relevant. Here the models are shown in order according to the number of parameters (K). However, this is only a convenience. This elaborates on the example in Table 1.1.

Model	Number of parameters[a]	AIC value	Δ_i	w_i	Adjusted R^2
ax^b	3	227.64	813.12	0.0000	0.962
$a + b\log(x)$	3	91.56	677.04	0.0000	0.986
$a\left(x/(b+x)\right)$	3	350.40	935.88	0.0000	0.903
$a(1 - e^{-bx})$	3	529.17	1114.65	0.0000	0.624
$a - bc^x$	4	223.53	809.01	0.0000	0.960
$(a + bx)/(1 + cx)$	4	57.53	643.01	0.0000	0.989
$a(1 - e^{-bx})^c$	4	−42.85	542.63	0.0000	0.995
$a\left(1 - [1 + (x/c)^d]^{-b}\right)$	5	−422.08	163.40	0.0000	0.999
$a[1 - e^{-(b(x-c))^d}]$	5	−585.48	0	1.0000	0.999

[a] K is the number of parameters in the regression model plus 1 for σ^2.

where $\hat{\sigma}^2 = \text{RSS}/n$ and K is the number of regression parameters plus 1 (for σ^2). AIC values for the 9 models are given in Table 2.2. The last model is clearly the best approximating model for these data. Values of $\Delta_i = \text{AIC}_i - \text{AIC}_{min} = \text{AIC}_i + 585.48$ are also given and allow the results to be more easily interpreted. Here, the second- and third-best models are quickly identified (corresponding to Δ_i values of 163.40 and 542.63, respectively); however, these Δ values are very large, and the inference here is that the final model is clearly the best of the candidate models considered for these specific data. This conclusion seems to be born out by Flather (1992), since he also selected this model based on a careful analysis of residuals for each of the 9 models and Mallows' C_p. The remaining question is whether a still better model might have been postulated with 6 or 7 parameters and increased structure. **Information criteria attempt only to select the best model from the candidate models available; if a better model exists, but is not offered as a candidate, then the information-theoretic approach cannot be expected to identify this new model.**

Adjusted R^2 values are shown in Table 2.2, and while these are useful as a measure of the proportion of the variation "explained," they are not useful in model selection (McQuarrie and Tsai 1998). In the case of Flather's data, the best 4 models all have an adjusted $R^2 \approx 0.99$, prompting one to conclude (erroneously) that all 4 models are an excellent fit to the data. Examination of the Δ_i values shows that models 6, 7 and 8 are incredibly poor, relative to model 9. The evidence ratio for the best model versus the second-best model is

$$w_9/w_8 = \frac{1}{\exp(-163.4/2)} \approx 3.0 \times 10^{35}.$$

There are additional reasons why adjusted R^2 is poor in model selection; its usefulness should be restricted to description.

2.15   Summary

Ideally, the investigator has a set of "multiple working hypotheses" and has thought hard about the background science of the issue at hand. Then, the science of the matter, experience, and expertise are used to define an a priori set of candidate models, representing each of these hypotheses. **These are important philosophical issues that must receive increased attention**. The research problem should be carefully stated, followed by careful planning concerning the sampling or experimental design. Sample size and other planning issues should be considered fully before the data-gathering program begins.

The basis for the information-theoretic approach to model selection and inference is **Kullback–Leibler information**,

$$I(f,g) = \int f(x) \log \left(\frac{f(x)}{g(x|\theta)} \right) dx.$$

$I(f,g)$ **is the "information" lost when the model g is used to approximate full reality or truth f**. An equivalent interpretation of $I(f, g)$ is a "distance" from the approximating model g to full truth or reality f. Under either interpretation, we seek to find a candidate model that minimizes $I(f, g)$, over the candidate models. This is a conceptually simple, yet powerful, approach. However, $I(f, g)$ cannot be used directly, because it requires knowledge of full truth or reality and the parameters in the approximating models g_i.

Akaike (1973), in a landmark paper, provided a way to *estimate* relative, expected $I(f, g)$, based on the empirical log-likelihood function. He found that the maximized log-likelihood value was a biased estimate of relative, expected Kullback–Leibler information and that under certain conditions this bias was approximately equal to K, the number of estimable parameters in the approximating model g. His method, *Akaike's information criterion* (AIC), allowed model selection to be firmly based on a fundamental theory and opened to door to further theoretical work. He considered AIC to be an extension of likelihood theory, the very backbone of statistical theory. Shortly thereafter, Takeuchi (1976) derived an asymptotically unbiased estimator of relative, expected Kullback–Leibler information that applies in general (i.e., without the special conditions underlying Akaike's derivation of AIC). His method (TIC for Takeuchi's information criterion) requires large sample sizes to estimate elements of two $K \times K$ matrices in the bias-adjustment term. TIC represents an important conceptual advance and further justifies AIC. Second order (i.e., small sample) approximations (AIC$_c$) were soon offered by Sugiura (1978) and Hurvich and Tsai (1989 and several subsequent papers). The three main approaches to adjusting for this bias (the bias-adjustment term is subtracted from the maximized log-likelihood) are summarized below:

Criterion	Bias adjustment term
AIC	K
AIC$_c$	$K + \frac{K(K+1)}{n-K-1}$
TIC	$\operatorname{tr}(J(\theta)I(\theta)^{-1}) \approx K.$

These information criteria are estimates of relative, expected K-L information and are an extension of Fisher's likelihood theory. AIC and AIC$_c$ are easy to compute, quite effective in many applications, and we recommend their use. When count data are found to be overdispersed, appropriate model selection criteria have been derived, based on quasi-likelihood theory (QAIC and QAIC$_c$). If overdispersion is found in the analysis of count data, the nominal log-likelihood function must be divided by an estimate of the overdispersion ($\hat{c}$) to obtain the correct log-likelihood. Thus, investigators working in applied data analysis have several powerful methods for selecting a "best" model for making inferences from empirical data to the population or process of interest. In practice, one need not assume that the "true model" is in the set of candidates (although this is sometimes mistakenly stated in the technical literature).

The AIC differences (Δ_i) and Akaike weights (w_i) are important in ranking and scaling the hypotheses, represented by models. The evidence ratios (e.g., w_i/w_j) help sharpen the evidence for or against the various alternative hypotheses. All of these values are easy to compute and simple to understand and interpret.

The principle of parsimony provides a philosophical basis for model selection, K-L information provides an objective target based on deep theory, and AIC, AIC$_c$, QAIC$_c$, and TIC provide estimators of relative, expected K-L information. Objective model selection is rigorously based on these principles. These methods are applicable across a very wide range of scientific hypotheses and statistical models. We recommend presentation of $\log(\mathcal{L}(\hat{\theta}))$, K, the appropriate information criterion (AIC, AIC$_c$, QAIC$_c$ or TIC), Δ_i, and w_i for various models in research papers to provide full information concerning the evidence for each of the models.

3
Basic Use of the Information-Theoretic Approach

3.1   Introduction

Model building and data analysis in the biological sciences somewhat presuppose that the investigator has some advanced education in the quantitative sciences, and statistics in particular. This requirement also implies that a researcher has substantial knowledge of statistical null hypothesis-testing approaches. Such investigators, including ourselves over the past several years, often find it difficult to understand the information-theoretic approach, only because it is conceptually so very different from the testing approach that is so familiar. Relatively speaking, the concepts and practical use of the information-theoretic approach are simpler than those of statistical hypothesis testing, and much simpler than some of the Bayesian approaches to data analysis (e.g., Laud and Ibrahim 1995 and Carlin and Chib 1995).

The prevailing philosophy has been to use some test or criterion or statistic to select a model, from a set of models, that is somehow "best" in some particular sense. Inference is then entirely conditional on this selected model. We believe that approach should be merely the beginning, and an inadequate or humble beginning at that. There is much more to the model selection problem than this initial solution. Substantive information is contained in the differences (Δ_i), since they are free from arbitrary (and unknown) constants and are directly interpretable in many cases. Both the Δ_i and Akaike weights (w_i) allow scientific hypotheses, carefully represented by models, to be ranked. The discrete likelihood of model i, given the data ($\mathcal{L}(g_i|x)$), provides a powerful way to assess the relative support for the alternative models. The w_i provide a strength of evidence for alternative models, given the set of models. Evidence ratios can be easily computed to promote understanding of the relative evidence for the second-, third-, and nth-best model, irrespective of other models in the set. These methods go well beyond just the selection of a best model and are very useful in assessing the empirical evidence for the alternatives in applied scientific problems.

It will be made clear in the next two chapters that even these extended analysis and inference philosophies are only a midway point in the information-theoretic paradigm. As we have struggled to understand the larger issues, it has become clear to us that inference based on only a single best model is often relatively poor for a wide variety of substantive reasons. Instead, we increasingly favor multimodel inference: procedures to allow formal statistical inference from all the models in the set. These procedures are simple to compute and interpret and are the subjects of Chapters 4 and 5. Such multimodel inference includes model averaging, incorporating model selection uncertainty into estimates of precision, confidence sets on models, and simple ways to assess the relative importance of variables.

The examples below focus on the selection of a single best model; extensions will appear in the following chapters. However, many methods illustrated go beyond this initial approach in terms of the evidence for each model in the set. Methods to assess model selection uncertainty (e.g., the differences Δ_i and Akaike weights w_i) are illustrated and discussed. Evidence ratios and relative likelihood of model i, given the data, provide additional evidence concerning inferences about the actual K-L best model. Still, these examples should be viewed as a halfway point in understanding the full information-theoretic approach where formal inferences are drawn from multiple models.

While the derivation of AIC (Chapter 7) lies deep in the theory of mathematical statistics, its application is quite simple. Our initial example is a simple multiple linear regression model of cement hardening and is a classic example in the model selection literature. The remaining examples in this chapter focus on more complex data sets and models. These examples will provide insights into real-world complexities and illustrate the ease and general applicability of AIC in model selection and inference. Several of these examples are continued in later chapters as additional concepts and methods are provided. Several examples deal with survival models, since that has been one of our research interests.

Given a model, likelihood inference provides a quantitative assessment of the strength of evidence in the data regarding the plausible values of the parameters in the model (Royall 1997). Given a well-developed set of a priori candidate models, information-theoretic methods provide a quantitative assessment of the strength of evidence in the data regarding the plausibility of which model is "best." Information criteria can be computed and interpreted without the aid of subjective judgment (e.g., α-levels or Bayesian priors) once a set of candidate models has been derived.

> **Importance of Modeling**
>
> Akaike (1981b) believed that the most important contribution of his general approach was the clarification of the importance of modeling and the need for substantial, prior information on the system being studied.

At some early point in the analysis of count data, the goodness-of-fit of the global model should be assessed using standard methods. Similar scrutiny should accompany continuous data (see Carrol and Ruppert 1988). There is generally no concept of overdispersion in continuous data; the modeling of residual variation should receive careful attention. One should examine outliers, highly leveraged points, symmetry, trends, and autocorrelations in the residuals (McCullagh and Nelder 1989). There are many standard diagnostic procedures that should be used to aid in the modeling of the residual variation. If, after proper attention to the a priori considerations, the global model still fits poorly, then information-theoretic methods will select only the best of the set of poor-fitting models. This undesirable situation probably reflects on the poor science that went into the modeling and definition of the set of candidate models. Lack of fit of the global model should be a flag warning that still more consideration must be given to the modeling, based on an understanding of the questions being asked and the design of the data collection. Perhaps the effort must be classed as exploratory and very tentative; this would allow some data dredging, leading perhaps to some tentative models and suggestive conclusions. Treated as the results of a pilot study, then new data could be collected and the analysis could proceed in a more confirmatory fashion using the techniques we outline in this book.

Computer programs for likelihood methods nearly always provide the value of the log-likelihood at its maximum, and the appropriate information criterion can be easily computed by hand, if necessary. Similarly, one can compute the MLE of σ^2 from standard output of LS programs and can compute the information criteria from this estimate in most cases. While many software packages currently print AIC, relatively few print the value of AIC_c or $QAIC_c$, and this is unfortunate (see Example 1 below, where AIC performs poorly, because the ratio n/K is small).

3.2   Example 1: Cement Hardening Data

The first example is a small set of data on variables thought to be related to the heat evolved during the hardening of Portland cement (Woods et al. 1932:635–649). These data represent a simple use of multiple linear regression analysis (see Section 1.2.2). This data set (the "Hald data") has been used by various authors (e.g., Hald 1952:635–649, Seber 1977, Daniel and Wood 1971, Draper and Smith 1981:294–342 and 629–673, Stone and Brooks 1990, George and McCulloch 1993, Hjorth 1994:31–33, Ronchetti and Staudte 1994, Laud and Ibrahim 1996, and Sommer and Huggins 1996) and will illustrate a variety of

TABLE 3.1. Cement hardening data from Woods et al. (1932). Four predictor variables (as a percentage by weight) [x_1 = calcium aluminate (3CaO · Al_2O_3), x_2 = tricalcium silicate (3CaO · SiO_2), x_3 = tetracalcium alumino ferrite (4CaO · Al_2O_3 · Fe_2O_3), x_4 = dicalcium silicate (2CaO · SiO_2)] are used to predict the dependent variable y = calories of heat evolved per gram of cement after 180 days of hardening.

x_1	x_2	x_3	x_4	y
7	26	6	60	78.5
1	29	15	52	74.3
11	56	8	20	104.3
11	31	8	47	87.6
7	52	6	33	95.9
11	55	9	22	109.2
3	71	17	6	102.7
1	31	22	44	72.5
2	54	18	22	93.1
21	47	4	26	115.9
1	40	23	34	83.8
11	66	9	12	113.3
10	68	8	12	109.4

important points. The data include 4 predictor variables and have a sample size of 13 (Table 3.1). The predictor variables (as a percentage of the weight) are x_1 = calcium aluminate (3CaO·Al_2O_3), x_2 = tricalcium silicate (3CaO·SiO_2), x_3 = tetracalcium alumino ferrite (4CaO · Al_2O_3 · Fe_2O_3), and x_4 = dicalcium silicate (2CaO · SiO_2), while the response variable is y = total calories given off during hardening per gram of cement after 180 days. Daniel and Wood (1971) provide further details on these data for the interested reader. "*What approximating model to use?*" is the primary focus of this example.

The small size of the sample necessitates the use of AIC_c (Section 2.4); however, we will present comparable values for AIC in this example. We will use an obvious notation for denoting what variables are in each candidate model. That is, if variables x_1 and x_3 are in a particular model, we denote this as model {13}; each model has an intercept (β_0).

3.2.1 Set of Candidate Models

Because only 4 variables are available, the temptation is to consider all possible models ($2^4 - 1 = 15$) involving at least one of the predictor variables. In view of the small sample size, we will consider this example as largely exploratory, and lacking any personal knowledge concerning the physics or chemistry of cement hardening, we will consider the full set of models, including the global model {1234} with $K = 6$ parameters. While we generally advise strongly against consideration of all possible models of the x_i (but no interactions or powers of the predictor variables), this approach will allow some comparisons with

TABLE 3.2. Summary of 15 models for the cement-hardening data, including the total number of estimable parameters (K), the ML estimated mean squared error $(\hat{\sigma}^2)$, and Δ_i values for both AIC and AIC_c followed by the Akaike weights (w_i), based on AIC_c. Models are ordered in terms of Δ_i for AIC_c.

Model	K	$\hat{\sigma}^2$	$\log(\mathcal{L})$	Δ_i AIC	Δ_i AIC_c	w_i
$\{12\}$[1]	4	4.45	-9.704	0.4346	0.0000	0.567
$\{124\}$	5	3.69	$-8,478$	0.0000	3.1368	0.118
$\{123\}$	5	3.70	-8.504	0.0352	3.1720	0.116
$\{14\}$	4	5.75	-11.370	3.7665	3.3318	0.107
$\{134\}$	5	3.91	-8.863	0.7528	3.8897	0.081
$\{234\}$	5	5.68	-11.290	5.6072	8.7440	0.007
$\{1234\}$	6	3.68	-8.469	1.9647	10.5301	0.003
$\{34\}$	4	13.52	-16.927	14.8811	14.4465	0.000
$\{23\}$	4	31.96	-22.519	26.0652	25.6306	0.000
$\{4\}$	3	67.99	-27.426	33.8785	31.1106	0.000
$\{2\}$	3	69.72	-27.586	34.2052	31.4372	0.000
$\{24\}$	4	66.84	-27.315	35.6568	35.2222	0.000
$\{1\}$	3	97.37	-29.760	38.5471	35.7791	0.000
$\{13\}$	4	94.39	-29.558	40.1435	39.7089	0.000
$\{3\}$	3	149.18	-32.533	44.0939	41.3259	0.000

[1] Here, $\log(\mathcal{L}) = -n/2 \cdot \log(\hat{\sigma}^2) = -9.7039$, $AIC_{min} = -2\log(\mathcal{L}) + 2K = 27.4078$, and $AIC_{c,min} = AIC + \frac{2K(K+1)}{n-K-1} = 32.4078$.

others in the published literature (e.g., Draper and Smith 1981, Hjorth 1994, and Hoeting and Ibrahim 1996). We note, however, that the 4 models with only a single variable might have been excluded on a priori grounds because cement involves a mixture of at least two compounds that react chemically. We will extend this example in Chapter 4 to examine the issue of model selection uncertainty and other issues.

3.2.2 Some Results and Comparisons

The use of AIC_c suggests model $\{12\}$ as the best approximating model for these data (Table 3.2). The estimated regression coefficients in the selected model are

$$\hat{E}(y) = 52.6 + 1.468(x_1) + 0.662(x_2),$$

where the estimated standard errors of the 3 estimated parameters (*given this model*) are 2.286, 0.121, and 0.046, respectively (this result is in agreement with Hald 1952). The adjusted $R^2 = 0.974$ and the MLE $\hat{\sigma} = 2.11$ for the AIC_c-selected model. The second-best model is $\{124\}$, but it is 3.14 AIC_c units from the best model (Table 3.2). Other candidate models are ranked, and clearly many of the models represent poor approximations to these (scant) data (at least the models in Table 3.2 with Δ_i values > 10). Note the differences in

Δ_i and associated rankings between AIC vs. AIC_c in Table 3.2; clearly, AIC_c is to be preferred over AIC, because the ratio $n/K (= 13/6)$ is only 2.2 for the global model (model {1234}).

The Akaike weight for the best model is not large, relative to the weight for the other models. The ratio of the weights for the best model versus the 4 next-best models ranges from only 4.8 to 7; this is not strong evidence that model {12} is likely best if other replicate samples were available.

Using a type of cross-validation criterion (Q_{cv}), Hjorth (1994:33) selected model {124} with $K = 5$ for these data. Here, his result is

$$\hat{E}(y) = 71.6 + 1.452(x_1) + 0.416(x_2) - 0.236(x_4),$$

where the estimated standard errors are 14.142, 0.117, 0.186, and 0.173, respectively. Model {124} has an adjusted $R^2 = 0.976$ and $\hat{\sigma} = 1.921$. Draper and Smith (1981:325–327) used cross-validation and the PRESS (Allen 1970) selection criterion, which is quite similar to Q_{cv}, and also selected model {124}. Note, had AIC been used, ignoring the ratio $n/K \approx 2$, model {124} would have been selected (Table 3.2); AIC_c should be used if this ratio is small (i.e., < 40).

Is there any basis to say that AIC_c selected a better approximating model than Hjorth's cross-validation procedure or AIC or the PRESS criterion? This is difficult to answer conclusively because truth is not known here. However, the regression coefficient on x_4 is not "significant" under the traditional hypothesis testing scenario ($t = 1.36$, 9 df), and the estimated standard error on the regression coefficient for x_2 increased by a factor of 4 from 0.046 to 0.186 compared to model {12}. The adjusted R^2 statistics for Hjorth's selected model is 0.976 (vs. 0.974), but it has one additional parameter. The correlation coefficient between x_1 and x_3 was -0.824, while the correlation between x_2 and x_4 was -0.973. Just on the basis of this latter correlation it seems unwise to allow both x_2 and x_4 in the same model (if n were 3,000 instead of only 13, perhaps there would be more support for including both x_2 and x_4). While not completely compelling, it would seem that AIC_c has selected the better parsimonious model in this case. An additional, negative, consideration is the computer-intensive nature of Hjorth's cross-validation algorithm (Q_{cv}) compared to the information-theoretic approach. With more reasonable sample sizes or more variables, or with more models to consider, the cross-validation approaches may often become computationally too "costly."

Draper and Smith (1981) used Mallows's C_p statistic and also selected model {12}, in agreement with AIC_c (this might be fortuitous, because no small sample version of C_p or Q_{cv} is available). They further point out that $\sum_{j=1}^{4} x_{ij} = a$ constant (approximately 98%) for any i; thus the $X'X$ matrix for model {1234} is theoretically singular. Small rounding errors were eventually introduced, since the percentage data were expressed as integers, leaving the $X'X$ matrix barely nonsingular. At best, model {1234} would be a poor model for the analysis of these data. They also warn against the unthinking use of all possible regressions and present a detailed analysis of forward, backward, and stepwise approaches, based on tests of hypotheses and arbitrary α levels. Draper and Smith (1981) also used the stepwise procedure (with $\alpha = 0.15$), which resulted in model {12}, after starting at step 1 with x_4, eventually dropping it, and retaining only x_1 and x_2. This represents an improvement over routines that merely add new variables, without looking to see whether a particular variable has become redundant. Draper and Smith (1981) provide a good discussion of the various older model selection alternatives and offer some useful recommendations (but do not discuss any of the information-theoretic approaches). They provide an intensive analysis of the cement data over several chapters and include detailed computer output in two large appendices.

Another analysis approach involves computation of the principal components on the (centered) $X'X$ matrix and examination of the correlation matrix for the 4 explanatory variables (see Draper and Smith 1981:327–332, Stone and Brooks 1990). The principal component eigenvalues here are 2.23570, 1.57607, 0.18661, and 0.00162. Approximately 95.3% of the total variance is contained in the first 2 eigenvectors, while 99.96% is in the first 3 eigenvectors. These results certainly suggest that the global model overfits these data (i.e., 4 predictor variables are redundant). In addition, it might suggest that 2 predictor variables will suffice (given $n = 13$). Critical interpretation of the percentage eigenvalues requires some judgment and subjectivity. Furthermore, relatively few biologists are knowledgeable about the concept of eigenvalues and eigenvectors. *We believe that the investigators should understand the methods leading to the results of their work; this is sometimes difficult with some advanced methods.* Such understanding seems relatively easy with the information-theoretic approaches.

One could ask whether there is a need for model selection when there are only 4 predictor variables (i.e., why not merely take the global model with 6 parameters and use it for inference?). This simple strategy is often very poor, as we illustrate here. First, note that this global model has $\Delta_i = 10.5301$, relative to model {12}, and is therefore a poor approximation to the meager data available. The estimates of parameters for the global model {1234} are

$$\hat{E}(\hat{y}) = 62.4 + 1.551(x_1) + 0.510(x_2) + 0.102(x_3) - 0.144(x_4),$$

where the estimated standard errors, given this model, are 70.071, 0.745, 0.728, 0.755, and 0.709, respectively. These standard errors are large because the $X'X$ is nearly singular (the percentage coefficients of variation for $\hat{\beta}_0$, $\hat{\beta}_1$, and $\hat{\beta}_2$ were 4.3, 8.2, and 6.9 under model {12}, compared to 112.3, 48.0, and 142.7, respectively, under model {1234} (see Wood and Thomas 1999). Only the regression coefficient for x_1 might be judged as "significant" in a hypothesis testing sense, and the model is clearly overfit (see Figure 1.4b). Model {1234} has an adjusted $R^2 = 0.974$ and $\hat{\sigma} = 1.918$. Surely a parsimonious model, such as {12}, would better serve the analyst in this case.

Loss of precision is expected in using an overfit global model; however, there is also a nonnegligible probability that even the sign of the estimated parameter may be incorrect in such cases. It seems somewhat compelling to withhold judgment if the information (data) is inadequate for reliable inference on a parameter or effect, because the estimate might be very misleading.

If all the predictor variables are mutually orthogonal (uncorrelated), model selection is not quite as critical, and the global model with $K = 6$ might not be so bad. Orthognality arises in controlled experiments where the factors and levels are *designed* to be orthogonal. In observational studies there is a high probability that some of the predictor variables will be mutually quite dependent. Rigorous experimental methods were just being developed during the time these data were taken (about 1930). Had such design methods been widely available and the importance of replication understood, then it would have been possible to break the unwanted correlations among the x variables and establish cause and effect.

With only a single data set, one could use AIC_c and select the best model for inference. However, if several other independent data sets were available, would the same model be selected? The answer is *perhaps* it would be; but generally there would be variation in the selected model from the data set, just as there would be variation in parameter estimates over data sets, given that the same model is used for analysis. The fact that other data sets might suggest the use of other models leads us to the issue of model selection uncertainty.

Based on simulation studies, we are usually surprised by how much variation there is in selecting a parsimonious model for a given problem. It is demonstrably the case that in many real-world problems there is substantial model selection uncertainty. We generated 10,000 bootstrap samples from these data to estimate model selection uncertainty. The parameters, in each of the 15 models shown in Table 3.2, were estimated and AIC_c was computed for each bootstrap sample. The following summary shows the relative model selection frequencies (π_i) from applying AIC_c (models not shown had zero selections) to each of the 10,000 bootstrap samples. Here, $\hat{\pi}$ are the estimated model selection probabilities. Also shown are the Akaike weights (w_i) from the original data:

Model	K	Bootstrap Sel. Freq. $\hat{\pi}_i$	Akaike weights w_i
{12}	4	0.5338	0.567
{124}	5	0.0124	0.118
{123}	5	0.1120	0.116
{14}	4	0.2140	0.107
{134}	5	0.0136	0.081
{234}	5	0.0766	0.007
{1234}	6	0.0337	0.003
{34}	4	0.0039	0.000.

As might be expected with such a small sample size, the selection frequencies varied substantially, and model {12} was selected as the best in only about 53% of the bootstrapped samples. Model {14} was selected 21% of the time; recall that the simple correlation between variables x_2 and x_4 was $r = -0.973$. Thus it is a quite reasonable result that models {12} and {14} are somewhat aliased.

Other models had much lower selection frequencies in this example. There is reasonable agreement between the $\hat{\pi}_i$ and the w_i, considering the sample size of 13 observations. Further results based on the bootstrap are given in the following chapter.

3.2.3 A Summary

In summary, the simple approach of using AIC_c appears to have given a good parsimonious model as the basis for inference from these data. The use of AIC_c sharpens the inference about which parsimonious model to use, relative to AIC. A priori information could have resulted in fewer candidate models and generally strengthened the process (note, that Hald (1952) first presented only an analysis of x_1 and x_2 and presented the analysis of the 2 additional variables several pages later). It seems likely that models with only a single variable might have been excluded from serious consideration based on what must have been known about cement in the late 1920s. Similarly, we suspect that Woods et al. (1932) had knowledge of the negative relationship between x_2 and x_4; after all, model {14} was their second-best model. AIC_c avoided use of both x_2 and x_4 in the same model (where the correlation was -0.973) and the over parametrized global model. An important feature of the information-theoretic approach is that it provides a ranking of alternative models, allowing some inferences to be made about other models that might also be useful. In addition, the rankings suggest that some models that remain very poor (e.g., models {24}, {1}, {13}, and {3} for the cement data). The Akaike weights serve to focus the evidence for or against the various models. **The importance of carefully defining a small set of candidate models, based on the objective and what is known about the problem, cannot be overemphasized**.

An investigator with, say, 10 explanatory variables cannot expect to learn much from the data and a multiple linear regression analysis unless there is some substantial supporting science that can be used to help narrow the number of models to consider. In this case, there would be $2^{10} = 1,024$ models (many more if transformations or interaction terms were allowed), and overfitting would surely be a risk. The analysis, by whatever method, should probably be considered exploratory and the results used to design further data gathering leading to a more confirmatory analysis, based on some a priori considerations.

3.3   Example 2: Time Distribution of an Insecticide Added to a Simulated Ecosystem

This example concerns the addition of the insecticide DURSBAN® to a laboratory system that simulates a pond of water. The original work was done by Smith (1966) and his colleagues; our main reference for this example was

Blau and Neely (1975), but also see Carpenter (1990) for a simplified Bayesian analysis of these data.

Blau and Neely note (1975) that the determination of the ultimate fate and distribution of this chemical introduced into an ecosystem is an important environmental issue. They go on to mention that "... a true mathematical model describing each step of the process would be extremely complex. It is important, however, to try to find a suitable model to identify the most important chemical, physical, and biological phenomena taking place and to predict the long-term environmental consequences." This view of modeling is consistent with Akaike's and the one recommended here. This example is used because it rests on systems of first-order differential equations whose parameters, given a model, are estimated by least squares. Such results can easily be used to compute AIC values to aid in selection of a parsimonious approximating model.

The active ingredient of DURSBAN® is 0,0-Diethyl 0-(3,5,6-trichloro-2-pyridyl) phosphorothioate, which was labeled with radioactive carbon 14 in the pyridyl ring and added at a level of 1 mg/6 gal in a 10-gallon glass jar (see Figure 3.1). This aquarium contained 2 inches of soil (13.3% organic matter), plants (salvinia, anacharis, milfoil, and water cucumber), and 45 goldfish. Samples of the various components were analyzed for radioactivity at 12 different time periods following the addition of DURSBAN®. Three samples at each time period yielded a sample size (n) of 36. The data (Table 3.3) are in percentages from the crude radioactivity measurements (Blau and Neely

FIGURE 3.1. Glass aquarium used in the studies of DURSBAN® (from Smith 1966).

TABLE 3.3. Distribution of radioactive carbon in DURSBAN® in a simulated ecosystem (from Blau and Neely 1975).

Time after DURSBAN® addition (hours)	Percent radioactivity		
	Fish	Soil & Plants	Water
0	0	0	100
1.5	15.2	35.2	49.7
3.0	19.0	46.0	28.3
4.0	19.3	56.0	24.5
6.0	20.7	61.0	18.3
8.0	23.0	60.5	17.0
10.0	24.2	59.3	18.2
24.0	21.2	51.5	26.5
48.0	23.0	38.3	34.5
72.0	22.7	38.3	39.5
96.0	20.5	36.3	43.0
120.0	17.3	38.3	44.5

1975:150). The authors of the study assumed that the model residuals were normally distributed, with zero means and a constant standard deviation of 1% (we take this to mean the actual measurement error of the instrument used).

3.3.1   Set of Candidate Models

Blau and Neely (1975) had a great deal of knowledge about this system, and they exploited this in a priori model building. They began by postulating that an equilibrium exists between DURSBAN® in the water (A), soil and plant components (B), and a direct uptake of the chemical by the fish (C). This led to their Model 1 (Figure 3.2), which was represented by a system of differential equations, where the rate parameters to be estimated are denoted by k_i,

$$dx_A(t)/dt = -k_1 x_A(t) + k_2 x_B(t) - k_3 x_C(t),$$
$$dx_B(t)/dt = k_1 x_A(t) - k_2 x_B(t),$$
$$dx_C(t)/dt = k_3 x_A(t),$$

with initial conditions $x_A(0) = 100$, $x_B(0) = 0$, and $x_C(0) = 0$. This is a type of compartment model (Brown and Rothery 1993) and is often used in some fields. Blau and Neely (1975) used $x_A(t)$, $x_B(t)$, and $x_C(t)$ as the percentages at time (t) of A, B, and C, respectively, with the restriction that

$$x_A(t) + x_B(t) + x_C(t) = 100.$$

They used nonlinear least squares to estimate model parameters (the k_i and σ^2), and their analytic methods were quite sophisticated. The parameter estimates for this model were $\hat{k}_1 = 0.510$, $\hat{k}_2 = 0.800$, $\hat{k}_3 = 0.00930$, and $\hat{\sigma}^2 = 149.278$

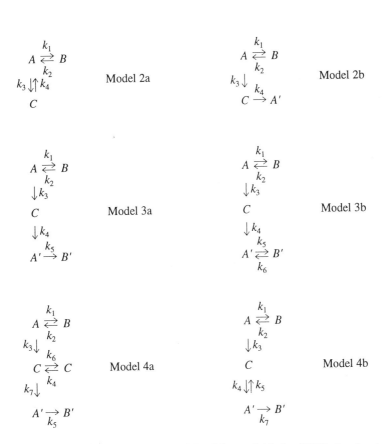

FIGURE 3.2. Summary of models used by Blau and Neely (1975) for the data on DURSBAN® in a simulated pond ecosystem.

(this is their residual sum of squares (RSS) divided by n to obtain the MLE of σ^2); thus, $K = 4$ for this model.

Blau and Neely (1975) built six other models, each based on their knowledge of the system, but also based on examination of the residuals from prior models (there are some inconsistencies here that we were unable to resolve; thus we will use the material from their paper). While some data dredging was evident, their main derivation of additional models seemed to stem primarily from hypotheses about the processes. They were well aware of the principle of parsimony and included a very nice discussion of LS and ML methods and their relationships in an introductory part of their paper. They computed goodness-of-fit tests and separated "pure error" from the remaining residual terms. Model selection was accomplished by statistical hypothesis tests (likelihood ratio tests) and examining the RSS. They found Model 4a (see Figure 3.2) to be the best and also found some support for Model 4b.

3.3.2   Some Results

Analysis of these data under an information-theoretic paradigm is simple, given Blau and Neely's (1975) Table II, since they provide values for $K - 1$ and RSS for each of their seven models. Due to the relationships between LS estimation and ML theory (see Section 1.2.2),

$$\log(\mathcal{L}(\hat{\underline{k}}, \hat{\sigma}^2 \,|\, data)) = -n/2 \cdot \log(\hat{\sigma}^2),$$

where $\hat{\sigma}^2 = $ RSS $/n$. Then,

$$\text{AIC} = -2 \cdot \log(\mathcal{L}(\hat{\underline{k}}, \hat{\sigma}^2 \,|\, data)) + 2K$$

and

$$\text{AIC}_c = \text{AIC} + \frac{2K(K+1)}{n - K - 1}.$$

These computations were done by hand on a simple calculator and took approximately 20 minutes. The results of this extended analysis are shown in Table 3.4 and suggest that Model 4a is the best to use for inference, in agreement with Blau and Neely (1975). Only Model 4b is a competitor, but it has a Δ_i value of 7.611 and seems relatively implausible for these data ($w_{4b} = 0.022$). The evidence ratio for model 4a vs. 4b is $0.978/0.022 = 44$; thus, there is strong support for 2-way transfer between the viscera and flesh in the fish (i.e., the essential difference between models 4 and 4b is $c \leftrightarrows c'$).

Carpenter (1990) used these data and seven models under a simplified Bayesian analysis with equal Bayesian prior probabilities on the models but

TABLE 3.4. Summary of model selection statistics (the first three columns taken from Blau and Neely 1975). Statistics for the AIC_c-selected model are shown in bold.

Model	RSS	$\log(\mathcal{L}(\hat{k}, \hat{\sigma}^2 \,|\, data))$	K	AIC	AIC_c	$\Delta_i \, \text{AIC}_c$	w_i
1	5374	−90.105	4	188.209	189.499	150.626	0.000
2a	1964	−71.986	5	153.972	155.972	117.099	0.000
2b	848	−56.869	5	123.737	125.737	86.864	0.000
3a	208.3	−31.598	6	75.196	78.094	39.221	0.000
3b	207.9	−31.563	7	77.127	81.127	42.254	0.000
4a	**58.6**	**−8.770**	**8**	**33.540**	**38.873**	**0.0**	**0.978**
4b	79.4	−14.238	7	42.475	46.475	7.602	0.022

with no prior probabilities specified on the model parameters in that same
semi-Bayesian context. He also concluded that Model 4a was the best, with
Model 4b a poor second. In this example, K ranged from only 4 to 8; thus
the various methods might be expected to be in somewhat close agreement.
This example illustrates that it is often easy to perform a reanalysis of data
on complex systems, based on information provided in published papers. The
analysis clearly shows that five of the seven models have essentially no support, and inferences from these models would likely be poor. For instance, the
third-best model (3a) has an evidence ratio of 3×10^9, while the worst model
(1) has an evidence ratio of 2×10^{33}. Clearly, these models are unsupported,
given the data available.

Blau and Neely's (1975) results are interesting, and well supported by the
best model. The evidence ratio for the second-best model is 44.7 and it seems
reasonable to base inference on just the best model in this case. Researchers
are often comfortable with the concept that inferences can be based on a proper
model; in a sense, the inference here is the model.

After a final model is chosen it is often wise to examine the residuals using
standard methods. Such examination may reveal issues that warrant further
study; in this sense, science never "stops."

Formal *statistical* inferences include the following: (1) there is a rapid equi-
libration between DURSBAN and the soil and plant system; (2) this is followed
by a shorter uptake of DURSBAN by the fish; (3) fish tend to metabolize and
excrete DURSBAN; (4) the liberated material (metabolized DURSBAN) is
again taken up by the soil and plants; (5) fish have two compartments, the viscera and the flesh; (6) the final sink for DURSBAN is the soil and the plants;
and (7) plants readily dissipate the metabolite as degraded CO_2, NH_3, and
H_2O. Estimates of the various transfer rates are given by the $\hat{k}_i$ and estimates
of precision are available as standard errors or confidence intervals.

3.4   Example 3: Nestling Starlings

We generated a set of Monte Carlo data to illustrate many of the points discussed with a much more complicated example of an experimental setting.
Thus, in a sense, the generating model is "truth"; we will accept this bit of un-
realism for the moment, but mitigate it by including many parameters ($K = 34$)
and a wide variety of tapering treatment effects. In addition, we will choose a
global model that has four fewer parameters than the generating model; thus
the generating model is not in the set of candidate models. Furthermore, this
example contains many so-called nuisance parameters (sampling probabilities). This is the only example in Chapter 3 where "truth" is known, and some
interesting insights can be gained from this knowledge. The essential question
is what parsimonious approximating model can be used for data analysis that
will lead to valid inference about the structure of the system, its parameters,

and the effects of the treatment. A second question relates to the strength of the evidence for the best model in relation to other models.

3.4.1 Experimental Scenario

We generated data to mimic the experiment conducted by Stromborg et al. (1988) (also see Burnham et al. 1987:343–348). The research question relates to the survival effects of an organophosphate pesticide administered to nestling European starlings (*Sturnus vulgaris*). We assume for illustration that a simple field experiment is designed using artificial nest boxes placed on a 5,000 hectare island. Fledgling birds are assumed not to leave the island during the summer and early fall months when the experiment is conducted (geographic closure). Nest boxes are monitored during the nesting season to determine the date of hatching. All nestlings are leg-banded with uniquely numbered bands 16 days following hatching, and half of those nestlings are randomly assigned to a treatment group and the remaining birds assigned to a control group. In total, we will assume that 600 nestling starlings are banded and returned to the nest box (i.e., the number of starlings originally released in each group is 300). All nest boxes contain 4 young birds (thus 2 treatment and 2 control), and we assume these to be of nearly uniform size and age and that once fledged, they move about and behave independently. Starlings randomly selected to be in the treatment group receive an oral dose of pesticide mixed in corn oil. Birds in the control groups are given pure corn oil under otherwise very similar conditions. Colored leg bands provide a unique identification for each starling and therefore its group membership, on each weekly resighting occasion. Data collection will be assumed to begin after a 4-day period following dosage, and for simplicity, we assume that no birds die due to handling effects following marking but before resighting efforts begin a week later. Surviving starlings are potentially resighted during the following 9 weeks; sampling covers the entire island and is done on each Friday for 9 weeks. Thus, the data are collected on 10 occasions; occasion 1 is the initial marking and release period, followed by 9 resighting occasions.

The pesticide is hypothesized to affect conditional survival probability (the parameters of interest) and resighting probabilities (the nuisance parameters); however, the pesticide industry's position is that only minor survival effects are likely, while environmental groups suspect that there are substantial acute (short-term) and chronic (long-term) effects on survival probabilities and worry that the resighting probabilities might also be affected by the treatment. Thus, the set of candidate models might span the range of the controversy. In practice, of course, one might design the experiment to include several "lots" of starlings, released at different, independent locations (islands), and these data would be the basis for empirical estimates of treatment effect and precision (see Burnham et al. 1987 for a discussion of experiments of this general type). Here we will focus on an example of the model selection issue and not on optimal design.

3.4.2   Monte Carlo Data

Monte Carlo data were generated using the following relationships for conditional survival probability (ϕ) and resighting probability (p) for treatment (t) and control (c) groups at week i:

$$\phi_{ti} = \phi_{ci} - (0.1)(0.9)^{i-1} \text{ for } i = 1, \ldots, 9,$$
$$p_{ti} = p_{ci} - (0.1)(0.8)^{i-2} \text{ for } i = 2, \ldots, 10,$$

using program RELEASE (Burnham et al. 1987). These relationships allow a smooth temporal tapering of effect size due to the treatment in both conditional survival and resighting probabilities. That is, each week the effect of the pesticide is diminished. We used the initial per-week survival and resighting probabilities for the control group as 0.9 and 0.8, respectively. Conditional survival and resighting probabilities for the control group did not differ by week (i.e., $\phi_{ci} \equiv \phi_c \equiv 0.9$ and $p_{ci} \equiv p_c \equiv 0.8$). The data are given in Table 3.5 for each treatment and control group.

3.4.3   Set of Candidate Models

Define ϕ_{vi} as the conditional probability of survival for treatment group v ($v = t$ for treatment and c for control) from week i to $i+1$ ($i = 1$ to 9) and p_{vi} as the conditional probability of resighting for treatment group v at week i (for $i = 2$ to 10). The set of models that seem reasonable might include one with no treatment effects (g_0), a model for an acute effect only on the first survival probability ($g_{1\phi}$), and a model for an acute effect on both the first survival probability and the first resighting probability (denote this by p_2, because it occurs at week 2) (model g_{2p}). This initial line of a priori consideration leads to three models:

Model	Parametrization
g_0	All $\phi_{ti} = \phi_{ci}$ and all $p_{ti} = p_{ci}$ (no treatment effect)
$g_{1\phi}$	g_0, except $\phi_{t1} \neq \phi_{c1}$ (an acute effect on ϕ_1)
g_{2p}	$g_{1\phi}$, except $p_{t2} \neq p_{c2}$ (acute effects on ϕ_1 and p_2)

Chronic effects might arise from starlings that are in poor health due to effects of the pesticide; these starlings might be more susceptible to predation (this would be revealed in lessened survival during the summer period) or might be less active in foraging (this might be revealed in differing probabilities of resighting compared to the control starlings, because sampling is done during the summer period). Chronic effects, if they exist, might be reduced with time. That is, one might expect chronic effects to diminish over time, relative to the starlings in the control group. Agreement is reached, based on biological evidence, that chronic effects, if they exist, should not last beyond the seventh week.

TABLE 3.5. Summary of the starling data as the matrix m_{vij}, where $v =$ treatment or control group, $i =$ week of release ($i = 1, \ldots, 9$), and $j =$ week of resighting ($j = 2, \ldots,$ 10). The data given for each group (v) are the number of starlings first captured in week j after last being released at time i. $R_i =$ the number of birds released at week i; note that all of those released in weeks $2, \ldots, 9$ were merely rereleased. Each row (i) plus the term $\left(R(i) - \sum_j m_{ij} \right)$ is modeled as a multinomial distribution with sample size $R(i)$.

Week	$R(i)$	Observed Recaptures for Treatment Group $m(i, j)$							
		$j = 2$	3	4	5	6	7	8	9
1	300	158	43	15	5	0	0	0	0
2	158		82	23	7	1	1	0	0
3	125			69	17	6	1	0	0
4	107				76	8	2	0	0
5	105					67	20	3	0
6	82						57	14	1
7	81							53	12
8	70								46

Week	$R(i)$	Observed Recaptures for Control Group $m(i, j)$							
		$j = 2$	3	4	5	6	7	8	9
1	300	210	38	5	1	0	0	0	0
2	210		157	20	8	2	0	0	0
3	195			138	24	2	1	0	0
4	163				112	24	2	0	0
5	145					111	16	6	0
6	139						105	16	4
7	124							93	12
8	115								89

Define $S_i = \phi_{ti}/\phi_{ci}$ for $i = 1$ to 7 as the measure of treatment effect on conditional survival probability, compared to the control group. (Starlings in the control group will experience some mortality as the summer progresses; here the interest is in any *additional* mortality incurred by starlings caused by the pesticide treatment.) The parameters S_i ($i = 1, 2, \ldots, 7$) are 0.889, 0.911, 0.929, 0.943, 0.954, 0.964, and 0.971, respectively. With dampened chronic effects, one expects $S_2 < S_3 < S_4 < \cdots < S_7 < 1$, as can be seen from the parameters above (of course, the unconstrained *estimates* of these parameters, based on some approximating model, might not follow these inequalities). Here, it seems reasonable to consider the presence of chronic effects only as additional impacts to the hypothesized acute effects. Thus, several models of chronic effects on both conditional survival and resighting probabilities are defined and might be included in the set of candidate models:

Model	Parametrization
$g_{2\phi}$	g_{2p}, except $\phi_{t2} \neq \phi_{c2}$ (chronic effect on ϕ_2)
g_{3p}	$g_{2\phi}$, except $p_{t3} \neq p_{c3}$ (chronic effect on p_3)
$g_{3\phi}$	g_{3p}, except $\phi_{t3} \neq \phi_{c3}$ (more chronic effects)
g_{4p}	$g_{3\phi}$, except $p_{t4} \neq p_{c4}$ (more chronic effects)
$\vdots$	
$g_{7\phi}$	All ϕ_{vi} and p_{vi} differ by treatment group for 7 weeks

This last candidate model ($g_{7\phi}$) allows chronic treatment effects on both conditional survival and resighting probabilities up through the 7th sampling week, in addition to the acute treatment effects on ϕ_{t1} and p_{t2}. This model will serve as our global model, and it has 30 parameters. The treatment effect extends through the ninth week; thus, the generating model is not in the set of candidate models and has more parameters than the global model (34 vs. 30).

Model g_0 has 17 parameters, while model $g_{7\phi}$ has 30 parameters. The simplest model would have a constant survival and resighting probability for each group ($g_{\phi,p}$) and thus no treatment or week effects on either conditional survival or resighting probabilities. This model would have only two parameters (ϕ and p). Alternatively, a four-parameter model could allow the time-constant parameters to differ by treatment group (ϕ_t, ϕ_c, p_t, and p_c). Considering the relatively large sample size in this example, these models seem to be too simple and unlikely to be useful based on initial biological information, and we might well exclude these from the set of candidate models. **Models without biological support should not be included in the set of candidate models**. However, as an example, we will include these simple models for consideration and note that these models might well be viewed as more viable models if the initial sample size released were 60 instead of 600.

The effective sample size in these product multinomial models is the number of starlings released (or rereleased) at each week. [The effective sample size in these product multinomial models is a complicated issue, but we will not divert attention to this matter here, except to say that here we used $n = \sum R_i$ in the context of AIC$_c$. Technical notes on this subject may be obtained from KPB.] In this example, $n = 2{,}583$ releases (a resighting is equivalent to being "recaptured and rereleased"). Because 600 starlings (300 in each group) were released at week 1 (the nest boxes), the remaining 1,983 starlings were resighted at least once. Because of the large effective sample size, the use of AIC$_c$ is unnecessary; however, if one chose always to use AIC$_c$ in place of AIC, no problems would be encountered because AIC$_c$ and AIC converge as n/K gets large.

A statistician on the research team suggests adding several models of the possible tapering treatment effects on conditional survival or resighting probabilities. This is suggested both to conserve the number of parameters (recognizing the bias–variance tradeoff, Figure 1.3) and to gain additional insights concerning possible long-term chronic treatment effects. Models employing a type of sine transformation on the parameters (ϕ_{vi} and p_{vi}) will be used here. In this transformation, the parameter (θ, representing either ϕ or p, assumed to be between 0 and 1) to be modeled as a function of an external covariate (e.g., X) is replaced by the expression $(\sin(\alpha + \beta X) + 1)/2$. The new parameters α and β are the intercept and slope parameters, respectively, in the covariate model. The transformation utilizes one-half of a sine wave to model increasing or decreasing sigmoid functions and is an example of a link function in generalized linear models. In particular, submodel $g_{\sin \phi_t}$ and submodel $g_{\sin p_t}$ were defined for the dynamics of starlings in the treatment group:

$$g_{\sin \phi_t} \qquad \sin(\phi) = \alpha + \beta(\text{week}),$$
$$g_{\sin p_t} \qquad \sin(p) = \alpha' + \beta'(\text{week}).$$

These submodels each have only 2 parameters (intercepts α and α' and slopes β and β') and assume that $\sin(\phi_t)$ or $\sin(p_t)$ is a linear function of week (e.g., conditional survival of starlings in the treatment group will gradually increase as the summer period progresses, eventually approximating that of starlings in the control group).

   These above two submodels for the treatment group can be crossed with four submodels below for the control group:

$g_{\phi ci}$ ϕ is allowed to differ for each week; hence
        ($i = 1, \ldots, 8$).
$g_{\phi c}$ ϕ is assumed constant across weeks.
$g_{p ci}$ p is allowed to differ for each week; hence
        ($i = 2, \ldots, 9$).
$g_{p c}$ p is assumed constant across weeks.

For example, a model can be developed using $g_{\sin \phi_t}$ for conditional survival of the treatment group and model $g_{\phi c}$ for the conditional survival of the control group. This part of the model has 3 parameters; α, β, ϕ_c, plus the parametrization of the resighting probabilities. Thus, one could consider model $g_{\sin p_t}$ for the treatment group and model $g_{p ci}$ for the control group as one parametrization for the resighting probabilities. This would add the parameters $\alpha', \beta', p_{c2}, p_{c3}, \ldots, p_{c10}$, for a total of $K = 14$ parameters. As an illustration, we consider a rich mixture of candidate models in Table 3.6 (a set of 24 candidate models). If this were a real situation, still other a priori models might be introduced and carefully supported with biological reason. If this experiment were based on only 60 nestlings, then several simple models should be included in the set, and high-dimensional models would be deleted. This set of 24 candidate models will serve as a first example where there is some substantial complexity.

TABLE 3.6. Summary of Akaike's information criterion (AIC) and associated statistics for 24 candidate models for the analysis of the simulated data on nestling starlings dosed with a pesticide. (All values are scaled by the additive constant $-4,467.779$; thus $\Delta_i = 0$ for the best model.) Akaike weights (w_i) are also shown.

Model	AIC	No. Parameters	Δ_i	w_i
$g_{7\phi}$ (global)	4,495.409	30	27.63	0.0000
g_{7p}	4,493.619	29	25.84	0.0000
$g_{6\phi}$	4,491.649	28	23.87	0.0000
g_{6p}	4,489.889	27	22.11	0.0000
$g_{5\phi}$	4,491.679	26	23.90	0.0000
g_{5p}	4,491.929	25	24.15	0.0000
$g_{4\phi}$	4,490.199	24	22.42	0.0000
g_{4p}	4,489.029	23	21.25	0.0000
$g_{3\phi}$	4,489.629	22	21.85	0.0000
g_{3p}	4,492.619	21	24.84	0.0000
$g_{2\phi}$	4,501.809	20	34.03	0.0000
g_{2p}	4,517.019	19	49.24	0.0000
$g_{1\phi}$	4,523.489	18	55.71	0.0000
g_0	4,532.599	17	64.82	0.0000
$g_{\sin\phi_t,\phi_{ci},\sin p_t,p_{ci}}$	4,485.669	21	17.89	0.0001
$g_{\sin\phi_t,\phi_{ci},\sin pt,pc}$	4,475.249	14	7.47	0.0217
$g_{\sin\phi_t,\phi_c,\sin p_t,p_{ci}}$	4,479.359	14	11.58	0.0028
$g_{\sin\phi_t,\phi_c,\sin pt,pc}$	**4,467.779**	**6**	**0.0**	**0.9014**
$g_{\sin\phi_t,\phi_{ci},p_{ti},p_{ci}}$	4,488.629	28	20.85	0.0000
$g_{\sin\phi_t,\phi_{ci},p_{ti},pc}$	4,478.209	21	10.43	0.0049
$g_{\sin\phi_t,\phi_c,p_t,p_{ci}}$	4,484.699	13	16.92	0.0002
$g_{\sin\phi_t,\phi_c,pt,pc}$	4,473.119	5	05.34	0.0629
$g_{\phi_t,\phi_c,p_t,pc}$	4,770.479	4	302.70	0.0000
$g_{\phi,p}$	5,126.609	2	356.13	0.0000

3.4.4   Data Analysis Results

As one would expect with simulated data, they fit the model used for their generation; $g_{9\phi}$ ($\chi^2 = 35.5$, 36 df, $P = 0.49$). [A large literature on goodness-of-fit testing in this class of models exists (e.g., Burnham et al. 1987 and Pollock et al. 1990); we will not pursue the details of such tests here.] These data were simulated such that no overdispersion was present, and an estimate of the overdispersion factor c could be computed under the generating model from the results of the goodness-of-fit test, $\hat{c} = \chi^2/\text{df} = 35.5/36 \approx 1$. The global model $g_{7\phi}$ has fewer parameters than the generating model, but also fits these data well ($\chi^2 = 35.4$, 30 df, $P = 0.23$). The value of $\hat{c}$ for the global model was 1.18, reflecting no overdispersion in this case, but some lack of fit (which is known to be true in this instance); after all, it, too, is only a model of "truth." In practice, one cannot usually distinguish between overdispersion and a structural lack of fit. One could consider a quasi-likelihood inflation of the variances and covariances of the estimates from the selected model by multiplying these by 1.18 (or the standard errors by the square root, 1.086). In particular, one might consider using the modifications to AIC given in Section 2.5 (i.e., QAIC $= -2\log(\mathcal{L})/1.18 + 2K$). We will mention these issues at a later point. The critical information needed for selection of a parsimonious model and ranking and scaling the other models is shown in Table 3.6.

The interpretation of the 24 models for the experimental startling data (Table 3.6) can be sharpened by examining the Akaike weights. Here the weight for the AIC-selected model ($g_{\sin\phi_t,\phi_c,\sin p_t,p_c}$) is 0.906, while the second-best model ($g_{\sin\phi_t,\phi_c,p_t,p_c}$) has a weight of 0.063 and the third-best model ($g_{\sin\phi_t,\phi_{ci},\sin p_t,p_c}$) has a weight of 0.022. The sum of the weights for the 21 remaining models is less than 0.01. In this case, one is left with strong support for the best model, with fairly limited support for the second-best model (evidence ratio of best vs. second-best $\geq$ 14). The evidence ratio for the best vs. third-best is about 41.2. Thus, the data support one model as covincingly best, and there seems to be little need to attempt model averaging or bootstrapping (Chapter 4) to gain further robustness in inferences from these data (for this set of models). In addition, the use of conditional standard errors, given the best model, will likely suffice. Note that bootstrapping in this example would be very, very difficult. Software development would be a very formidable task, and computer time on a Pentium 1PC would likely take several days. Thus, the Akaike weights provide a distinct advantage in complex problems such as this simulated starling experiment.

The model with the minimum AIC value was $g_{\sin\phi_t,\phi_c,\sin p_t,p_c}$ with $K = 6$ parameters (α, β, α', β', ϕ_c, and p_c). Using estimates of these 6 parameters one can derive MLEs of the survival and resighting parameters of interest; the MLEs for the treatment survival probabilities were as follows:

i	ϕ_{ti}	$\hat{\phi}_{ti}$	$\widehat{se}(\hat{\phi}_{ti})$
1	0.800	0.796	0.021
2	0.810	0.810	0.016
3	0.819	0.824	0.014
4	0.827	0.838	0.014
5	0.834	0.851	0.160
6	0.841	0.864	0.019
7	0.847	0.876	0.022
8	0.852	0.887	0.026
9	0.857	0.898	0.029

The survival parameter for the control group was 0.90, and its MLE from the selected model was 0.893 ($\widehat{se} = 0.008$). These estimates are reasonably close to the parameter values, and one can correctly infer the diminishing, negative effect of the treatment on weekly survival probabilities. On a technical note, the 9 estimates of survival probability for the treatment group (above) were derived from the MLEs of α and β in the submodel $\sin(\phi_{ti}) = \alpha + \beta(\text{week } i)$.

Model $g_{\sin\phi_t,\phi_c,\sin p_t,p_c}$ had the lowest AIC value (4,467.78, $\Delta_i = 0$); the AIC value is large because the sample size is large (Section 2.1.4). Here, the sine model estimates the acute and chronic effects of the treatment on both the conditional survival and resighting probabilities for birds in the treatment group. The conditional survival and resighting probabilities for birds in the control group were constant over weeks in this model, but differed from those in the treatment group. The AIC-selected model captures the main structure of the generated process. Figure 3.3 illustrates the similarities among the true values, the estimates from the global model ($g_{7\phi}$), and the estimates from the AIC-selected model in terms of the treatment effect, $1 - S_i$.

Part of the reason that this analysis was successful was the a priori reasoning that led to *modeling* the treatment effects, rather than trying to estimate the week-specific treatment effects (i.e., the S_i) individually. Such modeling allowed substantial insight into the tapering, chronic effects in this case. Note: The two simplest models ($g_{\phi_t,\phi_c,p_t,p_c}$ with $K = 4$ and $g_{\phi,p}$ with $K = 2$) were not at all plausible ($\Delta_i = 302.70$ and 356.13, respectively); recall that these models would not normally have been considered in a well-designed experiment, since they lacked any reasonable biological support, given the large sample size involved. Of course, had sample size been very small, then these models might have been more reasonable to include in the set of candidates.

If sample size is small, one must realize that relatively little information is probably contained in the data (unless the effect size if very substantial), and the data may provide few insights of much interest or use. Researchers routinely err by building models that are far too complex for the (often meager)

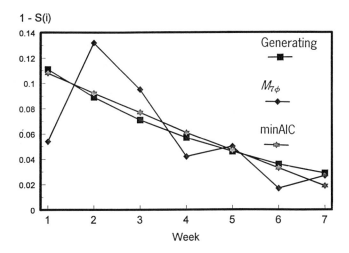

FIGURE 3.3. Treatment effect ($1 - S_i$, for week $i = 1, \ldots, 7$) for the starling data from the generating models ($g_{9\phi}$) with 34 parameters, compared with estimates of these parameters from the global model ($g_{7\phi}$) with 30 parameters and the AIC-selected model with 6 parameters.

data at hand. They do not realize how little structure can be reliably supported by small amounts of data that are typically "noisy." Some experience is required before analysts get a feeling for modeling based on sample size and what is known about the science of the problem of interest.

3.4.5   Further Insights into the First Fourteen Nested Models

If only the first 14 models (Table 3.6) had been defined a priori, the inference concerning which model to use would have been far less clear. First, the best of these 14 models is over 25 units from the AIC-selected model, but this would not have been known. Second, 7 models have AIC values within 4 units of the best of the 14. Thus, some additional steps would be necessary to incorporate model selection uncertainty into inference for these experimental data if the analysis was based on just the first 14 models.

We now examine further the results that would have been obtained had the set of candidate models included just the first 14 models in Table 3.6. Substantial theory (e.g., the estimators exist in closed form) and software (program RELEASE, Burnham et al. 1987) exist for this sequence of nested models, allowing the illustration of a number of deeper points. First, we must notice that these 14 models are clearly inferior to the models hypothesizing tapering treatment effects (a diminishing linear treatment effect embedded in a sine link function) for birds in the treatment group (e.g., the best model of the 14, model g_{4p}, is 21.25 AIC units above the selected model and has 23 parameters, compared to only 6 parameters for the AIC-selected model). Again, this points to the importance of a good set of candidate models. Second, many smaller chronic effects could not be identified by model g_{4p} (i.e., the relative treatment effects on survival in the later time periods, S_4, S_5, and S_6); however, the Δ_i values provide clues that at the very least, models $g_{4\phi}$ (therefore, S_4) and g_{6p} (therefore, S_5) are also somewhat supported by the data (Table 3.6). These models have AIC values within 1.17 and 0.86, respectively, of model g_{4p}. In fact, models g_{3p} through model $g_{6\phi}$ have fairly similar AIC values (Table 3.6 and Figure 3.4). Unless the data uniquely support a particular model, we should not take the resulting model as *the* answer for the issue at hand: just the best that the particular data set can provide. Perhaps more than one model should be considered for inference from the 14 models (Chapters 4 and 5).

The program RELEASE (Burnham et al. 1987) allows approximate expected values of estimators and theoretical standard errors to be computed easily for models in this class (i.e., the 14 appearing at the top of Table 3.6). These results allow insight into why the more minor chronic effects were not identified by model g_{4p} (the model estimated to be the best among the 14):

i	$1 - E(\hat{S}_i)$	$\widehat{se}(1 - \hat{S}_i)$	$(1 - E(\hat{S}_i))/\widehat{se}(1 - \hat{S}_i)$
4	0.057	0.053	1.08
5	0.046	0.055	0.84
6	0.036	0.057	0.63

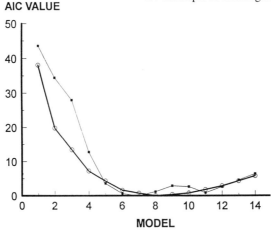

FIGURE 3.4. Estimated theoretical (heavy line) and sample Δ_i values for the 14 nested models used for the starling experiment. The estimated ($n = 50,000$ Monte Carlo reps) theoretical, expected AIC values (shown as open circles) are minimized ($\Delta_i = 0$) at model g_{5p}, while the realized AIC value from the sample data is minimized at model g_{4p} (see Table 3.6). Generally, there is good agreement between the theoretical and sample values, here plotted as Δ_i values).

The expected treatment effect size (i.e., $1 - E(\hat{S})$) was small (near 0), while the standard errors were of a similar magnitude or larger, as shown in the final two columns above. The larger effects (i.e., S_1 and S_2) are relatively easy to identify; however, at some point, the effect size is too small to detect directly with confidence from the information contained in the finite sample. Still, if one had only the first 14 models and had used AIC to select model g_{4p}, inference from the data in this example would have been fairly reasonable, but hardly optimal. The acute and larger chronic effects would have been convincingly identified. Comparison of AIC values for models $g_{4\phi}$ ($K = 24$) and g_{6p} ($K = 28$) would have provided reasonable evidence for some extended chronic treatment effects. Still, having to estimate 23–28 parameters would lead to imprecise estimators, compared to those under the best model ($g_{\sin \phi_t, \phi_c, \sin p_t, p_c}$). AIC, AIC$_c$, and QAIC$_c$ are fundamental criteria that provide a basis for a unified approach to the statistical analysis of empirical data in the biological sciences. Further details concerning this class of models are provided by Anderson et al. (1994), Burnham et al. (1994), Burnham et al. (1995a and b), and Anderson et al. (1998).

3.4.6   Hypothesis Testing and Information-Theoretic Approaches Have Different Selection Frequencies

At this point it is illustrative to examine briefly how information-theoretic selection compares to traditional approaches based on statistical hypothesis testing. Thus, Monte Carlo methods were employed to generate 50,000 inde-

TABLE 3.7. Selection percentages for six selection methods, based on 50,000 Monte Carlo repetitions. The hypothesis testing approaches use $\alpha = 0.05$. The data sets were generated under model $g_{9\phi}$ with 34 parameters, which was parametrized to reflect a tapering treatment effect on both conditional survival and resighting probabilities for the treatment group.

	Model	Hypothesis Testing			Information-Theoretic		
		Stepup	Stepdown	Stepwise	AIC	AIC_c	$QAIC_c$
1	g_0	0.6	0.0	0.0	0.0	0.0	0.0
2	$g_{1\phi}$	17.4	0.0	13.9	0.0	0.0	0.0
3	g_{2p}	14.8	0.3	13.5	0.2	0.2	0.2
4	$g_{2\phi}$	26.8	1.5	26.3	1.2	1.3	1.4
5	g_{3p}	16.3	2.8	16.9	2.7	2.8	2.6
6	$g_{3\phi}$	14.6	6.9	16.1	6.8	7.4	7.1
7	g_{4p}	5.9	7.5	7.1	8.5	9.0	8.3
8	$g_{4\phi}$	2.7	11.9	3.8	13.5	14.0	13.1
9	g_{5p}	0.8	10.3	1.3	12.0	12.3	11.4
10	$g_{5\phi}$	0.2	13.3	0.7	14.3	14.1	13.5
11	g_{6p}	0.0	10.9	0.2	11.3	10.9	10.6
12	$g_{6\phi}$	0.0	12.9	0.2	11.3	11.1	11.3
13	g_{7p}	0.0	10.2	0.1	8.8	8.1	8.9
14	$g_{7\phi}$	0.0	11.5	0.1	9.5	8.8	11.5

pendent samples (data sets) using the same methods as were used to generate the original set of simulated data on nestling starlings. That is, model $g_{9\phi}$ and the numbers released and all parameter values were identical to those used to generate the first set of data. Six methods were used to select a model for inference: The first 3 methods involve well-known selection methods based on hypothesis testing (stepup or forward selection, stepdown or backward selection, and stepwise selection), each using $\alpha = 0.05$. Three information-theoretic methods were also used on each of the data sets: AIC, AIC_c, and $QAIC_c$ (using $\hat{c}$ as a variance inflation factor, estimated for each simulated data set). The results (Table 3.7 and Figure 3.5) show substantial differences among model selection frequencies for the various methods.

Stepup selection, on average, selects model $g_{2\phi}$ with 20 parameters (the average was 20.3 parameters selected). These results are similar to the stepwise approach, which selects, on average, model g_{3p} with 21 parameters (here the average number of parameters was 20.6). Given that data were simulated under model $g_{9\phi}$ with 34 parameters, these methods seem to select relatively simple models that miss most of the chronic treatment effects. Stepdown testing resulted, on average, in model g_{6p} (mean 26.9 parameters) and resulted in quite different model selection frequencies than the other hypothesis testing approaches. Of course, the selection frequencies would differ substantially if a different (arbitrary) α level (say, 0.15 or 0.01) had been chosen or had the treatment impacts differed (i.e., a different model used to generate the data).

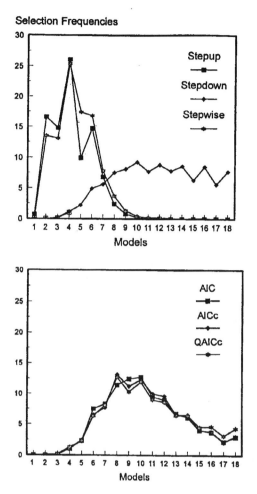

FIGURE 3.5. Model selection probabilities for three hypothesis testing approaches (top) and three information-theoretic approaches based on 50,000 Monte Carlo repetitions of the starling data, generated under model $g_{9\phi}$.

The practical utility of hypothesis testing procedures is of limited value in model identification (Akaike 1981b:722).

In this example, AIC selection averaged 25.4 parameters (approximately model g_{5p}). Both AIC and the stepdown testing did reasonably well at detecting the larger chronic effects. The differences between AIC, AIC_c, and QAIC_c are trivial, as one would expect from the large sample sizes used in the example (Table 3.7 and Figure 3.5). Even the use of QAIC_c made relatively little difference in this example because, the estimated variance inflation factor was near 1 ($\hat{c} = 1.18$).

While both AIC and likelihood ratio tests employ the maximized log-likelihood, their operating characteristics can be quite different, as illustrated in this simulated example. In addition, one must note that substantial uncertainty exists in model choice for all six approaches (Figure 3.5). This makes the material in Chapter 4 particularly important, since this component of uncertainty should be incorporated into estimates of precision of the parameter estimators. In this example the statistical hypothesis testing approach is a poor alternative to selection based on estimating the relative K-L information. In general, we recommend strongly against the use of null hypothesis testing in model selection.

3.4.7   Further Insights Following Final Model Selection

Selection of the best model and the relative ranking of all the candidate models is objective, given a set of candidate models, and can be implemented without the aid of subjective judgment. The formal data-based search for a best model is a key part of the analysis. In the example, model $g_{9\phi}$ was used to generate the data. Thus "truth" is known and serves here as a basis for comparisons. AIC does not try to select the model that generated these data; rather, it estimates which model is the "best approximating model" for analysis of these data in the sense of having the smallest K-L distance from approximating model to truth. Further information concerning the statistical properties of AIC-selected models in the $g_0, \ldots, g_{9\phi}$ class are given in Anderson et al. (1998).

The starling example illustrates an ideal a priori strategy; however, let us explore some potential realities *after* the analysis has been completed to this point. We select model $g_{\sin \phi_t, \phi_c, \sin p_t, p_c}$ as the best (with 6 parameters), but we must also perhaps consider models

$$g_{\sin \phi_t, \phi_c, p_t, p_c} \qquad (\Delta_i = 5.34 \text{ with only 5 parameters}),$$
$$g_{\sin \phi_t, \phi_{ci}, \sin p_t, p_c} \qquad (\Delta_i = 7.47, \text{ with 14 parameters}),$$

before making some final inferences (at this time, the analyst must address the variance component due to model uncertainty; Chapter 4). Model $g_{\sin \phi_t, \phi_{ci}, \sin p_t, p_c}$ is somewhat inconsistent. Why would there be unrestricted weekly variation in conditional survival for the birds in the control group, but only smooth time trends in the treatment group? Perhaps this finding might lead to a thorough review of field methods in an effort to detect some anomaly. Perhaps this model should not have been in the set of candidate models considered, since it may be picking up random variation in the data. Note, too, that Δ_i is 7.47, and this model has 8 additional parameters over the AIC-selected model. It would seem that this model is a relatively poor one for these data, although it might play a role in estimating the variance component due to model selection uncertainty.

After the analysis of the data to this point, suppose that one of the team members asks about models where there is no treatment effect on the resighting probabilities, just hypothesized treatment effects on conditional survival—either acute or chronic or both. Can new models reflecting these hypotheses be added to the set of candidate models and more AIC values computed? This question brings up several points. First, if this suggestion was made after examining the estimates of $p_{t2}, p_{t3}, p_{t4}, \ldots, p_{p7}$ vs. $p_{c_2}, p_{c3}, p_{c4}, \ldots, p_{c7}$ and noting that there seemed to be little difference between successive week-dependent pairs, then this is a form of data dredging, and any subsequent results should clearly detail the process by which the additional models were considered. **We encourage full investigation of the data to gain all possible insights; we only want investigators to reveal the extent of any data dredging that took place.** Second, if that suggestion was made on conceptual grounds rather than by studying the intermediate results, then the new class of models can be added to the list, AIC computed, the Δ_i, w_i, and evidence ratio values recomputed, and inferences made. However, in this second case, the team could be somewhat faulted for not considering the set of models more carefully in the first place.

3.4.8   Why Not Always Use the Global Model for Inference?

Some might argue that the global model (or another model with many parameters representing likely effects) should always be used for making inferences. After all, this model has been carefully defined by existing biological considerations and has all the effects thought to be reasonable. Why bend to the principle of parsimony and endure the various issues concerning model selection, selection uncertainty, etc.? We can illustrate problems with this approach using the starling data and the global model ($g_{7\phi}$) with 30 parameters. The key results are given here in detail for estimates of S_i for $i = 1, \ldots, 7$:

i	$\hat{S}_i$	$\widehat{se}(\hat{S}_i)$	CI_L	CI_U
1	0.946	0.044	0.858	1.033
2	0.868	0.052	0.766	0.969
3	0.905	0.058	0.792	1.017
4	0.958	0.057	0.847	1.069
5	0.950	0.047	0.859	1.042
6	0.983	0.051	0.883	1.084
7	0.973	0.059	0.858	1.088

The poor precision is illustrated by the upper confidence interval (CL_U), since 6 of the 7 include the value of 1 (i.e., no treatment effect; often such upper limits would be truncated at 1.0). The average coefficient of variation on the $\hat{\phi}_{ti}$ under model $g_{7\phi}$ is 4.75% vs. 2.08% under the AIC-selected model.

Attempts to select a properly parsimonious model for inference has its rewards, primarily an approximating model that has a reasonable tradeoff between bias and variance. The tradeoff between bias and variance is a byproduct of model selection where expected K-L information loss is minimized (i.e., select that model whose estimated distance to truth is the shortest). The routine reliance on the global model may have little bias, but will likely give estimates of model parameters that are unnecessarily imprecise (see Figure 1.3b and Section 3.2.2), and this weakens the inferences made. In fact, the estimates fail to show the real patterns that can be validly inferred from these data, such as the smooth decrease in ϕ_{ti} and thus in S_i. Sometimes the global model might have 50, 100, or even 200 parameters, and this makes interpretation difficult. One cannot see patterns and structure, since there are so many parameters, most estimated with poor precision. Thus, some analysts have tried to make analyses of these estimated parameters in order to "see the forest for the trees." This has rarely been done correctly, since the estimators usually have substantial sampling correlations, making simple analysis results misleading. It is far better to embed the reduced model in the log-likelihood function and use the information-theoretic criteria to select a simple, *interpretable* approximation to the information in the data.

3.5   Example 4: Sage Grouse Survival

3.5.1   Introduction

Data from sage grouse (*Centrocercus urophasianus*) banded in North Park, Colorado, provide insights into hypothesis testing and information-theoretic criteria in data analysis. The example is taken from Zablan (1993), and additional details are found there. Here we will use data on subadult (birds less than 1 year old) and adult (birds more than 1 year old) male grouse banded on leks during the breeding season (first week of March through the third week of May), from 1973 through 1987. Sage grouse are hunted in the fall, and nearly all of the band recoveries were from hunters who shot and retrieved a banded bird and reported it to the Colorado Division of Wildlife. During this time 1,777 subadult and 1,847 adult males were banded, and the subsequent numbers of band recoveries were 312 and 270, respectively (Table 3.8). The basic theory for modeling and estimation for these types of sampling data is found in Brownie et al. (1985).

Two types of parameters are relevant here: S_i is the conditional survival probability relating to the annual period between banding times i to $i + 1$, and r_i is the conditional probability of a band from a bird being reported in year i, given that the bird died in year i. In the model building it is convenient to use a as a subscript to denote age (subadult vs. adult) and t to denote annual variation (e.g., S_{a*t} denotes survival probabilities that vary by both age (a) and year (t); some models assume that r_i is a constant, thus resulting in identifiability of the parameter S_{15}).

TABLE 3.8. Summary of banding and recovery data for subadult (top) and adult male sage grouse banded in North Park, Colorado (from Zablan 1993).

Year Banded	Number banded	Recoveries by hunting season														
		73	74	75	76	77	78	79	80	81	82	83	84	85	86	87
1973	80	6	4	6	1	0	1	0	0	0	0	0	0	0	0	0
1974	54		6	5	2	1	0	0	0	0	0	0	0	0	0	0
1975	138			18	6	6	2	0	1	0	0	0	0	0	0	0
1976	120				17	5	6	2	1	1	0	0	0	0	0	0
1977	183					20	9	6	2	1	1	0	0	0	0	0
1978	106						14	4	3	1	0	0	0	0	0	0
1979	111							13	4	0	1	0	0	0	0	0
1980	127								13	5	3	1	0	0	0	0
1981	110									13	5	4	0	0	0	0
1982	110										7	1	3	1	1	0
1983	152											15	10	2	0	0
1984	102												12	4	0	0
1985	163													16	4	1
1986	104														5	2
1987	117															8

Year Banded	Number banded	73	74	75	76	77	78	79	80	81	82	83	84	85	86	87
1973	99	7	4	1	0	1	0	0	0	0	0	0	0	0	0	0
1974	38		8	5	1	0	0	0	0	0	0	0	0.	0	0	0
1975	153			10	4	2	0	1	1	0	0	0	0	0	0	0
1976	114				16	3	2	0	0	0	0	0	0	0	0	0
1977	123					12	3	2	3	0	0	0	0	0	0	0
1978	98						10	9	3	0	0	0	0	0	0	0
1979	146							14	9	3	3	0	0	0	0	0
1980	173								9	5	2	1	0	1	0	0
1991	190									16	5	2	0	1	0	0
1982	190										19	6	2	1	0	1
1983	157											15	3	0	0	0
1984	92												8	5	1	0
1985	88													10	1	0
1986	51														8	1
1987	85															10

3.5.2 Set of Candidate Models

The biological objective of this study was to increase understanding of the survival process of sage grouse. Zablan (1993) used model $\{S_{a*t}, r_{a*t}\}$ as the global model, with 58 parameters. For the purpose of this particular example, the set of candidate models includes the following:

Number/Model		K	Comment
Models with r constant:			
1	S, r	2	Constant S
2	S_t, r	16	Year-dependent S
3	S_a, r	3	Age-dependent S
4	S_{a+t}, r	17	Age- and year-dependent S, no interaction
5	S_{a*t}, r	31	Age- and year-dependent S, interaction
Models with r year-dependent (t):			
6	S, r_t	16	Constant S
7	S_t, r_t	29	Year-dependent S
8	S_a, r_t	17	Age-dependent S
9	S_{a+t}, r_t	30	Age- and year-dependent S, no interaction
10	S_{a*t}, r_t	44	Age- and year-dependent S, interaction
Models with r age-dependent (a):			
11	S, r_a	3	Constant S
12	S_t, r_a	17	Year-dependent S
13	S_a, r_a	4	Age-dependent S
14	S_{a+t}, r_a	18	Age- and year-dependent S, no interaction
15	S_{a*t}, r_a	32	Age- and year-dependent S, interaction

Models with interaction terms (denoted by the $a * t$) allow each age class to have its own set of time-dependent parameters. The additive models (denoted by "+") exclude interaction terms; e.g., for model S_{a+t} there is a constant difference between subadult and adult survival parameters, and the year-to-year estimates of survival probabilities for subadults and adults are parallel on a logit scale and separated by β_0 (see below). A logit transformation has been made on S, and age (a) and year (as a dummy variable, t_i) enter as a linear function,

$$S_{a+t} \text{ denotes logit}(S) = \beta_0 + \beta_1(a) + \beta_2(t_1) + \beta_3(t_2) + \cdots + \beta_{16}(t_{15}).$$

This approach is similar to logistic regression. However, this submodel is embedded in the log-likelihood function. Such models are often quite useful and can be biologically realistic. Models without interaction terms have fewer parameters; for example, model S_{a+t}, r_a has 18 parameters, compared to 32 parameters for model S_{a*t}, r_a.

These 15 models plus models $\{S_a, r_{a*t}\}$ and $\{S_a, r_{a+t}\}$ and the global model $\{S_{a*t}, r_{a*t}\}$ seem like sound initial choices; however, further biological considerations might lead one to exclude models with many parameters, in view of the relatively sparse data available (Table 3.8). We realize that this a priori set of models would ideally be fine-tuned in a real-world application. For example, if a long-term increase or decrease in survival was hypothesized, one might introduce submodels for survival such as

$$\text{logit}(S) = \beta_0 + \beta_1(a) + \beta_2(T) + \beta_3(a * T),$$

where T indexes the year of study as a continuous covariate $\{1, 2, \ldots, 15\}$. Alternatively, it was known that 1979 and 1984 had very severe winters; the survival probability in these years (say, S_s) could have been parametrized to differentiate them from the survival probability in more normal years (S_n). We assume that a great deal of thought has been put into the development of a set of candidate models.

A primary interest in banding studies is often to estimate survival probabilities and assess what external covariates might influence these. Thus, Zablan (1993) modeled sage grouse survival using 4 year-dependent environmental covariates (cov_t): winter precipitation (wp), winter temperature (wt), spring precipitation (sp), and spring temperature (st) (she provided operational definitions of these variables; we will not need to note the specific details here). Submodels with survival probabilities of the form

$$\text{logit}(S_t) = \beta_0 + \beta_1(\text{cov}_t) \quad \text{or} \quad \text{logit}(S_{t+a}) = \beta_0 + \beta_1(a) + \beta_2(\text{cov}_t)$$

could be constructed. Such submodels for survival have only 2 or 3 parameters (an intercept and one slope coefficient for the first submodel and an intercept, 1 age effect, and 1 slope coefficient for the second submodel), but also provide some insights into biological correlates, which themselves are time-dependent.

3.5.3 Model Selection

Zablan's analysis was done using the programs ESTIMATE and BROWNIE (Brownie et al. 1985) and SURVIV (White 1983). Zablan (1993) found that the global model $\{S_{a*t}, r_{a*t}\}$ fit the data well ($\chi^2 = 34.34$, 30 df, $P = 0.268$), and she computed a variance inflation factor from the global model as $\hat{c} = 34.34/30 = 1.14$. Her calculations are in agreement with ours (deviance of model $\{S_{a*t}, r_{a*t}\} = 87.85$, 80 df, $\hat{c} = 1.10$). There was little evidence of overdispersion; thus there was no compelling reason to use QAIC. The effective sample size for parameter estimation in these surveys is the sum of the number of birds banded, which equaled 3,624 in this case. Zablan's global model had 58 parameters, giving the ratio $n/K = 3{,}624/58 = 62$; thus AIC could have been safely used instead of AIC_c. We used the program MARK (White and Burnham 1999, White et al. 2001) to compute MLEs of the parameters and their conditional covariance matrix, the maximized value of the log-likelihood function, AIC_c, Δ_i, and w_i for each of the 17 candidate models without a weather covariate and 4 models with one of the weather covariates.

AIC_c selected model $\{S_a, r_a\}$ with 4 parameters (Table 3.9) among the models without a weather covariate on survival. This approximating model assumes that conditional survival and reporting probabilities are age-dependent (subadult vs. adult), but constant over years. Here, the ML estimate of adult survival probability was 0.407 ($\widehat{se} = 0.021$), while the estimated survival for subadults was higher, at 0.547 ($\widehat{se} = 0.055$). The respective percent coefficients of variation were 5.2 and 10.0. An inference here is that male subadult grouse survive at a higher rate than male adults; perhaps this reflects the cost

TABLE 3.9. Candidate models for male sage grouse, $\log(\mathcal{L})$, (K), AIC_c, Δ_i, and Akaike weights (w_i).

Number	Model	$\log(\mathcal{L})$	K	AIC_c	Δ_i	w_i
Without environmental covariates:						
1	S, r	$-2, 215.564$	2	4435.13	4.41	0.085
2	S_t, r	$-2, 205.074$	16	4442.30	11.57	0.002
3	S_a, r	$-2, 215.096$	3	4436.20	5.47	0.050
4	S_{a+t}, r	$-2, 203.797$	17	4441.76	11.04	0.003
5	S_{a*t}, r	$-2, 199.277$	31	4461.11	30.38	0.000
6	S, r_t	$-2, 204.893$	16	4441.94	11.21	0.003
7	S_t, r_t	$-2, 194.611$	29	4447.71	16.98	0.000
8	S_a, r_t	$-2, 204.526$	17	4443.22	12.50	0.001
9	S_{a+t}, r_t	$-2, 193.633$	30	4447.84	17.12	0.000
10	S_{a*t}, r_t	$-2, 188.531$	44	4466.17	35.44	0.000
11	S, r_a	$-2, 214.717$	3	4435.44	4.72	0.073
12	S_t, r_a	$-2, 204.544$	17	4443.26	12.53	0.001
13	$\boldsymbol{S_a, r_a}$	$\boldsymbol{-2, 211.357}$	**4**	**4430.72**	**0**	**0.772**
14	S_{a+t}, r_a	$-2, 204.544$	18	4439.96	9.23	0.008
15	S_{a*t}, r_a	$-2, 196.065$	32	4456.72	25.99	0.000
16	S_a, r_{a*t}	$-2, 197.572$	32	4459.73	29.01	0.000
17	S_{a*t}, r_{a*t}	$-2, 174.557$	58	4467.03	36.31	0.000

of breeding and increased predation on breeding males. Estimated reporting probabilities (the $\hat{r}$) for first-year subadult birds were also different from those for adult birds (0.227 ($\widehat{se} = 0.031$) and 0.151 ($\widehat{se} = 0.008$), respectively). The use of the AIC_c-selected model does not indicate that there was no year-dependent variation in the parameters, only that this variation was relatively small in the sense of a bias–variance tradeoff and K-L information loss. The estimated next-best model ($\Delta_i = 4.41$) without a covariate was model $\{S, r\}$ with only two parameters, while the third-best model ($\Delta_i = 4.72$) was $\{S, r_a\}$ with three parameters. There is relatively little structure revealed by these data; this is not surprising, since the data are somewhat sparse (Table 3.8).

The AIC-selected model assumed that the survival and reporting probabilities varied by age class. Thus we considered four models where logit(S) was a linear function of age (subadult vs. adult) and one of the weather covariates (wp, wt, sp, or st), while retaining the age-specific reporting probability. The results were interesting and, at first, suggest that each of the weather covariate models is nearly tied with the AIC_c-selected model $\{S_a, r_a\}$. This leads to an important point, illustrated below:

Number	Model	$\log(\mathcal{L})$	K	AIC_c	Δ_i
With environmental covariates:					
18	S_{a+wp}, r_a	$-2, 210.828$	5	4431.67	0.95
19	S_{a+wt}, r_a	$-2, 211.334$	5	4432.68	1.96
20	S_{a+sp}, r_a	$-2, 210.819$	5	4431.66	0.93
21	S_{a+st}, r_a	$-2, 210.802$	5	4431.62	0.90

Each of the four models with an environmental covariate has a Δ_i value less than 2 and seems to provide support for the hypotheses that annual survival probabilities are related to temperature or precipitation. However, upon closer examination, it can be seen that the value of the maximized log-likelihood is very similar to the best model in Table 3.9, without any covariates. Thus, the inclusion of models with a covariate has not improved the fit of the model to the data. The best model without a covariate has 4 parameters, whereas the covariate models have 5 parameters. This difference in the number of parameters explains most of the difference in the AIC_c values between the best model and the 4 models with a covariate. Hence, upon closer examination, there is virtually no support for any of the covariates from these data. This leads to a point that is important in general.

Models Within Two Units of the Best Model

Models having Δ_i within about 0–2 units of the best model should be examined to see whether they differ from the best model by 1 parameter *and* have essentially the same values of the maximized log-likelihood as the best model.

In this case, the larger model is not really supported or competitive, but rather is "close" only because it adds 1 parameter and therefore will be within 2 Δ_i units, even though the fit, as measured by the log-likelihood value, is not improved.

Further insights into the sage grouse data can be obtained from the Akaike weights for the first 17 models in Table 3.9 (ignoring here models 18–21). In this case, the weight for the AIC_{min} model $\{S_a, r_a\}$ is 0.773, while the second-best model $\{S, r\}$ has a weight of 0.085 (evidence ratio $= 9.1$). The third- and fourth-best models had weights of 0.073 and 0.050, while the weights for the other models were nearly zero (the sum of the Akaike weights for the 13 remaining models was < 0.02). The annual variation in conditional survival probabilities was small (temporal process variation $\hat{\sigma}_s = 0.0426$ for adults and 0.0279 for subadults); thus model $\{S_a, r_a\}$ seems reasonable. Models ranked 2–4 all had fewer parameters than the AIC_{min} model. Thus, conditional sampling variances from those models were smaller than from the AIC_{min} model. In addition, these three models had small Akaike weights. These considerations lead to some trust in the conditional sampling variances from the best model as a reasonable reflection of the precision of the parameter estimates.

3.5.4   Hypothesis Tests for Year-Dependent Survival Probabilities

Zablan (1993) computed a likelihood ratio test between models $\{S_a, r_{a*t}\}$ and $\{S_{a*t}, r_{a*t}\}$ (the global model) using the program BROWNIE and found strong evidence of year-dependent survival ($\chi^2 = 46.78$, 26 df, $P = 0.007$). Program MARK provides similar results ($\chi^2 = 46.03$, 26 df, $P = 0.009$). This test allowed a fairly general structure on the reporting probabilities and therefore seemed convincing and provided evidence that survival probabilities varied "significantly" by year. In fact, it might be argued that had a simpler structure been imposed on the reporting probabilities (e.g., r_a), the power of the test for year-dependent survival probabilities (i.e., S vs. S_t) would have increased and the test result been even more "significant." However, contrary to this line of reasoning, the test of $\{S_a, r_a\}$ vs. $\{S_{a*t}, r_a\}$ gives $\chi^2 = 30.58$, 28 df, and $P = 0.336$. Still other testing strategies are possible, and it is not clear which might be deemed the best.

Given a believed year-dependence in annual survival probabilities, Zablan (1993) asked whether this variability was partially explained by one of the four covariates, with or without an age effect. However, she was unable to find a relationship between annual survival probabilities and any of the four covariates using likelihood ratio tests (the smallest P-value for the four covariates was 0.194). She used model $\{S_{a*t}, r_{a*t}\}$ (the global model) as a basis for inference.

3.5.5 Hypothesis Testing Versus AIC in Model Selection

An apparent paradox can be seen in the results for the male sage grouse data, and this allows us to further compare alternative paradigms of statistical hypothesis testing and AIC for model selection. The test between the two models $\{S_a, r_{a*t}\}$ and $\{S_{a*t}, r_{a*t}\}$ attempts to answer the questions, "Given the structure $\{r_{a*t}\}$ on the recovery probabilities, is there evidence that survival is also (i.e., in addition to age) year-dependent?" The answer provided is yes ($P = 0.007$ or 0.009). But this answer is seemingly in contrast to the inferences from the AIC-selected model, where there is no hint of time-dependence in either S or r (Table 3.9). The Δ_i values for models $\{S_a, r_{a*t}\}$ and $\{S_{a*t}, r_{a*t}\}$ are 28.068 and 30.400, respectively. AIC lends little support for a best approximating model that includes year-dependent survival or reporting probabilities.

The answer to this paradox is interesting and important to understand. The null hypothesis that $S_1 = S_2 = S_3 = \cdots = S_{14}$ for a given age class is *obviously* false, so why test it? This is not properly a hypothesis testing issue (see Johnson 1995 and Yoccoz 1991 for related issues). The test result is merely

---

Model Interpretation

Sometimes, the selected model contains a parameter that is constant over time, or areas, or age classes (i.e., $\theta = \theta_1 = \theta_2 = \cdots = \theta_m$). This result should not imply that there is *no* variation in this parameter, rather that parsimony and its bias/variance tradeoff finds the actual variation in the parameter to be relatively small in relation to the information contained in the sample data. It "costs" too much in lost precision to add estimates of all of the individual θ_i. As the sample size increases, then at some point a model with estimates of the individual parameters would likely be favored.

Just because a parsimonious model contains a parameter that is constant across strata does not mean that there is no variation in that process across the strata.

telling the investigator whether there is enough information (data, sample size) to show that the null hypothesis is false. *However, a significant test result does not relate directly to the issue of what approximating model is best to use for inference.* One model selection strategy that has often been used in the past is to do likelihood ratio tests of each structural factor (e.g., a, t, $a + t$, $a * t$, for each of the parameters S and r) and then use a model with all the factors that were "significant" at, say, $\alpha = 0.05$. However, there is no theory that would suggest that this strategy would lead to a model with good inferential properties (i.e., small bias, good precision, and achieved confidence interval coverage at the nominal level).

Clearly, one must also worry about the multiple testing problem here and the fact that many such tests would not be independent. If overdispersion is present, then likelihood ratio tests are not chi-square distributed. Furthermore, the test statistics may not be chi-square distributed for nontrivial problems such as these, where sample sizes are far from asymptotic and, in particular, where many models contain nuisance parameters (the r_i). The choice of an α-level is arbitrary as well. Many of the models in Table 3.9 are not nested; thus likelihood ratio tests are not possible between these model pairs. We note a certain lack of symmetry (this is again related to the α-level) between the null and alternative hypotheses and how this might relate to selection of a "best approximating model" (see Section 2.7.2). A very general and important problem here is how the test results are to be incorporated into building a good model for statistical inference. This problem becomes acute when there are many (say, > 8–10) candidate models. Using just the set of 17 models for the sage grouse data, one would have 136 potential likelihood ratio tests; however, some of these models were not nested, prohibiting a test between these pairs. With 136 (or even 36) test results there is no theory or unique way to decide what the best model should be and no rigorous, general way to rank the models (e.g., which model is second-best? Is the second-best model virtually as good as the best, or substantially inferior?). Finally, what is to be done when test results are inconsistent, such as those found in Section 3.5.4?

The biological question regarding annual survival probabilities would better be stated as, "How much did annual survival vary during the years of study?" Has survival varied little over the 14 or 15 years, or has there been large variation in these annual parameters? Such questions are *estimation problems*, not ones of hypothesis testing (see Franklin et al. 2002). Here, the focus of inquiry should be on the amount of variation among the population parameters $(S_1, S_2, \ldots, S_{14})$ for each of the two age classes; we will denote this standard deviation among these parameters by σ_s. Of course, if we knew the parameters S_i, then $\hat{\sigma}_s = \left(\sum_{i=1}^{14}(S_i - \overline{S})^2/13\right)^{1/2}$.

We next ask why the AIC procedure did not pick up the "fact" that survival varied by year? The reason is simple; AIC attempts to select a parsimonious approximating model for the observed data. In the sense of K-L information loss or a tradeoff between bias and variance, it was poor practice to add some

54 additional parameters (the difference in parameters between models $\{S_a, r_a\}$ and $\{S_{a*t}, r_{a*t}\}$) or even 26 additional parameters (the difference between models $\{S_a, r_{a*t}\}$ and $\{S_{a*t}, r_{a*t}\}$) to model the variation in S_i or r_i (Table 3.9). Note that the difference in AIC values for model $\{S_a, r_a\}$ versus model $\{S_a, r_{a*t}\}$ is 29.01, suggesting that model $\{S_a, r_{a*t}\}$ is highly overfit. Whether differences among survival probabilities are large enough to be included in a model is a *model selection* problem, not one of hypothesis testing.

Estimates of the 30 survival probabilities under model $\{S_{a*t}, r_a\}$ are given in Table 3.10. The average of the 15 estimates of adult survival was 0.400, nearly the same as that from the 4-parameter model selected by AIC_c (0.407). However, the average percent coefficient of variation for each $\hat{S}_i$ was 20.4 for model $\{S_{a*t}, r_a\}$ compared to only 5.2 for $\hat{S}$ in the AIC_c-selected model. Thus, the AIC-selected model indicates that the best estimate of annual survival in a particular year is merely $\hat{S}$ (from model $\{S_a, r_a\}$).

The situation was similar for subadult survival; the average survival from model $\{S_{a*t}, r_a\}$ was 0.548, compared to 0.547 for the AIC_c-selected model. The respective average percent coefficients of variation were 20.6% and 10% for models $\{S_{a*t}, r_a\}$ and $\{S_a, r_a\}$. In summary, 54 (or even 26) additional parameters "cost too much" in terms of increased variability of the estimates (see Figure 1.3B and Table 3.10) and reflect substantial overfitting. The lack of precision illustrated in Table 3.9 for model $\{S_{a*t}, r_a\}$ was worse still when model $\{S_{a*t}, r_{a*t}\}$ was used; coefficients of variation were 35.6% for adult survival and 30.5% for subadult survival. The model suggested by the hypothesis testing approach had 58 parameters, while the AIC_c-selected model has only 4 parameters. This illustrates the cost of using overparametrized models, even though the results of hypothesis tests clearly show "significance" for year-dependent survival (and reporting) probabilities. Models $\{S_a, r_{a*t}\}$ and $\{S_{a*t}, r_{a*t}\}$ are very general for these data and lie far to the right of the bias–variance tradeoff region in Figure 1.2. Zablan recognized the problems in using model $\{S_{a*t}, r_{a*t}\}$ and commented, "While significant differences were found between survival and recovery rates of males and of both age classes, and between years, survival estimates had unacceptably wide confidence intervals."

3.5.6   A Class of Intermediate Models

The researcher could use the AIC_c-selected model $\{S_a, r_a\}$ to obtain estimates of parameters and then proceed to obtain an estimate σ_s for each of the two age classes, using model $\{S_{a*t}, r_{a*t}\}$ or $\{S_{a*t}, r_{a+t}\}$ if desired. It is not trivial to embed this parameter into the likelihood framework, allowing an ML estimate of σ_s; this becomes a "random effects" model. However, this is a problem in "variance components," and consistent estimates of σ_s can be computed using, say, model $\{S_{a*t}, r_a\}$, following, for example, Anderson and Burnham (1976: 62–66); and Burnham et al. (1987:260–269). This approach is conceptually based on the simple partitioning of the total variance $(\text{var}(\hat{S}_i))$ into its two

TABLE 3.10. MLEs of year-dependent survival probabilities under model $\{S_{a*t}, r_a\}$. The first 15 estimates relate to grouse banded as adults, while the second set of estimates relate to first-year survival of subadults. The model assumes that the subadults become adults the second year after banding and thus have the same year-dependent survival probabilities as birds banded as adults.

Year(i)	$\hat{S}_i$	Standard Error	95% Confidence Interval[a]	
			Lower	Upper
1	0.462	0.128	0.238	0.702
2	0.500	0.092	0.327	0.673
3	0.357	0.073	0.230	0.508
4	0.412	0.074	0.277	0.561
5	0.464	0.073	0.328	0.606
6	0.507	0.069	0.375	0.639
7	0.465	0.066	0.340	0.595
8	0.357	0.062	0.246	0.486
9	0.397	0.063	0.282	0.524
10	0.340	0.061	0.233	0.466
11	0.321	0.063	0.212	0.455
12	0.358	0.073	0.231	0.509
13	0.171	0.071	0.071	0.355
14	0.339	0.133	0.138	0.621
15	0.549	0.129	0.305	0.771
1	0.725	0.114	0.462	0.891
2	0.629	0.152	0.321	0.859
3	0.524	0.106	0.323	0.717
4	0.528	0.112	0.316	0.731
5	0.566	0.093	0.383	0.732
6	0.446	0.120	0.237	0.677
7	0.386	0.117	0.193	0.623
8	0.513	0.110	0.307	0.715
9	0.497	0.118	0.282	0.713
10	0.615	0.111	0.389	0.801
11	0.547	0.101	0.351	0.729
12	0.368	0.121	0.173	0.618
13	0.440	0.107	0.251	0.649
14	0.744	0.104	0.498	0.895
15	0.695	0.111	0.450	0.864

[a]Based on a back transformation of the interval endpoints on a logit scale (Burnham et al. 1987:214).

additive components: the variance in the population parameters σ_s^2 and the conditional sampling variance ($\text{var}(\hat{S}_i \mid model)$). The approach assumes that the true S_i are independently and identically distributed random variables (in this case, these assumptions are weak and the effect somewhat innocent). Here, one has ML estimates of the sampling covariance matrix and can estimate $\text{var}(\hat{S}_i)$ directly from the estimates $\hat{S}_i$; thus, by subtraction one can obtain an estimate of σ_s^2. Thus, inferences could be made from the AIC-selected model, hopefully after incorporating model selection uncertainty.

The estimate of σ_s would provide some insight into the variation in the survival parameters; this would be done in the context that one *knows* that the S_i vary. Exact details of the optimal methodology would take us too far afield; however, some unpublished results seem exciting. For the adult data, $\hat{\sigma}_s = 0.0426$, 95% likelihood interval $[0, 0.106]$, and cv $= 10.8\%$ on annual survival probability and for the subadult data $\hat{\sigma}_s = 0.0279$, 95% likelihood interval $[0, 0.129]$, and cv $= 4.9\%$ on S. Thus, one can infer that the relative variation in the true annual survival probabilities was fairly small (cv $\approx$ 5–10%). Thus, the large variation in the estimates of annual survival probabilities under model $\{S_{a*t}, r_a\}$ (Table 3.9) is due primarily to sampling variation, as the large estimated standard errors suggest. Additional details, including shrinkage estimates of annual survival probabilities, appear in Chapter 6 (from Burnham and White 2002).

Ideally, the number of parameters in the various candidate models would not have large increments (see Section 2.7.2). In the grouse models, a submodel for the survival probabilities without year-dependent survival might have one or two (if age is included) parameters, while a model with year-dependent survival would have as many as 30 parameters (15 for each of the two age groups). Large differences in the number of parameters between certain candidate models are not ideal, and one should consider intermediate models while deriving the set of candidate models. Zablan's (1993) various covariate models represent an example; here an intercept and slope parameter on one of the covariates would introduce 2 parameters (3 with age) instead of 15 (30 with age). In contrast with the hypothesis testing approach, AIC-selection showed the four weather covariate models to be essentially tied with the AIC-selected model $\{S_a, r_a\}$ (Table 3.9).

Hypothesis testing and AIC are fundamentally very different paradigms in model selection and in drawing inferences from a set of data. In the sage grouse example, AIC_c tries to select a model that well approximates the information in the data. That selected model then provides estimates of the parameters $S_1, S_2, \ldots, S_{15}$ for each age group in the sense of K-L information loss (or a bias versus variance tradeoff). That is, an estimate of average survival ($\hat{S}$) from model $\{S_a, r_a\}$ (i.e., 0.407 for adults) would be used to estimate, for example, S_5 for adult grouse (hence, $\hat{S}_5 = 0.407$, 95% confidence interval $[0.368, 0.448]$), and this estimate would have better inferential properties than that using model $\{S_{a*t}, r_a\}$, whereby S_5 would be estimated using the year-specific estimator $\hat{S}_5$ (see Table 3.10, where this estimate is given as 0.464 with 95% confidence interval of 0.328 to 0.606). If inference about the conditional survival in the fifth year is made from the general model $\{S_{a*t}, r_{a*t}\}$, then the estimate is 0.336, and the precision is worse yet (95% confidence interval of $[0.168, 0.561]$). In the last two cases, the precision is relatively poor (e.g., compare Figures 1.3B and C for further insights).

3.6   Example 5: Resource Utilization of *Anolis* Lizards

This example illustrates the use of information-theoretic criteria in the analysis of count data displayed as multidimensional contingency tables (see also Sakamoto 1982). Schoener (1970) studied resource utilization in species of lizards of the genus *Anolis* on several islands in the Lesser Antilles, in the Caribbean. Here we use his data collected on *Anolis grahami* and *A. opalinus* near Whitehouse, on Jamaica, as provided by Bishop et al. (1975). These data have been analyzed by Fienberg (1970), Bishop et al. (1975), and McCullagh and Nelder (1989) and Qian et al. (1996), and the reader is urged to compare the approaches given in these papers to that presented here.

In his general studies of species overlap, Schoener (1970) studied the two species of lizards in an area of trees and shrubs that had been cleared for grazing. The height (< 5 or ≥ 5 ft) and diameter (< 2 or ≥ 2 in) of the perch, insolation (sunny or shaded), and time of day (roughly early morning, midday, and late afternoon) were recorded for each of the two species of lizard seen. Data were taken on each individual only once per "census"; data were not recorded if the lizard was disturbed, and the census route was varied considerably from one observation period to the next. The data of interest for this example can be summarized as a $2 \times 2 \times 2 \times 3 \times 2$ contingency table corresponding to height (H), diameter (D), insolation (I), time of day (T), and species (S), shown in Table 3.11 (from McCullagh and Nelder 1989:128–135). The data on 546 observations of lizards appear in the table with 48 cells. [Note: the results here differ from those in the first edition due to errors (blunders, actually) in computing AIC$_c$.]

TABLE 3.11. Contingency table of site preference for two species of lizard, *Anolis grahami* and *A. opalinus* (denoted by g and o, respectively) on the island of Jamaica (from Schoener 1970).

			Time of Day (T); Species (S)					
Insolation	Diameter	Height	Early morning		Midday		Late afternoon	
(I)	(D) in	(H) ft	g	o	g	o	g	o
Sunny	≤ 2	< 5	20	2	8	1	4	4
		≥ 5	13	0	8	0	12	0
	> 2	< 5	8	3	4	1	1	3
		≥ 5	6	0	0	0	1	1
Shaded	≤ 2	< 5	34	11	69	20	18	10
		≥ 5	31	5	55	4	13	3
	> 2	< 5	17	15	60	32	8	8
		≥ 5	12	1	21	5	4	4

3.6.1   Set of Candidate Models

Because these data were collected nearly 30 years ago and we have little expertise in lizard ecology and behavior, decisions concerning an a priori set
of candidate models will necessarily be somewhat contrived. We will focus
attention on modeling and model selection issues as an example and comment
on several inference issues. In reading Schoener's (1970) paper and the literature he cited, it would seem that a model with all the main effects (i.e., H,
D, I, T, and S) might serve as a starting point for models to be considered.
Several second-order interactions might be suspected, e.g., H $*$ T and H $*$ I
and I $*$ T. If the two species are partitioning their resources, then models with
H $*$ S, D $*$ S, I $*$ S, and T $*$ S included should be reasonable. As the study was
designed and data were collected, it was probably evident that site occupancy
was affected by several variables as well as some interactions. This might suggest that a model with all main effects and second-order interactions might be
considered. Then issues remain concerning possible higher-order interactions.
On biological grounds, it might seem reasonable to consider third-order terms
such as H $*$ D $*$ I, H $*$ D $*$ T, and H $*$ D $*$ S; or further, to add H $*$ I $*$ S,
H $*$ T $*$ S, and I $*$ T $*$ S. Finally, the second-order term D $*$ T seems unlikely
to be important; thus some models without this term were considered. We will
use the short set of models in Table 3.12 for illustrative purposes here. Even
in the late 1960s, T. W. Schoener and his colleagues, including S. E. Fienberg,
could have developed a better set of a priori candidate models than ours.

3.6.2   Comments on Analytic Method

We used a loglinear model with Poisson errors following Agresti (1990: chapter
5 and pages 453–456), and the analysis was made conditional on the total of the
frequencies, $\sum n_j$. Specifically, we used the SAS program GENMOD (SAS
1985). The likelihood is, $\mathcal{L}(\mu|n_j, model) = \prod_{j=1}^{48} \frac{\mu^{n_j} e^{-\mu}}{(n_j)!}$. The form of the
log-likelihood for the global model can be expressed as

$$\log(\mathcal{L}) = \sum_{j=1}^{48} \left(n_j \cdot \log(\mu_j) - \mu_j\right) - \sum_{j=1}^{48} \log(n_j!),$$

where n_j is the number of observations in cell j with Poisson mean μ_j, where
$j = 1, 2, \ldots, 48$. The purpose of the modeling is to put some reduced structure
on the 48 means. Then one has the log-linear model $\log(\mu_j) = X\beta$ as in
analysis of variance. Thus, β is the vector of effects and the grand mean. The
final term in the log-likelihood is a constant; thus SAS GENMOD omits this
term, and the resulting log-likelihood is positive. AIC is computed in the usual
manner, even though the AIC values are scaled by $\sum_{j=1}^{48} \log(n_j!)$. Note that
such arbitrary, additive constants are not present in the Δ_i values. Several
software packages allow ML estimates from discrete data such as these and
provide a number of relevant analysis options (see summary in Agresti 1990:

484–488), and some also print AIC values (e.g., SAS). However, we do not know of any packages that print AIC_c. Section 2.5 provides additional guidance if overdispersion is thought to exist in count data. We note Agresti's (1990) comment, "In practice, we learn more from *estimating* descriptive parameters than from *testing hypotheses* about their values."

3.6.3 Some Tentative Results

A model with only the five main effects might be a starting point to represent the information in these data. Alternative models with various hypothesized interaction terms are shown in Table 3.12. None of the models with second- or third-order interactions are supported by the data; in fact, the only model with any support is the model with merely the main effects. The result is in agreement with McCullagh and Nelder (1989), but differs from those of Fienburg (1970) and Bishop et al. (1975).

Note that if another model were to be added, the Δ_i values would likely need to be recomputed (but not the AIC values). Such values are always with respect to the minimum AIC model, given a set of candidate models, and inferences derived from the data via a set of models are effectively conditional on the set of models considered. Strict experimentation might be expected to provide additional insights into the issue of resource utilization in these lizard species.

If our original interest had been only on differences in resource use by the two species, we could examine a model with all five main effects (H, D, I, T, S) vs. a similar model without a species effect (H, D, I, T). Here, there might have been only 2 candidate models considered. The difference Δ_i for these two models is 163.70, indicating strongly that the two species were using their resources differently. Alternatively, we could make a similar comparison of two models, but also include the relevant second-order interactions

H, D, I, T, S, H*D, H*I, H*T, H*S, D*I, D*T, D*S, I*T, I*S, T*S

vs.

H, D, I, T, H*D, H*I, H*T, D*I, D*T, I*T.

Here, the difference Δ_i is 84.63 and again clearly indicates that the two species are utilizing their habitat differently. Other alternative analysis might be pursued if more were known about the study design and field protocol.

Bishop et al. (1975) performed a number of hypothesis tests to build a model, and this resulted in a model with all the main effects plus the second-order interaction terms H∗D, H∗S, D∗S, T∗S and the third-order term H∗D∗S. If this study is considered to be merely exploratory, then one might consider a much wider class of candidate models, and many models with "small" Δ_i could be found. However, the number of possible models would be very large (perhaps 2,000, depending on what rules might be applied concerning the presence of lower-order effects if higher-order effects are included in the model). Even with powerful computing equipment, an exhaustive study of all models for

TABLE 3.12. Summary of a priori models of the lizard data (from Schoener 1970); log-likelihood; number of parameters (K), AIC_c, differences (Δ_i), and Akaike weights (w_i). The model with the minimum AIC_c is shown in bold.

Model	$\log(\mathcal{L})$	K	AIC_c	Δ_i	w_i
1 All main effects, H D I T S	**1,181.08**	**7**	**−2,347.95**	**0**	**1.00**
2 All main effects and					
second-order interactions	1,181.86	21	−2,319.96	27.99	0.00
3 Base[1] but drop DT	1,180.52	19	−2,321.59	26.35	0.00
4 Base[1] plus HDI, HDT					
and HDS terms	1,182.97	25	−2,313.44	34.51	0.00
5 Base[1] plus HDI, HDS, HIT,					
HIS, HTS, and ITS	1185.75	30	−2,307.89	40.06	0.00
6 Base[1] plus HIT, HIS, HTS,					
and ITS	1185.48	28	−2,311.82	36.13	0.00
7 Base[1] plus HIS, HTS,					
and ITS	1184.01	26	−2,313.32	34.64	0.00
8 Base[1] plus HIT, HIS,					
HTS, and ITS, but drop DT	1184.15	26	−2,313.60	34.36	0.00
9 Base[1] plus HIT, HIS, and ITS,					
but drop DT	1183.29	24	−2,316.28	31.67	0.00
10 Base[1] plus HIT and HIS,					
but drop DT	1182.18	22	−2,318.42	29.52	0.00

[1] "Base" is a model with the five main effects plus all second-order interaction terms.

this $2 \times 2 \times 2 \times 3 \times 2$ table is nearly prohibitive (see Agresti 1990:215). This illustrates again the importance of a set of good a priori models, even if the study is somewhat exploratory. Of course, data dredging could be very effective in finding a model that "fits" *these* data. However, the goals in data analysis usually stress an inference about the population or process, not merely data description. Thus, the results from intensive data dredging should be viewed as tenuous.

The rigorous analysis of multidimensional contingency tables remains problematic because of the large number of possible models. Gokhale and Kullback (1978:19) suggest that conclusions drawn from contingency tables should be only exploratory. While a good a priori set of models seems essential, it may be difficult to forecast higher-order interactions in many situations. Several other problematic issues are associated with contingency table analysis. Other analytic approaches exist (Manly et al. 1993), and alternative model formulations besides a log link and Poisson errors can be considered (see Santer and Duffy 1989 for additional information).

Solomon Kullback was born in 1907 in Brooklyn, New York. He graduated from the City College of New York in 1927, received an M.A. degree in mathematics in 1929, and completed a Ph.D. in mathematics at the George Washington University in 1934. Kully, as he was known to all who knew him, had two major careers: one in the Defense Department (1930–1962) and the other in the Department of Statistics at George Washington University (1962–1972). He was chairman of the Statistics Department from 1964–1972. Much of his professional life was spent in the National Security Agency, and most of his work during this time is still classified. Clearly, most of his studies on information theory were done during this time. Many of his results up to 1958 were published in his 1959 book *Information Theory and Statistics*. Additional details on Kullback may be found in Greenhouse (1994) and Anonymous (1997).

3.7 Example 6: Sakamoto et al.'s (1986) Simulated Data

Here we return briefly to the example used in Section 1.4.2 from Sakamoto et al.ś (1986) book. Ten data sets (each with $n = 21$) were generated from the simple model

$$\mathbf{y} = \mathbf{e}^{(x-0.3)^2} - 1 + \epsilon.$$

Sakamoto et al. used the simple polynomials from order 0 to 5 as the set of candidate models to visually illustrate the concepts of under- and overfitting (see Figure 1.4). Because the sample size ($n = 21$) is small in relation to the dimension of the largest model in the set ($K = 7$), AIC_c should be used for the analysis of data in this example. AIC_c was computed for each of the 10 data sets and then averaged for each of the 7 models. Then Δ_i and w_i values were derived from these averages:

Model	K	AIC_c	Δ_i	w_i
Mean	2	− 6.58	25.93	0.00
Linear	3	−18.94	13.57	0.00
Quadratic	4	−29.63	2.88	0.18
Cubic	5	−26.80	5.71	0.04
4th-order	6	−23.79	8.72	0.01
5th-order	7	−19.26	13.25	0.00
$\min f(x) = 0.3$	**2**	**−32.51**	**0.0**	**0.77.**

Clearly, the model based on the additional (hopefully a priori) information that $f(x)$ is 0 at $x = 0.3$ is the best of the set (compare these quantitative results with the plots in Figure 1.4). The Akaike weights (w_i) more clearly sharpen the inference and suggest that only the quadratic model (with $K = 4$) is a competitor to the special model for these simulated data. The evidence ratio between the best and second-best models is $0.77/0.18 = 4.3$, whereas this ratio between the best and third-best model is only $0.77/0.04 = 19.2$. These comparisons are in line with the visual images in Figure 1.4 and help to reinforce understanding of the information-theoretic quantities in a simple example. More complex data and models defy simple plots and a visual analysis; thus K-L information and various information criteria become essential.

3.8   Example 7: Models of Fish Growth

Shono (2000) presented a reanalysis of data on the growth of female masu salmon (*Oncorhynchus masou*) from Kiso et al. (1992). Shono (2000) presented a comparison of model selection under AIC, AIC_c, and BIC (a Bayesian criterion, Schwarz 1978); we will focus on model selection and extend the inferences made by using Δ_i, w_i, and evidence ratios.

   Kiso et al. (1992) estimated the parameters of three standard growth curves using length and age data over a period of 2 to 19 months; here we provide a brief review of the information; further details can be found in the original paper and, particularly, in Shono (2000). The sample size was not given explicitly, but was large relative to the number of parameters in the highest-dimensioned model (where $K = 5$). The models for length $L(t)$ as a function of time, in months, were

von Bertalanffy $L(t) = L_\infty[1 - \exp\{-\kappa(t - t_0,)\}]$,

Gompertz $L(t) = L_\infty \exp[-\exp\{-\kappa(t - t_0)\}]$,

Logistic $L(t) = \dfrac{L_\infty}{1 - \exp\{-\kappa(t - t_0)\}}$.

The basic model parameters are L_∞ and κ. Each model contains time (t) in the exponent, and this was itself modeled in four ways to reflect hypotheses concerning the seasonal pattern in growth. These submodels of time were

Basic $F(t) = t,$

Type 1 $F(t) = t + \dfrac{\theta_1}{2\pi} \sin 2\pi(t - t_1),$

Type 2 $F(t) = t + \dfrac{\theta_1}{2\pi} \sin 2\pi(t - t_1) + \dfrac{\theta_2}{4\pi} \sin 4\pi(t - t_1),$

Type 3 $F(t) = t + \dfrac{\theta_1}{2\pi} \sin 2\pi(t - t_1) + \dfrac{\theta_3}{6\pi} \sin 6\pi(t - t_1).$

The unknown parameters in these submodels are θ_1, θ_2, and θ_3, where in each case, $\theta_i \geq 0$. The θ_i are the amplitudes of sine curves with periods of one, two, and three cycles/year, respectively, and t_1 is the starting point of these sine curves. They define Type 1 as a modified type where the growth rate changes once per year. Type 2 allows growth rate changes in a combination of 1 and 2 cycles per year, whereas Type 3 allows growth rate changes in a combination of 1 and 3 cycles per year (Kiso et al. 1992:1780).

Each of the three growth models included four submodels of seasonality, giving a total of 12 models. The results, taken partially from Shono's Table 2, are given in Table 3.13. As shown by Shono (2000), the best model is the von Bertalanffy with the Type 1 seasonal effect. However, by examining the Δ_i values, we note that the Gompertz model with Type 1 seasonality is essentially tied ($\Delta = 0.72$); the evidence ratio is $0.27/0.19 = 1.42$. Actually, the Gompertz Type 3 ($\Delta = 1.13$) and the Gompertz Type 2 ($\Delta = 1.52$) are also close competitors. Three of the best four models involve the Gompertz form. Even the worst of the top four models is still quite good; its evidence ratio with the best model is $0.27/0.13 = 2.1$. Clearly, the logistic is a poor model, relative to the other types. In fact, all nine models with a nonconstant seasonal pattern have Δ_i values < 6, suggesting that some seasonality is very important in salmon growth. Beyond this, there is considerable model selection uncertainty. This is a clear case where inference based only on the selected best model is risky. Inference, including prediction, should probably be based on all twelve models or, at least, the nine models allowing seasonality in growth. In addition, estimates of precision should allow for the high uncertainty in model selection. These are subjects treated in Chapter 4.

3.9 Summary

The purpose of the analysis of empirical data is not to find the "true model"— not at all. Instead, we wish to find a best approximating model, based on the data, and then develop statistical inferences from this model. In some sense, the model is *the* inference from the available data. We search, then, not for a "true model," but rather for a *parsimonious model* giving an accurate approximation to the interpretable information in the data at hand. Data analysis involves the question, "What level of model complexity will the data support?" and both under- and overfitting are to be avoided. Larger data sets tend to support more

TABLE 3.13. Results of AIC_c-based model selection using data for mean length of masu salmon. AIC_c differences (Δ_i) and Akaike weights (w_i) are also shown (from Kiso et al. 1992).

Formulae	Type	K	AIC_c	Δ_i	w_i
von Bertalanffy	Basic	2	135.82	117.26	0.00
Gompertz	Basic	2	50.56	32.00	0.00
Logistic	Basic	2	54.05	35.49	0.00
von Bertalanffy	1	4	18.56	0.00	0.27
Gompertz	1	4	19.28	0.72	0.19
Logistic	1	4	21.02	2.46	0.08
von Bertalanffy	2	5	21.03	2.47	0.08
Gompertz	2	5	20.08	1.52	0.13
Logistic	2	5	24.52	5.96	0.01
von Bertalanffy	3	5	21.30	2.74	0.07
Gompertz	3	5	19.69	1.13	0.15
Logistic	3	5	24.25	5.69	0.02

complex models, and the selection of the size of the model represents a tradeoff between bias and variance.

The analysis of data under the information-theoretic approaches is relatively simple. That is, the computational aspects are simple, and the results are easy to understand and interpret if one has the value of the maximized log-likelihood ($\log(\mathcal{L})$) or the residual sum of squares (RSS) for each model in the set. These quantities are routinely printed by nearly any commercial data analysis software. Computation of AIC, AIC_c, or $QAIC_c$ from either of these values is simple to the point that it can easily be done by hand. Similarly, computation of the differences (Δ_i), Akaike weights (w_i), and evidence ratios is nearly trivial. The general approach is flexible enough to be used in a very wide variety of practical situations in the life sciences. These are all approaches that hark back to Kullback–Leibler information and have a deep theoretical basis. The easy part of the information-theoretic approaches includes both the computational aspects and the clear understanding of these results (the nature of the evidence).

The hard part, and the one where training has been so poor, is the a priori thinking about the science of the matter before data analysis—even before data collection. It has been too easy to collect data on a large number of variables in the hope that a fast computer and sophisticated software will sort out the important things—the "significant" ones (the "just the numbers" approach). Instead, a major effort should be mounted to understand the nature of the problem by critical examination of the literature, talking with others working on the general problem, and thinking deeply about alternative hypotheses. Rather than "test" dozens of trivial matters (is the correlation zero? is the effect of the lead treatment zero? are ravens pink?, Anderson et al. 2000), there must be a more concerted effort to provide evidence on *meaningful* questions that are

Richard Arthur Leibler was born in Chicago, Illinois, on March 18, 1914. He received bachelor's and master's degrees in mathematics from Northwestern University and a Ph.D. in mathematics from the University of Illinois and Purdue University (1939). After serving in the Navy during WWII, he was a member of the Institute for Advanced Study and a member of the von Neumann Computer Project 1946–1948. From 1948–1980 he worked for the National Security Agency (1948–1958 and 1977–1980) and the Communications Research Division of the Institute for Defense Analysis (1958-1977). He is the president of Data Handling Inc., a consulting firm for the Intelligence Community (1980–present).

to a discipline. This is the critical point: the common failure to address important science questions in a fully competent fashion. Thinking, reconsideration, synthesizing, and challenging the "known" lead to fresh hypotheses and mathematical models to carefully reflect these alternative hypotheses. We believe that science is likely to advance more quickly if the "hard part" of the information-theoretic approaches are given much more weight. We suspect that the Bayesians might "second" the cry for more a priori thinking before formal data analysis begins. A rereading of Platt (1964) and delving into Ford (2000) might often be good starting points in better understanding this philosophy of science.

Data from the simulated starling experiment and the banded sage grouse population both illustrate moderate complexity. In both examples, inference based on the global model would have been (needlessly) poor, inefficient, and difficult to interpret. The global model for the sage grouse data had 58 parameters, and the resulting MLEs had wide confidence intervals (Table 3.10) and dozens of substantial sampling correlations (e.g., $\widehat{\text{corr}}(\hat{S}_i, \hat{S}_{i+1})$ and $\widehat{\text{corr}}(\hat{S}_i, \hat{r}_i)$; see Brownie et al. 1985), making it difficult to examine patterns, trends, and associations (e.g., are age-specific survival probabilities linked across years? are there declines in survival probabilities over years?).

The compartment models of DURSBAN® in a simple ecosystem were sets of first-order differential equations, and they represent a higher degree of complexity. The real data on cement hardening and fish growth represent simple examples but provide insights into the interpretation of evidence. The extension of the analysis of the simulated data from Sakamoto et al. (1986) allows some comparisons with the graphical material in Section 1.4.2. The example of resource partitioning in *anolis* lizards must be considered only exploratory. Burnham et al. (1996) give a comprehensive example dealing with declining survival probabilities of the Northern Spotted Owl (*Strix occidentalis carina*), caused by widespread clear-cutting in national forests. This analysis involved a large data set over eleven geographic areas in a politically charged issue of national importance.

Researchers often attempt to perform some further, separate, analysis (e.g., multiple linear or logistic regression) of the estimates from a very general model in an effort to understand the structure of the process and gain insight into its behavior. However, such external analyses are not easily done correctly; the estimates for the starling and grouse data have a multinomial, not normal, variance structure; the variances are unequal (not constant); and successive estimates are dependent (not independent). When the analyst feels a need for such further "external" analysis of the parameter estimates of a fitted model, it is clear that a properly parsimonious, easily interpretable model has not been achieved, and hence the analysis has partially failed.

We recommend carefully developing a set of candidate models to explore the science of the issue (e.g., embed additive submodels for recovery probabilities into the log-likelihood using an appropriate link function) and obtain the MLEs under these models. Then one can focus on model selection to identify a properly parsimonious model(s) (the models in the set of candidates that are "close" to truth in the K-L information sense) that will serve as a basis for inference. In contrast, if only the high-dimensional global model is employed and estimates of parameters obtained by ML or LS, then the purpose of the analysis is virtually defeated, because a parsimonious interpretation of results may be impossible, patterns often cannot be found, and estimates are very imprecise. Zablan (1993) used a global model with 58 parameters in her analysis of the sage grouse data and correctly observed, "... survival estimates had unacceptably wide confidence intervals."

Statistical analysis of empirical data should not be just number crunching, given only a set of data (the numbers). The cement hardening data (Section 3.2) have too frequently been analyzed without examining the collection and treatment of the data: the important a priori considerations that we have stressed here. What was known about the chemistry of cement hardening (if not in the 1930s, then at least by the 1960s)? For example, can cement be expected to harden well with only a single ingredient? If not, this might have put the four single-variable models out of consideration. Given the chemical composition of variables x_2 and x_4 (3CaO · SiO$_2$ vs. 2CaO · SiO$_2$), is it surprising that these variables are highly correlated ($r = -0.973$)? The situation involving variables x_1 and x_3 is similar. Surely, these conditions were known a priori and could have affected the models in the set to be considered. How many analysts have known, or bothered to find out, that $\sum_{j=1}^{4} x_{ij}$ is a constant for each of the thirteen observations and therefore excluded the four-variable model from consideration, based on these a priori grounds? Instead, unthinking approaches have been the modus operandi, and "all possible models" have frequently been tried. "Let the computer find out" is a poor strategy for researchers who do not bother to think clearly about the problem of interest and its scientific setting. *The sterile analysis of "just the numbers" will continue to be a poor strategy for progress in the sciences.*

Researchers often resort to using a computer program that will examine all possible models and variables automatically. Here, the hope is that the computer will discover the important variables and relationships (a "just the numbers approach" void of any thinking or science). Cook et al. (2001:977) conducted stepwise linear regression analyses using AIC and Mallows's C_p (SAS Institute 1988:786) in a study of elk (*Cervus canadensis*) condition. They found that this approach "... provided results that often were biologically unrealistic, unstable due to multicolinearity, and overparameterized (≥ 5 variables)." The literature is full of such failed studies; just because AIC was used as a selection criterion does not mean that valid inference can be expected. The primary mistake here is a common one: the failure to posit a small set of a priori models, each representing a *plausible* research hypothesis.

The presentation of results in scientific publications should detail the logic used in arriving at a set of candidate models. The model set should have strong ties to study design and the alternative research hypotheses of interest. Presentation and discussion of the $\log(\mathcal{L})$ values, K, the appropriate information criterion for each model, Δ_i, and w_i is recommended (see Anderson et al. 2001d). Evidence ratios should be presented with other relevant values to allow a comprehensive assessment of the alternative hypotheses. Such information allows the merits of each model to be contrasted. If some exploratory data dredging was done following the formal analysis, this activity should be clearly noted and the tentative insights from these activities provided. We do

---

All Possible Models

Unthinking approaches have been the common modus operandi and using "all possible models" are frequently seen in the literature. "Let the computer find out" is a poor strategy and usually reflects the fact that the researcher did not bother to think clearly about the problem of interest and its scientific setting.

The sterile analysis of "just the numbers" will continue to be a poor strategy for progress in the sciences.

not encourage the use of the word "significant" in publication of scientific results, since this word is so tied to statistical null hypothesis testing, the arbitrary
α level, and the resulting problematic P-values and misinterpretations.

4

Formal Inference From More Than One Model: Multimodel Inference (MMI)

4.1   Introduction to Multimodel Inference

Model selection is most often thought of as a way to select just the best model, then inference is conditional on that model. However, information-theoretic approaches are more general than this simplistic concept of model selection. Given a set of models, specified independently of the sample data, we can make formal inferences based on the entire set of models. Here, the conditioning is on all the models in the set and this has several advantages; however, it does reinforce the importance of having a good set of models to carefully represent the scientific hypotheses of interest. Part of multimodel inference includes ranking the fitted models from best to worst, based on the Δ_i values, and then scaling to obtain the relative plausibility of each fitted model (g_i) by a weight of evidence (w_i) relative to the selected best model. Using the conditional sampling variance ($\mathrm{var}(\hat{\theta}|x, g_i)$) from each model and the Akaike weights (w_i), unconditional inferences about precision can be made over the entire set of models. Model-averaged parameter estimates and estimates of unconditional sampling variances can be easily computed. Model selection uncertainty is a substantial subject in its own right, well beyond just the issue of determining the best model.

By *unconditional*, we mean not conditional on any particular model; however, inference is still conditional on the full set of models. This is the reason for spending the time to arrive at a good set of models, based on what is known or hypothesized about the science underlying the study. Ideally, this set of models should be small and well justified, at least for confirmatory studies.

We address four main issues in this chapter. First, how can parameter estimates be made using all the models in the set, and what are the advantages in so doing? Second, how can model selection uncertainty be quantified and incorporated into estimates of precision? Third, how can the relative importance of predictor variables in analyses such as linear or logistic regression be assessed? Finally, how can a confidence set of models be established for the K-L best model? More research is required to develop and understand general methods for these issues, but we provide several approaches that are useful. Four examples are provided to illustrate the use of multiple models in making formal inference.

4.2   Model Averaging

4.2.1   Prediction

Consider model-based inference for prediction, where R models are considered, each having the parameter θ as the predicted value of interest. Each model i allows an estimate of the parameter, θ_i. If one of the models was clearly the K-L best (e.g., if its $w \geq 0.90$), then inference could probably be made, conditionally, on the selected best model. However, it is often the case that no single model is clearly superior to some of the others in the set. If the predicted value ($\hat{\theta}$) differs markedly across the models (i.e., the $\hat{\theta}_i$ differ across the models $i = 1, 2, \ldots, R$), then it is risky to base prediction on only the selected model. An obvious possibility is to compute a weighted estimate of the predicted value, weighting the predictions by the Akaike weights (w_i).

Model Averaging

This concept leads to the model averaged estimates,

$$\hat{\bar{\theta}} = \sum_{i=1}^{R} w_i \hat{\theta}_i,   \text{(4.1)}$$

where $\hat{\bar{\theta}}$ denotes a model averaged estimate of θ. Alternatively, if the bootstrap is used to provide the estimated model selection frequencies ($\hat{\pi}_i$), model averaging can be done using,

$$\hat{\bar{\theta}} = \sum_{i=1}^{R} \hat{\pi}_i \hat{\theta}_i.   \text{(4.2)}$$

This type of model averaging is useful for prediction problems or in cases where a particular parameter (e.g., γ an immigration probability) occurs in all the models in the set. Prediction is an ideal way to view model averaging, because each model in a set, regardless of its parametrization, can be used to make a predicted value.

Hirotugu Akaike was born in 1927 in Fujinomiya-shi, Shizuoka-jen, in Japan. He received B.S. and D.S. degrees in mathematics from the University of Tokyo in 1952 and 1961, respectively. He worked at the Institute of Statistical Mathematics for over 30 years, becoming its Director General in 1982. He has received many awards, prizes, and honors for his work in theoretical and applied statistics (deLeeuw 1992, Parzen 1994). The three-volume set, *"Proceedings of the First US/Japan Conference on the Frontiers of Statistical Modeling: An Informational Approach* (Bozdogan 1994) commemorated Professor Hirotugu Akaike's 65th birthday. Bozdogan (1994) records that the idea of a connection between the Kullback–Leibler discrepancy and the empirical log-likelihood function occurred to Akaike on the morning of March 16, 1971, as he was taking a seat on a commuter train.

4.2.2 Averaging Across Model Parameters

If one has a large number of closely related models, such as in linear-regression based variable selection (e.g., all subsets selection), designation of a single best model is unsatisfactory because that "best" model is often highly variable. That is, the model estimated to be best would vary from data set to data set, where replicate data sets would be collected under the same underlying process. In this situation, model averaging provides a relatively much more stabilized inference.

The concept of inference being tied to all the models can be used to reduce model selection bias effects on linear regression coefficient estimates in all subsets selection. For the linear regression coefficient β_j associated with predictor variable x_j there are two versions of model averaging. First, we have the estimate $\hat{\bar{\beta}}_j$ where β_j is averaged over all models in which x_j appears (i.e., when j is not zero):

$$\hat{\bar{\beta}}_j = \frac{\sum_{i=1}^{R} w_i I_j(g_i) \hat{\beta}_{j,i}}{w_+(j)},$$

$$w_+(j) = \sum_{i=1}^{R} w_i I_j(g_i),$$

and

$$I_j(g_i) = \begin{cases} 1 & \text{if predictor } x_j \text{ is in model } g_i, \\ 0 & \text{otherwise.} \end{cases}$$

Here, $\hat{\beta}_{j,i}$ denotes the estimator of β_j based on model g_i. The notation $w_+(j)$ is merely the sum of the Akaike weights over all models in the set where predictor variable j is explicitly in the model. Note, $w_+(j)$ is itself a model-average value about whether variable x_j is in (or not in) a particular model. Thus, $\hat{\bar{\beta}}_j$ is a "natural" average to consider, as it only averages $\hat{\beta}_j$ over models where an unknown β_j parameter appears. Note, however, that the estimator $\hat{\bar{\beta}}_j$ ignores evidence about models g_i wherein $\beta_{j,i} \equiv 0$.

An alternative way to average over linear regression models is to consider that variable x_j is "in" every model, it is just that in some models the corresponding β_j is set to zero, rather than being considered unknown. Conditional on model g_i being selected, model selection has the effect of biasing $\hat{\beta}_{j,i}$ away from zero (Section 1.6). Thus, a second model-averaged estimator, denoted $\tilde{\beta}_j$, is suggested:

$$\tilde{\beta}_j = w_+(j) \hat{\bar{\beta}}_j.$$

This $\tilde{\beta}_j$ actually derives from model averaging over all R models. In cases where x_j is not in a particular model, it is because $\beta_{j,i} \equiv 0$ is used instead of the estimate $\hat{\beta}_{j,i}$. The resultant average is identical to $w_+(j) \hat{\bar{\theta}}_j$. Heuristically, $w_+(j)$ serves to *shrink* the conditional $\hat{\bar{\theta}}_j$ back towards zero, and this shrinkage serves to ameliorate much of the model selection bias of $\hat{\bar{\theta}}_j$ (Section 1.6). Investigation of this general idea and its extensions are an open research area.

One point here is that while $\hat{\bar{\beta}}_j$ can be computed ignoring models other than those where x_j appears, $\tilde{\beta}_i$ does require fitting all R of the a priori models. Improved inference requires fitting all the a priori models and then using a type of model averaging. When possible, one should use inference based on all the models, via model averaging and selection bias adjustments, rather than a "select the best model and ignore the others" strategy.

There are several advantages, both practical and philosophical, to model-averaging, when it is appropriate. Where a model averaged estimator can be used it often has reduced bias and, sometimes has better precision, compared to

$\hat{\theta}$ from the selected best model. Hoeting et al. (1999) provides an introduction to model averaging from a Bayesian viewpoint (also see Leamer 1978 for motivating ideas). Bayesian model averaging is easy to understand, but can be difficult to implement in practice. Information-theoretic methods for model averaging are easy both to understand and implement, even when there is a large number of models, each with potentially many parameters.

While there are many cases where model averaging is useful, we warn against model averaging structural parameter estimates in some types of nonlinear models. While it is often appropriate to average slope parameters in linear regression models, structural parameters in nonlinear models such as

$$E(y) = (a + b_x)/(1 + cx) \quad \text{or} \quad E(y) = a(1 - [1 + (x/c)^d]^{-b})$$

should not be averaged. For example, a weighted average across these two models of any of the parameters $a, b, c,$ or d would not be appropriate. Instead, model averaging the predicted expected response variable $\hat{E}(y)$, for a given value of x, across models, is advantageous in reaching a robust inference that is not conditional on only a single model.

It is important to realize that the expected value of the model-averaged estimate, $E(\hat{\bar{\theta}})$, is not necessarily the same as θ from absolute truth. Under classical sampling theory the estimator $\hat{\theta}$ ($\equiv \hat{\theta}_i$ for the selected model g_i which varies by sample), arrived at in the two-stage process of model selection followed by parameter estimation given the model, is by definition an unbiased estimator of $E(\hat{\bar{\theta}})$ as given by (4.1 or 4.2). Therefore, the unconditional sampling variance of $\hat{\theta} \equiv \hat{\bar{\theta}}$ is to be computed with respect to $E(\hat{\bar{\theta}})$. Any remaining bias, $E(\hat{\bar{\theta}}) - \theta$, in $\hat{\bar{\theta}}$ cannot be measured or allowed for in model selection uncertainty. However, part of the intent of having a good set of models and sound model selection is to render this bias negligible with respect to the unconditional $se(\hat{\theta})$.

Model-averaging ideas are well developed from the Bayesian perspective (see Madigan and Raftery 1994, Draper 1995, Raftery 1996a and (particularly) Hoeting et al. 1999; Newman 1997 provides an application). Model averaging has not yet been commonly adapted into applied frequentist inferences. Some theoretical basis for these approaches and ideas appears in Chapter 6 (also see Buckland et al. 1997 and the Bayesian references just above).

4.3 Model Selection Uncertainty

An understanding of statistical inference requires that one consider the process that generates the sample data we observe. For a given field, laboratory, or computer simulation study, data are observed on some process or system. If a second, independent data set could be observed on the same process or system under nearly identical conditions, the new data set would differ somewhat from the first. Clearly both data sets would contain information about the process, but the information would likely be slightly different, by chance. An obvious goal of data analysis is to make an inference about the process based on the data observed. However, inferences must not be overly specific with respect to the (single) data set observed. That is, we would like our inferences to be robust, with respect to the particular data set observed, in such a way that we tend to avoid problems associated with both underfitting and overfitting (overinterpreting) the limited data we have. Thus, we would like some ability to make inferences about the process as if a large number of other data sets were also available.

With only a single data set, one could use AIC, and select the best model for inference. However, if several other independent data sets were available, would the same model be selected? The answer is that perhaps it would be; but generally, there would be variation in the selected model from data set to data set, just as there would be variation in parameter estimates over data sets, given that the same model is used for analysis. The fact that other data sets might suggest the use of other models leads us to model selection uncertainty and hence another variance component that should be included in measures of precision of parameter estimates (Section 1.7).

If an analyst selects a model using AIC_c (or using some other procedure such as cross-validation) and makes estimates of the sampling variance of an estimated parameter in that model, he invariably does so conditional on the selected model. The estimated precision will then likely be overestimated, because the variance component due to model selection uncertainty has been omitted. The standard errors computed conditional on the model will be too small, confidence intervals will be too narrow, and achieved coverage will be below the nominal level. Chatfield (1995b) reviews this issue in some detail; also see Rencher and Pun (1980), Chow (1981), Hurvich and Tsai (1990), Pötscher (1991), Goutis and Casella (1995), and Kabaila (1995).

This section presents a variety of methods that can be used to (1) measure the uncertainty associated with model selection, either what is the actual best model, or regarding uncertainty about selected variables; and (2) provide measures of unconditional precision (e.g., sampling variance, standard errors, and confidence intervals) for parameter estimators, rather than just using the usual measures of sampling uncertainty conditional on a selected model. Additional research is encouraged to better understand the properties and limitations of these approaches.

There are three general approaches to assessing model selection uncertainty: (1) theoretical studies, mostly using Monte Carlo simulation methods; (2) the bootstrap applied to a given set of data; and (3) utilizing the set of AIC differences (i.e., Δ_i) and model weights w_i from the set of models fit to data. Useful insights can be obtained about model selection and associated uncertainties by extensive Monte Carlo simulations of model selection (e.g., McQuarrie and Tsai 1998). Use of the bootstrap and $\hat{\pi}_i$ values applies directly to a single data set; hence they represent our focus here. The bootstrap may require 10,000 samples for reliable results, and it could take many hours of computer time to apply the bootstrap to complex data-analysis cases. In contrast, the third method (i.e., use of w_i values) is easily computable and merits more development and understanding.

Monte Carlo investigations generate 1,000 to 10,000 independent data sets (sometimes, 100,000 or even a million samples are needed) from a stated generating model. These data sets are then analyzed, the log-likelihoods maximized to obtain the MLEs, and model selection is done to identify the best model for each sample. Finally, one can summarize resultant relative frequencies of models selected and other information of interest, such as variation of the Δ_i as well as conditional and unconditional variances of parameter estimators over models. For results to apply fully to the K-L model selection paradigm envisioned for real data, the generating model (which is truth, f, in the simulation study) should be complex and not contained in the set of approximating models, $g_1, \ldots, g_R$ (i.e., g_i or $g_i(x \mid \theta)$). We present some interesting examples of Monte Carlo simulation results in Chapters 5 and 6.

Many, if not most, simulation studies on model selection do not at all meet these conditions (e.g., Wang et al. 1996). Rather, they are far too simplistic because (1) a simple generating model is used (so no tapering effects and only small K), (2) the set of models considered contains the generating model (i.e., contains "truth"), and (3) the model selection goal is usually to select the generating model (hence to select "truth"). None of these features are realistic of real data-analysis problems; hence we discount the results of such simulation studies as appropriate guides to real-world model selection issues (see Chapter 6 for more details on this common error).

The fundamental idea of the model-based sampling theory approach to statistical inference is that the data arise as a sample from some conceptual probability distribution, f, and hence the uncertainties of our inferences can be measured if we can estimate f. There are ways to construct a nonparametric estimator of (in essence) f from the sample data. The fundamental idea of the bootstrap method (Section 2.13) is that we compute measures of our inference uncertainty from that estimated sampling distribution of f.

4.3.1   Concepts of Parameter Estimation and Model Selection Uncertainty

Statistical science should emphasize estimation of parameters and associated measures of estimator uncertainty. Given a correct model (most theory assumes $g = f$), an MLE is reliable, and we can compute a reliable estimate of its sampling variance and a reliable confidence interval (such as a profile likelihood interval; see Royall 1997). If the model is selected entirely independently of the data at hand, and is a good approximating model, and if n is large, then the estimated sampling variance is essentially unbiased, and any appropriate confidence interval will essentially achieve its nominal coverage. This would be the case if we used only one model, decided on a priori, and it was a good model, g, of the data generated under truth, f. However, even when we do objective, data-based model selection (which we are advocating here), the selection process is expected to introduce an added component of sampling uncertainty into any estimated parameter; hence classical theoretical sampling variances are too small: They are conditional on the model and do not reflect model selection uncertainty. One result is that conditional confidence intervals can be expected to have less than nominal coverage.

Consider a scalar parameter θ, which may be used in all or only some of the models considered, but is in the selected model, and therein has unknown value θ_i given model g_i. Here, the subscript i denotes the model used to estimate θ, with the understanding that this parameter means the same thing in all models in which it appears. There is a conceptual true value of θ for the given study. However, the value of θ that we would infer, in the sense of $E_f(\hat{\theta}_i \mid g_i) = \theta_i$ (for large sample size) from model g_i applied to the data, may vary somewhat by model. Given model g_i, the MLE, $\hat{\theta}_i$, has a conditional sampling distribution, and hence a conditional sampling variance $\text{var}(\hat{\theta}_i \mid g_i)$. We mean the notation $\text{var}(\hat{\theta}_i \mid g_i)$ to be functionally and numerically identical to $\text{var}(\hat{\theta}_i \mid \theta_i)$. The latter notation is more traditional; we use the former notation when we want to emphasize the importance of assuming the model in its totality when in fact other models are also being considered.

There is a concept of the true value of θ: It is the value of θ we would compute based on knowing truth, f, even though θ need not literally appear in f. To the extent a model, g_i, is wrong (i.e., $g_i \neq f$), θ_i may not equal θ when this K-L best value is determined for θ under assumed model g_i. That is, even when data are generated by f, if those data are interpreted under model g_i, we will infer (for large n) that the value of θ is θ_i. For a good model this possible "bias" (i.e., $\theta_i - \theta$) is not of great concern, because it will be dominated by the conditional sampling standard error of $\hat{\theta}_i$ (in essence, this domination is one feature of a "good" model). The bias $\theta_i - \theta$ induces one source of model selection uncertainty into $\hat{\theta}$; that is, this bias varies over models in an unknown manner. In many situations the model, as such, means something to us, and we will then take $\hat{\theta}_i$ derived only from the selected model, g_i, as the most meaningful estimator of θ. This is what has been commonly done, and seems sensible, so much so that the alternative of model averaging seems at first strange, but is an alternative to getting an estimator of θ based on multiple models. Model averaging arises in a natural way when we consider the unconditional sampling variance of $\hat{\theta}_i$.

Another problem arising from model selection uncertainty concerns cases where the estimate of sampling variance is derived from the residuals from the fitted model (e.g., multiple linear regression). In this case, overfitting produces a negatively biased estimator of conditional sampling variance, $\text{var}(\hat{\beta}_i \mid g_i)$. Two related concepts are required to understand this issue. First, when the model structure is Poisson, binomial, multinomial (includes contingency-table-based models), or negative binominal models, there is a known theoretical sampling variance, and this can be inferred from the estimates of the model parameters (e.g., for the binomial, the model parameter is p and the theoretical sampling variance is $\text{var}(\hat{p}) = \frac{p(1-p)}{n}$).

Underestimation of sampling variance due to structural overfitting does not seem to be a serious problem in cases where such a theoretical sampling variance is known. In particular, this is true if we also use a variance inflation factor, $\hat{c}$, applied for all models in the set of R models (so selection is not an issue as regards $\hat{c}$) to adjust for any modest structural lack of fit of the global model.

The second situation occurs often in regression models where the residual sampling variance, σ^2, is functionally unrelated to the model structure and there is no true replication. Then σ^2 must be estimated only from residuals to the fitted model. In this case there is neither true replication nor a theoretical basis to infer σ^2, such as there is in models for count data. If we overfit the structural component of the model to the data, we will get $\hat{\sigma}^2$ biased low, and hence estimated sampling standard errors of any $\hat{\theta}_i$ will be biased low (there is likely to be a compensating increase in the factor $(\hat{\theta}_i - \theta)^2$). True replication at the level of the regressor values can eliminate this problem, but often we do not have such true replication.

The ideas of classical sampling theory can be used to derive the theoretical sampling variance of $\hat{\theta}$ resulting from the two-stage process of (1) model selection, then (2) using $\hat{\theta} \equiv \hat{\theta}_i$ given that model g_i was selected. Imagine this process carried out many times, m, each time on an independent sample. For sample j we get $\hat{\theta}_j$ as our estimator of θ. This conceptual $\hat{\theta}_j$ comes from the selected model in repetition j, but we do not need, hence avoid using, a doubly indexed notation (such as $\hat{\theta}_{i,j}$) to denote both sample j and selected model i given sample j.

The estimated unconditional sampling variance, $\widehat{\text{var}}(\hat{\theta})$, from these m replicates would be $\sum (\hat{\theta}_j - \bar{\theta})^2/(m-1)$; $\bar{\theta}$ is the simple average of all m estimates (hence the $\hat{\theta}_j$ have been averaged over selected models). This variance estimator represents the total variation in the set of m values of $\hat{\theta}$; hence both within and between-model variation is included. This set of m values of $\hat{\theta}$ can be partitioned into R subsets, one for each model wherein the ith subset contains all the $\hat{\theta}$'s computed under selected model i. Then one can compute from the ith subset of the $\hat{\theta}$ values an estimate of the conditional sampling variance of $\hat{\theta}$ when model g_i was selected. Formal mathematics along this line of partitioning the above $\widehat{\text{var}}(\hat{\theta})$ into R components and taking expectations to get a theoretical unconditional sampling variance gives the result for $\text{var}(\hat{\theta})$ as a weighted combination of conditional variances, $\text{var}(\hat{\theta}_i \mid g_i)$ plus a term for variation among $\theta_1, \ldots, \theta_R$. The weights involved are the model selection probabilities. Relevant formulas are given in the next section, but first we mention one more issue.

The above heuristics were presented as if the parameter of interest appeared in every model. However, a given parameter may appear only in some of the models. In this case the basis for unconditional inference about that parameter can be (but need not be) made based on just those models in which that parameter appears. An example is variable selection in linear regression, say y versus p regressors, $x_1, \ldots, x_p$ (plus an intercept). There are 2^p possible models, but each regressor appears in only half of these models (i.e., in 2^{p-1} models). Thus if regressor variable x_j, hence parameter β_j, is in the selected AIC best model, we could restrict ourselves to just that subset of models that contain β_j in order to directly estimate the unconditional sampling variance of $\hat{\beta}_j$. All the above (and below) considerations about conditional and unconditional variances with regard to a particular parameter can be interpreted to apply to just the subset of models that include the parameter.

We have emphasized models as approximations to truth. A model being "wrong" is technically called model misspecification (see White 1994), and the usual theoretical sampling variances of MLEs, $\text{var}(\hat{\theta}_i \mid g_i)$, may be wrong, but only trivially so if the model is a good approximation to truth. There is theory that gives the correct conditional (on the model) sampling variance of $\hat{\theta}_i$ in the event of model misspecification (Chapter 7 gives some of this theory). However, the correct estimator of $\text{var}(\hat{\theta}_i \mid g_i)$ is then so much more complex and variable (a type of instability) that it generally seems better to use the theoretical estimator supplied by the usual model-specific information matrix (which assumes that the model is correct). This simplified approach seems especially defensible when done in conjunction with sound model selection procedures intended to minimize both serious overfitting and underfitting. We believe AIC is suitable for this selection purpose and that the only additional consideration is thus to get reliable unconditional sampling variances (and confidence intervals) for MLEs after model selection.

4.3.2   Including Model Selection Uncertainty in Estimator Sampling Variance

We continue to assume that the scalar parameter θ is common to all models considered. This will often be the case for our full set of a priori specific models, and is always the case if our objective is prediction with the fitted model, such as interpolation or extrapolation with a generalized linear model. Alternatively, if our focus is on a model structural parameter that appears only in a subset of our full set of models, then we can restrict ourselves to that subset in order to make the sort of inferences considered here about the parameter of interest. In the latter case we simply consider the relevant subset as the full set of models under consideration.

In repeated (conceptual) samples there is a probability π_i of selecting each model. Presentation of a defensible way to augment $\widehat{\text{var}}(\hat{\theta}_i \mid g_i)$ with model selection uncertainty involves the idea of model averaging in that we must define a model-averaged parameter value, $\bar{\theta}$, as

$$\bar{\theta} = \sum_{i=1}^{R} \pi_i \theta_i, \text{ and its estimator } \hat{\bar{\theta}} = \sum_{i=1}^{R} \hat{\pi}_i \hat{\theta}_i.$$

In some theory development we use π_i rather than $\hat{\pi}_i$, but we still use notation such as $\bar{\theta}$ and $\hat{\bar{\theta}}$.

The theoretical, unconditional sampling variance of the estimator of θ is given by

$$\text{var}(\hat{\bar{\theta}}) = \sum_{i=1}^{R} \pi_i \left[\text{var}(\hat{\theta}_i \mid g_i) + (\theta_i - \bar{\theta})^2 \right]. \tag{4.3}$$

This result follows directly from frequentist sampling theory. It is noted in Section 4.3.1 that if we had m independent samples and then applied model selection to each sample to get $\hat{\theta}_j$, $j = 1, \ldots, m$, then an estimator of $\widehat{\text{var}}(\hat{\bar{\theta}})$ would be

$$\widehat{\text{var}}(\hat{\bar{\theta}}) = \sum (\hat{\theta}_j - \bar{\bar{\theta}})^2 / (m - 1).$$

Here, j indexes the sample that $\hat{\theta}_j$ came from (whatever the model used), whereas i indexes that $\hat{\theta}_i$ arose from model i (whatever the sample was). This notation allows us to focus on different aspects of the model selection problem without a notation so complex that it hinders understanding of concepts. Letting m become infinite, the above estimator of $\text{var}(\hat{\bar{\theta}})$ converges to the theoretical unconditional sampling variance of $\hat{\bar{\theta}}$. By first grouping the set of m different $\hat{\theta}$ values by model and then taking the needed limit as $m \to \infty$ we get (4.3).

Readers less interested in the derivation of an estimator of the variance of the model-averaged estimate may want to skip to the following box. The quantity $\text{var}(\hat{\theta}_i \mid g_i) + (\theta_i - \bar{\theta})^2$ is just the mean square error of $\hat{\theta}_i$ given model i. Thus, in one sense the unconditional variance of $\hat{\bar{\theta}}$ is just an average mean square error. Specifically,

$$E[(\hat{\theta}_i - \hat{\bar{\theta}})^2 \mid g_i] = \text{var}(\hat{\theta}_i \mid g_i) + (\theta_i - \bar{\theta})^2,$$

and we recommend thinking of the above quantity as the sampling variance of $\hat{\theta}_i$, given model i, when $\hat{\theta}_i$ is being used as an estimator of $\bar{\theta}$. The incorporation of model selection uncertainty into the variance of $\hat{\theta}_i$ requires some new thinking like this. The matter arises again when we must consider a type of covariance, $E[(\hat{\theta}_i - \bar{\theta})(\hat{\theta}_j - \hat{\bar{\theta}}) \mid g_i]$, that also allows for model selection uncertainty.

One might think to estimate the augmented sampling variance of $\hat{\theta}_i$ by $\widehat{\text{var}}(\hat{\theta}_i \mid g_i) + (\hat{\theta}_i - \hat{\bar{\theta}})^2$. Such an estimator is not supported by any theory and is likely to be both biased (it could be bias-corrected) and quite variable; however, we have not investigated this possible estimator. Rather, to get an estimator of $\text{var}(\hat{\theta})$ we could plug estimated values into (4.3) to get $\widehat{\text{var}}(\hat{\theta}) = \sum \hat{\pi}_i [\widehat{\text{var}}(\hat{\theta}_i \mid g_i) + (\hat{\theta}_i - \hat{\bar{\theta}})^2]$. Ignoring that the π_i and $\text{var}(\hat{\theta}_i \mid g_i)$ are estimated (they are not the major source of estimation variation in this variance estimator), we can evaluate $E(\widehat{\text{var}}(\hat{\theta}))$ to bias-correct $\widehat{\text{var}}(\hat{\theta})$. The result involves the sampling variance, $\text{var}(\hat{\bar{\theta}})$, of the model-averaged estimator and is

$$E(\widehat{\text{var}}(\hat{\theta})) = \text{var}(\hat{\theta}) + \sum \pi_i \, \text{var}(\hat{\theta}_i \mid g_i) - \text{var}(\hat{\bar{\theta}}),$$

which leads to

$$\text{var}(\hat{\theta}) = \text{var}(\hat{\bar{\theta}}) + \sum \pi_i E(\hat{\theta}_i - \hat{\bar{\theta}})^2. \tag{4.4}$$

It seems that we cannot avoid estimating $\hat{\bar{\theta}}$ and $\text{var}(\hat{\bar{\theta}})$ even though it is $\hat{\theta}$ and $\widehat{\text{var}}(\hat{\theta})$ that we are seeking. First, note that efforts to evaluate $\sum \pi_i E(\hat{\theta}_i - \hat{\bar{\theta}})^2$ are circular, thus useless. Anyway, at worst we would only need to estimate this quantity without (much) bias, which we can clearly do. Second, (4.4) shows us that as one might expect, if our goal is to estimate $\bar{\theta}$, then the model-averaged $\hat{\bar{\theta}}$ is to be preferred to $\hat{\theta}_i$ because it will have a smaller sampling variance. However, given that our goal is to estimate θ, there is no theoretical basis to claim that $\hat{\bar{\theta}}$ is the superior estimator as compared to $\hat{\theta} \equiv \hat{\theta}_i$.

From Buckland et al. (1997) we will take the needed $\text{var}(\hat{\bar{\theta}})$ as

$$\text{var}(\hat{\bar{\theta}}) = \left[\sum_{i=1}^{R} \pi_i \sqrt{\text{var}(\hat{\theta}_i \mid g_i) + (\theta_i - \bar{\theta})^2} \right]^2, \tag{4.5}$$

with the estimator as

$$\widehat{\text{var}}(\hat{\bar{\theta}}) = \left[\sum_{i=1}^{R} \hat{\pi}_i \sqrt{\widehat{\text{var}}(\hat{\theta}_i \mid g_i) + (\hat{\theta}_i - \hat{\bar{\theta}})^2} \right]^2. \tag{4.6}$$

Formula (4.5) entails an assumption of perfect pairwise correlation, ρ_{ih}, of $\hat{\theta}_i - \hat{\bar{\theta}}$ and $\hat{\theta}_h - \hat{\bar{\theta}}$ for all $i \neq h$ (both i and h index models). Such pairwise correlation of $\rho_{ih} = 1$ is unlikely; however, it will be high. The choice of a value of $\rho_{ih} = 1$ is conservative in that $\text{var}(\hat{\bar{\theta}})$ computed from (4.5) will tend to be too large if this assumption is in error. Also, by just plugging estimators into (4.6) a further upward bias to (4.5) results. Thus from (4.4) the use of $\widehat{\text{var}}(\hat{\theta}) = \widehat{\text{var}}(\hat{\bar{\theta}}) + \sum \pi_i (\hat{\theta}_i - \hat{\bar{\theta}})^2$ with $\widehat{\text{var}}(\hat{\bar{\theta}})$ from (4.6) seems to risk too much positive bias. Hence, we are now suggesting just using $\widehat{\text{var}}(\hat{\theta}) = \widehat{\text{var}}(\hat{\bar{\theta}})$ from (4.6).

All simulations we have done so far, in various contexts, have supported use of this estimator:

$$\widehat{\text{var}}(\hat{\bar{\theta}}) = \left[\sum_{i=1}^{R} \hat{\pi}_i \sqrt{\widehat{\text{var}}(\hat{\theta}_i \mid g_i) + (\hat{\theta}_i - \hat{\bar{\theta}})^2} \right]^2. \tag{4.7}$$

These simulations have used θ (i.e., truth), not $\hat{\bar{\theta}}$, as the target for confidence interval coverage, and this may be another reason that the dual usage of (4.6) and (4.7) is acceptable. Improved estimation of unconditional sampling variances under model selection may be possible. However, our objective is to give practical solutions to some problems under model selection with the expectation that improvements will be further explored.

The $\hat{\pi}_i$ in (4.7) (and the equivalent (4.6)) will usually be taken as the Akaike weights, w_i. In general, $w_i \neq \pi_i$; rather, w_i can be considered to approximate π_i, but (4.7) seems robust to slightly imprecise values of the weights. Alternatively, one can use the bootstrap estimates, $\hat{\pi}_i = b_i/B$; however, given that one has bootstrap samples, the analytical formulas above are not needed.

As a final part of this section we give some details of the derivation of (4.5) in a more restricted context than was used in Buckland et al. (1997). Specifically, we do not assume that the R models are randomly selected from all possible models. Rather, we just condition on the set of R models that have been provided; hence, inferences are conditional on just this set of models. Formally, each $\hat{\theta}_i$ is considered as an estimator of $\bar{\theta}$, and it is this conceptualization that is critical to getting a variance formula that includes model selection uncertainty.

Ignoring that the π_i in (4.5) need to be estimated, the variance of $\hat{\bar{\theta}}$ can be expressed as

$$\mathrm{var}(\hat{\bar{\theta}}) = \sum_{i=1}^{R} (\pi_i)^2 \mathrm{E}[(\hat{\theta}_i - \bar{\theta})^2 \mid g_i]$$

$$+ \sum_{h \neq i}^{R} \sum^{R} \pi_i \pi_h \left[\mathrm{E}(\hat{\theta}_i - \bar{\theta})(\hat{\theta}_h - \bar{\theta}) \mid g_i, g_h) \right].$$

From above we know that

$$\mathrm{E}[(\hat{\theta}_i - \bar{\theta})^2 \mid g_i] = \mathrm{var}(\hat{\theta}_i \mid g_i) + (\theta_i - \bar{\theta})^2.$$

In order to coherently allow for model selection uncertainty and to be consistent with the definition of a correlation, we must interpret the covariance term in this expression for $\mathrm{var}(\hat{\bar{\theta}})$ as

$$\mathrm{E}(\hat{\theta}_i - \bar{\theta})(\hat{\theta}_h - \bar{\theta}) \mid g_i, g_h) = \rho_{ih} \sqrt{\mathrm{E}[(\hat{\theta}_i - \bar{\theta})^2 \mid g_i]\mathrm{E}[(\hat{\theta}_h - \bar{\theta})^2 \mid g_h]};$$

hence,

$$\mathrm{E}(\hat{\theta}_i - \bar{\theta})(\hat{\theta}_h - \bar{\theta}) \mid g_i, g_h)$$
$$= \rho_{ih} \sqrt{[\mathrm{var}(\hat{\theta}_i \mid g_i) + (\theta_i - \bar{\theta})^2][\mathrm{var}(\hat{\theta}_h \mid g_h) + (\theta_h - \bar{\theta})^2]}.$$

Thus we have

$$\text{var}(\hat{\bar{\theta}}) = \sum_{i=1}^{R} (\pi_i)^2 \left[\text{var}(\hat{\theta}_i \mid g_i) + (\theta_i - \bar{\theta})^2 \right]$$

$$+ \sum_{h \neq i}^{R} \sum^{R} \pi_i \pi_h \rho_{ih} \sqrt{[\text{var}(\hat{\theta}_i \mid g_i) + (\theta_i - \bar{\theta})^2][\text{var}(\hat{\theta}_h \mid g_h) + (\theta_h - \bar{\theta})^2]}.$$

We have no basis to estimate the across-model correlation of $\hat{\theta}_i - \bar{\theta}$ with $\hat{\theta}_h - \bar{\theta}$ (other than the bootstrap, but then we do not need theory for $\text{var}(\hat{\bar{\theta}})$). The above simplifies if we assume that all $\rho_{ih} = \rho$:

$$\text{var}(\hat{\bar{\theta}}) = (1 - \rho) \left[\sum_{i=1}^{R} (\pi_i)^2 \left[\text{var}(\hat{\theta}_i \mid g_i) + (\theta_i - \bar{\theta})^2 \right] \right]$$

$$+ \rho \left[\sum_{i=1}^{R} \pi_i \sqrt{\text{var}(\hat{\theta}_i \mid g_i) + (\theta_i - \bar{\theta})^2} \right]^2. \tag{4.8}$$

From (4.8), if we further assume $\rho = 1$, then we get (4.5):

$$\text{var}(\hat{\bar{\theta}}) = \left[\sum_{i=1}^{R} \pi_i \sqrt{\text{var}(\hat{\theta}_i \mid g_i) + (\theta_i - \bar{\theta})^2} \right]^2.$$

Unconditional Variance Estimator

Then, using the Akaike weights (w_i) instead of the model selection frequencies (π_i) and using estimates instead of parameters, we obtain a very useful result,

$$\widehat{\text{var}}(\hat{\bar{\theta}}) = \left[\sum_{i=1}^{R} w_i \sqrt{\widehat{\text{var}}(\hat{\theta}_i \mid g_i) + (\hat{\theta}_i - \hat{\bar{\theta}})^2} \right]^2. \tag{4.9}$$

where $\hat{\bar{\theta}}$ is model-averaged estimate (4.1).

This estimator of the unconditional variance can be used for either the MLE $\hat{\theta}$ from the selected model or for the model averaged estimator $\hat{\bar{\theta}}$.

If only a subset of the R models in used, then the w_i must be recalculated, based on just these models (thus these new weights must satisfy $\sum w_i = 1$). If one has the estimated model selection frequencies (π_i) from the bootstrap, the estimator

$$\widehat{\text{var}}(\hat{\bar{\theta}}) = \left[\sum_{i=1}^{R} \hat{\pi}_i \sqrt{\widehat{\text{var}}(\hat{\theta}_i \mid g_i) + (\hat{\theta}_i - \hat{\bar{\theta}})^2} \right]^2$$

is useful. In either case, $\widehat{\text{se}}(\hat{\bar{\theta}}) = \sqrt{\widehat{\text{var}}(\hat{\bar{\theta}})}$.

The concept of an unconditional variance can be extended to deriving an expression for an unconditional sampling covariance between two different parameter estimators in a model, or (what is the same thing) for an unconditional covariance between two estimators as based on a generalized linear model. A formula for the unconditional $\text{var}(\hat{\bar{\theta}})$ values is

$$\text{var}(\hat{\bar{\theta}}) = \left[\sum_{i=1}^{R} \pi_i \sqrt{\text{var}(\hat{\theta}_i|g_i) + (\theta_i - \hat{\bar{\theta}})^2} \right]^2 ;$$

we need something useful for the analogous

$$\text{cov}(\hat{\bar{\theta}}_1, \hat{\bar{\theta}}_2),$$

$$\hat{\bar{\theta}}_1 = \sum_{i=1}^{R} \pi_i \hat{\theta}_{1i},$$

$$\hat{\bar{\theta}}_2 = \sum_{i=1}^{R} \pi_i \hat{\theta}_{2i}.$$

For a useful estimation formula the Akaike weight, w_i, will replace π_i.

We propose the formula below for $\widehat{\text{cov}}(\hat{\bar{\theta}}_1, \hat{\bar{\theta}}_2)$:

$$= \bar{r}_{1,2} \left[\sum_{i=1}^{R} w_i \sqrt{\widehat{\text{var}}(\hat{\theta}_{1i}|g_i) + (\hat{\theta}_{1i} - \hat{\bar{\theta}}_1)^2} \right] \left[\sum_{i=1}^{R} w_i \sqrt{\widehat{\text{var}}(\hat{\theta}_{2i}|g_i) + (\hat{\theta}_{2i} - \hat{\bar{\theta}}_2)^2} \right]$$

$$= \bar{r}_{1,2} \, \widehat{\text{se}}(\hat{\bar{\theta}}_1) \, \widehat{\text{se}}(\hat{\bar{\theta}}_2),$$

where $\bar{r}_{1,2}$ is

$$\bar{r}_{1,2} = \sum_{i=1}^{R} w_i \times r_{1,2|i},$$

and $r_{1,2|i}$ is the estimated sampling correlation between $\hat{\theta}_{1i}$ and $\hat{\theta}_{2i}$ given model i. This model-conditional sampling correlation can be obtained from the conditional-on-model-i sampling variance-covariance matrix (or indirectly by large sample means).

Now if we are considering the difference $d = \hat{\bar{\theta}}_1 - \hat{\bar{\theta}}_2$ (see Conner et al. 2001), then

$$\widehat{\text{var}}(d) = A^2 + B^2 - 2\bar{r}_{1,2}AB,$$

where

$$A = \sum_{i=1}^{R} w_i \sqrt{\widehat{\text{var}}(\hat{\theta}_{1i}|g_i) + (\hat{\theta}_{1i} - \hat{\bar{\theta}}_1)^2},$$

$$B = \sum_{i=1}^{R} w_i \sqrt{\widehat{\text{var}}(\hat{\theta}_{2i}|g_i) + (\hat{\theta}_{2i} - \hat{\bar{\theta}}_2)^2}.$$

The average correlation $\bar{r}_{1,2}$ will be between -1 and 1. Therefore, $\widehat{\text{var}}(d)$ will be between $(A - B)^2$ and $(A + B)^2$; hence, $\widehat{\text{var}}(d) \geq 0$ for any value of $\bar{r}_{1,2}$. More work and experience are needed with the formula versus simulation results and formula performance.

4.3.3   Unconditional Confidence Intervals

The matter of a $(1 - \alpha)100\%$ unconditional confidence interval is now considered. We have two general approaches: the bootstrap (see, e.g., Buckland et al. 1997), or analytical formulas based on analysis results from just the one data set. The analytical approach requires less computing; hence we start with it.

The simplest such interval is given by the endpoints $\hat{\theta}_i \pm z_{1-\alpha/2} \, \widehat{\text{se}}(\hat{\theta}_i)$, where $\widehat{\text{se}}(\hat{\theta}_i) = \sqrt{\widehat{\text{var}}(\hat{\theta}_i)}$. One substitutes the model-averaged $\hat{\bar{\theta}}$ for $\hat{\theta}_i$ if that is the estimator used. A common form used and recommended as a conditional interval is $\hat{\theta}_i \pm t_{\text{df}, 1-\alpha/2} \, \widehat{\text{se}}(\hat{\theta}_i \mid g_i)$. When there is no model selection, or it is ignored, it is clear what the degrees of freedom (df) are for the t-distribution here. For (4.7) it is not clear what the degrees of freedom should be. Note, however, that we are focusing on situations where sample size is large enough that the normal approximation will be applicable. These simple forms for a confidence interval are based on the assumption that $\hat{\theta}_i$ has a normal sampling distribution.

We will hazard a suggestion here; it has not been evaluated in this context, but a similar procedure worked in a different context. If for each fitted model we have degrees of freedom df_i for the estimator $\widehat{\text{var}}(\hat{\theta}_i \mid g_i)$, then for generally small degrees of freedom one might try using the interval $\hat{\theta}_i \pm z_{1-\alpha/2} \, \widehat{\text{ase}}(\hat{\theta}_i)$, where the adjusted standard error estimator is

$$\widehat{\text{ase}}(\hat{\theta}_i) = \sum_{i=1}^{R} \hat{\pi}_i \sqrt{\left(\frac{t_{\text{df}_i, 1-\alpha/2}}{z_{1-\alpha/2}} \right)^2 \widehat{\text{var}}(\hat{\theta}_i \mid g_i) + (\hat{\theta}_i - \hat{\bar{\theta}})^2}.$$

In cases where $\hat{\theta}_i \pm z_{1-\alpha/2} \, \widehat{\text{se}}(\hat{\theta}_i)$ is not justified by a normal sampling distribution (as judged by the conditional distribution of $\hat{\theta}_i$), intervals with improved coverage can be based on a transformation of $\hat{\theta}_i$ if a suitable transformation is known. Log and logit transforms are commonly used, often implicitly in the context of general linear models. In fact, in general linear models the vector parameter $\underline{\theta}$ will be linked to the likelihood by a set of transformations, $\underline{\theta} = \underline{W}(\beta)$. Then it is β that is directly estimated, and it is often the case that the simple normal-based confidence limits on components of β can be reliably used. An interval constructed from a component of $\hat{\beta}$ and its unconditional sampling variance ((4.5) applies) can be back-transformed to an interval on the corresponding component of $\underline{\theta}$.

The above methods are justified asymptotically, or if a normal sampling distribution applies to $\hat{\theta}$. However, "asymptotically" means for some suitable large sample size n. We do not know when to trust that n is suitably large in nonlinear and nonnormal random variation models. A general alternative when there is no model selection is the profile likelihood interval approach (Leonard and Hsu 1999, Sprott 2000). We suggest here an adaptation of that approach that widens the likelihood interval to account for model selection uncertainty.

Let the vector parameter $\underline{\theta}$ be partitioned into the component of interest, θ, and the rest of the parameters, denoted here by $\underline{\gamma}$. Then the profile likelihood, as a function of θ_i (the subscript denotes the model used) for model g_i is given by

$$\mathcal{PL}(\theta_i \mid \underline{x}, g_i) = \max_{\underline{\gamma}_i \mid \theta_i} \left[\mathcal{L}(\theta_i, \underline{\gamma}_i \mid \underline{x}, g_i) \right];$$

almost always $\mathcal{PL}(\theta_i \mid \underline{x}, g_i)$ has to be computed numerically. We define a profile deviance as

$$\mathcal{PD}(\theta_i) = 2 \left[\mathcal{PL}(\hat{\theta}_i \mid \underline{x}, g_i) - \mathcal{PL}(\theta_i \mid \underline{x}, g_i) \right]. \tag{4.10}$$

The large sample profile likelihood interval ignoring model selection uncertainty is the set of θ_i that satisfy the condition $\mathcal{PD}(\theta_i) \leq \chi^2_{1,1-\alpha}$. Here, $\chi^2_{1,1-\alpha}$ is the upper $1 - \alpha$ percentile of the central chi-squared distribution on 1 df. This interval is approximately a $(1 - \alpha)100\%$ confidence interval.

We propose an interval that is a version of (4.9) adjusted (widened) for model selection uncertainty: the set of all θ_i that satisfy

$$\mathcal{PD}(\theta_i) \leq \left[\frac{\widehat{\text{var}}(\hat{\theta})}{\widehat{\text{var}}(\hat{\theta}_i \mid g_i)} \right] \chi^2_{1,1-\alpha}. \tag{4.11}$$

It suffices to solve (numerically) this inequality for the confidence interval endpoints, $\hat{\theta}_{i,\text{L}}$ and $\hat{\theta}_{i,\text{U}}$. In the event that we are not doing model averaging, it seems logical to use the resultant confidence interval from (4.10).

All of the above was assuming that the parameter of interest occurred in each of the R models in the full set of models. Often this will not be the case. Rather, there will be a subset of size $Q < R$ of the models in which the parameter θ occurs. Conceptually, the parameter θ does not occur in the other $R - Q$ models, even as a value equal to zero. In this event we suggest applying all the above theory to just that subset of Q models. The $R - Q$ models in which θ does not appear seem totally uninformative about the value of θ; hence they cannot play a direct role in inference about θ. The situation excludes variable selection as in linear all subset selection because there we can usefully consider that every structural parameter is in each model, but sometimes we have set $\theta \equiv 0$ in a model.

In the case that $Q = 1$ (θ is unique to one model in the set of R models), none of the above results can be used. In this case it seems that there may not be a direct way to include model selection uncertainty into the uncertainty about the value of θ. An approach we can envision here is to adjust upward the conditional sampling variance estimator, $\widehat{\text{var}}(\hat{\theta} \mid g_i)$, by some variance inflation factor. What the variance inflation factor would be is not clear because we have looked at the variance inflation factor

$$\frac{\widehat{\text{var}}(\hat{\theta})}{\widehat{\text{var}}(\hat{\theta}_i \mid g_i)}$$

and found that it can vary greatly by parameter. Thus, estimation of a variance inflation factor for $\hat{\theta}$ based on a different parameter in a subset of different models (from the one that model θ appears in) seems very problematic. Fundamentally, it is not clear that we should inflate the conditional sampling variance of a parameter unique to just one model in the set of models. Perhaps all we can, and should, do in this case is note the uncertainty about whether that model is likely to be the K-L best model in the full set of models and use the model-specific conditional sampling variance for that $\hat{\theta}$. Confidence intervals for θ are then constructed based on just the one model in which θ appears (e.g., profile likelihood interval, or other parametric methods such as $\hat{\theta} \pm z_{1-\alpha/2}\,\widehat{\text{se}}(\hat{\theta} \mid g)$).

Bootstrap construction of an unconditional confidence interval on a parameter in the selected model is not fundamentally different from such bootstrap-based interval construction without model selection. The latter is much discussed in the statistical literature (see, e.g., Efron and Tibshirani 1993, Mooney and Duval 1993, Hjorth 1994).

First one generates a large number, B, of bootstrap samples from the data and applies model selection to each bootstrap sample. All R models in the original set are fit to each bootstrap sample, and one of these models will be selected as best. Only the estimated parameters from that selected best model are kept in an output set of parameter estimates for each bootstrap sample (plus the index of the selected model for each sample). Hence if a parameter θ is not in the selected model g_i for bootstrap sample b, the value of $\hat{\theta}_b^*$ is missing for bootstrap sample b (the subscript b and $\hat{\theta}_b^*$ denote that $\hat{\theta}^*$ is from the bth bootstrap sample; the model used to get this $\hat{\theta}^*$ varies over bootstrap samples). For any parameter in common over all models, there will be B values of $\hat{\theta}_b^*$ in the output data set. In either case the variation in the output set of (not missing) values of $\hat{\theta}_b^*$, $b = 1, \ldots, m\ (\leq B)$, reflects both model selection variation in $\hat{\theta}$ and within-model sampling variation of $\hat{\theta}$ given a model.

As noted in Section 4.2.2, the model selection frequencies can be estimated from these bootstrap results; but our focus here is on unconditional estimator uncertainty and confidence intervals (Efron and Tibshirani 1993). The average of all m values of $\hat{\theta}_b^*$, $\overline{\hat{\theta}^*}$ is an estimator of the model-averaged parameter $\hat{\bar{\theta}}$. Hence the empirical variance of the set of m values of $\hat{\theta}_b^*$,

$$\text{var}(\hat{\theta}^*) = \sum(\hat{\theta}_b^* - \overline{\hat{\theta}^*})^2 \big/ (m-1),$$

is the bootstrap estimator of the unconditional sampling variance of $\hat{\theta}$ ($\equiv \hat{\theta}_i$ for the parameter θ estimated from the selected best model, i). Given this bootstrap $\mathrm{var}(\hat{\theta}^*) = \widehat{\mathrm{var}}(\hat{\theta})$, the simplest confidence interval is $\hat{\theta} \pm z_{1-\alpha/2}\,\widehat{\mathrm{se}}(\hat{\theta})$. However, such an interval fails to make full use of the value of the bootstrap method in finding upper and lower interval estimates that allow for a nonnormal sampling distribution for $\hat{\theta}$ under model selection.

The simple, direct bootstrap-based confidence interval on θ is the percentile interval (Efron and Tibshirani 1993). Order the bootstrap values $\hat{\theta}^*_b$ from smallest to largest and denote these ordered values by $\hat{\theta}^*_{(b)}$, $b = 1, \ldots, m$. For a $(1 - \alpha)100\%$ confidence interval select the $\alpha/2$ lower and $1 - \alpha/2$ upper percentiles of these ordered bootstrap estimates as $\hat{\theta}_L$ and $\hat{\theta}_U$. These percentiles may not occur at integer values of b, but if m is large, it suffices to use $\hat{\theta}_L = \hat{\theta}^*_{(l)}$ and $\hat{\theta}_U = \hat{\theta}^*_{(u)}$, where $l = [m \cdot \frac{\alpha}{2}]$ and $u = [m \cdot (1 - \frac{\alpha}{2})]$. More complex, but possibly better, unconditional intervals after model selection based on the bootstrap are considered by Shao (1996) for regression problems. Bootstrapping can be done with model averaging in a straightforward way. Here, interest is in $\hat{\bar{\beta}}$ or $\tilde{\beta}$ and model averaging should be done for each bootstrap sample. Thus, one obtains $\hat{\bar{\beta}}^*$ or $\tilde{\beta}^*$ for each of the B bootstrap samples and then computes the standard error and confidence intervals from these results.

Note that B needs to be at least several hundred for the bootstrap method to begin to work well, and we recommend 10,000 (and at least use $B = 1,000$). If the parameter of interest is in every model, then $m = B$, which is user selected. If the parameter is truly not in every model, m is random, and B may need to be made larger to ensure that a sufficient sample size, m, of relevant bootstrap samples is obtained.

4.4 Estimating the Relative Importance of Variables

Data analysis is sometimes focused on the variables to include versus exclude in the selected model (e.g., imporatant vs. unimportant). Variable selection is often the focus of model selection for linear or logistic regression models. Often, an investigator uses stepwise analysis to arrive at a final model, and from this a conclusion is drawn that the variables in this model are important, whereas the other variables are not important. While common, this is poor practice and, among other issues, fails to fully consider model selection uncertainty. Here, we provide simple methods to quantify the evidence for the importance of each variable in the set.

Consider 10 models based on combinations of a number of regressor variables. Assume that the selected best model includes x_1 and has an Akaike weight of only 0.3. There is considerable model selection uncertainty here, and hence there would seem to be only weak evidence for the importance of variable x_1 based on the selected best model. But one must consider the Akaike weights of all other models that include x_1 in order to quantify the importance of x_1. It might be that all models that exclude x_1 have very low Akaike weights; that situation would suggest that x_1 is a very important predictor. The measure of this importance is to sum the Akaike weights (or the bootstrap $\hat{\pi}_i$) over the subset of models that include variable x_1. This idea is applicable in general to model selection whenever it is equated to variable selection, for linear or nonlinear models of any type.

Consider the hypothetical example of three regressors, x_1, x_2, and x_3, and a search for the best of the eight possible models of the simple linear regression type: $y = \beta_0 + \beta_1 x_1 + \beta_2 x_2 + \beta_3 x_3 + \epsilon$. The possible combinations of regressors that define the eight possible models are shown below, along with hypothetical Akaike weights w_i (a 1 denotes that x_i is in the model; otherwise, it is excluded):

x_1	x_2	x_3	w_i
0	0	0	0.00
1	0	0	0.10
0	1	0	0.01
0	0	1	0.05
1	1	0	0.04
1	0	1	0.50
0	1	1	0.15
1	1	1	0.15.

While the selected best model has weight of only 0.5, i.e., a probability of 0.5 of being the actual K-L best model here, the sum of the weights for variable x_1 is 0.79. This is evidence of the importance of this variable, across the models considered. Variable x_2 was not included in the selected best model; but this should not suggest that it is of zero importance. Actually, its relative weight of evidence support is 0.35. Finally, the sum of the Akaike weights for predictor variable x_3 is 0.85. Thus the evidence for the importance of variable x_3 is substantially more that just the weight of evidence for the best model. We can order the three predictor variables in this example by their estimated importance: x_3, x_1, x_2 with importance weights of 0.85, 0.79, and 0.35. As with other methods recommended here, we see that we are able to use model selection to go well beyond just noting the best model from a set of models.

Relative Variable Importance

Estimates of the relative importance of predictor variables x_j can best be made by summing the Akaike weights across all the models in the set where variable j occurs. Thus, the relative importance of variable j is reflected in the sum $w_+(j)$.

The larger the $w_+(j)$ the more important variable j is, relative to the other variables. Using the $w_+(j)$, all the variables can be ranked in their importance.

The direction and magnitude of effect size should often be based on model-averaged estimates with appropriate measures of precision.

This idea extends to subsets of variables. For example, we can judge the importance of a pair of variables, as a pair, by the sum of the Akaike weights of all models that include the pair of variables. For the pair x_1 & x_2, the weight of evidence for the importance of this pair is 0.19. For pair x_2 & x_3, the weight of evidence for importance is 0.23, while for the pair x_1 & x_3, the weight of evidence is 0.65 (compared to 0.5 for the selected model as such). Similar procedures apply when assessing the relative importance of interaction terms.

When assessing the relative importance of variables using sums of the w_i, it is important to achieve a balance in the number of models that contain each variable j. For example, in the numerical example above, each of the three variables appeared in four models. This balancing puts each variable on equal footing.

To summarize, in many contexts the AIC selected best model will include some variables and exclude others. Yet this inclusion or exclusion by itself does not distinguish differential evidence for the importance of a variable in the model. The model weights, w_i or $\hat{\pi}_i$, summed over all models that include a given variable provide a better weight of evidence for the importance of that variable in the context of the set of models considered.

4.5   Confidence Set for the K-L Best Model

4.5.1   Introduction

There exists a concept of a confidence set for the K-L best model based on the data, just as there is a confidence interval for a parameter based on a model and data. For a 95% confidence set on the actual K-L best model, a rational (but not unique) approach is to sum the Akaike weights from largest to smallest until that sum is just ≥ 0.95; the corresponding subset of models is a type of confidence set on the K-L best model. In this example (assuming that we have indexed the models as 1 to 7 in order of decreasing weights), the confidence set is models {1, 2, 3, 4, 5}, which has sum of weights = 0.966. In using this approach to a confidence set of models we are interpreting the Akaike weight as a posterior probability (i.e., given the data and the set of a priori models) that model i is the K-L best model (see Section 6.4.5). This is not the best approach but it is easy to understand.

There is another approach to developing a confidence set of models based on the idea of a Δ_i being a random variable with a sampling distribution. In particular, let index value *best* correspond to the actual expected K-L best model in the set. There is always a K-L best model in the set of models (ignoring that ties might occur). It is thus model g_{best} that we should use for the data analysis; we just do not happen to know a priori the value of *best*. Then the Δ_i of conceptual interest is

$$\Delta_p = \text{AIC}_{best} - \text{AIC}_{min} . \tag{4.12}$$

This unobservable random variable (Δ_p) is analogous to $\theta - \hat{\theta}$, which can often be used (after normalization by $\widehat{se}(\hat{\theta})$) as a pivotal value for construction of a confidence interval on θ. A pivotal quantity is one whose sampling distribution is independent of any unknown parameters, a t-distributed pivotal, for example. The "p" in the Δ defined by (4.12) denotes that this Δ is a conceptual pivotal value rather than an actual Δ_i that we can compute from real data.

It is not exact to consider $\Delta_p = AIC_{best} - AIC_{min}$ as a pivotal quantity, but it seems a useful approximation in some contexts. The context it seems useful in is one of complex truth; tapering effect sizes; many models, some being good approximations to truth, with full truth not in the set of models used; and a lot of nested sequences of models (as in the starling experiment example in Chapter 3). Monte Carlo studies on the above Δ_p can be done; we have done many of these and results support the conclusion that in this context, the sampling distribution of this Δ_p has substantial stability and the 95th percentile of the sampling distribution of Δ_p is generally much less that 10, and in fact generally less than 7 (often closer to 4 in simple situations). This means that an alternative rule of thumb for an approximate 95% confidence set on the K-L best model is the subset of all models g_i having $\Delta_i \leq$ some value that is roughly in the range 4 to 7. In fact, the Δ value to use when a model is not competitive as a candidate for the K-L best model is variable, but is probably somewhere between 2 and 10 in many situations. Thus, a Δ_i of 2 is not large, while a $\Delta_i = 10$ is strong evidence against model g_i being the K-L best model in the set of models considered, if sample size is not small. These guidelines, rough as they are, are useful.

We review this interpretation of evidence from the Δ_i when observations are independent, sample sizes are large, and models are nested:

Δ_i	Level of Empirical Support
0–2	Substantial
4–7	Considerably less
> 10	Essentially none

Models with $\Delta > 10$ represent very strong evidence that the model is not the K-L best model. The reader should not take these guidelines as inviolate since as there are situations to which they do not apply well (such as when there is a small sample size or dependent observations). Likewise, if there are thousands of models, these guidelines may not hold.

We had these guidelines well in mind when we encountered similar guidelines for the Bayes factor. The Bayes factor is a Bayesian-based ratio for the relative data-based likelihood of one model versus another model, but without considering any priors on the set of models (Berger and Pericchi 1996, Raftery 1996a); it is somewhat analogous to $\exp(-\frac{1}{2}\Delta_i)$. Raftery (1996a:252, 1996b:165) presents a similar scale for interpretation of 2 log(Bayes factor) as evidence for the simpler model being considered.

A third reasonable basis for a confidence set on models is motivated by likelihood-based inference (see e.g., Edwards 1992, Azzalini 1996, Royall

1997), hence is analogous to a profile likelihood interval on a parameter given a model. Here we would simply agree on some value of the relative likelihood of model i versus the estimated K-L best model *best* as a cutoff point for a set of relatively more plausible models. Thus our confidence set of models is all models for which the ratio

$$\frac{\mathcal{L}(g_i|x)}{\mathcal{L}(g_{min}|x)} > \textbf{cutoff},$$

where the cutoff value might be $\approx 1/8$. Models where this evidence ratio is greater than 1/8 are in the confidence set and are deemed plausible. There is no direct sampling theory interpretation required and no necessary appeal to the idea of the selected subset of models including the K-L best model with a preset, known, long-run inclusion relative frequency such as 95%. This procedure has the advantage that the cutoff remains unchanged by the addition or deletion of a model (of course, a new model will be either in or out of the confidence set). Thus, a confidence set based on the evidence ratio has a desirable invariance property. In contrast, any change in the set of R models can alter the confidence set when summing the Akaike weights.

We have presented three approaches to finding a confidence set on models: (1) base it directly on the Akaike weights, interpreted as approximate probabilities of each model being the actual best model, given the data; (2) use a cutoff Δ_i motivated by the idea of the sampling distribution of the approximate pivotal, Δ_p (using, say, the 95th percentile of this distribution as the cutoff Δ); or (3) think in terms of relative likelihood and hence (for *min* indexing the selected AIC best model) use a cutoff value of Δ for which $\mathcal{L}(g_i|x)/\mathcal{L}(g_{min}|x) \equiv \exp(-\frac{1}{2}\Delta_i)$ is small, say $0.135 \, (\Delta_i = 4), 0.082 \, (\Delta_i = 5)$, or $0.050 \, (\Delta_i = 6)$. In general we favor this third approach.

The use of intervals based purely on relative likelihood is soundly supported by statistical theory (cf. Berger and Wolpert 1984, Edwards 1992, Azzalini 1996, Royall 1997), but rarely taught or used. Rather, most users of statistics have been taught to think of confidence intervals in terms of coverage probability; hence they might feel more at home with methods (1) and (2), both of which are motivated by the sampling theory idea of a 95% confidence interval on a parameter. The approach based on simple evidence ratios seems quite useful. More needs to be known about the properties of these three methods to construct a confidence set of models before we would be comfortable recommending just one approach.

4.5.2 Δ_i, Model Selection Probabilities, and the Bootstrap

For a given set of data we can estimate the sampling distribution of model selection frequencies and the distribution of $\Delta_p = \text{AIC}_{best} - \text{AIC}_{min}$ (i.e., formula 4.12) using the bootstrap method. In this method the role of the actual (unknown) K-L best model is played by the model selected as best from the data analysis; denote that model by model g_{best}. For example, if model g_5 is selected by AIC, this means that $best = 5$. For each bootstrap sample we fit each of the R models, compute all R of the AIC_i^*, and then find the single $\Delta_p^* = AIC_{best}^* - AIC_{min}^*$; $best$ does not change over bootstrap samples. The model producing AIC_{min}^* varies by bootstrap sample. However, it will often be model g_{best} in which case $\Delta_p^* = 0$. When it is not model g_{best} that produces AIC_{min}^*, then $\Delta_p^* > 0$.

The B bootstrap samples provide B values of Δ_p^* that are independent and conditional on the data. The percentiles of the empirical probability distribution function of Δ_p^* provide the estimate of the percentiles of the sampling distribution of Δ_p, and hence provide a basis for a confidence set on the K-L best model for the actual data. For a $(1 - \alpha)100\%$ confidence set on the K-L best model, order the $\Delta_{p,(b)}^*$ (smallest to largest) and find the $\Delta_{p,(b)}^*$ value for $b = [(1 - \alpha)B]$. For the actual data analysis results, the subset of the R models g_i having $\Delta_i \leq \Delta_{[(1-\alpha)B]}^*$ is the desired confidence set. For reliable results on the upper tail percentiles of Δ_p, B needs to be 10,000.

Other information can be gained from these bootstrap results about model selection uncertainty, in particular, the frequency of selection of each of the R models. Let b_i be the number of samples in which model i was selected as the K-L best model. Then an estimator of the relative frequency of model selection in the given situation is $\hat{\pi}_i = b_i/B$. These estimated selection probabilities are useful for assessing how much sampling variation there is in the selection of the best model: they directly quantify model selection uncertainty. These estimated selection probabilities are similar to, but not identical in meaning to, the Akaike weights, which also quantify strength of evidence about model selection uncertainty.

Also, for each bootstrap sample we can compute the Akaike weights,

$$w_i^* = \frac{\exp(-\frac{1}{2}\Delta_i^*)}{\sum_{r=1}^{R} \exp(-\frac{1}{2}\Delta_r^*)}$$

and then average these over the B samples to get $\overline{w}_i^*$. Comparison of the w_i, $\overline{w}_i^*$, and $\hat{\pi}_i$ is informative as to the coherence of these methods, each of which provides information about the sampling uncertainty in model selection. The theoretical measure of model selection sampling uncertainty is the set of true, unknown selection probabilities, $\pi_1 \ldots, \pi_R$. Either the $\hat{\pi}_i$ (from the bootstrap) or the Akaike weights, w_i (we often prefer the latter because they do not require computer-intensive calculations and they relate more directly to strength of evidence based on the data at hand), may be taken as the estimated inference uncertainty about model selection. Note that in this usage of the term "model" we mean the structural form of the model (such as which variables are included vs. excluded) without consideration of the specific parameter values required in each model. Parameter-estimation uncertainty is conceptually separable from (but influenced by) model selection uncertainty.

4.6   Model Redundancy

Consider a set of three models, in which models g_2 and g_3 are identical because a mistake was made in setting up the problem on the computer. Thus, models g_2 and g_3 are 100% redundant in the set of models; the model set should contain only models g_1 and g_2. Assume $\Delta_1 = 0$ and $\Delta_2 = \Delta_3 = 4$. For the redundant set of three models we get $\mathcal{L}(g_1)/\mathcal{L}(g_2) = \mathcal{L}(g_1)/\mathcal{L}(g_3) = 7.4$. Similarly, for the correct set of two models, $\mathcal{L}(g_1)/\mathcal{L}(g_2) = 7.4$. The unfortunate model redundancy has not affected the Δ_i nor the likelihood evidence ratios of models. However, the (normalized) Akaike weights (Section 2.9) are affected: For the set of two models, $w_1 = 0.881$ and $w_2 = 0.119$; whereas for the set with model redundancy, $w_1 = 0.787$ and $w_2 = w_3 = 0.106$. Note that for either model set we still have $w_1/w_2 = 7.4$ ($= w_1/w_3$): likelihood ratios are not affected by model redundancy.

The difference between a w_1 of 0.881 and one of 0.787 is not large. However, our point is that this, clearly erroneous, model redundancy in the three-model set has affected the Akaike weights. The weights for the model set with a redundant model included are not correct because the value Δ_2 shows up twice (one time "disguised" as Δ_3). The effect on the weights is not dramatic here (but it could be), but they are wrong, and this could affect (presumably adversely) calculations using the w_i (as $\hat{\pi}_i$), as for example in model averaging and unconditional variance calculations.

If the model redundancy was recognized, and we wanted to retain it (we should not), we could correct the situation by considering the set of models as having two subsets: Model g_1 is one subset; a second subset contains models g_2 and g_3. Given that we know that models g_2 and g_3 are 100% redundant, we allocate prior weights, about which model is the expected K-L best model, as 1/2 to each subset, and the 1/2 is further divided equally for each model in a subset. Thus, $\tau_1 = 0.5$, $\tau_2 = 0.25$, and $\tau_3 = 0.25$. Now we use

$$w_i = \frac{\mathcal{L}(g_i|\underline{x})\tau_i}{\sum_{r=1}^{R} \mathcal{L}(g_r|\underline{x})\tau_r}$$

from Section 2.9 to compute correct Akaike weights for the set of three models; thus

$$w_1 \propto 1.0 \cdot \frac{1}{2}, \quad w_2 \propto 0.135335 \cdot \frac{1}{4}, \quad w_2 \propto 0.135335 \cdot \frac{1}{4}.$$

The normalized (to add to 1) weights are 0.8808, 0.0596, and 0.0596. Now the sum of the weights for models g_2 and g_3 correctly add up to what they ought to be, for model averaging and unconditional sampling variance estimation will produce correct results when applied to the redundant set of three models.

This hypothetical example presumably would not occur deliberately, but it serves to introduce the concept, and issue, of model redundancy in the set of models considered. It is possible to have actual model redundancy if one is not careful in constructing the set of models considered. For example, in analysis of distance sampling data (Buckland et al. 2001 and Chapter 5), the program DISTANCE (Laake et al. 1994) can consider, in effect, the full set of models structured into two or more subsets. Different subsets of models may be specified in such a way that they have one key model in common, to which adjustment terms are applied to get a sequence of models. The same key model can be used with different types of adjustment terms. Schematically, we can have this situation: The full set of models is given as two subsets of models, $\{g_1, g_2, g_3, g_4\}$ and $\{g_1, g_5, g_6, g_7\}$. If the full set of models is considered as just 8 different models, then the redundancy of model g_1 is not being recognized. The situation can easily be rectified if it is recognized: either label the models 1 to 7, or compute the w_i from the Δ_i, for the models labeled as 1 to 8, based on differential priors, τ_i, as $\left\{\frac{1}{14}, \frac{1}{7}, \frac{1}{7}, \frac{1}{7}\right\}$ and $\left\{\frac{1}{14}, \frac{1}{7}, \frac{1}{7}, \frac{1}{7}\right\}$. It is not clear whether there might be a partial model redundancy in the models g_1 to g_7.

Failure to completely understand the models used can result in model redundancy. For example, a logistic regression may be used for the structural model of the probability of success,

$$p(x) = \frac{\exp(a + bx)}{1 + \exp(a + bx)},$$

where the parameters are a and b. However, this model (stucture) can be expresed as,

$$p(x) = \frac{1}{1 + \exp[-\{(x - d)/c\}]},$$

where $a \equiv d/c$ and $b = -1/c$. The second model is just a combination of a different model representation and a 1-to-1 reparametrization. If both model forms where included in the model set, total redundancy results. Users should avoid such model redundancy.

To further illustrate model redundancy we consider some models for capture–recapture data, obtained on k capture occasions, to estimate animal population size. The parameters of such models are population size (N) and capture probabilities (denoted by p), by occasion, animal, or by both factors. One possible type of model is model g_b under which there are only two different capture probabilities: for first capture or for recapture. This model is for the case where animals have a behavioral response to first capture, but no other factors affect capture probability ($K = 3$).

A different model (g_t) allows capture probabilities to vary, but only by occasion; so we have $p_1, p_2, \ldots, p_k$ ($K = k + 1$). Thus, we can have two very different models. However, the model under which capture probabilities can vary by time allows for many submodels ($2^k - k$ possible models, including the most general case). Some example (sub) models are

$$g_{t1} : p_1 = p_2, \text{ other } p_i \text{ all differ } (K = k),$$
$$g_{t2} : p_1 = p_2 = p_3, \text{ other } p_i \text{ all differ } (K = k - 1),$$
$$g_{t3} : \text{all } p_i = p \ (K = 2),$$
$$g_{t4} : \text{all } p_i \text{ are different } (K = k + 1).$$

If we now take as our model the set $\{g_{t1}, g_{t2}, g_{t3}, g_{t4}, g_b\}$ (so these are in order, as models 1 to 5), we have model redundancy that if ignored could cause problems. If we were to get the Δ_i, in order 1 to 5, as {15, 10, 16, 20, 0}, then model redundancy becomes irrelevant because g_b is overwhelmingly the best model: The usual Akaike weight for that model is here $w_5 = 0.992$. We do claim, however, that the correct weights here should be based on model priors as $\left\{\frac{1}{8}, \frac{1}{8}, \frac{1}{8}, \frac{1}{8}, \frac{1}{2}\right\}$ not $\left\{\frac{1}{5}, \frac{1}{5}, \frac{1}{5}, \frac{1}{5}, \frac{1}{5}\right\}$. If the Δ_i are {2, 0, 1, 5, 20}, then again model redundancy is irrelevant (redundancy is only between model 5 and the others; there is no redundancy in models 1 to 4 if model 5 is ignored). But if the result for the Δ_i is {2, 2, 2, 2, 0}, then model redundancy matters a great deal as regards the proper w_i. For wrong priors $\left\{\frac{1}{5}, \frac{1}{5}, \frac{1}{5}, \frac{1}{5}, \frac{1}{5}\right\}$, $w_5 = 0.40$, but under correct priors $\left\{\frac{1}{8}, \frac{1}{8}, \frac{1}{8}, \frac{1}{8}, \frac{1}{2}\right\}$, $w_5 = 0.73$.

By adding submodels of the general time-specific model to our set of models, we dilute the absolute strength of evidence for model g_b as measured by Akaike weights; and we must use such absolute weights in certain formulas (e.g., model averaging). Inasmuch as these added models deal only with time variation in capture probabilities, they are all of a type (hence, redundant as regards their evidence *against* model g_b), so they unfairly "gang up" against model g_b, which is a totally different type of model.

The appropriateness of unequal priors if submodels of the general time model, g_t, are included is justified here on a theoretical basis. It is well documented in the capture–recapture literature that there is no practical advantage, as regards estimating N, of considering constrained versions of the general time-specific model. Thus, the original set of two models, $\{g_t, g_b\}$, should not be augmented as above. Hence, in the last example we should really have only these two models, and they have Δ_1 and Δ_2 as {2, 0}. Now, for model g_b, $w_2 = 0.73$. A key point here is that when we did have model redundancy, the use of the unequal priors did produce the correct Akaike weights. Thus, we think that model redundancy can be coped with analytically by appropriate modification of the otherwise equal model priors, τ_i.

Even more important than accepting model redundancy, and therefore modifying model priors, is to construct the set of models to be considered so that there is no model redundancy. As the above example illustrates, all suitable knowledge about the correct formulation and use of models for the problem at hand should be utilized in defining the a priori set of models to consider. Another point worth repeating is that neither the Δ_i nor the relative likelihoods of the models will be affected by model redundancy. Thus confidence sets on models based on all models with Δ_i less than some cutoff value may be the safest type to use. Our ideas on the cutoff value to use can be obtained from the distribution of Δ_p, but only for situations with no model redundancy. The recommendations already made on this matter were so developed.

The concept and issue of model redundancy was brought to our attention by S. T. Buckland (personal communication); the above ideas are our own. Professor Buckland suggested that the bootstrap would automatically be a solution to model redundancy as long as for a given bootstrap sample it is forced to select one best model. This seems reasonable; but we still perceive a need for analytical formulas, and we now think that the analytical solution to model redundancy lies in construction of unequal model priors. However, then we must be able to recognize model redundancy in our set of models. If we can do that (we can and should), we think that redundancy can be eliminated. If model redundancy operates at a more subtle level than considered here, the bootstrap would have an advantage. We are currently disinclined to think that there will be a model redundancy problem as regards Akaike weights as long as the set of models considered is carefully constructed. (More research on the issue would be helpful.)

4.7   Recommendations

If data analysis relies on model selection, then inferences should acknowledge model selection uncertainty. If the goal is to get the best estimates of a set of parameters in common to all models (this includes prediction), model averaging is recommended. If the models have definite, and differing, interpretations as regards understanding relationships among variables, and it is such understanding that is sought, then one wants to identify the best model and make inferences based on that model. Hence, reported parameter estimates should then be from the selected model (not model averaged values). However, even when selecting a best model, also note the competing models, as ranked by their Akaike weights. Restricting detailed comparisons to the models in a 90% confidence set on models should often suffice. If a single model is not strongly supported, $w_{min} \geq 0.9$, and competing models give alternative inferences, this should be reported. It may occur that the basic inference(s) will be the same from all good models. However, this is not always the case, and then inference based on a single best model may not be sound if support for even the best model is weak (in all-subsets selection when $R > 1,000$, w_{min} can be very small, e.g., < 0.01).

We recommend that investigators compute and report unconditional measures of precision based on (4.9) when inference is based on a best model, unless the Akaike weight w_i for the selected model is large (e.g., ≥ 0.9). For an unconditional confidence interval, often the form $\hat{\theta} \pm 2\,\widehat{se}(\hat{\theta})$ will suffice, or an interval of this type back-transformed from a function of $\hat{\theta}$ such as occurs via the link function in general linear models. If such a simple interval has clear deficiencies, or in general if the computation can be done, use inflated profile likelihood intervals based on formulas (4.10) and (4.11).

If interest is really just on some parameters in common to all models, then we recommend using model-averaged parameter estimates from (4.1). The sampling variance estimate to use is then (4.9). Again, often the form $\hat{\theta} \pm 2\,\widehat{se}(\hat{\theta})$ will suffice for a confidence interval.

We think that these analytical procedures can suffice, so we would not initially use the bootstrap to evaluate model selection uncertainty. The bootstrap can produce robust estimates of unconditional sampling variances and confidence intervals, as by the percentile confidence intervals, especially for $\tilde{\tilde{\theta}}$. The bootstrap provides direct, robust estimates of model selection probabilities π_i, but we have no reason now to think that use of bootstrap estimates of model selection probabilities rather than use of the Akaike weights will lead to superior unconditional sampling variances or model-averaged parameter estimators. The primary purpose of the bootstrap is to assess uncertainty about inferences; therefore, we recommend that the point estimates used be the actual MLEs from the selected model (not the bootstrap means). In analyses that are very complex, where there may be no suitable analytical or numerical estimators of conditional (on model) sampling variances, the bootstrap could be used to get conditional and unconditional measures of precision. We recommend that more bootstrap samples be used than is commonly the case; use 10,000 for really reliable results, but even 400 would be better than no assessment of model selection uncertainty (no assessment has often been the default).

Be mindful of possible model redundancy. A carefully thought-out set of a priori models should eliminate model redundancy problems and is a central part of a sound strategy for obtaining reliable inferences. Do not regress to statistical tests of post hoc null hypotheses, the associated P-values, and decisions concerning supposed "significance" are not valid.

The theory here applies if the set of models is a priori to the data analysis. If any models considered have been included after some analyses, because said model(s) are suggested by the data, then theoretical results (such as variance formulas) might fail to properly apply (in principle, the bootstrap can still be used). Even for such data-driven model selection strategies we recommend assessing model selection uncertainty rather than ignoring the matter.

4.8   Cement Data

We return to the cement data of Section 3.2 to compare bootstrap estimates of model selection frequencies (π_i), Δ_i values, Akaike weights (w_i), and unconditional estimation of sampling variances. These quantities are summarized in Table 4.1; the AIC$_c$-selected model is shown there in bold. The remaining seven models are not shown in Table 4.1 because they were never selected in the 10,000 bootstrap samples (also, they have virtually zero Akaike weights). The three simple approaches shown in Table 4.1 provide useful insights into model selection uncertainty for this very small ($n = 13$) data set. Clearly, model {12} is indicated as the best by all approaches. However, substantial model selection uncertainty is evident because that best model has an Akaike weight of only 0.57 and a bootstrap selection probability of 0.53. All three approaches cast substantial doubt concerning the utility of the final three or four models in Table 4.1. Model {34} is particularly unsupported, with $\hat{\pi}_i < 0.004$ and $w_i = 0.0004$.

TABLE 4.1. Bootstrap selection probabilities, $\hat{\pi}_i$, for the models of the cement data used in Section 3.2; $B = 10{,}000$ bootstrap samples were used; also shown are AIC differences Δ_i and derived Akaike weights computed from the data.

Model	K	$\hat{\pi}_i$	Δ_i	w_i
{12}	4	0.5338	0.0000	0.5670
{124}	5	0.0124	3.1368	0.1182
{123}	5	0.1120	3.1720	0.1161
{14}	4	0.2140	3.3318	0.1072
{134}	5	0.0136	3.8897	0.0811
{234}	5	0.0766	8.7440	0.0072
{1234}	6	0.0337	10.5301	0.0029
{34}	4	0.0039	14.4465	0.0004

Evidence for the importance of each variable can be obtained by using the bootstrap and tallying the percentage of times that each variable occurred in the AIC_c selected model (Section 4.4). For the 10,000 bootstrap samples, x_1 occurred in 93% of the models, followed by x_2 (76%), x_3 (23%), and x_4 (36%). Again, this simple approach indicates the importance of x_1 and x_2 relative to x_3 and x_4. Similar evidence can be obtained by summing the Akaike weights over those models with a particular variable present. Using that simple approach, the relative support of the four variables is as follows: x_1 (99%), x_2 (81%), x_3 (21%), and x_4 (32%). Considering the small sample size ($n = 13$), the bootstrap and Akaike weights seem to give similar results.

Using the idea of the pivotal Δ_p (Section 4.5) to obtain the bootstrap distribution as Δ_p^*, we find that an approximate 90% confidence set occurs for $\Delta_i < 8.75$, while a 95% set is achieved if $\Delta_i < 13.8$. These bootstrap based percentile values of Δ_i are quite extreme here because the sample size in this example is so small ($n = 13$).

Using the bootstrap selection frequencies ($\hat{\pi}_i$), models {12}, {14}, and {123} represent an approximate 86% confidence set, while adding model {234} reflects an approximate 94% confidence set of models. Using Akaike weights (w_i), an approximate 90% confidence set includes models {12}, {124}, {123}, and {14}. The Δ_i values suggest that the final three models in Table 4.1 have little utility. These types of ranking and calibration measures have not been available under a hypothesis testing approach or cross-validation.

We now illustrate the computation of unconditional estimates of precision, first for a parameter in common to all models. What if one wanted to predict the value $\hat{E}(Y_0)$, denoted for simplicity by $\hat{Y}_0$, given the values $x_1 = 10$, $x_2 = 50$, $x_3 = 10$, and $x_4 = 20$ (cf. Table 3.1)? The prediction under each of the eight models of Table 4.1 is shown in Table 4.2; we used PROC REG (SAS 1985) in SAS to easily compute predicted values and their conditional standard errors, $\widehat{se}(\hat{Y}_0 \mid g_i)$. Clearly, $\hat{Y}_0$ is high for model {234}, relative to the other models. The estimated standard error for model {1234} is very high, as might be expected

TABLE 4.2. Some analysis results for the cement data of Section 3.2; $\hat{Y}_0$ is a predicted expected response based on the fitted model (see text for x_i values used); conditional-on-model measures of precision are given for $\hat{Y}_0$; $\overline{\hat{Y}}$ denotes a model-averaged predicted value; and $\hat{Y}_0 - \overline{\hat{Y}}$ is the estimated bias in using a given model to estimate Y_0.

Model	K	$\hat{Y}_0$	$\widehat{se}(\hat{Y}_0 \mid g_i)$	$\widehat{var}(\hat{Y}_0 \mid g_i)$	$(\hat{Y}_0 - \overline{\hat{Y}}_{bootstrap})^2$	$\left(\hat{Y}_0 - \overline{\hat{Y}}\right)^2$
{12}	4	100.4	0.732	0.536	4.264	1.503
{124}	5	102.2	1.539	2.368	0.070	0.329
{123}	5	100.5	0.709	0.503	3.861	1.268
{14}	4	105.2	0.923	0.852	7.480	12.773
{134}	5	105.2	0.802	0.643	7.480	12.773
{234}	5	111.9	2.220	4.928	89.019	105.555
{1234}	6	101.6	5.291	27.995	0.748	0.001
{34}	4	104.8	1.404	1.971	5.452	10.074

because the X matrix is nearly singular. Both of these models have relatively little support, as reflected by the small relative weights, so the predicted value under these fitted models is of little credibility.

The predicted value for the AIC_c-selected model is 100.4 with an estimated conditional standard error of 0.73. However, this measure of precision is an underestimate because the variance component due to model selection uncertainty has not been incorporated. Model averaging (4.2) results in a predicted value of 102.5 using the bootstrap estimated weights ($\hat{\pi}_i$) and 101.6 using the Akaike weights (w_i) (4.1). The corresponding estimated unconditional standard errors are 3.0 using the bootstrap-based weights and 1.9 using the Akaike weights. These unconditional standard errors are substantially larger than the conditional standard error of 0.73. In Monte Carlo studies we have done we find that the unconditional standard errors better reflect the actual precision of the predicted value, and conditional confidence interval coverage is often quite near the nominal level (Chapter 5).

Study of the final three columns in Table 4.2 above shows that the variation in the model-specific predictions (i.e., the $\hat{Y}_0$) from the weighted mean (i.e., $(\hat{Y}_0 - \overline{\hat{Y}}_{bootstrap})^2$ or $(\hat{Y}_0 - \overline{\hat{Y}})^2$) is substantial relative to the estimated variation, conditional on the model (i.e., the $\widehat{var}(\hat{Y}_0 \mid g_i)$). Models {124} and {1234} are exceptions because they overfit the data (i.e., more parameters than are needed). The Akaike weights are relatively easy to compute compared to the effort required to obtain the bootstrap estimates $\hat{\pi}_i$; w_i seem preferred for this reason, in this example. That is, we perceive no advantage here from the bootstrap-based results, compared to the Akaike-weight-based results, that compensates for the computational cost of the bootstrap (we do not claim that the bootstrap-based results are any worse, just not better).

The investigator has the choice as to whether to use the predicted value from the AIC_c-selected model (100.4) or a model-averaged prediction (102.5 for the bootstrap weights, $\hat{\pi}_i$, or 101.6 for the Akaike weights w_i). In this example, the differences in predicted values are small relative to the unconditional standard errors (3.0 for the bootstrap and 1.9 for Akaike weights); thus here the choice of weights makes no great difference. However, there is considerable model uncertainty associated with this data set, and we would suggest the use of model-averaged predictions (when prediction is the objective), based on the Akaike weights. Thus, we would use 101.6 as the predicted value with an unconditional standard error of 1.9. If the AIC_c-selected model was much more strongly supported by the data, then we might suggest use of the prediction based on that (best) model (i.e., $\hat{Y}_0 = 100.4$) combined with use of the estimate of the unconditional standard error (1.9), based on the Akaike weights.

The selected model includes only regressor variables x_1 and x_2. For that model the estimated partial regression coefficients and their conditional standard errors are $\hat{\beta}_1 = 1.4683$ (conditional $\widehat{se} = 0.1213$) and $\hat{\beta}_2 = 0.6623$ (conditional $\widehat{se} = 0.0459$). Each of these parameters appears in eight models. To compute the estimate of unconditional sampling variation for $\hat{\beta}_1$ we first find each model containing β_1, its estimate and conditional variance in that model, and the model's Akaike weight:

Model	$\hat{\beta}_1$	$\widehat{se}(\hat{\beta}_1 \mid g_i)$	w_i
{12}	1.4683	0.1213	0.5670
{124}	1.4519	0.1170	0.1182
{123}	1.6959	0.2046	0.1161
{14}	1.4400	0.1384	0.1072
{134}	1.0519	0.2237	0.0811
{1234}	1.5511	0.7448	0.0029
{1}	1.8687	0.5264	0.0000
{13}	2.3125	0.9598	0.0000.

The first step is to renormalize the w_i so they sum to 1 for this subset of models. Here that sum is 0.9925 before renormalizing, so we will not display the renormalized w_i, but they are the weights to use in applying (4.1) and (4.9). The model-averaged estimate of β_1 is 1.4561 (from 4.1). Now apply (4.9) as

$$\widehat{var}(\hat{\theta}) = \left[\sum_{i=1}^{8} w_i \sqrt{\widehat{var}(\hat{\theta}_i \mid g_i) + (\hat{\theta}_i - \hat{\bar{\theta}})^2} \right]^2.$$

For example, the first term in the needed sum is $0.069646 = 0.5713 \times \sqrt{(0.1213)^2 + (0.0122)^2}$. Completing the calculation, we get $\widehat{var}(\hat{\theta}) = (0.1755)^2$, or an estimated unconditional standard error on $\hat{\beta}_1$ of 0.1755, compared to the conditional standard error given the selected model of 0.1213.

For the same calculations applied for $\hat{\beta}_2$ we start with

Model	$\hat{\beta}_2$	$\widehat{se}(\hat{\beta}_2 \mid g_i)$	w_i
{12}	0.6623	0.0459	0.6988
{124}	0.4161	0.1856	0.1457
{123}	0.6569	0.0442	0.1431
{1234}	0.5102	0.7238	0.0035
{234}	−0.9234	0.2619	0.0089
{23}	0.7313	0.1207	0.0000
{24}	0.3109	0.7486	0.0000
{2}	0.7891	0.1684	0.0000

When all 16 models are considered, the Akaike weights for just the eight models above add to 0.8114. However, to compute results relevant to just these eight models we must renormalize the relevant Akaike weights to add to 1. Those renormalization Akaike weights are what are given above. The model-averaged estimator of β_2 is 0.6110, and the unconditional estimated standard error of $\hat{\bar{\beta}}_2$ is 0.1206 (compared to the conditional standard error of 0.0459). It is important here to compute and use unconditional standard errors in all inferences after data-based model selection. Note also that (to be conservative) confidence intervals on β_1 and β_2, using results from model {12}, should be constructed based on a t-statistic with 10 df ($t_{10,0.975} = 2.228$ for a two-sided 95% confidence interval). Such intervals here will still be bounded well away from 0; for example, the 95% confidence interval for β_2 is 0.39 to 0.93.

We generated 10,000 bootstrap samples of these data and applied AIC_c selection to all 16 models fit to each bootstrap sample. Then as per Section 4.3.3 (and common belief about the bootstrap) it should be acceptable to estimate the unconditional standard error of an estimated partial regression coefficient based on the standard deviation of the set of realized estimates, such as $\hat{\beta}_{1,b}^*$, over all bootstrap samples, b, wherein the selected model included variable x_1. The results are given below, along with the average value of the parameter estimate over all relevant bootstrap samples:

parameter	bootstrap results	
	average	st. error
β_1	1.461	0.760
β_2	0.453	0.958
β_3	−0.420	1.750
β_4	−0.875	1.237

From the selected model, {12}, we get $\hat{\beta}_1 = 1.47$ and $\hat{\beta}_2 = 0.66$ with estimated unconditional standard errors of 0.18 and 0.12, respectively, as determined by analytical methods using Akaike weights.

Based on the above, and other comparisons not given, we conclude that the bootstrap failed here when all 16 models were allowed to be considered. As noted in Section 3.2, the full design matrix, X, for this regression example is essentially singular: The first three eigenvalues (in a principal components analysis on X) sum to 99.96% of the total of all four eigenvalues; the first two eigenvalues sum to 95.3% of the total. Also, the pairwise correlation of x_2 and x_4 is $r = -0.973$. This information, to us, strongly justifies (virtually forces) one to drop model {1234} from consideration and to drop all other models in which both x_2 and x_4 appear. Thus without any model fitting we can, and should, reduce the 16 possible models to 12 by eliminating models {24}, {124}, {234}, and {1234}. These sorts of considerations should be done routinely and do not compromise an a priori (as opposed to exploratory) model selection strategy.

With the reduced set of 12 models we computed the Δ_i and the w_i, and ran 10,000 new bootstrap samples (to get $\hat{\pi}_i$ and bootstrap estimates, $\hat{\theta}_b^*$), getting the results below (models not shown were never selected in the bootstrap samples):

Model	K	$\hat{\pi}_i$	Δ_i	w_i
{12}	4	0.5804	0.0000	0.6504
{123}	5	0.1315	3.1720	0.1332
{14}	4	0.2340	3.3318	0.1229
{134}	5	0.0465	3.8897	0.0930
{34}	4	0.0076	14.4465	0.0005

Applying here the method of Section 4.4 based on the sum of the Akaike weights, we get the relative importance for the four variables as follows: x_1 (0.9995), x_2 (0.7836), x_3 (0.2267), and x_4 (0.2164). Using the methods of Section 4.3.2, especially (4.7) with the above Akaike weights, we computed unconditional standard errors as $\widehat{se}(\hat{\beta}_1) = 0.18$ (for $\hat{\beta}_1 = 1.47$, $\widehat{se}(\hat{\beta}_1 \mid g_{\{12\}}) = 0.12$) and $\widehat{se}(\hat{\beta}_2) = 0.046$ (for $\hat{\beta}_2 = 0.66$, $\widehat{se}(\hat{\beta}_2 \mid g_{\{12\}}) = 0.046$). The bootstrap estimates of unconditional standard errors are 0.34 and 0.047 for $\hat{\beta}_1$ and $\hat{\beta}_2$, respectively.

The two different methods (analytical Akaike weights and bootstrap) now agree for $\widehat{se}(\hat{\beta}_2)$ but not for $\widehat{se}(\hat{\beta}_1)$. The resolution of this discrepancy hinges on two items. First, the correlation in the data of x_1 and x_3 is $r = -0.82$; second, the sample size is only $n = 13$. As a result, fitted models {123} and {134} are very unstable over bootstrap samples as regards the estimate of β_1. For example, the sampling standard deviation (this estimates $se(\hat{\beta}_1 \mid g_{\{12\}})$) of the 1,315 bootstrap values of $\hat{\beta}_{1,b}$ that resulted when model {123} was selected by AIC_c was 0.65 (the average of the bootstrap estimates $\hat{\beta}_{1,b}$ was 1.78). The theory-based estimate is $\widehat{se}(\hat{\beta}_1 \mid g_{\{12\}}) = 0.12$.

There are several points we wish to make with this example. **Results are sensitive to having demonstrably poor models in the set of models considered; thus it is very important to exclude models that are a priori poor.** The analytical method (vs. the bootstrap method) of assessing unconditional standard errors seems more stable, as regards having or excluding poor models from the set of models considered. In fact, the bootstrap approach failed when all 16 models were (erroneously) used. However, the analytical approach seemed reasonable even with all 16 models considered (the results were more precise when only the 12 models were used). With the reduced set of models the bootstrap results are still suspect, but now only because sample size is so small ($n = 13$). Monte Carlo evaluation and comparison of both methods is needed before definitive statements about reliability will be possible.

4.9   Pine Wood Data

We consider here an example of only two simple linear regression models, neither one nested in the other. This example has been used by Carlin and Chib (1995) on Bayesian model choice using Markov chain Monte Carlo methods. The data also appear elsewhere, such as in Efron (1984). The data (see Table 4.3) can be considered a trivariate response vector $(y, x, z)'$ for sample size $n = 42$. Variable y is the measured strength of a piece of pine wood, x is the measured density of that wood, and z is the measured density after adjustment for the measured resin content of the wood. The scientific question is which of x or z is a better predictor of the wood strength y, based on a linear model, either $y = a + bx + \epsilon$ or $y = c + dz + \delta$ (ϵ or δ are random residuals from the expected linear model structure). Residuals are taken to be normally distributed and homogeneous under either model. Scientifically, it is thought that wood density adjusted for resin content should be a better predictor of wood strength, but it takes more time and cost to measure variable z.

TABLE 4.3. Pine wood strength data y, wood density x, and wood density adjusted for resin content z (from Carlin and Chib (1995)); $n = 42$.

y	x	z	y	x	z	y	x	z
3040	29.2	25.4	2250	27.5	23.8	1670	22.1	21.3
2470	24.7	22.2	2650	25.6	25.3	3310	29.2	28.5
3610	32.3	32.2	4970	34.5	34.2	3450	30.1	29.2
3480	31.3	31.0	2620	26.2	25.7	3600	31.4	31.4
3810	31.5	30.9	2900	26.7	26.4	2850	26.7	25.9
2330	24.5	23.9	1670	21.1	20.0	1590	22.1	21.4
1800	19.9	19.2	2540	24.1	23.9	3770	30.3	29.8
3110	27.3	27.2	3840	30.7	30.7	3850	32.0	30.6
3160	27.1	26.3	3800	32.7	32.6	2480	23.2	22.6
2310	24.0	23.9	4600	32.6	32.5	3570	30.3	30.3
4360	33.8	33.2	1900	22.1	20.8	2620	29.9	23.8
1880	21.5	21.0	2530	25.3	23.1	1890	20.8	18.4
3670	32.2	29.0	2920	30.8	29.8	3030	33.2	29.4
1740	22.5	22.0	4990	38.9	38.1	3030	28.2	28.2

Simple linear regression can be used to find the MLE of σ^2 ($\hat{\sigma}^2$ = residual sum of squares divided by n) for each model. This gives us $\hat{\sigma}^2 = 109{,}589$ for model g_x (i.e., y vs. x) and $\hat{\sigma}^2 = 73{,}011$ for model g_z; both models have $K = 3$. The AIC_c values are 493.97 for model g_x and 476.96 for model g_z. The latter model being the estimated K-L best model, we select z as the best predictor. The two Δ_i are 0 and 17.01 for models g_z and g_x, respectively. The corresponding Akaike weights are 0.9998 and 0.0002.

The context in which we developed guidelines about interpreting Δ_i is one having more complexity and more models than here. Therein, a Δ_i of 17 would be considered overwhelming evidence for the superiority of model g_z. But the matter of interpretation of the strength of evidence is uncertain here. Therefore, for this model selection problem we recommend applying the bootstrap as well as using the above analytical results. The bootstrap is quite feasible here, much more so than in more complex model selection situations.

Based on 10,000 bootstrap samples of the data in Table 4.3 followed by AIC model selection we obtained results as follows. The bootstrap sampling distribution of Δ_p^* gives us estimated percentiles for the sampling distribution of Δ_p as 1.29 (95th percentile), 2.56 (96th percentile), 4.13 (97th percentile), 9.84 (99th percentile), and 17 is at about percentile 99.85. The Akaike weights are thus consistent here with the bootstrap sampling distribution of Δ_p^*, which gives us a basis to interpret Δ_i as regards the plausibility that model g_i is actually the K-L best model for the data. However, when we tabulate the model selection relative frequencies from the bootstrap we find model g_z selected in 93.8% of the 10,000 bootstrap samples. This is still strong evidence in favor of variable z as the better predictor.

It is clear that we select, based on strong evidence, the model structure $E(y) = c + dz$ as the better model. The estimates of the structural parameters of this model, and their conditional standard errors, are $\hat{c} = -1917.6$ ($\widehat{se} = 252.9$), $\hat{d} = 183.3$ ($\widehat{se} = 9.3$). We do not have a way we would consider reliable to compute unconditional standard errors when a parameter is unique to a single model (which d definitely is). However, when the evidence is strongly in favor of the selected model, such as here, it is reasonable to act as if we considered only that model, hence act as if that model would always be the one fit to such data, in which case conditional standard errors apply. So here we accept use of the conditional standard errors as a measure of estimator precision. As a rule of thumb we will hazard the suggestion that if the selected model has Akaike weight ≥ 0.90, it is acceptable to use the conditional standard errors. The exact value (i.e., 0.90) is not critical; the concept is that if (and only if) the data support the selected model strongly enough ($w_{min} \geq 0.9$ seems also be a safe rule of thumb; also see Royall (1997), where a similar rule is proposed), then conditional and unconditional standard errors will be nearly the same. In a case like this if one is bothered by the issue here of using conditional inference after data-based model selection, a modest simulation study can be done to explore that issue, as well as other matters.

Monte Carlo simulation methods are very useful for exploring model selection issues (and much more will be done with this idea in Chapter 5). We will introduce one use of simulation here, namely generating simulated "data" that closely mimic the apparent nature of the real data. This allows us to explore model selection in a case like these pine wood data when we know what model generated the data. Generally, we would be against such simple simulations as having relevant applicability to AIC model selection issues. However, here the issue is clearly one of just deciding between two linear models, so more complex data simulation models than used below do not seem needed.

We proceed by considering $(y, x, z)'$ as a trivariate normal random variable with mean vector μ and variance–covariance matrix $\Sigma = DCD$ where the diagonal matrix D has as its diagonal the marginal standard deviations of y, x, and z, and C is the matrix of correlations

$$\begin{bmatrix} 1 & \rho_{yx} & \rho_{yz} \\ \rho_{xy} & 1 & \rho_{xz} \\ \rho_{yz} & \rho_{zx} & 1 \end{bmatrix}.$$

From the data we obtain $\hat{\mu}' = (2992, 27.86, 26.79)$; the estimated marginal standard deviations are $894.60 (= \hat{\sigma}_y)$, 4.4946 and 4.6475; and the correlation estimates are $\hat{\rho}_{yx} = 0.9272$, $\hat{\rho}_{yz} = 0.9521$, and $\hat{\rho}_{xz} = 0.9584$. To generate a simulated observation mimicking the data we generate three independent standard normal random variables (i.e., normal(0,1)), say $\underline{v}$, then compute

$$(y, x, z)' = \underline{\mu} + DC^{0.5}\underline{v}$$

for some parameter choices "near" the estimated parameters. There are many software packages that will find the needed "square root" of matrix C (we used MATLAB, Anonymous 1994).

The best model here is the one that has the smaller residual standard error of y, given the predictor. Those true residual variances are $\sigma_{y|x}^2 = \sigma_y^2(1 - \rho_{yx}^2)$ and $\sigma_{y|z}^2 = \sigma_y^2(1 - \rho_{yz}^2)$. Therefore, in the simulation, the best variable for predicting y is the one with the biggest correlation coefficient with y (z, here). What we cannot determine without simulation is performance aspects of the model selection method.

We can tell from theory that only the values of ρ_{yx}, ρ_{yz}, and ρ_{xz} affect model selection performance, including the distribution of Δ_p and Δ_i values, and selection frequencies, hence Akaike weights. Hence, parameter values for μ and D are irrelevant to that aspect of the problem (we might be more concerned about values for μ and D if we wanted a realistic evaluation of model selection bias on parameter estimators). Therefore, in the simulations here it sufficed to set $\mu = 0$ and $D = I$. Using these values results in $a = c = 0$ and $b = \rho_{yx}$, and $\overline{d} = \rho_{yz}$, and knowing these as truth, we can infer relative degrees of model selection bias that might occur.

We generated 10,000 simulated observations for the three correlation coefficients being at their estimated values (as truth) and then did AIC_c model selection. To look some at the sensitivity of results to correlation coefficients, this process was repeated for four more sets of correlations with the same value of $\rho_{yz} - \rho_{yx} = 0.025$ as in the real data and a final case wherein all three true correlation coefficients used in the simulation were set to 0.95. Our primary objective was to determine the relative frequency of model selection as π_z (this is without loss of generality since $\pi_z + \pi_x = 1$) and the expected Akaike weight $E(w_z)$ (also without loss of generality since $E(w_x) + E(w_z) = 1$) and the 95th and 99th percentiles of Δ_p (denoted below by $\Delta_{p,0.95}$ and $\Delta_{p,0.99}$). Results are given below by assumed sets of correlation coefficients (estimated proportions have coefficients of variation of about 1%; the estimated Δ have coefficients of variation more like 2.5%, and this for 10,000 samples):

ρ_{yx}	ρ_{yz}	ρ_{xz}	π_z	$E(w_z)$	$\Delta_{p,0.95}$	$\Delta_{p,0.99}$
0.927	0.952	0.958	0.97	0.96	0.0	4.5
0.927	0.952	0.900	0.90	0.89	4.6	12.8
0.927	0.952	0.980	0.99	0.99	0.0	0.0
0.900	0.925	0.958	0.92	0.90	1.8	7.1
0.900	0.925	0.900	0.84	0.83	7.6	15.8
0.950	0.950	0.950	0.50	0.50	18.3	26.7

In these cases it is clear that $\pi_z \approx E(w_z)$, and that the sampling distribution of Δ_p is quite variable. This is a worse case as regards variability of the distribution of Δ_p (only two models, and they are nonnested). We can see that if the first case above were reality, we would expect to select the correct model in about 96% of all samples (for $n = 42$). These Monte Carlo results give us added faith in the usefulness of the bootstrap results based on the actual data and add faith in the strength of evidence deduced here from the data using $\Delta_x (= 17)$ and from $w_z = 0.9998$.

By looking at the more detailed results (not given here) on average values of estimated parameters, $\hat{\theta}$, and their averaged estimated standard errors $\widehat{se}(\hat{\theta} \mid g)$, given the selected model we can assess model selection bias in both point estimates and standard errors. If case 1 above were truth, then in the simulations suggested there would be little model selection bias here when g_z was selected and conditional standard errors applied (to be expected if model g_z is selected 96% of the time). When model g_x was selected, no strong biases in parameter estimators were suggested, but the sample size for this inference was only $m = 314$.

There is another interesting question we can explore with these simulation results. When a model is selected, right (g_z) or wrong (g_x), how frequently do we then judge the weight of evidence to be strongly in favor of that model? Our interest in such a question is mostly focused on when we make the wrong choice (we will not know this to be the case): Having picked the model that is not the K-L best model, will the data appear strongly to support the selected model as being best, or will the evidence be weak? For case 1 above we make the wrong choice with sampling probability only about 0.03 (314 of 10,000 samples). If we considered $w_x > 0.9$ as strong evidence in favor of the model g_x, when it was selected, we find that 101 of the 314 samples produced strong evidence in favor of the wrong model. Thus only 1% of all 10,000 samples would be strongly misleading in this simulated scenario. Conversely, for the 9,688 samples wherein model g_z was selected, 9,223 produced strong evidence ($w_z > 0.9$) in favor of the selected model. Hence we expect that in 92% of all samples (in this particular scenario) we would select the correct model and do so with convincing evidence. Note however, that we cannot tell from the actual data whether it is one if the "1%" strongly misleading samples. We can say, again just for this simulated scenario, that the *estimated* odds are 92 : 1 that we have reached a correct conclusion for these pine wood data.

4.10   The Durban Storm Data

Linhart and Zucchini (1986:176–182) apply AIC to storm frequency data from the Botanical Gardens in Durban, South Africa. The detailed data are given in their Table 10.1. By seven-day periods in the year ("weeks"), beginning with 1 January, they obtained the frequency of weeks with at least one storm event occurring. For example, in 47 consecutive years of data, for January 1–7 there were 6 years with at least one storm event. The data are based on a rigorous definition of a storm: "a rainfall event of at least 30 mm in 24 hours" (Linhart and Zucchini 1986:176). We use here their period I data y_i (i denotes week, 1 to 52), wherein for the first 22 weeks the sample size of years is $n_i = 47$; for weeks $23 \leq i \leq 52$, $n_i = 48$. Thus, under a fixed effects approach sample size here is 2,474. The data are from January 1932 to December 1979. We ignore, as did Linhart and Zucchini (1986), the minor matter of a few weeks needing to have 8 days (such as 26 February to 4 March when a leap year occurs). Listed in order $i = 1$ to 52, the data y_i are

6, 8, 7, 6, 9, 15, 6, 12, 16, 7, 9, 6, 8, 2, 7, 4, 4, 3, 3, 10, 3, 3, 0, 5, 1, 2,

4, 0, 2, 0, 3, 1, 1, 5, 4, 3, 6, 1, 8, 3, 4, 6, 9, 5, 8, 6, 5, 7, 5, 8, 5, 4.

Conceptually, there exists a probability p_i of a storm at the Durban Botanical Gardens in week i. Based on these data, what is a "good" estimate of p_1 to p_{52}? That was the analysis objective of Linhart and Zucchini, and it will be one of our objectives. Our other objective is to reliably assess the uncertainty of our $\hat{p}_i$. A simple estimator is $\hat{p}_i = y_i/n_i$; it is very nonparsimonious, lacks precision, and (most seriously) fails to be a smooth, hence informative, estimator of time trends in the true p_i. We expect that anyone considering this problem would strongly believe that the p_i would have a considerable degree of smoothness as a function over the 52 weeks. Therefore, we want to fit some model $p_i(\underline{\theta})$ for a not-large number of parameters represented by $\underline{\theta} = (\theta_1, \ldots, \theta_K)'$.

4.10.1 Models Considered

We agree with the general approach taken by Linhart and Zucchini (1986); they construct a likelihood by treating the y_i as a set of independent binomial random variables on sample sizes n_i for parameters p_i, and use the structural model as

$$\text{logit}(p_i) = \log(p_i/(1 - p_i)) = \sum_{j=1}^{K} \theta_j z_{ji}$$

being some suitable linear model on $\underline{\theta}$, for known "covariates" z_{ji}. Essentially, this is a type of logistic regression (we consider theory for AIC model selection in this situation in Section 7.6.6). Linhart and Zucchini used a finite Fourier series model for the z_{ji} and used TIC for model selection (which here became essentially the same as AIC). We extend their example by using QAIC and model averaging; also, we compute unconditional confidence intervals on the p_i. Here, K is the number of structural parameters in the model, plus 1 (for $\hat{c}$).

The structure of the simplest model, model g_1, is given by

$$\text{logit}(p_i) = \theta_1, \qquad i = 1, \ldots, 52.$$

For model g_2:

$$\text{logit}(p_i) = \theta_1 + \theta_2 \cos\left(\frac{2\pi(i - 1)}{52}\right) + \theta_3 \sin\left(\frac{2\pi(i - 1)}{52}\right), \qquad i = 1, \ldots, 52.$$

For model g_3:

$$\text{logit}(p_i) = \theta_1 + \theta_2 \cos\left(\frac{2\pi(i - 1)}{52}\right) + \theta_3 \sin\left(\frac{2\pi(i - 1)}{52}\right)$$
$$+ \theta_4 \cos\left(\frac{4\pi(i - 1)}{52}\right) + \theta_5 \sin\left(\frac{4\pi(i - 1)}{52}\right), \qquad i = 1, \ldots, 52.$$

In general, the structure for model g_r (wherein $K = 2r$) is given by

$$\text{logit}(p_i) = \theta_1 + \sum_{j=1}^{r-1}\left[\theta_{2j} \cos\left(\frac{2j\pi(i - 1)}{52}\right)\right.$$
$$\left. + \theta_{2j+1} \sin\left(\frac{2j\pi(i - 1)}{52}\right)\right], \qquad i = 1, \ldots, 52.$$

Assuming marginal binomial variation and independence, the form of the likelihood for any model is

$$\mathcal{L}(\underline{\theta}) \propto \Pi_{i=1}^{52}(p_i)^{y_i}(1 - p_i)^{n_i - y_i}.$$

Given the model for $\text{logit}(p_i)$ as a function of $\underline{\theta}$, say $h_i(\underline{\theta}) = \text{logit}(p_i)$, we compute p_i as

$$p_i = \frac{1}{1 + \exp[-h_i(\underline{\theta})]}.$$

TABLE 4.4. Some basic results from models g_1 to g_9 fitted to the weekly storm incidence data (Linhart and Zucchini 1986); QAIC is based on taking g_7 as the global model (hence $\hat{c} = 1.4$); the df of the goodness-of-fit χ^2 is $52 - K + 1$.

Model	K^a	$\log(\mathcal{L})$	Δ-AIC	χ^2	P	$\hat{c}$	Δ-QAIC
1	2	-863.24	62.66	131.4	0.000	2.57	40.67
2	4	-833.83	7.85	76.5	0.007	1.56	2.66
3	6	-829.17	2.53	69.3	0.019	1.47	0.00
4	8	-826.37	0.93	61.2	0.054	1.36	**0.00**
5	10	-823.91	**0.00**	55.6	0.094	1.29	0.49
6	12	-823.89	3.95	55.6	0.064	1.36	4.45
7	14	-823.40	7.04	54.7	0.049	1.40	7.76
8	16	-822.76	9.70	54.0	0.035	1.46	–
9	18	-822.47	13.11	53.8	0.022	1.54	–

[a] The number of structural parameters plus 1 for $\hat{c}$ for QAIC; the number of parameters for AIC in this example is $K - 1$.

The independence assumption may not be true, but it seems likely to be not badly wrong. Similarly, the count y_i may not be the sum of exactly homogeneous Bernoulli events over the n_i years. Truth may correspond more closely to having varying year-to-year weekly probabilities of a storm. A useful way to cope with these types of model inadequacies is to use ideas from quasi-likelihood theory, hence to use a variance inflation factor $\hat{c} = \chi^2/\text{df}$. This $\hat{c}$ is computed from the global model goodness-of-fit chi-square (χ^2) on degrees of freedom df. Then we use QAIC, rather than AIC; also, conditional sampling variances based on assumed models are multiplied by $\hat{c}$ (Section 2.5).

Following Linhart and Zucchini (1986) we consider seven models as our set over which model uncertainty and model averaging are computed. For model g_7, $K = 14$. We obtained MLEs for these models by using SAS PROC NLIN (SAS Version 6.12); it is easy to adapt PROC NLIN to produce ML estimates (see, e.g., Burnham 1989). In this example, it is not clear as to which model should serve as the basis for estimation of the variance inflation factor ($\hat{c}$). Thus, several models were explored (Table 4.4), and estimates of $\underline{c}$ were relatively stable at about 1.40. For each fitted model we also computed the usual chi-square goodness-of-fit statistic, its significance level (P-value), and $\hat{c}$. For the purpose of a more thorough consideration of model fit we also fit models g_8 ($K = 16$) and g_9 ($K = 18$). Table 4.4 gives basic results from these fitted models: K, $\log(\mathcal{L})$, Δ-AIC, χ^2 goodness-of-fit, and corresponding P-values, $\hat{c}$, and Δ-QAIC. The values of Δ-QAIC are for when model g_7 is taken as the global model.

The $\log(\mathcal{L})$ values in Table 4.4 for models 1 through 7 match the values of Linhart and Zucchini (1986) in their Table 10.3 (they did not fit models g_8 and g_9). The AIC-selected model has 9 parameters, and our MLEs $\hat{\underline{\theta}}$ match the results of Linhart and Zucchini (on their page 182). The parameters, $\underline{\theta}$, in the likelihood of these models do not have intrinsic meaning and are not of direct interest. Therefore, we do not present values of $\hat{\theta}$ from any fitted models, nor their estimated conditional standard errors (standard likelihood theory was applied to obtain the large-sample variance–covariance matrix of the MLE $\hat{\theta}$). Rather, our goal is to estimate well the set of p_1 to p_{52}, which in effect are parameters in common to all models.

4.10.2 Consideration of Model Fit

Before we accept the AIC-selected model, we must consider whether the global model fits. Based on the results in Table 4.4, the global model, g_7, is a not a good fit to the data: $P = 0.049$. More importantly, $\hat{c} = 1.4$ on 39 df is sufficiently greater than 1 that we should not accept results of AIC-selection that here require $c \approx 1$. Even the AIC-selected model has $\hat{c} = 1.29$ (and $P = 0.094$ even though this model is deliberately selected to fit well). To explore this issue further we fit two more models; models g_8 and g_9 also fit the data poorly. If the problem was an inadequate structural model, we would expect the fit to g_8 and g_9, compared to model g_7, to improve. The results for $\hat{c}$ in Table 4.4 strongly suggest that there is extrabinomial variation in these count data. Such a result is common for real count data such as these, as is the value of $\hat{c}$ (i.e., $1 < \hat{c} <\approx 2$).

However, before automatically resorting here to QAIC, there is another issue worth noting. The expected counts from the models fitted here are often small (i.e., the data are sparse in the sense of being small counts). For example, for model g_5, $\hat{E}(y_{26})$ to $\hat{E}(y_{32})$ are about 1.5; these are the smallest estimated expected count values here; the largest estimated expected values are about 10. Perhaps even if the global model is structurally true, the plethora of small count values will invalidate the usual central chi-square null distribution of the goodness-of-fit statistic.

We explored this matter by Monte Carlo methods (also called the parametric bootstrap method). We generated data based on truth being $\hat{p}_i$ from the AIC-selected model g_5. That is, independent y_i^* were generated as binomial$(n_i, \hat{p}_i)$ based on fitted model g_5. For each such data set we then fitted model g_5 and computed the chi-square goodness-of-fit statistic to see whether its distribution was noticeably different from that of a central chi-square on 43 df. We used only a sample of 100 such generated data sets because we were looking for a big effect: for this situation is $c = 1$ or 1.29?

The answer was clear: If the model truly fits, then on average we will get $\hat{c} = 1$; i.e., the usual null distribution holds well here despite small counts. The average of the 100 χ^2 goodness-of-fit values was 41.9 (theoretically it is 43). The largest and smallest of the 100 values were 71.6 ($P = 0.004$) and 21.8 ($P = 0.997$); these are not unusual for a sample of 100 such test statistics. When each test statistic was converted to a P-value, the set of 100 P-values fit a uniform $(0, 1)$ distribution. Finally, the average of the set of 100 values of $\hat{c}$ was 0.98 ($\widehat{se} = 0.21$). While the possibility remains that for these data $c = 1$

is appropriate (and we just happened to get an unusual realized sample), this is quite unlikely based on the Monte Carlo evidence. Moreover, experience-based general statistical wisdom for real data supports the belief that we should accept that extra binomial variation often exists in count data. We therefore will use QAIC, not AIC, with $\hat{c} = 1.4$ as our basis for model selection.

The number of estimable parameters must be augmented by 1 to reflect the need to estimate the variance inflation factor (c); if more than one variance inflation factor is estimated, then the number of such estimated factors should be included in K. If the estimate of c is close to 1, then no variance inflation is necessary and K should not be increased. If one believes that there is no concern about overdispersion and, therefore takes $c = 1$, then K should also not be incremented.

When sufficient precision is used in the calculations, we find that model g_4 is the QAIC best model, although for practical purposes models g_3 and g_4 are tied for best (and model g_5 is almost as good, based on QAIC). Figure 4.1 gives a plot of the fitted $\hat{p}_i$ for both models g_3 and g_4. Also shown are the approximate 95% confidence bands on p_i based on $\hat{p}_{L,i}$ and $\hat{p}_{U,i}$ for each week i. We next explain the calculation of these confidence intervals.

4.10.3 Confidence Intervals on Predicted Storm Probability

Basically, $\hat{p}_{L,i}$ and $\hat{p}_{U,i}$ arise as back-transformed lower and upper confidence limits on logit(p_i). However, we used SAS PROC NLIN to directly gives us the estimated MLE-based theoretical $\widehat{se}_t(\hat{p}_i \mid g)$ that is computed assuming $c = 1$. The first step is then to form the correct (inflated) estimated standard error: $\sqrt{\hat{c}} \cdot \widehat{se}_t(\hat{p}_i \mid g) = 1.183\,\widehat{se}_t(\hat{p}_i \mid g) = \widehat{se}(\hat{p}_i \mid g)$. The interval $\hat{p}_i \pm 2\,\widehat{se}(\hat{p}_i \mid g)$ could be used. However, it is better to use here what is basically an appropriate back-transformed logit-based interval (Burnham et al. 1987:214):

$$\hat{p}_{L,i} = \frac{\hat{p}_i}{\hat{p}_i + (1 - \hat{p}_i)C},$$

$$\hat{p}_{U,i} = \frac{\hat{p}_i}{\hat{p}_i + (1 - \hat{p}_i)/C},$$

where

$$C = \exp\left[\frac{t_{\alpha/2,df}\,\widehat{se}(\hat{p}_i \mid g)}{\hat{p}_i(1 - \hat{p}_i)}\right]$$

(acceptable as long as $\hat{p}_i$ does not get too close to 0 or 1). The confidence bands in Figure 4.1 were computed in this manner and are thus conditional on the model. We used $t_{\alpha/2}$, df $= 39$, because the df that apply here are those of $\hat{c}$; thus here df $= 39$.

Estimates of unconditional standard errors require the Akaike weights (or the bootstrap), in this case based on Δ_i from QAIC (Table 4.4). We find that $w_1, \ldots, w_7$ are (from Section 2.6).

0.0000, 0.0833, 0.3149, 0.3149, 0.2465, 0.0340, 0.0064.

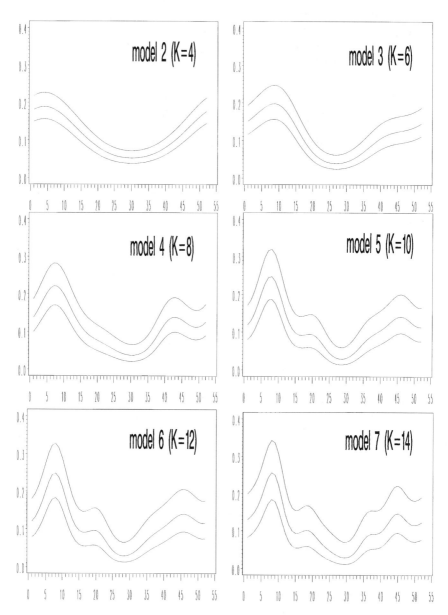

FIGURE 4.1. Plots of the predicted probability of one or more storms per week, $\hat{p}_i$ (y-axis), from models g_2 to g_7 fitted to the Durban storm data from Linhart and Zucchini (1986) (see text for details). Also shown are approximate 95% confidence bands on p_i; these bands are conditional on the model (see text for details).

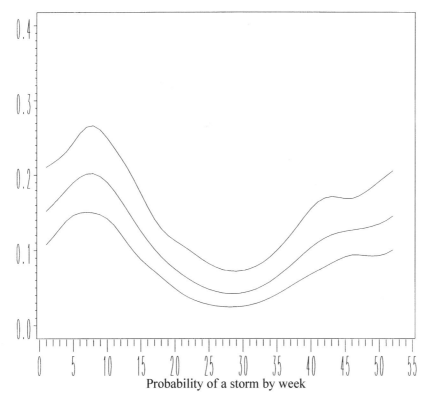

FIGURE 4.2. Plot of the model-averaged (from models g_2 to g_7) predicted probability of one or more storms per week, $\hat{p}_{a,i}$, from the Durban storm data from Linhart and Zucchini (1986) (see text for details). Also shown are approximate 95% confidence bands on p_i; these bands include model selection uncertainty.

Then for each week we find $\hat{p}_i$ under models g_1 to g_7 and apply (4.1) to find the model-averaged $\hat{p}_{a,i}$. Next we apply (4.9) to obtain the unconditional $\widehat{\text{se}}(\hat{p}_{a,i})$; this unconditional standard error also applies to $\hat{p}_i$ from the selected model as well as to $\hat{p}_{a,i}$. In cases like this where the $\hat{\underline{\theta}}$ are not of direct interest we recommend that the $\hat{p}_i$ to use are the model-averaged values $\hat{p}_{a,i}$. Moreover, here we use the unconditional $\widehat{\text{se}}(\hat{p}_{a,i})$, and we base confidence bands on the above formula for $\hat{p}_{L,i}$ and $\hat{p}_{U,i}$, however, for C based on $\widehat{\text{se}}(\hat{p}_{a,i})$. The resulting $\hat{p}_{a,i}$, $\hat{p}_{L,i}$, and $\hat{p}_{U,i}$ are shown in Figure 4.2.

4.10.4 Comparisons of Estimator Precision

We now give some considerations about (estimated) standard errors for different versions of the $\hat{p}_i$. Common practice would be to select a model and use the standard errors conditional on the model. In this case that would mean using model g_4 (by the slimmest of margins). We computed and examined the ratios

$$r_i(\text{se}) = \frac{\widehat{\text{se}}(\hat{p}_i \mid g_4)}{\widehat{\text{se}}(\hat{p}_{a,i})}, \qquad i = 1, \ldots, 52$$

and

$$r(\overline{\text{se}}) = \frac{\sum \widehat{\text{se}}(\hat{p}_i \mid g_4)}{\sum \widehat{\text{se}}(\hat{p}_{a,i})}.$$

These ratios are less than 1 if the unconditional is larger than the conditional standard error (the notation used for these ratios has no special meaning; we just need to represent them somehow).

We obtained $0.78 \leq r_i(\text{se}) \leq 1.02$ and $r(\overline{\text{se}}) = 0.90$. Thus the proper unconditional standard errors are on average 1.11 times the standard errors that are conditional on the model, and hence ignore model uncertainty. Also, we note that the average of the 52 values of $\widehat{\text{se}}(\hat{p}_{a,i})$ was 0.0214 $(0.012 \leq \widehat{\text{se}}(\hat{p}_{a,i}) \leq 0.035)$. This is good absolute precision; the actual average width of the 52 confidence intervals was 0.084.

An alternative that avoids model selection is to use $\hat{p}_i = y_i/n_i$. This parameter-saturated model is not very useful. Estimated standard errors under this model are given by $\widehat{\text{se}}_s(\hat{p}_i) = \sqrt{\hat{p}_i(1 - \hat{p}_i)/n_i}$ (no adjustment by any $\hat{c}$ is used here since there is no basis on which to compute a variance inflation factor given this model). We computed and examined the ratio $\left[\sum \widehat{\text{se}}_s(\hat{p}_i)\right] / \left[\sum \widehat{\text{se}}(\hat{p}_{a,i})\right]$ and considered the separate $\widehat{\text{se}}_s(\hat{p}_i)$ and $\widehat{\text{se}}(\hat{p}_{a,i})$. We obtained

$$\frac{\sum \widehat{\text{se}}_s(\hat{p}_i)}{\sum \widehat{\text{se}}(\hat{p}_{a,i})} = 2.31;$$

so on average the unconditional standard errors of the model-averaged $\hat{p}_{a,i}$ were more precise by a multiplicative factor $0.433 = 1/2.31$ compared to the much less useful parameter-saturated model estimates. Also, we observed that $0 \leq \widehat{\text{se}}_s(\hat{p}_i) \leq 0.082$ (and a variance estimate of 0 is quite wrong), whereas $0.012 \leq \widehat{\text{se}}(\hat{p}_{a,i}) \leq 0.035$; thus the standard errors for the model-averaged $\hat{p}_{a,i}$ are much more stable than is the case for the parameter-saturated model.

Linhart and Zucchini (1986) used TIC for model selection, not AIC. The only difference between the two methods is the use of $\text{tr}(JI^{-1}) = K$ rather than estimating this trace term, which Linhart and Zucchini denote by $\text{tr } \Omega_n^{-1} \Sigma_n$. In fact, in Section 7.6.6 we demonstrate by theory and example that $\text{tr}(JI^{-1})$ is very near K unless the structural model is truly terrible. In fact, we think that it is better for count data (and generally simpler) to use $\text{tr}(JI^{-1}) = K$ rather than estimate this quantity. Linhart and Zucchini (1986:181) give $\text{tr}(JI^{-1})$ in their Table 10.3. For models g_1 to g_7 the ratios of the estimated trace term to K are 0.95, 0.97, 0.97, 0.98, 0.98, 0.98, 0.98. Not only are these values close to 1, but they are all less than 1 (see Section 7.6).

4.11 Flour Beetle Mortality: A Logistic Regression Example

Young and Young (1998, 510–514) give as an example the analysis of acute mortality of flour beetles (*Tribolium confusu*) caused by an experimental five hour exposure to gaseous carbon disulfide (CS_2); the data are originally from Bliss (1935). Table 4.5 gives the basic data, as dose levels, number of beetles tested, and the number that died as an immediate causal result of exposure (note, we take sample size here as total beetles tested, hence $n = 471$). Observed mortality rate (Table 4.5) increases with dosage in a roughly, but not totally smooth, sigmoid form. It is typical to fit a parametric model to effectively smooth such data, hence to get a simple estimated dose-response curve and confidence bounds, and to allow predictions outside the dose-levels used. An extrapolation beyond the range of applied dose levels requires a fitted dose-response model; classically only one model was used. As an example, we will examine multimodel prediction of mortality at dose-level 40 mg/L.

A generalized linear models approach may easily, and appropriately, be used to model the probability of mortality, π_i, as a function of dose level x_i. The probability distribution assumed for the data is binomial: for n_i beetles tested at dose level x_i the response random variable y_i (number killed) is assumed to be distributed as binomial(n_i, π_i). Therefore, the likelihood is given by

$$\mathcal{L} = \prod_{i=1}^{8} (\pi_i)^{y_i} (1 - \pi_i)^{n_i - y_i}.$$

Within this setting a model means some parametric form for $\pi_i \equiv \pi(x_i)$ such that $0 < \pi(x) < 1$ is maintained. Moreover, in the context of generalized linear models there must be a nonlinear transformation (i.e., link function) of $\pi(x)$ to give a linear structural model in the parameters. There is one more constraint here: as dose increases, modeled mortality must not decrease (i.e., monotonicity). There are several commonly used forms for such a link-function-based

TABLE 4.5. Flour beetle mortality at eight dose levels of CS_2 (from Young and Young 1998, their Table 14.1).

	Number of Beetles		Observed
Dose (mg/L)	tested	killed	mortality rate
49.06	49	6	0.12
52.99	60	13	0.22
56.91	62	18	0.29
60.84	56	28	0.50
64.76	63	52	0.83
68.69	59	53	0.90
72.61	62	61	0.98
76.54	60	60	1.00

linear model but no single model-form that is theoretically the correct, let alone true, one.

We consider three commonly used generalized linear models and associated link functions: logistic, hazard, and probit (all are implemented in SAS PROC LOGISTIC, SAS Institute Inc., 1985). The logistic model form is

$$\pi(x) = \frac{1}{1 + e^{-(\alpha + \beta x)}},$$

with link function

$$\log\left(\frac{\pi(x)}{1 - \pi(x)}\right) = \text{logit}(\pi(x)) = \alpha + \beta x.$$

The hazard model and associated complementary log-log link function are

$$\pi(x) = 1 - e^{-e^{(\alpha + \beta x)}},$$

and

$$\log[-\log(1 - \pi(x))] = \text{cloglog}(\pi(x)) = \alpha + \beta x.$$

The cumulative normal model and associated probit link (SAS denotes it as NORMIT) are

$$\pi(x) = \int_{-\infty}^{\alpha + \beta x} \left[\frac{1}{\sqrt{2\pi}} e^{-\frac{1}{2}z^2}\right] dz \equiv \Phi(\alpha + \beta x),$$

and

$$\Phi^{-1}(\pi(x)) = \text{probit}(\pi(x)) = \alpha + \beta x.$$

Here, $\Phi(\cdot)$ denotes the standard normal cumulative probability distribution, which does not exist in closed form.

Traditionally, the predictor variable x can be either dose or log(dose). When dosages range over an order of magnitude or more, hence the lowest dose is relatively near 0 compared to the largest dose, then log(dose) is more commonly used because then $\hat{\pi}(0) = 0$ will apply. Otherwise, with the model forms above one risks getting $\hat{\pi}(0) > 0$, when in fact $\pi(0) = 0$ applies when we are recording acute mortality over a time interval so short that no natural mortality will occur. However, dose levels here are tightly clustered far away from 0 so either form of x is plausibly appropriate. One approach a person might try is to use models with both forms of x, thus six models.

A priori we do not know whether the simple model structure $\alpha + \beta x$ will suffice to fit the data. Thus one might be motivated to extend the basic linear model to be, say, link($\pi(x)$) $= \alpha + \beta x + \gamma f(x)$ where $f(x)$ can be such as x^2 (or log(dose) if $x = $ dose). Clearly, there is no unique model here. Thus, let us consider six more models: the three link functions combined with the form $\alpha + \beta x + \gamma x^2$, where x can be either dose or log(dose). One might proceed now without further thought. We do not recommend doing so but will first give the results for these 12 fitted models and then note further simple a priori thoughts

TABLE 4.6. AIC results for the 12 fitted models; a model can be recognized by its link function and the predictors included. Models are ordered by Δ.

AIC	Δ	weight	K	predictors		link
366.536	0.000	0.19356	2	dose		cloglog
366.641	0.105	0.18366	2	logdose		cloglog
367.608	1.072	0.11325	3	dose,	$dose^2$	logit
367.698	1.162	0.10826	3	logdose,	$logdose^2$	logit
367.804	1.268	0.10267	3	logdose,	$logdose^2$	probit
367.998	1.462	0.09318	3	dose,	$dose^2$	probit
368.420	1.884	0.07546	3	dose,	$dose^2$	cloglog
368.430	1.894	0.07508	3	logdose,	$logdose^2$	cloglog
370.246	3.710	0.03028	2	dose		probit
371.313	4.777	0.01776	2	dose		logit
374.138	7.602	0.00433	2	logdose		probit
375.222	8.686	0.00252	2	logdose		logit

that, in fact, allow an a priori restriction to just the three link functions and $link(\pi(x)) = \alpha + \beta x$, for $x = $ dose.

Table 4.6 gives AIC results for the 12 generalized linear models mentioned above. The models are identifiable by the predictors they contain and the link function used. There is considerable model uncertainty, as reflected by the Akaike weights. However, a first thing to note is that the eight plausible models ($\Delta < 2$) are "paired" on dose and log(dose), by link function and number of predictors, in the sense that the two models of a pair have almost the same Δ value. There is a logical reason for this, and it should be determined a priori, which would have led to consideration of only 6 models, based on either log(dose) or dose.

For these data (Table 4.5) dosage is between 49 and 77 mg/L. Over such a restricted interval log(dose) is almost perfectly linearly correlated with dose: $\log_e(dose) = 1.35376 + 0.007001 \cdot dose$, with $r^2 = 0.9960$. This near perfect correlation justifies using either dose or log(dose), but not both, as the basis for our models. In fact, to include here the six models based on dose and the corresponding six models based on log(dose) is a form of model redundancy (Section 4.6), and we recommend against it. Thus, a priori we would have only six models; it is our choice to keep things simple and just base models directly on dose.

Of the remaining six models, a further a priori consideration suffices to eliminate the quadratic models such as $logit(\pi(dose)) = \alpha + \beta dose + \gamma dose^2$. These three models cannot be monotonic increasing in dose because they are quadratic. They might fit the data well (they do), but they will either increase at lower doses or decrease at high doses. Because the quadratic models cannot be monotonic increasing in dose (unless $\hat{\gamma} = 0$ occurs), they should not have been used. A plot of a fitted model will reveal this fact, at which point one can feel justified in eliminating that model.

TABLE 4.7. AIC results for three fitted models justifiable a priori.

AIC	Δ	weight	K	predictor	link
366.536	0.000	0.80114	2	dose	cloglog
370.246	3.710	0.12534	2	dose	probit
371.313	4.777	0.07352	2	dose	logit

Table 4.7 shows the three models, of the 12, that can be justified here a priori. The associated Akaike weights change, relative to the 12 models, but the model evidence ratios are invariant. For example, $0.80114/0.07352 = 10.90 = 0.19356/0.01776$ for the hazard vs. logistic model based on dose as the predictor. Figure 4.3 shows plots of the three fitted models of Table 4.7.

Common analysis practice for these data would be to fit just one of the models of Table 4.7: historically probit, but in recent decades, logit. A very recent analysis might have looked at all three models of Table 4.7 (or all 12 models of Table 4.6) and used AIC to select the best model and then would have based inferences conditional on just that model. This practice ignores model uncertainty. Such uncertainty can be greater for extrapolations outside the range of doses used than at the actual doses. To illustrate multimodel inference (model averaging, here) we predict mortality at a dose of 40 mg/L ($\hat{\pi}(40)$).

Because there is a causal relationship of dose and mortality, extrapolation beyond the data, while risky, is both reasonable and of interest. However, when extrapolation is done model deficiencies can be important, such as non-monotonicity. Figure 4.4 gives plots of four fitted models: the three models of Table 4.7 and the third model in Table 4.6, the logit link on dose and dose^2. This latter model predicts increasing mortality as dose goes below about 44 mg/L, even though it provides a quite acceptable fit within the range of the actual data. Even if only discovered after the fact, we would use this non-monotonicity as a basis for eliminating this model.

Table 4.8 gives the basic conditional and unconditional results for the three models considered. The estimated best model has a weight of 0.80, which is not overwhelming (the evidence ratio for the best vs. second-best model is 6.39). The point estimates of $\pi(40)$ from the three models range from 0.0031 to 0.0308 and produce, at these extremes, nonoverlapping confidence intervals (Table 4.8). This information should be reported if knowledge of $\pi(40)$ is critical. If a single best-point estimate is acceptable, use the model-averaged $\hat{\pi}(40) = 0.0257$ with unconditional 95% confidence interval 0.0094 to 0.0680. Note that the unconditional standard error is 35% larger than the conditional standard error for the AIC best model. For a confidence interval on $\pi(40)$, based on the model-averaged $\hat{\pi}(40)(= 0.0257)$ and its unconditional standard error (0.01274), we used the logit-based interval of Burnham et al. (1987:214) (see also Section 4.10.3).

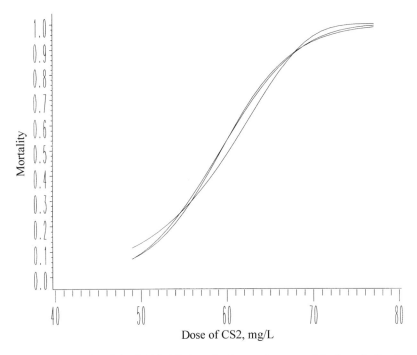

FIGURE 4.3. Fitted models of Table 4.7 plotted only over the range of the data; the fitted probit and logistic models are nearly identical compared to the slightly different complementary log-log model.

TABLE 4.8. Predicted mortality probability at dose 40mg/L; shown are results conditional on each of the three models and the unconditional model-averaged inference.

Link function	Akalke weight	$\hat{\pi}(40)$	$\widehat{se}$	95% Confidence Interval lower	upper
cloglog	0.8011	0.0308	0.00945	0.0168	0.0560
probit	0.1253	0.0031	0.00226	0.0007	0.0117
logit	0.0735	0.0085	0.00382	0.0035	0.0204
model averaged:		0.0257	0.01274	0.0094	0.0680

All three models used for Tables 4.7 and 4.8 fit the data well, so there was no need for any overdispersion adjustment. A simple Pearson observed vs. expected chi-square comparison suffices:

$$\chi^2 = \sum \frac{(O_j - \hat{E}_j)^2}{\hat{E}_j},$$

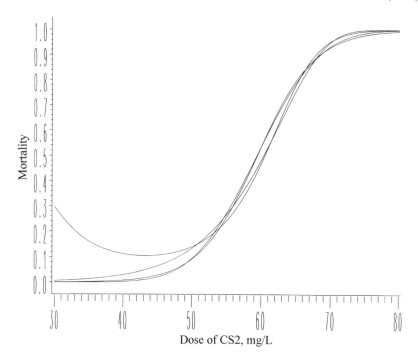

FIGURE 4.4. Four fitted models plotted beyond the range of dose levels used: complementary log-log, probit and logit models of Table 4.7 (and Figure 4.3), and the logit model on dose and dose^2 (Table 4.6), which wrongly predicts increasing mortality as dose goes below 44 mg/L.

on 6 degrees of freedom ($= 8 - 2$ since each model has 2 estimated parameters). For these binomially distributed data this chi-square statistic is

$$\chi^2 = \sum_{i=1}^{8} \frac{(y_i - n_i\hat{\pi}_i)^2}{n_i\hat{\pi}_i(1 - \hat{\pi}_i)}.$$

Goodness-of-fit results are as follows

model	χ^2	P
cloglog	3.49	0.74
probit	7.06	0.32
logit	7.65	0.26.

A possible outcome, not observed here, is the best model fitting but the worst model not fitting, say with $P = 0.01$ and $\hat{c} = 2.8$. If, as here, that worst-fitting model is not a global model, but is just an alternative of the same dimension as the other models, we should attribute the failure of fit to model structural inadequacies, not overdispersion. Therefore, we keep $c = 1$ (i.e., no overdispersion adjustments). Overdispersion would result in a lack of fit of all three models, i.e., even the best model would not fit well.

A key feature of this beetle mortality example, as compared, for example, to the Durban storm data (Section 4.10), is causality. The experimentally applied dose caused the observed mortality, whereas for the storm data there is only prediction. For that reason it sufficed with the storm data to have a single link function plus a series of increasingly general nested models; this also allowed a global model for that example. For the beetle data we can establish a priori that (1) the only predictor needed, or useful, is dose and (2) monotonicity of expected response must be imposed by any model. The issue about a model is thus reduced to one of an appropriate functional form, hence, in a generalized linear models framework, to what is the appropriate link function. However, as a result, we have no global model, but rather several (3 were used) alternatives for a best causal-predictive model (this situation could also arise in observational studies). We saw that this has implications for goodness-of-fit and overdispersion evaluation.

Moreover, an important issue here is prediction outside the range of doses applied. It is well known that many models might fit the observed data well, as occurred here, but give quite different extrapolated predictions, as also occurred here. It is thus not a sound idea to pick a single model and unquestioningly base extrapolated predictions on it when there is model uncertainty. This example illustrates how easy it is to compute model-averaged predictions, given the information from the fitted models, and that so doing may produce more realistically cautious predictions.

4.12   Publication of Research Results

We provide an outline of how results under the information-theoretic approach might be presented in papers submitted for publication (taken largely from Anderson et al. 2001d). The *Introduction* is the place to clearly state the study objectives and note the degree to which the paper is exploratory versus confirmatory.

Chamberlin's (1965) concept of "multiple working hypotheses" should underlie the *Methods* section. This is the place to describe and justify the a priori hypotheses and models in the set and how these relate specifically to the study objectives. Ideally, one should be able to justify why a particular model is in the set, as well as support the decision to exclude another model from the set. Avoid the routine inclusion of a trivial hypothesis or models in the model set; all the models considered should have some reasonable level of interest and support. The *Methods* section should always provide sufficient detail so that the reader can understand what was done.

A common mistake is the use of AIC rather than the second-order criterion AIC_c. Use AIC_c unless the number of observations is at least 40 times the number of explanatory variables (i.e., $n/K > 40$) for the most highly parametrized model in the set. If using count data, some detail should be given as to how goodness of fit was assessed and, if necessary, an estimate of a variance inflation factor (c), and its degrees of freedom. If evidence of overdispersion is found, the log-likelihood should be computed as $\log(\mathcal{L})/\hat{c}$, QAIC$_c$ should be used in model selection, and the covariance matrix should be multiplied by $\hat{c}$. When using QAIC or QAIC$_c$, one parameter must be added to the count of the number of estimated parameters in the model (K); this accounts for the estimation of the overdispersion parameter c. When the appropriate criterion has been identified (AIC, AIC$_c$, or QAIC$_c$), it should be used for all the models in the set. The adequacy of the global model should also be addressed when the response variable is continuous (e.g., regression); there are a host of procedures to aid in this task and many general ways to model the residual variation.

Discuss or reference the use of other aspects of the information-theoretic approach, such as model averaging, confidence sets on models, and examination of the relative importance of variables. Define or reference the notation used (e.g., K, Δ_i and w_i). Ideally, the variance component due to model selection uncertainty should be included in estimates of precision (i.e., unconditional vs. conditional standard errors), unless there is strong evidence favoring the best model, say an Akaike weight (w_i) > 0.9.

For well-designed, strict experiments in which the number of effects or factors is small and factors are orthogonal, use of the full model is usually most appropriate (rather than considering more parsimonious models). If an objective is to assess the relative importance of variables, the models should be selected to provide a balance among variables (Section 4.4). Inference about the relative importance of variables can then be based on the sum of the Akaike weights for each variable, across models, and these sums should be reported. Avoid the implication that variables not in the selected (estimated "best") model are unimportant. Give estimates of the important parameters (e.g., effect size) and measures of precision (preferably a confidence interval). Further evidence can be assessed using the Akaike weights (w_i) and evidence ratios. Provide quantities of interest from the best model, or others in the set (e.g., $\hat{\sigma}^2$, coefficient of determination, estimates of model parameters and their standard errors, insights from an analysis of the residuals). In other words, all the evidence should be gathered, interpreted, and presented objectively.

The *Results* section should be easy to report if the *Methods* section outlines convincingly the hypotheses and associated models of interest. Show a table of the value of the maximized log-likelihood function ($log(\mathcal{L})$), the number of estimated parameters (K), the appropriate selection criterion (AIC, AIC$_c$ or QAIC$_c$), the simple differences (Δ_i), and the Akaike weights (w_i) for models in the set (or at least the models with some reasonable level of support, say where $\Delta_i < 10$). It is often helpful to report the models in order, by ranking, based on the differences (Δ_i); this makes it easy to see which model is best, second best, and so on.

Do not include test statistics and *P*-values when using the information-theoretic approach as this inappropriately mixes differing analysis

paradigms. For example, do not use AIC_c to rank models in the set and then test whether the best model is "significantly better" than the second-best model. The classical tests one would use for this purpose are invalid, once the model pairs have been ordered by an information criterion. Do not imply that the information-theoretic approaches are a test in any sense. **Avoid the use of terms such as significant and not significant, or rejected and not rejected**; instead, view the results in a strength of evidence context (Royall 1997).

If some data dredging and modeling were done after the a priori effort, then be sure that this is clearly explained when such results are mentioned in the *Discussion* section. It is important to separate analysis results based on questions and hypotheses formed before examining the data from results found sequentially from examining the results of data analyses. The first approach tends to be more confirmatory, while the second approach tends to be more exploratory. In particular, if the analysis of data suggests a particular pattern leading to an interesting hypothesis, then at this midway point, few statistical tests or measures of precision remain valid. That is, an inference concerning patterns or hypotheses as being an actual feature of the population or process of interest are not well supported. Conclusions reached after repeated examination of the results of prior analyses, while interesting, cannot be taken with the same degree of confidence as those from the more confirmatory analysis. Often such post hoc results, while somewhat likely to be spurious, may represent intriguing hypotheses to be readdressed with a new, independent set of data. This is often an important part of good science.

4.13   Summary

Model selection should not be considered just the search for the best model. Rather, the basic idea ought to be to make more reliable inferences based on the entire set of models that are considered a priori. This means that we rank and scale the set of models and determine, perhaps, a confidence subset of the R models, for the K-L best model. Parameter estimation should also make use of all the models when appropriate (e.g., model averaging for prediction) and attempt to use unconditional variances unless the selected best model is strongly supported (say its $w_{min} > 0.9$).

In general there is a substantial amount of model selection uncertainty in many practical problems (but see the simulated starling experiment, and real pine wood data, for exceptions). Such uncertainty about what model structure (and associated parameter values) is the K-L best approximating model applies whether one uses hypothesis testing, information-theoretic criteria, dimension-consistent criteria, cross-validation, or various Bayesian methods. Often, there is a nonnegligible variance component for estimated parameters (this includes prediction) due to uncertainty about what model to use, and this component should be included in estimates of precision.

Model selection uncertainty can be quantified in two basic ways: based on the w_i values for the set of models considered or based on use of the bootstrap method. The simple computation of the relative likelihood of each model considered, given the data, and rescaling these model likelihoods to the Akaike weights, w_i, is often effective and easy to understand and interpret. (The w_i provide a basis for a measure of relative support of the data for the models, what we can call model scaling. The use of evidence ratios (Section 2.10) is effective and convenient.) The bootstrap, while computationally intensive, provides estimates of model selection probabilities, $\hat{\pi}_i$.

If there is substantial model selection uncertainty and if the sampling variance is estimated conditionally on the selected model, the actual precision of estimated parameters will likely be overestimated, and the achieved confidence interval coverage will be below the nominal level (e.g., perhaps 80% rather than 95%). Estimates of unconditional standard errors can be made using either Akaike weights or bootstrap selection probabilities. The full set of results for the R models considered over all bootstrap samples may also be useful to compute unconditional standard errors of estimated parameters.

Relatively little work has been done to understand and lessen model selection bias (Section 1.6). Shrinkage estimators such as

$$\tilde{\tilde{\theta}} = w_+ \hat{\tilde{\theta}}$$

deserve investigation. In particular, the estimation of the precision of $\tilde{\tilde{\theta}}$ needs further work.

Conditional inference can be relatively poor. There is a need for additional research on general methodology to incorporate model selection uncertainty as a variance component in the precision of parameter estimators. Model averaging has potential in some applications where interest is concentrated on parameters that appear (implicitly, at least) in every model in the set. Prediction might often be approached as a problem in (weighted) model averaging where the weights are either Akaike weights (w_i) or bootstrap selection probabilities ($\hat{\pi}_i$).

The importance of a *small* number (R) of candidate models, defined prior to detailed analysis of the data, cannot be overstated. Small is best conceptualized in contrast to what is not small: If the number of models considered exceeds the sample size, hence $R > n$, then there is *not* a small number of candidate models. This condition often occurs in the commonly used case of considering all possible linear models in variable selection. In that all-models case, when one has p variables, then $R = 2^p$; hence $R = 1024$ for $p = 10$, or $R = 1,048,576$ for $p = 20$. One should have R much smaller than n. MMI approaches become increasingly important in cases where there are many models to consider. If the background science is lacking, so that needed a priori considerations are deemed impossible, then the analysis should probably be considered to be only exploratory.

Finally, investigators should explain what was actually done in the model selection (i.e., data analysis). Was it objective model selection and assessment based on an a priori set of models? Alternatively, was the selected model a result of a subjective strategy of seeking a model that fits the data well by introducing new models into consideration as data analysis progresses? In the former case we recommend AIC (or its variants as needed). In the latter case, if the strategy can be implemented as a computer algorithm, then use the bootstrap to assess model selection uncertainty for such subjective or iterative searches. If the model selection strategy cannot be represented as an explicitly defined algorithm, one cannot determine model selection uncertainty.

5

Monte Carlo Insights and Extended Examples

5.1 Introduction

This chapter gives results from some illustrative exploration of the performance of information-theoretic criteria for model selection and methods to quantify precision when there is model selection uncertainty. The methods given in Chapter 4 are illustrated and additional insights are provided based on simulation and real data. Section 5.2 utilizes a chain binomial survival model for some Monte Carlo evaluation of unconditional sampling variance estimation, confidence intervals, and model averaging. For this simulation the generating process is known and can be of relatively high dimension. The generating model and the models used for data analysis in this chain binomial simulation are easy to understand and have no nuisance parameters. We give some comparisons of AIC versus BIC selection and use achieved confidence interval coverage as an integrating metric to judge the success of various approaches to inference.

Section 5.3 focuses on variable selection (equivalent to all-subsets selection) in multiple regression for observational data assuming normally distributed homogeneous errors. A detailed example of AIC_c data analysis and inference is given, and it is shown how to extend the example to a relevant Monte Carlo investigation of methodology. The same Monte Carlo methods are used to generate additional specific illustrative results. We also discuss and illustrate ways to reduce a priori the number of variables. A discussion subsection provides details of model selection bias and other ideas with emphasis on problems of, and better approaches to, all-subsets selection.

Detailed examples of data analysis are given in Sections 5.3 and 5.4. In both cases we make a number of general points about K-L–based model selection. One real example uses distance sampling data on kangaroos in Australia. The reader is encouraged to read these examples, because each is used to convey some general information about data analysis under a model selection approach.

In the following material it is important to distinguish between sampling variances, or standard errors, that are conditional on one particular model, versus those that are unconditional, hence not based on just one specific model. Conditional measures of precision based on a restrictive model g are often denoted by $\text{var}(\hat{\theta}_j \mid \theta_j)$ or $\text{se}(\hat{\theta}_j \mid \theta_j)$; however, they are more properly denoted by $\text{var}(\hat{\theta}_j \mid g)$ or $\text{se}(\hat{\theta}_j \mid g)$. Corresponding unconditional values may be denoted by $\text{var}(\hat{\theta}_j)$ or $\text{se}(\hat{\theta}_j)$, and have (as needed) an added variance component due to model selection uncertainty. Even such "unconditional" sampling variances do depend on the full set of models considered.

The "as needed" phrase is appropriate because sometimes inference is design-based rather than model-based (see, e.g., Schreuder et al. 1993: Chapter 6, and Edwards 1998). This applies, for example to the use of the sample mean from a random sample; or here to the use of just $\hat{S}_i = n_{i+1}/n_i$: There is no restrictive model assumed for this inference. However, model-based inference is usually necessary and can be very effective for parameter estimation based on complex data.

5.2 Survival Models

5.2.1 A Chain Binomial Survival Model

We consider here tracking a cohort of animals through the entire survival process of the cohort. Thus at time 1 there are a known number n_1 of animals alive. Usually, we are thinking that these are all young of the year, hatched or born at time 1. One year later there are n_2 survivors; in general, at annual anniversary dates $i > 1$ there are n_i survivors. Eventually, we have for some last anniversary year ℓ, $n_\ell > 0$ but $n_{\ell+1} = 0$. Given the initial cohort size n_1, the survival probability in year one (i.e., for the first year of life) is conceptually defined as $S_1 = E(n_2 \mid n_1)/n_1$. In general, for year i, $S_i = E(n_{i+1} \mid n_i)/n_i$. Thus the obvious general estimator is $\hat{S}_i = n_{i+1}/n_i$; this estimator is not based on any assumed model structure imposed on the set of basic survival parameters. If n_i is large enough, then $\hat{S}_i$ is an acceptable estimator in terms of precision. However, as the cohort dies out, n_i becomes small, and then $\hat{S}_i$ is not reliable. In particular, in the year ℓ that the cohort dies out we always get $\hat{S}_\ell = (0/n_\ell) = 0$. This is a terrible estimator of age ℓ survival probability. To avoid this latter problem, and to improve age-specific survival estimates for ages i at which n_i

is small, we must impose a model structure on the set of parameters S_1, S_2, ..., S_ℓ. First, we consider what to use here for the global model.

The quantity n_1 is known at the study initiation, while the subsequent counts $n_2, n_3, n_4, \ldots$, being not then known, are treated sequentially as both random variables and known ancillaries in the eventual data analysis. Given n_1 we will generate, in the Monte Carlo study, n_2 as binomial(n_1, S_1). Because the survival process is sequential in time, once the survivors (hence n_2) at time 2 are known, we can model n_3, given n_2, as binomial(n_2, S_2). In general, at time i we know n_i; hence we can generate n_{i+1} as a binomial(n_i, S_i) random variable. Thus we generate our probability model, hence likelihood, for the data as a chain binomial model with conditional independence of random variable n_i given the temporally preceding n_1 to n_{i-1}. The likelihood function for the most general possible model is

$$\mathcal{L}(\underline{S} \mid \underline{n}) = \prod_{i=1}^{\ell} \binom{n_i}{n_{i+1}} (S_i)^{n_{i+1}} (1 - S_i)^{n_i - n_{i+1}}, \tag{5.1}$$

where $\underline{S}$ and $\underline{n}$ are vectors of the survival parameters and data counts, respectively. The underlying parameters explicitly in (5.1) are S_1, S_2, ..., S_ℓ. In this context a (restricted) model for the parameters is some imposed smoothness such as $S_i \equiv S$ for all ages $i \geq 1$ (model g_1), or perhaps the S_i are restricted only after age three; hence $S = S_4 = \cdots = S_\ell$, and S_1, S_2, and S_3 are unrestricted (model g_4). If no structural restrictions are imposed, we have the most general possible global model, g_i, for which the MLEs are

$$\hat{S}_i = \frac{n_{i+1}}{n_i}, \quad i \leq \ell,$$

with conditional sampling variances

$$\mathrm{var}\left(\hat{S}_i \mid g_i\right) = \frac{S_i(1 - S_i)}{n_i}.$$

If such data were from moderately long-lived species such as owls, elk, or wolves, one might expect substantial differences in survival for the first few age classes, hence $S_1 < S_2$ and so on for several age classes. There would then be near-equal survival probabilities for adults until a decrease occurs in yearly survival as the surviving animals approach old age (i.e., senescence). Thus, there are some large age-specific effects in survival followed by smaller tapering effects (this conceptual model is not universally applicable; for example, it does not apply to salmon). In real populations, the age-specific survival rates would be confounded with annual environmental effects and also, most likely, with some individual heterogeneity. We will not pursue all of these realities here and instead focus on some points related to conditional and unconditional inference in model selection and some comparison of information-theoretic approaches versus dimension-consistent criteria (e.g., BIC). This examination will illustrate the amount of variability in model selection and the effects of sample size n_1.

Ritei Shibata was born in Tokyo, Japan, in 1949. He received a B.S. in 1971, an M.S. in 1973, and a Ph.D. in mathematics in 1981 from the Tokyo Institute of Technology. He is currently a professor in data science, Department of Mathematics, at Keio University, in Yokohama, Japan. He has been a visiting professor at the Australian National University, University of Pittsburgh, Mathematical Science Research Institute (Berkeley, CA), and Victoria University of Wellington, New Zealand. He is interested in statistical model selection, time series analysis, S software, and data and description (D&D).

Because of the paucity of data at older ages, some constraint (i.e., model) must be assumed, at least for the older ages, to get reliable results. One simple solution is to pool data beyond some particular age and assume that the pooled ages have a constant survival probability. Thus, for animals of age r and older we assume that they all have the same survival probability, $S \equiv S_i, i \geq r$. This is not to say that truth is no longer age-specific, but rather that a parsimonious *model* of the survival process is not age-specific after some age r. An alternative is to do some modeling (e.g., logistic models) of the survival process and then do model selection. We will address both approaches in this section.

Our first sequence of models is defined below in terms of restrictions on the age-specific annual survival probabilities:

Model	K	Parameters	
1	1	$S \equiv S_i,$	$i \geq 1$
2	2	$S_1, S \equiv S_i,$	$i \geq 2$
3	3	$S_1, S_2, S \equiv S_i,$	$i \geq 3$
$\vdots$	$\vdots$	$\vdots$	$\vdots$
R	R	$S_1, S_2, \ldots, S_{R-1}, S \equiv S_i,$	$i \geq R$

The pattern is obvious: Model g_r has r parameters with different survival probabilities for ages 1 to $r - 1$, and constant annual survival probability for age r and older. The global model is g_R; model g_r has $K = r$ parameters.

This particular set of models is convenient for Monte Carlo simulations because all MLEs exist in closed form. Let the tail sum of numbers of animals alive at and after age r be denoted by $n_{r,+} = n_r + n_{r+1} + \cdots + n_j + \cdots$. Then the MLEs for model g_r ($r \leq \ell$) are

$$\hat{S}_i = \frac{n_{i+1}}{n_i}, \quad i = 1, \ldots, r - 1,$$

$$\hat{S} = \frac{n_{r+1,+}}{n_{r,+}} = \frac{n_{r+1} + n_{r+2} + \cdots}{n_r + n_{r+1} + \cdots}.$$

Also, the likelihood for the r parameters given model g_r is (from (5.1) constrained by model g_r)

$$\mathcal{L}_r(\underline{S} \mid \underline{n}) = \left[\prod_{i=1}^{r-1} \binom{n_i}{n_{i+1}} (S_i)^{n_{i+1}} (1 - S_i)^{n_i - n_{i+1}} \right] \left[\binom{n_{r,+}}{n_r} (S)^{n_{r,+}} (1 - S)^{n_r} \right].$$

$$(5.2)$$

5.2.2 An Example

The first example compares the performance of AIC_c and BIC for sample size $n_1 = 150$. BIC is the Bayesian information criterion (also similar to SIC) developed independently and from somewhat differing viewpoints by Schwarz (1978) and Akaike (1978, 1979). The form of BIC is

$$-2\log(\mathcal{L}) + K \cdot \log(n).$$

This is a type of criterion that Bozdogan (1987) calls "dimension consistent," since such criteria are often based on the assumption that a true model exists and this model is in the set of candidate models. Then, such criteria lead to an estimate of the dimension of this true model with probability 1 as sample size increases asymptotically. Other interpretations weaken this assumption to "quasi-true models" (see Cavanaugh and Neath (1999)) and avoid the notion that BIC provides an estimate of the true model's dimension. At best, BIC and

SIC are barely *information criteria*, since they are not estimates of Kullback–Leibler information. Theoretical issues regarding BIC are discussed in Chapter 6 in some detail. Here, we only present some comparisions for those with some exposure to BIC or SIC.

The true survival probabilities S_1, S_2, and so forth were 0.5, 0.7, 0.75, 0.8, 0.8, and then beyond S_5, the age-specific annual survival decreased by 2% per year (e.g., 0.784, 0.768, 0.753, 0.738, and so forth for S_6 and beyond). Thus, in the data-generating model the survival parameters vary smoothly, increasing to a maximum at ages 4 and 5, then simulating senescence by a 2% per year decrease in S_j. For each Monte Carlo sample the cohort is followed until all n_1 animals are dead. For example, the simulated animal counts for one repetition $(n_i, i = 1, 2, \ldots, 16 = \ell + 1)$ were

$$150 \text{ (fixed)}, 74, 49, 34, 27, 21, 13, 9, 6, 6, 6, 4, 4, 1, 1, 0.$$

The set of models considered here is g_1 to g_{10}; thus $R = 10$. This is a nested set of models within the global model g_{10}. Hence, the global model here allows separate estimates of S_1, S_2, $\ldots$, S_9 but a single "pooled" estimate of S after age 9. The results for 10,000 Monte Carlo samples are given in Table 5.1, where the two selection approaches yield substantial differences in the model selected. On average, AIC_c selected an approximating model with $K = 3.3$ parameters, while BIC selected a model with an average of 2.1 parameters.

The thinking underlying BIC is that the true model exists and is in the set of candidate models; this condition is not met here: Truth is not in the set of ten models (the model closest to truth is model g_{10}). Asymptotically BIC attempts to estimate the dimension of the true model, a concept that is not even well-defined here because there are nominally an unbounded number of age-

TABLE 5.1. Comparison of model selection relative frequencies for AIC_c vs. BIC using models g_1 to g_{10} and data from the chain binomial generating model with parameters S_1 to S_{10} as 0.5, 0.7, 0.75, 0.8, 0.8, 0.784, 0768, 0753, 0738, and 0.723 (i.e., after $j = 5$, $S_{j+1} = 0.98 \cdot S_j$); results are for sample size $n_1 = 150$.

Model	K	AIC_c model selection		BIC model selection	
		percent	cumul. %	percent	cumul. %
1	1	0	0	0.2	0.2
2	2	52.0	52.0	88.6	88.8
3	3	22.4	74.4	10.0	98.8
4	4	5.6	80.0	0.5	99.3
5	5	5.5	85.5	0.5	99.8
6	6	5.0	90.5	0.1	99.9
7	7	3.4	93.9	0	100.0
8	8	2.4	96.3	0	100.0
9	9	2.2	98.5	0	100.0
10	10	1.5	100.0	0	100.0

TABLE 5.2. Summary of Monte Carlo results for estimated age-specific survival probabilities under AIC_c and BIC model selection (see text for details of data generation); the
models used were g_1 to g_{10} (see text); these results are based on 10,000 samples; results are
for sample size $n_1 = 150$.

Age	S_i	$E(\hat{S}_i)$		$E(\widehat{se})$, AIC		$E(\widehat{se})$, BIC		Coverage, AIC_c	
i		AIC_c	BIC	cond.	unc.	cond.	unc.	cond.	unc.
1	0.500	0.500	0.501	0.041	0.041	0.041	0.041	0.959	0.959
2	0.700	0.711	0.732	0.039	0.053	0.029	0.052	0.731	0.959
3	0.750	0.751	0.749	0.035	0.053	0.026	0.038	0.916	0.966
4	0.800	0.772	0.751	0.033	0.052	0.026	0.031	0.639	0.921
5	0.800	0.769	0.750	0.032	0.050	0.026	0.029	0.630	0.906
6	0.784	0.760	0.749	0.032	0.048	0.026	0.029	0.806	0.958
7	0.768	0.754	0.749	0.032	0.048	0.026	0.029	0.895	0.976
8	0.753	0.750	0.749	0.032	0.048	0.026	0.029	0.914	0.977
9	0.738	0.745	0.749	0.032	0.047	0.026	0.029	0.870	0.962
10	0.723	0.740	0.749	0.032	0.047	0.026	0.029	0.770	0.912
Average								0.813	0.950

specific survival rates. However, there will not be enough data ever to estimate
all those parameters, and increasing n_1 does not change this conundrum.

In sharp contrast, AIC_c attempts to select a parsimonious approximating
model as a basis for inference about the population sampled. It does not assume
that full truth exists as a model, nor does it assume that such a "true model"
is in the set of candidates. AIC_c estimates relative expected Kullback–Leibler
distance and then selects the approximating model that is closest to unknown
truth (i.e., the model with the smallest value of AIC_c). Based on the evaluated
$E(\Delta_i)$ (good to about two significant digits), it is model g_2 that is the AIC_c best
model here, on average, but only by a minute winning margin over model g_3.
Therefore, without loss of generality we can set $E(\Delta_2) = 0$ and thus give the
other values of $E(\Delta_i)$ compared to this minimum value. In order, $i = 1$ to 10,
we have the $E(\Delta_i)$ as 25.0, 0.0, 0.09, 1.2, 2.0, 2.7, 3.6, 4.6, 5.6, and 6.7.

Further results from the Monte Carlo study are presented in Table 5.2. Specifically, we look at the properties of the estimated age-specific survival rates
under AIC_c and BIC model selection strategies. In Table 5.2, $E(\hat{S}_i)$ is the estimated expected value of $\hat{S}_i$, where $\hat{S}_i$, by sample, is the estimate of age i
survival rate based on whatever model was selected as best for that sample;
this estimator is computed for both AIC_c and BIC model selection methods.
For example, if model g_4 is selected, then $\hat{S}_i = n_{i+1}/n_i$, $i = 1, 2, 3$, and
$\hat{S}_i = n_{5,+}/n_{4,+}$ for $i \geq 4$.

From Table 5.2 the estimators of age-specific survival probabilities were
nearly unbiased; however, bias was slightly smaller for the AIC_c-selected
models than for the BIC-selected models for 8 of the 9 ages where the $E(\hat{S}_i)$
differed. The expected estimated standard errors, conditional (cond.) on the

selected model, were generally smaller for BIC selection compared to AIC_c selection, as was expected. The conditional (on selected model) standard errors under both AIC_c and BIC selection are too small, since they do not account for any model selection uncertainty.

Various comparisons are possible from these simulations, such as achieved empirical standard errors versus expected estimated standard errors based on theory. However, confidence-interval coverage is an integrating measure of how well the methodology is performing. Therefore, we focus on achieved confidence interval coverage for nominal 95% intervals, all computed (in this section) as $\hat{S} \pm 2\,\widehat{se}$. Use of this simple form may have resulted in an additional 1 to 3% failure in coverage of S_i for older ages, but we judge this as irrelevant to the contrast of conditional versus unconditional coverage.

Conditional confidence interval coverage in Table 5.2 of true S_i under AIC_c model selection is generally below the nominal level, ranging from 0.731 to 0.959 over S_1 to S_{10}. Adjusting the conditional standard errors of $\hat{S}_i$ to be unconditional (unc.) using (4.9) provides much improved coverage (averaging 95.0%), ranging from 0.906 to 0.977 (Table 5.2).

Achieved coverage using conditional standard errors for BIC model selection averaged 77.8% across the 10 age classes (Table 5.2). Use of (4.6) with BIC model selection (there is no theoretical basis to justify doing this) improved the achieved coverage of true S_i for BIC model selection to 86.4% (range 60.1 to 98.7%); coverage generally remained below the nominal level (95%). Buckland et al. (1997) present results on a survival model that is similar to the one used here.

This example shows the large amount of uncertainty associated with AIC_c model selection when sample size is small ($n_1 = 150$) and the generating model has tapering effects. This simple simulation exercise also demonstrates that BIC selection cannot be recommended. It requires very large sample sizes to achieve consistency; and typically, BIC results in a selected model that is underfit (e.g., biased parameter estimates, overestimates of precision, and achieved confidence interval coverage below that achieved by AIC_c-selected models). Conditional estimates of precision, under either AIC_c or BIC, exclude model selection uncertainty, and this is often an important omission. Incorporating model selection uncertainty can bring achieved confidence interval coverage up to approximately the nominal level for model selection under AIC_c (using (4.7)) (this applies to AIC also).

The results in Table 5.2 are typical of many similar simulation cases we have examined (with different sample size, various sets of S_i, and numbers of models R). To make more use of this particular example, in terms of the set of true S_i, we obtained confidence interval coverage on S_i, averaged over all S_i, $i = 1$ to R, for AIC_c model selection for some additional sample sizes. We also tabulated the 90th, 95th, and 99th percentiles of Δ_p; this information is useful for interpretation of the Δ_i. As sample size increases, the AIC_c best

TABLE 5.3. Some Monte Carlo results for AIC_c model selection, for the same example set of S_i used in Tables 5.1, 5.2 (see text for details), for varying sample size n_1; R is the number of models considered; achieved confidence interval coverage (nominally 95%) has been averaged over conditional (cond.) and unconditional (unc.) intervals on S_1 to S_R.

Sample size n_1	R	Best model K	Confidence interval coverage, %		percentiles of Δ_p		
			cond.	unc.	0.90	0.95	0.99
100	10	2	84.4	95.5	4.0	5.9	10.7
250	10	3	77.8	93.3	4.8	7.1	12.7
500	10	3	75.5	92.2	7.6	10.3	16.8
1,000	10	8	78.6	92.8	5.9	7.3	10.4
1,000	12	8	76.0	90.6	6.4	7.8	12.0
10,000	10	10	86.2	93.0	1.0	1.9	2.0
10,000	20	13	78.3	88.9	5.4	7.2	11.2

model, should include more parameters. Thus one needs also to increase the size of the model set, i.e., R.

Unconditional confidence interval coverage is much superior to conditional coverage (Table 5.3). When R is large, such as 20 for $n_1 = 10,000$, average (over S_1 to S_{20}) confidence interval coverage suffers as a result of the often extrapolated estimates of, say, S_{20} based on a fitted model averaging around g_{13}. The S_i have a shallow peak at $i = 4$ and 5, and this feature of truth requires a large sample size to detect reliably. Thus the theoretically best AIC_c model, though increasing as sample size increases, stalls at model g_3 until a threshold sample size is passed, after which that theoretically best AIC_c model is more responsive to increasing sample size. It is this feature that here causes the percentiles of Δ_p to be so big (e.g., for $n_1 = 500$, a 95th percentile of 10.3 and a 99th percentile of 16.8).

The other notable feature in Table 5.3 is found when $n_1 = 10,000$ and $R = 10$. The theoretically best AIC_c model is then model g_{10}, the most general model in the set considered. Model selection probabilities here include $\pi_{10} = 0.8$ and $\pi_9 = 0.18$; hence, Δ_p is 0 in 80% of the samples and small in the other 20% of the samples.

Use of BIC is theoretically inappropriate for model selection in the example of Table 5.3 because the set of models used does not include a simple generating model with fixed, small K and with all effects being big as sample size increases. However, BIC performs best at large sample size, so one might ask whether BIC performs well here when $n_1 = 10,000$ and $R = 10$ or 20. The achieved coverage (for nominally 95% intervals) for BIC model selection for these two cases is given below, along with average (over S_1 to S_R) MSE for

TABLE 5.4. Some Monte Carlo results for AIC_c model selection under the chain binomial generating model $S_i = 0.5+0.3/i$; K for the theoretical AIC_c best model is given; achieved confidence interval coverage (nominally 95%) has been averaged over conditional (cond.) and unconditional (unc.) intervals on S_1 to S_R.

Sample size n_1	R	Best model K	Confidence interval coverage, %		percentiles of Δ_p		
			cond.	unc.	0.90	0.95	0.99
100	10	3	78.0	93.4	2.7	4.3	8.2
250	10	3	74.6	92.2	3.6	5.4	9.6
500	10	3	76.4	92.3	4.8	7.0	11.5
1,000	10	4	76.3	92.6	3.5	5.2	9.4
1,000	12	4	72.0	91.5	3.6	5.3	9.6
10,000	10	6	80.1	92.4	3.5	4.6	8.3
10,000	20	6	59.6	85.2	3.7	5.3	9.0

both BIC and AIC (AIC coverage is in Table 5.3):

n_1	R	coverage, BIC		average MSE	
		cond.	unc.	BIC	AIC
10,000	10	77.8	83.7	0.000245	0.000195
10,000	20	45.9	51.5	0.003077	0.001484

Average mean square error is lower for AIC, and confidence interval coverage is much better. Using BIC, model g_8 is selected with probability 0.37 ($R = 10$) or 0.36 ($R = 20$), and the probability of selecting one of models g_7, g_8, g_9, g_{10} is 0.96 ($R = 10$) or 0.93 ($R = 20$).

For some additional results and comparison to Table 5.3, we mimicked Table 5.3 results for a model wherein true $S_i = 0.5+0.3/i$. Table 5.4 gives confidence interval coverage results and percentiles of Δ_p for this example. Because of the monotonicity of the true S_i, the percentiles of Δ_p are much more stable, being *about* 4 (90th percentile), 5 (95th percentile), and 9 (99th percentile). Again, for the last case ($n_1 = 10,000$ and $R = 20$) the poor coverage is due to averaging coverage over the intervals on all of S_1 to S_{20}, yet $\hat{S}_{11}$ to $\hat{S}_{20}$ are considerable extrapolations. Generally, AIC_c does not select a model beyond model g_{10}. If for this last case we look only at intervals on S_1 to S_{10}, the coverage levels under AIC_c model selection are 80.7% (cond.) and 93.3% (unc.).

5.2.3   An Extended Survival Model

A second Monte Carlo example is given here to add additional complexity and realism in the set of approximating models. This example uses the same chain binomial generating model and associated parameters presented in Section 5.2.2; only the set of approximating models is different. Rather than assuming for model g_r the constraint $S \equiv S_i$ for all $i \geq r$, the age-specific survival probabilities at and beyond age r are assumed to follow a logistic model. The 10 models used here are given below:

Model 1	$\text{logit}(S_i) = \alpha + \beta \cdot i$,	$i \geq 1$,
Model 2	$S_1, \text{logit}(S_i) = \alpha + \beta \cdot i$,	$i \geq 2$,
Model 3	$S_1, S_2, \text{logit}(S_i) = \alpha + \beta \cdot i$,	$i \geq 3$,
$\vdots$	$\vdots$	$\vdots$
Model 10	$S_1, S_2, \ldots, S_9$, and $\text{logit}(S_i) = \alpha + \beta \cdot i$,	$i \geq 10$.

Each model has an intercept and slope parameter for the logistic regression fitting of age-specific survival rates S_i for ages $i \geq r$. Model g_r, for $r \geq 2$, also fits unconstrained age-specific survival rates for ages 1 to $r - 1$. Thus, model g_r has $K = r + 1$ parameters.

The initial population size was $n_1 = 150$ animals, as before, and all animals were followed until the last one died. The true parameters were $S_1 = 0.5$, $S_2 = 0.7$, $S_3 = 0.75$, $S_4 = 0.8$, $S_5 = 0.8$, and for $r > 5$, $S_{r+1} = 0.98 S_r$. The computer generated 10,000 independent repetitions under the true S_r, for analysis using the 10 approximating models. In some samples not all 10 models could be fit because all animals died before reaching age 10. As before, let ℓ be the last age at which $n_\ell > 0$. If $\ell - 2 \geq 10$, then all ten logistic-based models were fit. Otherwise, only $\ell - 2$ models were fit (if $\ell - 2 < 10$). Thus, the number of models that could be fit to a sample varied somewhat. This did not matter to the overall strategy of either selecting a best model or using model averaging, because for each fitted model it was always possible to compute the derived $\hat{S}_i$ (and its estimated conditional variance) for any value of i.

Generally, the logistic approximating models were superior in this situation as compared to the models used in Section 5.2.2. The theoretically best model (under AIC_c selection) was g_3 ($K = 4$) with model g_4 a close second ($\text{E}(\Delta_4) = 0.17$ relative to setting $\text{E}(\Delta_3) = 0$). The model selection relative frequencies are shown in Table 5.5. In 84.5% of samples either model g_2, g_3, or g_4 was selected. Model selection uncertainty in this example is similar to that of the example in Section 5.2.2 (compare Tables 5.1 and 5.5), but here the best model to use has 4 parameters, rather than 2, as was the case with the models previously considered. **Use of a better set of models leads to more informative inferences from the data.**

For the simulation underlying Table 5.5 results, all 10 models were fitted in 9,491 samples. In 18 samples, only models g_1 to g_7 could be fit (under the protocol noted above). Only models g_1 to g_8 were fitted in 128 samples; and only up to model g_9 were fitted in 363 samples.

Table 5.6 summarizes the estimated expected values of $\hat{S}_j$ for AIC_c-selected models and under model averaging (MA) using (4.2), with $\hat{\pi}_i = w_i$ being the Akaike weights. Also shown in Table 5.6 is the achieved coverage of confidence intervals based on estimated unconditional sampling variances (i.e., using (4.9)). Model averaging provided a slightly less biased estimator of conditional survival. However, both methods performed relatively well.

TABLE 5.5. Summary of AIC_c model selection relative frequencies, based on 10,000 Monte Carlo samples, for the logistic models. The generating model allowed survival probabilities to increase to a maximum at ages 4 and 5 and then decrease slowly with age, exactly as in Section 5.2.2 (see text for details). Model g_3 is the theoretically best model; results are for sample size $n_1 = 150$.

Model	Percent	Cumul. %
1	00.0	00.0
2	35.2	35.2
3	29.5	64.7
4	19.8	84.5
5	6.3	90.8
6	3.5	94.3
7	2.2	96.4
8	1.5	98.0
9	1.2	99.2
10	0.8	100.0

TABLE 5.6. Summary of Monte Carlo results, based on 10,000 samples, of age-specific survival estimation under AIC_c selection and model averaging (MA). The generating model had parameters S_i; 10 logistic models ($g_1, \ldots, g_{10}$) were fit to the simulated age-specific data; results are for sample size $n_1 = 150$.

Age i	S_i	$E(\hat{S})$ AIC_c	$E(\hat{S})$ MA	Coverage AIC_c	Coverage MA
1	0.500	0.499	0.499	0.957	0.957
2	0.700	0.708	0.707	0.907	0.925
3	0.750	0.758	0.757	0.920	0.934
4	0.800	0.789	0.790	0.926	0.946
5	0.800	0.785	0.786	0.917	0.943
6	0.784	0.772	0.773	0.951	0.964
7	0.768	0.759	0.760	0.964	0.974
8	0.753	0.744	0.745	0.962	0.974
9	0.738	0.727	0.728	0.962	0.970
10	0.723	0.709	0.709	0.957	0.964
Average				0.942	0.955

The achieved confidence interval coverage for these intervals, which includes model selection uncertainty, is very close to the nominal level (0.95) for both approaches; both approaches use the same estimate of unconditional standard error.

From Table 5.6 the unconditional confidence interval coverage averaged over the intervals on S_1 to S_{10}, and over all 10,000 samples, was 0.942 under the strategy of inference based on the selected AIC_c best model. Confidence intervals based on the estimated sampling variances conditional on the selected model ("conditional intervals") produced a corresponding overall coverage of 0.860. In the worst case, which was at age 5 (i.e., for S_5), conditional coverage was only 0.717, versus unconditional of 0.917.

In this example the simulation program also computed the mean squared error (MSE) for estimators of S_i at each age for AIC_c model selection, for model averaging, and for the simple (almost model-free) estimator of age-dependent survival $\hat{S}_i = n_{i+1}/n_i$, $i = 1, 2, \ldots, 10$. The MSE for the model-averaged estimator was smaller than that of the AIC_c-based estimator (ranging from 11% to 25% smaller, except for $\hat{S}_1$, where the two approaches had the same MSE). In this example, at least, model averaging, using Akaike weights, has important advantages. Both AIC_c model selection and model averaging are quite superior to the use of the simple estimator of S_i. For example, at age nine, the MSEs for $\hat{S}_i$ from AIC_c model selection, model averaging, and the simple estimator were 0.0041, 0.0035, and 0.0186, respectively. This illustrates the advantage of some appropriate *modeling* of the underlying survival probabilities (this is similar to the lesson learned from the starling experiment in Section 3.4).

We computed the pivotal quantities $\Delta_p = AIC_{best} - AIC_{min}$ (4.12) for each of the 10,000 samples to better understand these values for nested models of this type. Some percentiles of the sampling distribution of Δ_p for this example are shown below:

Percentile	Δ_p
50.0	1.20
75.0	2.09
80.0	2.29
85.0	3.11
90.0	4.21
95.0	6.19
97.5	8.33
98.0	8.80
99.0	10.63

Thus, in approximately 90% of the simulated data sets the difference, on average, between AIC_c for the selected model, for that sample, and the AIC_c for the theoretically best model was ≤ 4.21. If we adopted the idea of using an approximate 90% confidence set on what is the actual AIC_c theoretically best model, then in this example, fitted models that had Δ_i values below about 4 should be in that set, and hence might be candidates for some further consideration in making inferences from an individual data set.

To make further use of the generating model and set of approximating models used here, we determined several quantities for differing sample sizes: the theoretically best model, conditional and unconditional coverage under AIC_c model selection (as opposed to model averaging), and 90th, 95th, and 99th percentiles of Δ_p. Regardless of this variation in sample size from 50 to 100,000 the unconditional confidence interval coverage stays between about 93 and 95% (Table 5.7). Also, there is considerable stability of the shown percentiles of the (approximate) pivotal Δ_p. These same results hold fairly well for other generating models and fitted models that we have examined by Monte Carlo methods. Often the conditional coverage averages are less than what occurs in Table 5.7 (more like 70% to 85%).

The true S_i do not exactly fit any of the approximating logistic models. However, model g_4 turns out to be an excellent approximation to truth, at least for $i \leq 15$ (we did not plan this to be the case). Here, even if we could know S_1 to S_{15} exactly, but we needed a mathematical model to represent these numbers, we would likely fit and use model g_4. We would do so because the fitted model then provides a concise, yet almost exact, summary of truth. The lack of fit is both trivial and statistically "significant" given a huge sample size. We do not generally have such huge sample sizes, but if we did, then the usual concerns of small-sample-size statistics would not apply (practical significance then replaces statistical significance), and we might very well select model g_4 rather than model g_8 as a preferred approximation to truth. This illustrates a philosophical point: Even if we knew truth, we would often prefer to replace said truth by a low-order parsimonious fitted model because of the advantages this confers about understanding the basic structure of full truth.

5.2.4 Model Selection if Sample Size Is Huge, or Truth Known

The results in Table 5.7 provide a motivation for us to mention some philo-sophical issues about model selection when truth is essentially known, or equivalently in statistical terms, when we have a huge sample size. With a

TABLE 5.7. Some Monte Carlo results (10,000 repetitions per sample size) for AIC$_c$ model selection, for the same generating model and set of models g_i as used for Table 5.6. Sample size is n_1; R is the maximum number of models considered; K is shown for the theoretical AIC$_c$ best model g_k; achieved confidence interval coverage (nominally 95%) has been averaged over conditional (cond.) and unconditional (unc.) intervals on S_1 to S_R.

Sample size n_1	R	Best model K	Confidence interval coverage, %		percentiles of Δ_p		
			cond.	unc.	0.90	0.95	0.99
50	10	2	84.8	92.6	3.8	5.8	10.1
100	10	3	86.2	94.1	3.7	5.4	9.8
150	10	3	86.0	94.2	4.2	6.2	10.6
200	10	4	86.1	94.2	3.6	4.1	7.8
500	10	4	88.3	94.6	2.7	4.3	8.1
1,000	12	4	90.4	95.4	2.6	4.2	7.9
10,000	15	4	90.3	95.5	4.9	7.1	12.3
100,000	15	8	86.4	93.3	5.0	6.6	11.2

sample size of $n_1 = 10,000$ model g_4 is theoretically the best model to fit to the survival data. At $n_1 = 100,000$ the optimal model, based on information-theoretic statistical model selection, is g_8. However, at a sample size of 100,000 we find that (1) the parameter estimates under either model g_4 or g_8 are precise to at least two significant digits (so $\hat{S}_i \approx E(\hat{S}_i)$ for practical purposes), and (2) the difference (i.e., bias) under either model between $E(\hat{S}_i)$ and the true S_i is also trivial (but not always 0), hence ignorable for $i = 1, \ldots, 15$.

Therefore, at a sample size of 100,000, or if we literally knew the true S_i, it is more effective from the standpoint of understanding the basic pattern of variation in the survival probabilities S_i to fit and use model g_4 (fit to data if $n_1 = 100,000$, or to truth if truth is in hand). In this situation nothing practical is gained by using model g_8 rather than model g_4. On the contrary, using the less parsimonious, more complicated model g_8 has a cost in terms of ease with which we perceive the basic pattern in the true S_i. In essence, we are also saying that even if we knew the true S_i, we would be better able to understand the pattern of information therein by fitting (to the S_i) and reporting model g_4 and using that simple five-parameter model, rather than using either model g_8 or the actual S_i. (In practice one might find the simple truth here that $S_{i+1} = 0.98 \cdot S_i$, $i \geq 5$).

The point is that once complex truth, as a set of parameters, is known quite well, such as to two or three leading significant digits (which is quite beyond what we can usually achieve with real sample sizes), we may not need statistical model selection anymore. Instead, other criteria would be used: Either use truth as is, or use a simple, parsimonious, interpretable model in place of absolute truth when that model explains almost all the variation in truth. The ability to interpret, understand, and communicate one's model is important in all uses of models, and all numerical descriptions of reality should be considered as just a model if for no other reason than we will never know all digits of an empirical parameter.

Another view of this matter is that model g_8 is actually no better than model g_4 once we can fit either model to truth. Either model fitted to truth would correspond to producing a correlation coefficient ≥ 0.999 between true S_i and model-predicted survival probability. Once you can fit truth that well with a simple model, the gains in understanding and communicating the model, as truth, override that fact that the model is not exactly truth: It is close enough.

As a corollary of this philosophy, model goodness-of-fit based on statistical tests becomes irrelevant when sample size is huge. Instead, our concern then is to find an interpretable model that explains the information in the data to a suitable level of approximation, as determined by subject-matter consider-ations. Unfortunately, in life sciences we probably never have sample sizes anywhere near this large; hence statistical considerations of achieving a good fit to the data are important as well as subject-matter considerations of having a fitted model that is interpretable.

Data that are deterministic, as may arise in some physical sciences, correspond to huge sample sizes, hence model selection therein need not (but could) be based on statistical criteria. There is no theory of how well a deterministic model must fit deterministic truth in order for that model to be useful. Similarly, if truth as a probability distribution is known, but a one seeks an approximating model, then K-L can be used, but there is no theory about how much information loss is tolerable. That is a subject-matter decision.

5.2.5   A Further Chain Binomial Model

Here we consider a generating model with damped oscillations in survival as age increases. It is not intended to be considered as a biological example; it is just a case of a complex truth against which we can examine some aspects of model selection. We let the initial population size be $n_1 = 1,000$. We generated 10,000 independent samples using the chain binomial data-generating model with true survival probabilities as

$$S_i = 0.7 + (-1)^{i-1}(0.2/i),$$

where i is year of life ("age" for short). The data $(n_1, n_2, \ldots, n_{18}; \ell = 17$, as $n_{18} = 0$ but $n_{17} > 0)$ for one repetition were

1000 (fixed), 890, 520, 414, 270, 202, 130, 93, 67, 50, 35, 26, 19, 8, 7, 4, 2, 0.

Animals were followed until all were dead. The models used for analysis were the logistic models g_i of Section 5.2.3. The model set was $g_1, \ldots, g_{15}$ (i.e., $R = 15$). It is because sample size was 1,000 rather than 150 that we increase the size of the model set used. It is a general principle that as sample size increases, one can expect to reliably estimate more parameters, so the size of the set of models considered should depend weakly on sample size.

Results for AIC_c and BIC model selection are very different (Table 5.8). On average, AIC_c selected a model with 7.6 parameters, while BIC selected a model on average with 5.1 parameters. The theoretically best model to use, in terms of the expected K-L criterion, in this example is g_6, which has $K = 7$ parameters. Under BIC selection 95% of the models selected contained 4, 5, or 6 parameters. In contrast, under AIC_c selection approximately 95% of the models selected had between 5 and 13 parameters, inclusive.

That BIC selection produces a more concentrated distribution of selected models certainly seems to be an advantage. However, useful comparisons must focus on bias and precision of parameter estimates, hence on confidence interval coverage (given that we have essentially the shortest intervals possible under the different model selection strategies). From Table 5.9, AIC_c selection produces average conditional and unconditional coverage of 84.9% and 92.7%, respectively. Under BIC selection we had average conditional and unconditional coverage of 73.9% and 78.5%, respectively. BIC selection often (but not always) did achieve a smaller MSE in these examples. For the example of Table 5.9, MSEs averaged over S_1 to S_{15} were 0.00143 under BIC model

TABLE 5.8. Comparison of model selection relative frequencies for AIC_c vs. BIC based on truth as the chain binomial survival model with damped oscillations, $S_i = 0.7 + (-1)^{i-1}(0.2/i)$, and the model set for data analysis as g_1 to g_{15}; results are for sample size $n_1 = 1,000$.

| Model | K | AIC_c model selection | | BIC model selection | |
i		percent	cumul. %	percent	cumul. %
1	2	0	0	0	0
2	3	0	0	0	0
3	4	0.6	0.6	23.0	23.0
4	5	11.5	12.1	50.3	73.3
5	6	26.3	38.4	21.7	95.0
6	7	20.5	58.9	4.1	99.2
7	8	16.0	75.0	0.8	99.9
8	9	7.8	82.8	0.1	100.0
9	10	6.0	88.8	0	100.0
10	11	3.0	91.8	0	100.0
11	12	2.5	94.4	0	100.0
12	13	1.6	96.0	0	100.0
13	14	1.6	97.6	0	100.0
14	15	1.1	98.7	0	100.0
15	16	1.3	100.0	0	100.0

selection, whereas for AIC that average MSE was 0.00209. However, part of this cost of smaller MSEs (by BIC) for the $\hat{S}_i$ is poor confidence interval coverage; for S_1 to S_{15}, unconditional BIC coverage varied from 45% to 95% (see Table 5.9 for AIC coverage).

Comparison of the expected estimated standard errors illustrates the magnitude of the variance component due to model selection uncertainty under AIC_c. For example, for age 10, the expected conditional standard error is 0.028 versus the expected unconditional standard error of 0.037. Table 5.9 gives these results for ages 1 to 15.

We computed Δ_p values for each of the 10,000 repetitions of this case where there were damped oscillations in the S_i parameters. Some percentiles of the sampling distribution of Δ_p are shown below:

Percentile	Δ_p
50.0	1.7
75.0	3.0
80.0	3.5
85.0	4.2
90.0	5.7
95.0	7.9
97.5	10.1
98.0	10.9
99.0	13.3

TABLE 5.9. Some Monte Carlo results, based on 10,000 samples, on estimation of age-specific survival probabilities under AIC_c model selection; sample size was $n_1 = 1,000$; the generating model allowed damped oscillations in survival as age increased (see text for details). Conditional (cond.) and unconditional (unc.) standard errors and confidence intervals were also evaluated.

Age	S_i	$E(\hat{S}_i)$	$E(\widehat{se}(\hat{S}_i))$		Coverage	
i			cond.	unc.	cond.	unc.
1	0.900	0.900	0.009	0.009	0.950	0.950
2	0.600	0.600	0.016	0.016	0.955	0.955
3	0.767	0.767	0.018	0.018	0.951	0.951
4	0.650	0.651	0.023	0.024	0.904	0.932
5	0.740	0.734	0.023	0.027	0.821	0.882
6	0.667	0.678	0.025	0.031	0.701	0.860
7	0.729	0.716	0.024	0.032	0.671	0.880
8	0.675	0.690	0.024	0.034	0.711	0.899
9	0.722	0.706	0.025	0.035	0.767	0.926
10	0.680	0.693	0.028	0.037	0.838	0.937
11	0.718	0.699	0.030	0.039	0.871	0.944
12	0.683	0.691	0.035	0.043	0.887	0.941
13	0.715	0.692	0.040	0.048	0.900	0.949
14	0.686	0.685	0.046	0.054	0.902	0.942
15	0.713	0.681	0.052	0.061	0.908	0.955
Average					0.849	0.927

These results are consistent with percentile values for Δ_p that we have found for other examples.

Table 5.10 gives expected Akaike weights $E(w_i)$ and model selection probabilities π_i from the 10,000 Monte Carlo samples for this example. To compare results from this case of a large sample size ($n_1 = 1,000$, g_6 is theoretically best), Table 5.10 also gives $E(w_i)$ and π_i for $n_1 = 50$ (g_3 is theoretically best) under the same data-generating model. For sample size 1,000 we see from Table 5.10 that on average models g_4 to at least g_9 must be considered in making inference about the population under AIC_c model selection. The Akaike weights give some support to models g_{10} and g_{11}, while π_{10} and π_{11} provide somewhat less support to these models.

There is a high degree of model selection uncertainty for this example, particularly for the larger sample size. The underlying process (damped oscillations in the S_i) is complicated, and the set of logistic approximating models was relatively poor. Had some science been brought to bear on this (artificial) problem, hopefully the set of approximating models would have included some models with at least some oscillating features. Thus, model selection uncertainty would likely have been greatly reduced.

Use of the bootstrap to estimate model selection probabilities is effective but computationally intensive. Akaike weights can be easily computed and offer a

TABLE 5.10. Expected Akaike weights $E(w_i)$ compared to AIC_c model selection relative frequencies π_i based on Monte Carlo simulation (10,000 samples) for the generating model allowing damped oscillations in survival as age increases; models used for analysis were g_1 to g_{15}; two sample sizes are considered.

Model	$n_1 = 50$		$n_1 = 1,000$	
i	$E(w_i)$	π_i	$E(w_i)$	π_i
1	0.050	0.038	0.000	0.000
2	0.293	0.431	0.000	0.000
3	0.283	0.325	0.011	0.006
4	0.160	0.096	0.091	0.115
5	0.104	0.064	0.177	0.263
6	0.056	0.026	0.176	0.205
7	0.033	0.014	0.153	0.160
8	0.015	0.004	0.109	0.078
9	0.006	0.002	0.084	0.060
10	0.000	0.000	0.058	0.030
11	0.000	0.000	0.044	0.025
12	0.000	0.000	0.032	0.016
13	0.000	0.000	0.026	0.016
14	0.000	0.000	0.020	0.011
15	0.000	0.000	0.019	0.013

simple and effective alternative. From Table 5.10 we can observe the general agreement between the expected Akaike weights and the model selection probabilities π_i. Note, however, that these are not estimates of the same quantity, so that exact agreement is not expected. Akaike weights, normalized to add to 1, reflect the relative likelihood of each fitted model in the set. These weights provide information about the relative support of the data for the various candidate models. Finally, as all the examples in this section (and other sections) show, the weights w_i are very useful in model averaging and computing estimates of unconditional sampling variances, hence obtaining unconditional confidence intervals that do substantially improve coverage after model selection.

5.3   Examples and Ideas Illustrated with Linear Regression

The model selection literature emphasizes applications in regression and time series, often as selection of variables in regression; this is the same thing as model selection. McQuarrie and Tsai (1998) is devoted to, and entitled, *Regression and Time Series Model Selection*. Our explorations do not overlap much with those of McQuarrie and Tsai (1998), because we emphasize model selection uncertainty and, in general, multimodel inference. In this section we focus on all-subsets model selection by presenting an extensive example and some Monte Carlo results for a small member of predictor variables. Also given are other results and thoughts about K-L–based model selection for all-subsets regression.

5.3.1 All-Subsets Selection: A GPA Example

We use an example based on four regressors (Table 5.11) and a sample size of 20 (a larger number of regressors makes it too demanding of space to present the full results). The example of Table 5.11 comes from Graybill and Iyer (1994). They use these example data extensively to illustrate model selection, including all-subsets selection based on several criteria, but not using AIC or the other ideas we use herein. Also, we note that the Table 5.11 data are "realistic but not real" (H. Iyer, personal communication).

The full model to fit is

$$y = \beta_0 + \beta_1 x_1 + \beta_2 x_2 + \beta_3 x_3 + \beta_4 x_4 + \epsilon.$$

This model, g_{15}, is also denoted in tables by $\{1234\}$ because all four predictor variables are used (see, for example, Table 5.12). As another example of this notation, model g_6 uses only the predictors x_1 and x_3 and is thus also denoted by $\{1 \cdot 3 \cdot\}$:

$$y = \beta_0 + \beta_1 x_1 + \beta_3 x_3 + \epsilon.$$

Including the intercept-only model g_{16}, $\{\cdot\cdot\cdot\cdot\}$, there are 16 models here in the standard all-subsets approach. We use AIC_c for selection from this set of 16 models, and we also apply the bootstrap (10,000 samples) to these data with full AIC_c model selection applied to all 16 models for each bootstrap sample.

The selected best model (g_{11}) includes predictors x_1, x_2, and x_3 (Table 5.12). However, support for g_{11} as the only useful model to use here is weak, because its Akaike weight is just 0.454. Model g_5, $\{12 \cdot\cdot\}$, is also credible, as is, perhaps, model g_6, $\{1 \cdot 3 \cdot\}$. The Akaike weights w_i and bootstrap-selection probabilities $\hat{\pi}_i$ agree to a useful extent, and both demonstrate the model selection uncertainty in this example. For a confidence set on models here (if we felt compelled in practice to provide one in this sort of application; we do not) we would use a likelihood ratio (evidence) criterion: the set of all models g_i for which $w_{11}/w_i \leq 8$ (cf. Royall 1997). Thus, we have the set $\{g_{11}, g_5, g_6, g_{15}\}$. This criterion is identical to using as a cutoff models for which $\Delta_i \leq 4.16$ (4 would suffice).

We can explore for this example the coherence (with the likelihood approach) of such a rule by determining the bootstrap distribution of the pivotal Δ_p as $\Delta_p^* = AIC_{c,11}^* - AIC_{c,min}^*$. Here, $AIC_{c,min}^*$ is the minimum AIC_c value for the given bootstrap sample, and $AIC_{c,11}^*$ is the AIC_c value in that same bootstrap sample for model g_{11}. We computed the 10,000 bootstrap values of $\Delta_{p,b}^*$ and

TABLE 5.11. First-year college GPA and four predictor variables from standardized tests administered before matriculation. This example comes from Graybill and Iyer (1994: Table 4.4.3).

First-year GPA	SAT math	SAT verbal	High School math	High School English
y	x_1	x_2	x_3	x_4
1.97	321	247	2.30	2.63
2.74	718	436	3.80	3.57
2.19	358	578	2.98	2.57
2.60	403	447	3.58	2.21
2.98	640	563	3.38	3.48
1.65	237	342	1.48	2.14
1.89	270	472	1.67	2.64
2.38	418	356	3.73	2.52
2.66	443	327	3.09	3.20
1.96	359	385	1.54	3.46
3.14	669	664	3.21	3.37
1.96	409	518	2.77	2.60
2.20	582	364	1.47	2.90
3.90	750	632	3.14	3.49
2.02	451	435	1.54	3.20
3.61	645	704	3.50	3.74
3.07	791	341	3.20	2.93
2.63	521	483	3.59	3.32
3.11	594	665	3.42	2.70
3.20	653	606	3.69	3.52

give below some percentiles of this random variable:

percentile	Δ_p^*
50	1.00
80	3.21
90	3.62
95	5.18
98	7.00
99	8.79

Thus, we have a good basis to claim that the confidence set $\{g_{11}, g_5, g_6, g_{15}\}$ on the expected K-L best model is about a 95% confidence set. As with any statistical inference, this result depends on sample size, and hence could change if n increased.

Because model selection in regression is often thought of as seeking to identify the importance of each predictor variable, we next computed the variable importance weights of each variable as the sum of the Akaike weights w_i for each model in which the predictor variable appears (Table 5.13). We also

TABLE 5.12. All-subsets selection results based on AIC_c for the example GPA data in Table 5.11. GPA is regressed against all 16 combinations of four predictor variables as indicated by the set notation code such as {1234} for the full model. Here, the estimated model selection probabilities $\hat{\pi}_i$ are based on 10,000 bootstrap samples.

Model i	Predictors used	AIC_c	Δ_i	w_i	$\hat{\pi}_i$
11	{123·}	−43.74	0.00	0.454	0.388
5	{12··}	−42.69	1.05	0.268	0.265
6	{1·3·}	−40.77	2.97	0.103	0.152
15	{1234}	−39.89	3.85	0.066	0.022
12	{12·4}	−39.07	4.67	0.044	0.017
13	{1·34}	−38.15	5.58	0.028	0.064
1	{1···}	−38.11	5.62	0.027	0.045
8	{1··4}	−35.19	8.55	0.006	0.012
14	{·234}	−32.43	11.30	0.002	0.027
10	{··34}	−31.59	12.15	0.001	0.004
7	{·23·}	−29.52	14.21	0.000	0.001
9	{·2·4}	−25.87	17.86	0.000	0.003
3	{··3·}	−25.71	18.02	0.000	0.000
2	{·2··}	−23.66	20.07	0.000	0.000
4	{···4}	−21.70	22.03	0.000	0.000
16	{····}	−15.33	28.41	0.000	0.000

TABLE 5.13. Evidence for the importance of each regressor variable j in the GPA example, based on sums of Akaike weights $w_+(j)$ (denoted by $\sum w_i$ here) over models in which the variable occurs and based on relative frequency of occurrence of the variable in the selected model based on 10,000 bootstrap samples.

Predictor	Σw_i	$\Sigma \hat{\pi}_i$
x_1	0.997	0.965
x_2	0.834	0.722
x_3	0.654	0.658
x_4	0.147	0.148

computed the relative frequency of models selected containing each variable (hence, the same sum of the bootstrap-based $\hat{\pi}_i$). The results (Table 5.13) show that at this sample size, x_4 is not important, whereas x_1 is very important and x_2 and x_3 are at least moderately important. Indeed, the selected best model was g_{11}, {123·}.

Standard theory gives us the MLEs for each β_j included in a model and the estimated conditional standard error $\widehat{se}(\hat{\beta}_j \mid g_i)$. We go a step further here and compute the model-averaged estimate of each regression parameter $\hat{\bar{\beta}}_j$, whether or not β_j is in the selected best model. However, $\hat{\overline{\beta}}_j$ is computed only over those models in which variable x_j appears. Also computed is the corresponding estimated unconditional standard error $\widehat{se}(\hat{\overline{\beta}}_j)$, which we also use as the estimated unconditional standard error of any estimated β_j from the AIC_c-selected best model. The needed formulas are given in Section 4.3.2, especially (4.9) with $\hat{\pi}_i = w_i$. The numerical inputs to these calculations are given in Table 5.14 for each of the four parameters: Note that we can compute a model-averaged estimate of a parameter independently of issues of selecting a single best model. In Table 5.14 we show for each β_j its estimate for the eight models it appears in, and the corresponding conditional standard error and Akaike weight. These weights must be renormalized to sum to 1 to apply (4.1) and (4.9). In Table 5.14 the normalizing constant is shown along with the model-averaged estimate and its estimated unconditional standard error. Note that for a given regression coefficient β_j, its estimate can be quite different by model. Part of this variation is due to model bias, as discussed in Section 5.3.5. For example, $\hat{\beta}_3$ is 0.18 ($\widehat{se} = 0.09$) for model g_{11}, $\{123 \cdot\}$; however, for model g_3, $\{\cdot \cdot 3 \cdot\}$, we have $\hat{\beta}_3 = 0.51$ ($\widehat{se} = 0.12$). It is because the predictors are correlated here that β_3 actually varies over the different models (i.e., due to inclusion/exclusion of other predictor variables), and this source of variation contributes to model uncertainty about β_3 under model selection, or model averaging, of the set of $\hat{\beta}_3$ values in Table 5.14.

We also used the 10,000 bootstrap samples to estimate the model-averaged parameter $\overline{\beta}_j$ and its unconditional sampling variation $se(\hat{\overline{\beta}}_j)$, $j = 1, 2, 3, 4$. The results from the analytical approach using model selection and Akaike weights, and the bootstrap results, are shown in Table 5.15. The three estimated unconditional standard errors based on Akaike weights and model averaging (i.e., from (4.9)) are each larger than the corresponding conditional standard error for the selected best model, g_{11}, $\{123 \cdot\}$. Note especially the case for β_1 where based on analytical methods $\widehat{se}(\hat{\overline{\beta}}_1) = 0.000535$, whereas $\widehat{se}(\hat{\beta}_1 \mid g_{11}) = 0.000455$, and from the bootstrap, $\widehat{se}(\hat{\overline{\beta}}_1) = 0.000652$. This and the other bootstrap results are precise to essentially two decimal places.

The three different point estimates (bootstrap, AIC_c-based model-averaged, and best model) for each of β_1, β_2, and β_3 are quite similar, given their standard errors. However, this is not true for the two estimates of β_4; we do not know whether the bootstrap result (point estimate 0.29, $\widehat{se} = 0.31$) or analytical result (point estimate 0.09, $\widehat{se} = 0.20$) is better in this example.

Note also that here the bootstrap-estimated unconditional standard errors are less than the estimated conditional standard errors for $\hat{\overline{\beta}}_2$ and $\hat{\overline{\beta}}_3$. Unfortunately, this example constitutes only a sample of size 1 as regards comparing bootstrap and information-theoretic analytical methods. It will take a very large Monte Carlo study to make a reliable general comparison of these two approaches to assessing model selection uncertainty.

TABLE 5.14. Parameter estimates $\hat{\beta}_1, \ldots, \hat{\beta}_4$ by models that include each x_j (hence β_j), conditional standard errors given the model, and Akaike weights; also shown is the model-averaged $\hat{\bar{\beta}}_j$ (4.1), its unconditional standard error (Section 4.3.2), and the sum of Akaike weights over the relevant subset of models.

Model	Predictors	Results by model		
i	used	$\hat{\beta}_1$	$\widehat{se}(\hat{\beta}_1 \mid g_i)$	w_i
11	{123·}	0.002185	0.0004553	0.454
5	{12··}	0.002606	0.0004432	0.268
6	{1·3·}	0.002510	0.0004992	0.103
15	{1234}	0.002010	0.0005844	0.066
12	{12·4}	0.002586	0.0005631	0.044
13	{1·34}	0.002129	0.0006533	0.028
1	{1···}	0.003178	0.0004652	0.027
8	{1··4}	0.002987	0.0006357	0.006
$\hat{\bar{\beta}}_1$ and unc. se:		0.002368	0.0005350	0.997

Model	Predictors	Results by model		
i	used	$\hat{\beta}_2$	$\widehat{se}(\hat{\beta}_2 \mid g_i)$	w_i
11	{123·}	0.001312	0.0005252	0.454
5	{12··}	0.001574	0.0005555	0.268
14	{·234}	0.001423	0.0007113	0.002
15	{1234}	0.001252	0.0005515	0.066
12	{12·4}	0.001568	0.0005811	0.044
7	{·23·}	0.002032	0.0007627	0.000
9	{·2·4}	0.002273	0.0008280	0.000
2	{·2··}	0.003063	0.0008367	0.000
$\hat{\bar{\beta}}_2$ and unc. se:		0.001405	0.0005558	0.834

5.3.2 A Monte Carlo Extension of the GPA Example

Simulation is a very useful way to gain insights into complex model selection issues. In particular, here we can assume that the five-dimensional vector $(y, x_1, x_2, x_3, x_4)'$ is multivariate normal, MVN(μ, Σ) (this is now "truth"), generate 10,000 independent simulated sets of data under this generating model, and do full model selection to learn about selection performance issues. Given the matrix Σ we can determine the true regression coefficients (and their approximate true conditional standard errors) under any of the 16 regression models. The β_j (and other needed quantities) given a regression model depend only on elements of Σ, so it suffices to set the general scale to zero: $\mu = \underline{0}$. The needed 5×5 variance–covariance matrix is taken here as the sample variance–covariance matrix from the GPA data. Thus, our simulation will be under an assumed truth that is close enough to the truth underlying

TABLE 5.14. (*Continued*)

Model	Predictors	Results by model		
i	used	$\hat{\beta}_3$	$\widehat{se}(\hat{\beta}_3 \mid g_i)$	w_i
11	{123·}	0.1799	0.0877	0.454
7	{·23·}	0.3694	0.1186	0.000
6	{1·3·}	0.2331	0.0973	0.103
15	{1234}	0.1894	0.0919	0.066
3	{··3·}	0.5066	0.1236	0.000
13	{1·34}	0.2474	0.0990	0.028
14	{·234}	0.3405	0.1045	0.002
10	{··34}	0.4171	0.1054	0.001
$\hat{\bar{\beta}}_3$ and unc. se:		0.1930	0.0932	0.654

Model	Predictors	Results by model		
i	used	$\hat{\beta}_4$	$\widehat{se}(\hat{\beta}_4 \mid g_i)$	w_i
15	{1234}	0.08756	0.1765	0.066
12	{12·4}	0.01115	0.1893	0.044
13	{1·34}	0.17560	0.1932	0.028
8	{1··4}	0.09893	0.2182	0.006
14	{·234}	0.45333	0.1824	0.002
10	{··34}	0.57902	0.1857	0.001
9	{·2·4}	0.51947	0.2207	0.000
4	{···4}	0.77896	0.2407	0.000
$\hat{\bar{\beta}}_4$ and unc. se:		0.09024	0.1989	0.147

TABLE 5.15. Bootstrap (10,000 samples) and Akaike weight-based results for the GPA example for model-averaged estimated regression coefficients and associated estimated unconditional standard errors, which include model selection uncertainty. Also shown are the estimate from the selected model g_{11} and its estimated conditional standard error.

	Bootstrap results		w_i model-averaged		AIC_c best model	
j	$\hat{\bar{\beta}}_j$	$\widehat{se}(\hat{\bar{\beta}}_j)$	$\hat{\bar{\beta}}_j$	$\widehat{se}(\hat{\bar{\beta}}_j)$	$\hat{\beta}_j$	$\widehat{se}(\hat{\beta}_j \mid g_{11})$
1	0.00236	0.000652	0.00237	0.000535	0.00219	0.000455
2	0.00156	0.000508	0.00141	0.000556	0.00131	0.000525
3	0.2296	0.0684	0.1930	0.0932	0.1799	0.0877
4	0.2938	0.3056	0.0902	0.1989		

these GPA data to provide useful results and insights about the analysis of this GPA example.

This particular use of Monte Carlo simulation is also called the paramet-ric bootstrap: We use as the generating model the parametric model whose parameters are estimated from the actual data. Consequently, we can expect some of the results here to be about the same as those already obtained from the (nonparametric) bootstrap. The advantage of this parametric approach is that we can specify true values of parameters and hence evaluate confidence interval coverage.

Symbolically, the full variance–covariance matrix is partitioned as below for model g_i:

$$\Sigma_i = \begin{bmatrix} \sigma_y^2 & \underline{c}' \\ \underline{c} & \Sigma_x \end{bmatrix}. \tag{5.3}$$

The marginal variance of the response variable y is σ_y^2. For whatever m predictors x_j are in the regression model, the vector $\underline{c}$ ($m \times 1$, $1 \leq m \leq 4$) gives their covariances with y ($\text{cov}(y, x_j)$). The variance–covariance matrix of just the predictors considered (i.e., for any of the 15 models excluding the intercept-only model, $\{\cdots\}$) is given by matrix Σ_x ($m \times m$). The vector of true regression parameters, other than the intercept β_0, is given by

$$\underline{\beta}' = \underline{c}'(\Sigma_x)^{-1} \tag{5.4}$$

(we ignore β_0). The approximate sampling variance–covariance matrix of $\hat{\underline{\beta}}'$ is given by $\sigma_{y|x}^2 (\Sigma_x)^{-1}$, where

$$\sigma_{y|x}^2 = \sigma_y^2 - \underline{c}'(\Sigma_x)^{-1}\underline{c} \tag{5.5}$$

is the true residual variance in the regression (a good reference for this multivariate theory is Seber 1984).

The actual simulation process generates the rows of the design matrix X as random, but then we condition on them in the regression model $\underline{y} = X\underline{\beta} + \underline{\epsilon}$. Conditionally on X, $\hat{\underline{\beta}}$ is unbiased; so it is also unconditionally unbiased; hence $E(\hat{\underline{\beta}}') = \underline{c}'(\Sigma_x)^{-1}$. This same argument applies to $\hat{\sigma}_{y|x}^2$. However, conditionally (by sample), the variance–covariance matrix of $\hat{\underline{\beta}}$ is $\sigma_{y|x}^2 (X'X)^{-1}$, and $E(X'X)^{-1} = (\Sigma_x)^{-1}$ holds only asymptotically as sample size gets large. Thus for the simulations, $\sigma_{y|x}^2 (\Sigma_x)^{-1}$ is only an approximation to the true average variance–covariance matrix of $\hat{\underline{\beta}}$.

Rather than show Σ, we show the derived correlation matrix, upper elements only:

	x_1	x_2	x_3	x_4
y	0.850	0.653	0.695	0.606
x_1		0.456	0.559	0.663
x_2			0.434	0.417
x_3				0.272

(the ordered x_i are SATmath, SATverbal, HSmath, HSenglish). No pairwise x_i, x_j correlations are so high that we would need to eliminate any x_i. This

TABLE 5.16. True values of $\sigma^2_{y|x}$ and β's and approximate conditional standard errors and cv's for some models when they are fit to data, for $n = 20$, from the true generating model as g_{15} (as detailed in the text).

Model g_{15}, {1234}: $\sigma^2_{y|x} = 0.05692$

j	β_j	$se(\hat{\beta}_j \mid g_{15})$	$cv(\hat{\beta}_j \mid g_{15})$
1	0.002010	0.0005061	0.252
2	0.001252	0.0004776	0.381
3	0.1895	0.0796	0.420
4	0.0875	0.1528	1.745

Model g_{11}, {123·}: $\sigma^2_{y|x} = 0.05785$

j	β_j	$se(\hat{\beta}_j \mid g_{11})$	$cv(\hat{\beta}_j \mid g_{11})$
1	0.002185	0.0004072	0.186
2	0.001312	0.0004698	0.358
3	0.1799	0.0784	0.436

Model g_5, {12··}: $\sigma^2_{y|x} = 0.07307$

j	β_j	$se(\hat{\beta}_j \mid g_5)$	$cv(\hat{\beta}_j \mid g_5)$
1	0.002606	0.0004086	0.157
2	0.001574	0.0005121	0.325

Model g_1, {1···}: $\sigma^2_{y|x} = 0.10759$

j	β_j	$se(\hat{\beta}_j \mid g_1)$	$cv(\hat{\beta}_j \mid g_1)$
1	0.003178	0.0004413	0.139

Model g_3, {··3·}: $\sigma^2_{y|x} = 0.20001$

j	β_j	$se(\hat{\beta}_j \mid g_3)$	$cv(\hat{\beta}_j \mid g_3)$
3	0.5066	0.1173	0.117

is an important consideration. In addition, a principal-components analysis of the covariance matrix of the predictor variables is a reasonable approach.

The data-generating model used here (g_{15}, {1234}), based on the GPA data, has the pairwise correlations given above and the residual variance $\sigma^2_{y|x} = 0.05692$ (from (5.5)). From (5.4) we compute the true β's for the generating model g_{15}, and from $\sigma^2_{y|x}(\Sigma_x)^{-1}$ we compute the approximate conditional standard errors of the $\hat{\beta}_j$ under model g_{15}. We also give the approximate conditional coefficient of variation of each $\hat{\beta}_j$ (Table 5.16). These same quantities are computable for any submodel fitted to the generated data; Table 5.16 shows these theoretical values for models {1234}, {123·}, {12··}, {1···}, and {··3·}.

The expected results of fitting any model to data reflect both the truth as contained in empirical data and the adequacy of the model. For example, from Table 5.16 if we use model g_3, then $E(\hat{\beta}_3) = 0.5066$ (cv $= 0.117$), whereas when all four predictors are included, $\beta_3 = E(\hat{\beta}_3) = 0.1895$ (cv $= 0.420$). For model g_{11}, $E(\hat{\beta}_3) = 0.1799$ (cv $= 0.436$). Results in Table 5.16 apply when the specified model is always fit to the data; hence no data-based model selection occurs. When the inference strategy is to first select a model based on the data, then use it for inference, the properties of estimators and other inferences are affected (model selection biases and uncertainties occur).

Examination of results in Table 5.16 demonstrates that in general, as measured by its coefficient of variation, precision of a parameter estimator increases as the number of other parameters in the models decreases. That a given β_j varies by model is because the predictors are correlated. This effect (i.e., model variation in $E(\hat{\beta}_j)$) will get more pronounced if correlations get stronger; it does not occur if all predictors are uncorrelated with each other. Leaving an x_j out of fitted model g_i has little effect if that predictor is unimportant as measured by a large $cv(\hat{\beta}_j \mid g_i)$. For example, x_4 (HSenglish) can be left out of the fitted model $\{1234\}$; hence one uses $\{123\cdot\}$. Indeed, in this Monte Carlo example the expected K-L best model is g_{11} (based on 10,000 simulation samples).

What we want to illustrate with simulation here (and in Section 5.3.4) are some results under all-subsets AIC_c model selection in regression. First, we focus on unconditional vs. conditional confidence interval coverage on true β_j, i.e., the value of β_j in the generating model g_{15} in Table 5.16. For confidence intervals we used $\hat{\beta} \pm 2\,\widehat{se}$; hence, we ignored the issue of a t-distribution-based multiplier. This affects coverage a little, but the focus is really on the difference between conditional and unconditional coverage.

Second, we look at induced model selection bias in $\hat{\sigma}^2_{y\mid x}$. A selected model g_i, out of the 16 models fitted (especially at a small sample size, even using AIC_c), tends to have a better fit for that data set, hence a smaller residual sum of squares, than would occur on average if model g_i were always fitted. Thus, data-based selection in regression will tend to result in fitting the data a little too well; as a result, we get $E(\hat{\sigma}^2_{y\mid x}) < \sigma^2_{y\mid x}$ (this also has an effect on confidence interval coverage that is not correctable by using unconditional intervals). Confidence intervals depend on $\hat{\sigma}_{y\mid x}$; however, the true value of $\sigma_{y\mid x}$ varies by model, so what we report to assess selection bias is the relative bias

$$\text{RB} = \frac{E(\hat{\sigma}_{y\mid x}) - \sigma_{y\mid x}}{\sigma_{y\mid x}}. \tag{5.6}$$

Other quantities of interest include the expected value of the model-averaged estimator $\overline{\hat{\beta}}_j$ and unconditional interval coverage based on this estimator, model selection variation, and percentiles of Δ_p. From the Monte Carlo results of this example we find the percentiles below:

percentile	Δ_p
50	0.9
80	3.1
90	3.6
95	5.1
98	7.3
99	9.4

When a parameter (hence $\hat{\beta}_j$) appeared in the selected model we computed several quantities: the model-averaged estimate $\hat{\bar{\beta}}_j$, the unconditional standard error $\widehat{se}(\hat{\bar{\beta}}_j)$, and three confidence intervals (nominally 95%). The conditional interval (cond.) is based on $\hat{\beta}_j$ and its estimated conditional standard error given the selected model. The unconditional interval (unc.) is based on $\hat{\beta}_j$ and $\widehat{se}(\hat{\beta}_j)$. The interval based on model averaging (MA) uses $\hat{\bar{\beta}}_j$ and $\widehat{se}(\hat{\bar{\beta}}_j)$. Finally, it needs to be clearly understood that the coverage we refer to is on the true parameter from the actual data-generating model.

In this example, the achieved coverage of the unconditional interval is better than that of the conditional intervals (Table 5.17), especially for β_1 (89% versus 80%; both coverage percentages increase by about 0.02 if a t-distribution-based interval is used). A source of lowered coverage comes from bias due to model selection. For example, here model selection results in the bias $E(\hat{\beta}_1) - \beta_1 = 0.00229 - 0.00201 = 0.00028$. This bias is important only in relation to the unconditional standard error, which is here $se(\hat{\beta}_1) = 0.000514$. Thus, the bias/se ratio is $\delta = 0.54$; this value of δ will result in a coverage decrease to 92.1% if coverage would be 95% at $\delta = 0$ (see Cochran 1963:Table 1.1).

An unexpected result in this example is that the conditional coverage for β_2 and β_3 was as high as (about) 0.9. The unconditional coverage then does improve and without exceeding 95% coverage.

The model selection bias induced in $\hat{\sigma}_{y|x}$ is negative and depends on the probability that the model will be selected (Table 5.18). Shown in Table 5.18

TABLE 5.17. Expected values of estimators of β_j and confidence interval coverage on true β_j under AIC$_c$-based all-subsets model selection from the Monte Carlo generated data (10,000 samples) mimicking the GPA example; occurrence frequency is the number of samples in which the selected model included the indicated β_j.

Occur. freq.	j	β_j	$E(\hat{\beta}_j)$	$E(\hat{\bar{\beta}}_j)$	Achieved coverage cond.	unc.	MA
9544	1	0.00201	0.00229	0.00232	0.801	0.879	0.889
7506	2	0.00125	0.00156	0.00157	0.906	0.927	0.930
6506	3	0.190	0.248	0.253	0.887	0.913	0.916

TABLE 5.18. The relative bias (RB) in $\hat{\sigma}_{y|x}$ (5.6) induced by AIC_c model selection for the Monte Carlo example (10,000 samples) based on the GPA data; $E(\hat{\sigma}^2_{y|x} \mid g_i)$ is the average of the 10,000 values $\hat{\sigma}^2_{y|x}$ when the model is fit to every generated data set (i.e., no selection occurs).

Model i	Predictors used	$E(\hat{\sigma}^2_{y|x} \mid g_i)$ no selection	π_i	Selection RB
11	{123·}	0.0576	0.3786	−0.0618
5	{12··}	0.0728	0.2730	−0.0762
6	{1·3·}	0.0803	0.1451	−0.1161
13	{1·34}	0.0762	0.0458	−0.1854
15	{1234}	0.0566	0.0389	−0.1878
1	{1···}	0.1076	0.0351	−0.1724
12	{12·4}	0.0728	0.0301	−0.1693
14	{·234}	0.1017	0.0216	−0.2664
10	{··34}	0.1272	0.0161	−0.2681
8	{1··4}	0.1063	0.0078	−0.2121
7	{·23·}	0.1420	0.0064	−0.2843
9	{·2·4}	0.1695	0.0012	−0.3634
2	{·2··}	0.2227	0.0001	−0.3837
3	{··3·}	0.2012	0.0001	−0.5496
4	{···4}	0.2446	0.0001	−0.3913

is the average value of $\hat{\sigma}^2_{y|x}$ when the model is fit to all 10,000 generated data sets (compare to theoretical results in Table 5.16). The relative bias of $\hat{\sigma}_{y|x}$ under model selection is given by RB from (5.6). Good models (under the K-L paradigm) do not correspond to very bad levels of RB. As the model becomes less acceptable (in terms of expected K-L value), it is selected only when the data are an unusually good fit to that model.

While general in their qualitative nature, these numerical results are more extreme for a sample size of 20 than would be the case at a large sample size. In fact, for this generating model the selection bias in $\hat{\sigma}_{y|x}$ is trivial at sample size $n = 50$, and confidence interval coverage is nearly 95% for each type of interval. Even though model selection can induce biases, under information-theoretic selection and associated unconditional inferences results can be quite good and certainly better (for the sample size) than use of an unnecessarily high-dimensional global model that includes all predictors.

5.3.3   An Improved Set of GPA Prediction Models

An even better way to improve on the all-predictors global model is first to reduce one's models to a smaller set of a priori meaningful models suggested by subject matter or logical considerations. Basically, this means using logical transformations of the predictors (consideration of meaningful model forms is also important) and dropping predictors that are very unlikely to be related to the response variable of interest (investigators often record a variable simply because it is easy to measure). To illustrate this idea we conceived of five different GPA prediction models based on simple derived predictors that make some sense.

The original four predictors are each just indices to general academic ability, and they are measured with error. That is, a person's test grade would surely vary by circumstances (and luck), such as if they had a cold the day of the exam. Viewing these predictor variables as just semicrude indices, why not just compute a single averaged index? In so doing we average over math and verbal (English) ability, but grades in many courses depend upon both abilities anyway. With a large sample size (say $n > 1,000$) it makes sense to let the regression fit calibrate the relative importance of the four indices. However, with only 20 observations some combining of indices may be advantageous.

The SAT and HS scores are on very different scales. There are several ways to allow for this, such as first to normalize each predictor variable to have a mean of 0 and a standard deviation of 1 and then just average all four adjusted predictors to get a total (*tot*) predictor index. To circumvent that minor nuisance we used geometric means to cope with the scale issue. Thus the five new variables that replace the original four variables are

$$sat = (x_1 \times x_2)^{0.5},$$
$$hs = (x_3 \times x_4)^{0.5},$$
$$math = (x_1 \times x_3)^{0.5},$$
$$engl = (x_2 \times x_4)^{0.5},$$
$$tot = (sat \times hs)^{0.5} = (math \times engl)^{0.5} = (x_1 \times x_2 \times x_3 \times x_4)^{0.25}.$$

These variables are interpretable and seem just as adequate as the original four variables as indices to first-year college GPA.

Next, we would not use, in an a priori analysis, any of the original 16 models. The only linear regression models we would (did) consider with these derived predictors are given below, in terms of predictor variables in the model (all models have an intercept and σ^2). We numbered these as models 17 to 21 in order to compare results to the original 16 models:

model	K	variables included
g_{17}	3	tot
g_{18}	4	sat hs
g_{19}	4	math engl
g_{20}	3	sat
g_{21}	3	hs

We conceptualized these models before examining fit of the original 16 models to the GPA data, and no other derived models were considered.

The AIC_c best model of the above is g_{17} (Table 5.19). In fact, model g_{17} is best in the full set of all 21 models (results are not shown for all 21 models in Table 5.19). In adding new models to an existing set, no earlier AIC_c values

TABLE 5.19. Results of fitting, to the GPA data, the five new models (g_{17}–g_{21}) based on transformations of the original test scores (x_1–x_4); also given are results for some of the 16 models originally considered (see text for details of the new predictors).

Model i	Predictors used	AIC_c	Δ_i	w_i
17	tot	−48.20	0.00	0.590
18	sat hs	−45.97	2.23	0.193
20	sat	−43.96	4.24	0.071
11	{123·}	−43.74	4.47	0.063
5	{12··}	−42.69	5.52	0.037
6	{1·3·}	−40.77	7.44	0.014
15	{1234}	−39.89	8.32	0.009
19	math engl	−39.23	8.97	0.007
...	...	...	...	
21	hs	−33.98	14.22	0.000
...	...	...	...	

need to be recomputed: Just reorder the full set from smallest to largest AIC_c. The full set of Δ_i values may need to be recomputed if the best model changes (as here, from g_{11} to g_{17}). Given the new set of Δ_i, recompute the Akaike weights w_i.

The results in Table 5.19 illustrate that the best model in a set of models is relative only to that set of models. Kullback–Leibler model selection does not provide an absolute measure of how good a fitted model is; model g_{11} is best only in the set of 16 all-subsets models. Compared to model g_{17}, model g_{11} (and the entire original set of 16 models) can almost be discarded as contenders for expected K-L best model for these data. Correspondingly, we emphasize that any model-based inference is conditional on the set of models considered. The specifics of inferences and computable uncertainties are conditional on the models formally considered.

The models used here are useful only for prediction; they do not relate to any causal process. Hence, we illustrate inclusion of model uncertainty into prediction based on models g_{17}–g_{21} (standard aspects of prediction inference given a fitted linear model are assumed here; see, e.g., Graybill and Iyer 1994). As computed in Graybill and Iyer (1994:244), under model g_{15}, {1234}, the prediction of expected GPA at $x_1 = 730$, $x_2 = 570$, $x_3 = 3.2$, and $x_4 = 2.7$ is $\hat{E}(y) = 3.185$ with conditional (on model) standard error 0.172. For comparison we note that given model g_{11}, {123·}, the corresponding results are $\hat{E}(y) = 3.253$, $\widehat{se} = 0.102$.

The model-averaged predicted expected GPA is $\hat{\bar{E}}(y) = 3.06$ with estimated unconditional standard error of 0.11 (Table 5.20). These results are computed using (4.1) and (4.9). To construct a confidence interval here that allows for the small degrees of freedom of $\hat{\sigma}^2_{y|x}$ we suggest that it suffices in this example to

TABLE 5.20. Quantities needed in the computation of model-averaged prediction of expected GPA $\hat{E}(y)$ ($= 3.056$) and its unconditional standard error ($= 0.1076$), under models g_{17}–g_{21} for the predictors $x_1 = 730$, $x_2 = 570$, $x_3 = 3.2$, and $x_4 = 2.7$.

Model i	Predictors used	Δ_i	w_i	$\hat{E}(y)$	$\widehat{se}(\hat{E}(y) \mid g_i)$
17	tot	0.00	0.685	3.016	0.0738
18	sat hs	2.23	0.224	3.095	0.1210
20	sat	4.24	0.082	3.177	0.1993
19	math engl	8.97	0.008	3.271	0.1045
21	hs	14.22	0.001	2.632	0.0832
		Weighted results:		3.056	0.1076

use $3.06 \pm t \times \widehat{se}(\hat{E}(y))$, where the multiplier $t = 2.10$ is from the t-distribution on 18 df. Model g_{17} has 18 df for $\hat{\sigma}^2_{y \mid x}$, and the weight on that model is $w_{17} = 0.685$ (more sophisticated procedures will not make a practical difference here). The model-averaged result is distinctly more precise than the prediction based on the fitted global model (standard errors of 0.108 versus 0.172, model-averaged versus global model-based). Also, the inclusion of model uncertainty increases the standard error as compared to the result conditional on model g_{17} ($\widehat{se} = 0.074$).

5.3.4   More Monte Carlo Results

The theory for Monte Carlo generation of regression data, with random regressors, was presented in Section 5.3.2. Using that approach we computed a few more simulations. Our motivation was firstly to see whether anything bad occurred in using model averaging and unconditional confidence intervals (it did not), and secondly to see what biases might result from model selection and what confidence interval coverage could be achieved.

This is far from a full-scale simulation study because we greatly restricted the many factors to consider in the design of an all-subsets model selection study. For example, we used only $m = 4$ predictors here (but much larger values of m need to be explored). Thus, the global model is model g_{15}, $\{1234\}$. Given a sample size n, one generates a sample from the $(m+1)$-dimensional MVN(μ, Σ) generating model. Without loss of generality we can set $\mu = 0$. However, there are still (in general) $(m + 1) \times (m + 2)/2$ parameters to specify in Σ. To make this design problem tractable we used the following structure on the generating model: either

$$\underline{c}' = [0\,0\,0\,0] \quad \text{or} \quad \underline{c}' = [0.8\,0.6\,0.4\,0.2],$$

mostly the latter; and

$$\Sigma_x = \begin{bmatrix} 1 & \rho & \rho & \rho \\ \rho & 1 & \rho & \rho \\ \rho & \rho & 1 & \rho \\ \rho & \rho & \rho & 1 \end{bmatrix}.$$

The regression parameters of the generating model are given by $\beta' = \underline{c}'(\Sigma_x)^{-1}$.

Another design factor was taken to be the regression residual variance $\sigma_{y|x}^2 = \sigma_y^2 - \underline{c}'(\Sigma_x)^{-1}\underline{c}$. Given values for $\sigma_{y|x}^2$ (we used only 1 and 25) we find the marginal variance of y, σ_y^2. These quantities (i.e., $\underline{c}$, Σ_x, and σ_y^2) suffice to compute the full 5×5 variance–covariance matrix of (5.3). The conditional variance–covariance of $\underline{\hat{\beta}}$ is given by

$$\frac{\sigma_{y|x}^2}{n}(\Sigma_x)^{-1}.$$

We see that factors $\sigma_{y|x}^2$ and n are redundant in their effect on the sampling variance. Therefore, for not-small sample sizes it is much more economical to fix n (say $n \geq 28 + K$, for the global model value of K) and directly lower $\sigma_{y|x}^2$ to gain precision, rather than to fix $\sigma_{y|x}^2$ and simulate greater precision by increasing sample size. We did not do so here; it is still necessary to have small actual n to explore small-sample-size effects.

Because of the choice of the form of Σ_x, the estimators $\hat{\beta}_1$, $\hat{\beta}_2$, $\hat{\beta}_3$, and $\hat{\beta}_4$ all have the same conditional variance. In fact, for any ρ it suffices to present the constant diagonal element v of $(\Sigma_x)^{-1}$, because

$$\text{var}(\hat{\beta}_i \mid g_{15}) = v \times \sigma_{y|x}^2 \, / \, n.$$

For $\underline{c}' = [0.8, \ 0.6, \ 0.4, \ 0.2]$ we simulated 10,000 samples at each combination of $\rho = 0, 0.2, 0.4, 0.6$, and 0.8 crossed with $n = 20, 50$, and 100. We focused on $\hat{\beta}_1$, but looked at other parameters in a few cases (based then on another set of 10,000 samples). To these cases 1 to 5 for ρ we added cases 6 and 7, as noted in Table 5.21. In total we looked at 29 simulated "populations."

We tabulated some basic results (Table 5.22) wherein full AIC$_c$ model selection was applied to all 16 possible models (the labeling of models is the same as in Table 5.12). In particular, we tabulated the 90th, 95th, and 99th percentiles of Δ_p. There is one variation here; with no models more general than the generating global model (g_{15}) if sample size gets too large, selection converges on model g_{15}, and all percentiles of Δ_p go to 0. This "boundary" effect (i.e., some degree of reduction in percentiles of Δ_p) will not normally occur in real data analysis, so we flagged populations where a boundary effect is occurring. Our recommendations about interpreting Δ_i are for when no boundary effect occurs. For the 15 populations where no boundary effect occurred, the mean percentiles of Δ_p in Table 5.22 are 4.7, 6.4, and 10.6 (90th, 95th, and 99th, respectively).

TABLE 5.21. Values of design factors used in simulation exploration of model selection (see text for details); sample size sets used are n-set, a $= \{20, 50, 100\}$, b $= \{20, 50, 100, 200, 500, 1,000\}$; β_i are rounded to two decimal places.

Case	ρ	n-set	$\sigma^2_{y \mid x}$	υ	β_1	β_2	β_3	β_4
1	0	a	1	1.00	0.8	0.6	0.4	0.2
2	0.2	a	1	1.09	0.69	0.44	0.19	-0.06
3	0.4	a	1	1.36	0.73	0.39	0.06	-0.27
4	0.6	a	1	1.96	0.93	0.43	−0.07	−0.57
5	0.8	a	1	3.82	1.65	0.65	−0.35	−1.35
6	0	a	1	1.00	0	0	0	0
7	0	b	25	1.00	0.8	0.6	0.4	0.2

Sample size also has an effect on the distribution of Δ_p; it is not a strong effect for n greater than about 20. For case 7 in Table 5.22 the effect of sample size ($20 \leq n \leq 500$) is about 2 units at the 90th and 95th percentiles and about 3 units at the 99th percentile. These are typical of sample size effects we have observed.

There is considerable model selection uncertainty in these 29 simulated populations (Table 5.22), for example, as indexed by how low the selection probability π_k is even for the expected AIC$_c$ best model. The other index of model selection uncertainty used in Table 5.22 is simply a count of the number of models, of the 16, that have selection probabilities ≥ 0.01 (often $\geq$ one-half of the possible models). For the all too typical application of variable selection with 10 or more variables ($R \geq 1,024$) and not-large sample size, the selection process will be highly unstable (cf. Breiman 1996) as to what model is selected. That is, selection probabilities can be expected to be very low, even for the actual K-L best model, and not exhibit a strong mode. As a result the selected model itself is not at all the basis for a reliable inference about the relative importance of the predictor variables, even if the selected model provides reliable predictions.

A confidence interval for a parameter β_i was computed only when that parameter was in the selected model, in which case point estimates computed were the MLE $\hat{\beta}_i$ and the model-averaged $\bar{\hat{\beta}}_i$. Three types of intervals were computed: the classical conditional interval $\hat{\beta}_i \pm 2 \, \widehat{se}(\hat{\beta}_i \mid g_r)$, the corresponding unconditional interval $\hat{\beta}_i \pm 2 \, \widehat{se}(\bar{\hat{\beta}}_i)$, and the interval based on the model-averaged point estimate $\bar{\hat{\beta}}_i \pm 2 \, \widehat{se}(\bar{\hat{\beta}}_i)$. One result was that the coverage for interval types two and three was barely different, but was slightly better for the interval $\bar{\hat{\beta}}_i \pm 2 \, \widehat{se}(\bar{\hat{\beta}}_i)$ (ratio of coverages: 0.995). Hence, we present results only for this latter interval (Table 5.23). We focused on the coverage for β_1 (i.e., parameter index 1); Table 5.23 gives coverage results for a few instances of a difference parameter (β_3 or β_4).

TABLE 5.22. Monte Carlo results (10,000 samples used) for simulated populations; column four is the AIC_c best model; column five is the corresponding selection probability, π_k; column six is the number of models for which $\pi_i \geq 0.01$; also given are percentiles for Δ_p.

Case	ρ	n	Best Model	π_k best	# π_i ≥ 0.01	Percentiles of Δ_p 90	95	99	Boundary Effect
1	0	20	11	0.243	13	5.37	6.74	10.84	no
1	0	50	15	0.368	4	2.55	2.59	4.01	yes
1	0	50	15	0.378	4	2.56	2.59	4.25	yes
1	0	100	15	0.668	2	1.81	2.12	2.26	yes
2	0.2	20	5	0.342	12	4.65	6.51	11.37	no
2	0.2	20	5	0.334	12	4.58	6.41	10.77	no
2	0.2	50	11	0.264	7	2.67	3.78	7.10	yes
2	0.2	100	11	0.462	4	2.22	3.55	7.14	yes
3	0.4	20	1	0.398	12	7.03	9.35	14.80	no
3	0.4	50	12	0.353	8	3.35	4.29	7.02	yes
3	0.4	100	12	0.611	6	2.17	3.36	7.06	yes
4	0.6	20	8	0.227	12	6.38	8.73	13.53	no
4	0.6	50	12	0.544	8	2.47	3.65	6.41	yes
4	0.6	100	12	0.743	4	1.74	2.89	6.60	yes
5	0.8	20	8	0.499	10	5.43	7.68	12.58	no
5	0.8	50	15	0.305	4	3.32	4.03	4.77	yes
5	0.8	50	15	0.308	4	3.24	4.03	4.83	yes
5	0.8	100	15	0.595	4	2.04	2.23	2.88	yes
5	0.8	100	15	0.597	4	2.03	2.23	2.94	yes
6	0	20	16	0.594	11	3.43	5.11	8.92	no
6	0	50	16	0.535	11	3.35	4.96	8.71	no
6	0	100	16	0.518	11	3.41	4.90	8.65	no
7	0	20	16	0.493	11	4.55	6.47	10.55	no
7	0	50	1	0.180	13	5.62	7.48	11.42	no
7	0	100	5	0.131	16	5.11	6.90	10.86	no
7	0	200	11	0.143	15	4.08	5.20	8.42	no
7	0	200	11	0.140	15	4.10	5.20	8.43	no
7	0	500	11	0.391	8	3.42	4.93	8.92	no
7	0	1000	15	0.380	4	2.00	2.04	3.30	yes

Confidence interval coverage is affected by bias in either the point estimator or its standard error estimator. Therefore, we tabulated information on these biases for the interval based on the model-averaged estimator, $\hat{\bar{\beta}}_i \pm 2\,\widehat{se}\!\left(\hat{\bar{\beta}}_i\right)$. In this context bias is important only in relation to standard error, which Table 5.23 shows,

$$\delta = \frac{E(\hat{\bar{\beta}}_i) - \beta_i}{E(\widehat{se}(\hat{\bar{\beta}}_i))},$$

which is bias of the model-averaged estimator of β_i divided by the expected unconditional standard error of the estimator. The effect on confidence interval coverage is trivial for $|\delta| \leq 0.25$ and ignorable for $|\delta| \leq 0.5$.

The other factor that can affect coverage is bias in the estimator $\widehat{se}(\hat{\bar{\beta}}_i)$; hence we show the ratio

$$se\text{-}r = \frac{E\left(\widehat{se}\left(\hat{\bar{\beta}}_i\right)\right)}{MC\text{-}se\left(\hat{\bar{\beta}}_i\right)}$$

in Table 5.23. Here, $MC\text{-}se(\hat{\bar{\beta}}_i)$ is the actual achieved standard error of $\hat{\bar{\beta}}_i$ over the Monte Carlo samples (out of 10,000) wherein $\hat{\bar{\beta}}_i$ is computed. A value of $se\text{-}r = 1$ is desirable. Coverage would be reduced (other factors being equal) if $se\text{-}r$ becomes much less than 1. The effect on coverage is ignorable for $0.9 \leq se\text{-}r \leq 1.1$. For all these results, the relevant sample size is denoted by "Freq." in Table 5.23: the number of samples wherein $\hat{\bar{\beta}}_i$ is computed because the parameter is in the AIC_c-selected model.

The biggest surprise was the high achieved coverage of the traditional (conditional) confidence interval (Table 5.23). When that coverage was poor (for example, case 2, $\rho = 0.2$, $n = 20$, unconditional coverage of 0.755 on β_3), it was because of severe bias in either the point estimator or its standard error estimator. Moreover, these biases are clearly a form of model selection bias, and they occurred when the reference parameter was infrequently selected (i.e., infrequent selection of any model containing β_i), which itself is a result of the predictor variable x_i being unimportant at the given sample size. Confusing the matter, however, not all instances of small frequency of selecting models including β_i resulted in deleterious effects on coverage (for example, case 5, $\rho = 0.8$, $n = 50$, unconditional coverage of 0.925 on β_3). On the positive side, if a predictor variable was important (as judged by high selection frequency), its unconditional (and conditional) coverage was always good.

In all 29 simulated populations the unconditional interval coverage was greater than or equal to the conditional coverage and provided improved coverage when the conditional coverage was less than 0.95 (for example, case 5, $\rho = 0.8$, $n = 20$, conditional and unconditional coverage of 0.858 versus 0.937 on β_1). For the 23 populations where the parameter was important to the selected model, the average conditional and unconditional confidence interval coverage was 0.930 versus 0.947 (and $\bar{\delta} = 0.28$, $\overline{se\text{-}r} = 1.1$). The improvement in coverage is not dramatic, but is generally worthwhile.

For the other six populations, the bias in coverage is seen to be caused by strong biases in point estimates or their standard errors, as reflected by δ and $se\text{-}r$ ($abs(\delta) = 0.64$ and $\overline{se\text{-}r} = 0.63$). These biases are a direct result of model selection, i.e., they are model selection bias. In those cases where the selected model is not the expected AIC_c best (i.e., expected K-L best) model, rather it includes a variable x_s that is rarely included in the selected model, then

TABLE 5.23. Confidence interval coverage from the Monte Carlo simulations; coverage is for β_i (hence Parm. index i), and is from all selected models containing β_i: The number of such selected models is denoted by "Freq."; coverage is for the traditional conditional interval and the interval based on the model-averaged estimator $\hat{\bar{\beta}}_i$; see text for explanation of δ and se-r.

Case	ρ	n	Parm. index	Freq.	Coverage cond.	Coverage MA	Bias/se δ	Se-ratio se-r
1	0	20	1	8864	0.932	0.948	0.22	1.07
1	0	50	1	9994	0.943	0.948	−0.01	0.98
1	0	50	4	4337	0.934	0.934	0.90	1.43
1	0	100	1	10000	0.948	0.950	0.01	0.98
2	0.2	20	1	8309	0.923	0.947	0.42	1.12
2	0.2	20	3	2106	0.694	0.755	1.37	0.82
2	0.2	50	1	9970	0.945	0.955	0.07	1.02
2	0.2	100	1	10000	0.944	0.950	0.02	0.99
3	0.4	20	1	8026	0.908	0.945	0.41	1.13
3	0.4	50	1	9912	0.918	0.947	0.03	1.00
3	0.4	100	1	9999	0.932	0.944	0.02	0.98
4	0.6	20	1	8178	0.905	0.957	0.27	1.13
4	0.6	50	1	9953	0.905	0.942	0.00	0.97
4	0.6	100	1	10000	0.937	0.945	−0.03	0.98
5	0.8	20	1	9175	0.858	0.937	0.17	1.03
5	0.8	50	1	10000	0.918	0.945	−0.02	0.99
5	0.8	50	3	3572	0.913	0.925	−0.93	1.22
5	0.8	100	1	10000	0.930	0.946	−0.07	0.97
5	0.8	100	3	6055	0.957	0.960	−0.61	1.41
6	0	20	1	1220	0.396	0.495	0.13	0.45
6	0	50	1	1450	0.643	0.677	0.06	0.51
6	0	100	1	1501	0.658	0.676	−0.06	0.51
7	0	20	1	2014	0.726	0.759	1.19	0.73
7	0	50	1	3624	0.904	0.913	1.01	1.32
7	0	100	1	5594	0.952	0.955	0.73	1.47
7	0	200	1	7911	0.967	0.969	0.37	1.31
7	0	200	4	2225	0.785	0.796	1.00	0.78
7	0	500	1	9828	0.969	0.969	0.06	1.06
7	0	1000	1	9998	0.956	0.956	0.02	1.01

inference on β_s (an unimportant variable) can be very misleading because of resultant model selection bias for $\hat{\beta}_s$. Fortunately, this scenario is uncommon, almost by definition, since it is a case of x_s being commonly excluded from the selected model. Also, even then inference on an important parameter was generally sound in these simulations and others we have done even if inference on an unimportant β_s was slightly (but not strongly) misleading when its unconditional confidence interval was considered.

5.3.5   *Linear Regression and Variable Selection*

We present here some thoughts on aspects of model selection with specific reference to linear regression and so-called variable selection. This is arguably the most used and misused application area of model selection. In particular, every conceivable type of model (variable) selection method seems to have been tried in the context of having m predictors and using linear multiple regression models (see, for example, Hocking 1976, Draper and Smith 1981, Henderson and Velleman 1981, Breiman and Freedman 1983, Copas 1983, Miller 1990, Hjorth 1994, Breiman 1995, Tibshirani 1996, Raftery et al. 1997). However, almost always the statistical literature approaches the problem as if it is only a matter of "just-the-numbers" data analysis methodology. Firstly, in fact there is always a subject-matter scientific context, and a possible limitation of sample size, that must be brought into the problem, and so doing will make an enormous difference as compared to any naive model selection approach that does not consider context, prior knowledge, and sample size.

Secondly, there is always a goal of either (1) selecting a best model (this should include ranking competitor models) because one seeks understanding of the relationships (presumably causal) between $\underline{x}$ (independent variables) and y, or (2) prediction of $E(y \mid \underline{x})$ at values of $\underline{x}$ (predictors) not in the sample (prediction of $E(y \mid \underline{x})$ for $\underline{x}$ in the sample can be considered just parameter estimation). These goals really are different. That is, if there is considerable model selection uncertainty, then selecting the best model under goal (1) and using it for goal (2), prediction, is not optimal.

We recommend that prior to any data analysis full consideration be given to how the problem (i.e., set of models) should be structured and restricted. This means dropping variables that cannot reasonably be related to y for prediction, or cannot reasonably be causally related at detectable effect levels given the sample size. From the literature and our experience, investigators are far too reluctant to drop clearly irrelevant variables and otherwise apply a priori considerations based on logic and theory. This is the "measure everything that is easy to measure and let the computer sort it out" syndrome, and it does not work. Even a good exploratory analysis needs input of investigator insights to reach useful results. In this regard we quote Freedman et al. (1988):

> A major part of the problem in applications is the curse of dimensionality: there is a lot of room in high-dimensional space. That is why investigators need model specifications tightly derived from good theory. We cannot expect statistical modeling to perform at all well in an environment consisting of large, complicated data sets and weak theory. Unfortunately, at present that describes many applications.

An important a priori aspect is to consider reducing the number of independent variables by functionally combining them into a smaller set of more useful variables. This may be as simple as computing, by observation, an average of some of the predictor variables (such as in the GPA example of this section), and then that average replaces all the variables that went into it. Use of such derived variables is common (for example, wind chill factor, relative humidity, density as mass per unit volume in physics or animals per resource in ecology, rates of all sorts, and so forth). Consider also any bounds on y. For example, often college GPA is bounded on 0 to 4; hence, we do not want our model to be able to make a prediction of 4.2. We could model GPA/4 using a logistic link function and rescale predictions by 4.

As a rule of thumb, the maximum number of structural parameters to allow in a regression (or other univariate) model should be $n/10$. It is not possible to reliably estimate anything like $n/2$ (or $n/3$) parameters from "noisy" data. Mistakenly, models of such size are often fit to data, and may be selected based on an invalid criterion such as minimum residual variance, or an inappropriate to the situation criterion like adjusted R^2 or even AIC (AIC$_c$ correctly adjusts for either small sample size or large K).

To illustrate some of this thinking we consider another somewhat classic example of variable selection (Hocking 1976; see also Hendersen and Velleman 1981 for the actual data; we did not read this latter paper before the thinking below): automobile gas mileage (y) as MPG (miles per gallon) versus $m = 10$ independent variables (there is clearly causation involved here; when this is not so we use the term predictor variable). Note that y is already a derived variable. The 10 x_i are:

1   Engine shape (straight or V)
2   Number of cylinders (4, 6, or 8)
3   Transmission type (manual or automatic)
4   Number of transmission speeds
5   Engine size (cubic inches)
6   Horsepower
7   Number of carburetor barrels
8   Final drive ratio
9   Weight
10 Quarter mile time

The data arose from testing 32 ($= n$) different types of automobiles under standardized conditions. We independently generated a priori considerations (it would be better to get an automotive engineer involved). We did not first look at the dependent variable (either as y alone, or as y versus the x_i). In general, given that one has decided that the analysis will be only to look at models for a response variable y based on $x_1, \ldots, x_m$ as predictor variables (upon which the models are all conditional), one next reduces the number of such variables as much as possible by logical and subject-matter considerations. Given the resultant reduced set of predictors, we recommend looking at the correlations of those independent variables to be assured that there are no remaining pairs having an extremely high pairwise correlation.

A more comprehensive examination would be a principal components eigenvalue evaluation of the design matrix X. Such results are given in Hocking (1976) for the full set of 10 predictors (but the data are not given there). The eigenvalue analysis suggests that a two-variable model might fit as much as 90% of the variation in the MPG response variable (because the first two eigenvalues add to about 90% of the total of all 10 eigenvalues). This much explained variation is often all we can hope for without overfitting the data at sample size 32.

The x_i are highly intercorrelated in this observational study because of car design: Big cars have bigger engines; are more likely to have 8, not 4 or 6, cylinders; are therefore more likely to have a V-engine design; and so forth. For such observational studies if the issue of interest is causality, there are substantial inference problems (see Draper and Smith 1981, page 295, for some sage cautionary comments on such problems with observational data).

One of us (KPB, who is automotively challenged) proceeded as follows. Because $n = 32$, do not include more than three structural parameters. Gas mileage is strongly dependent on car weight, so always include x_9. Given that an intercept will be used here, this leaves room for only one more variable. As a first thought, then, consider the nine models

$$y = \beta_0 + \beta_1 x_9 + \beta_i x_i + \epsilon, \qquad i \neq 9,$$

plus $y = \beta_0 + \beta_1 x_9 + \epsilon$. However, bearing in mind the intercorrelated nature of these variables, consider dropping some on a priori grounds. Do not drop any x_i based just on a high correlation unless it is extreme such as $|r| \geq 0.95$, since then there is a variable redundancy problem (near colinearity). Do eliminate (near) colinearity problems; and do eliminate variables based on knowledge, reasoning, and experience.

As a type of thought experiment (because the data did not arise from an experiment) consider whether engine shape (x_1) is causally related to MPG. Do we really think that if all car features were held fixed except whether the engine is a straight or V8 that there would be any effect on MPG? Probably either not at all, or at a level we will never care about and could not detect except with an experiment and a huge sample size. Conclusion: Drop variable x_1 (we surmise that it was recorded because it is easy to determine—this is not justification for including a variable). **Recommendation: Use thought experiments in conjunction with observational studies**.

The same reasoning leads KPB to drop variable 2: number of cylinders. Again the thought is that if all else (horsepower, total cylinder displacement, etc.) were fixed, would just number of cylinders (as 4, 6, or 8) alone affect MPG? And again, no; at least not in these data. Variables 2 through 8 would be retained on fundamental grounds. Quarter mile time is in effect a complex derived variable; it might predict MPG well, but it is not causally related. Instead, variable 10 might itself be well predicted based on variables x_3–x_9. Conclusion: drop x_{10}. Thus KPB would consider only models with variable x_9 always included and at most one of variables 3–8 (seven models). This is very different from an all-subsets selection over $10^m = 1{,}024$ models, and this sort of thinking can, and should, always be brought to bear on a variable-selection problem.

One advantage of doing thinking such as this is that it focuses one's attention on the problem and basic issues. In this case the realization arose (still for KPB) that horsepower itself is a derived variable but one that can be engineered. To some extent horsepower might replace variables 5 and 7, and might be more important than variables 3, 4, 6, and 8. Thus we have the a priori hypothesis that the best two-variable model might be based on x_9 and x_6. If this model was not best, but nearly tied for best with a less-interpretable model, this a priori thinking would justify objective selection of these two variables as most important.

Another issue is the suitability of the linear model form. Because MPG is bounded below by zero but weight can be unbounded, a linear model could predict negative MPG. Surely, over a big enough weight range the relationship is curvilinear, such as $E(y \mid x_9) = \beta_0 e^{-\beta_1 x_9}$ or $E(1/y \mid x_9) = \beta_0 + \beta_1 x_9$. We use MPG only by convention; hence a priori KPB would fit all models as linear but based on an inverse link function to MPG. One could just fit $1/y$ to several linear models and select a best model. However, AIC_c is then not comparable from models fit to y versus models fit to $1/y$.

Less time was afforded to this exercise by DRA (who is much more auto-motively knowledgeable), who independently put forth two a priori models. Both include weight (x_9);

$$y = \beta_0 + \beta_1 (x_9)^2 + \beta_2 z + \epsilon$$

and

$$y = \beta_0 + \beta_1 (x_9)^2 + \beta_2 x_{10} + \epsilon.$$

The variable z is a derived variable meant to reflect the combined effect of several variables on MPG:

$$z = \frac{x_2 \times x_5}{x_6}.$$

Similarly, x_{10} is used here as a predictor that summarizes many features of the ability of the car to consume fuel.

Considerations like these based on reasoning and theory must be thought about before data analysis if reliable uncertainty bounds are to be placed on an inference made after model selection. One can always do data-dependent exploratory analyses after the a priori analysis. We just recommend separating the two processes, because results of exploratory analyses are not defensible as reliable inferences in the sense that the data cannot both suggest the question (the model) and then reliably affirm the inferential uncertainty of the answer (the same model). There is a saying from the USA western frontier days: "Shoot first, ask questions later." This strategy often precludes obtaining desired information. Similarly, "compute first, then create models" (or "compute first, think later") is also not a strategy for making reliable inference. The result can be a model that describes the data very well (because it overfits the data), but is a poor model as an inference to independent data from the same generating process. Overfitting, if it occurs, cannot be diagnosed with that same data set that has been overfit.

5.3.6   Discussion

A variety of comments and opinions are given here, some because they do not fit well elsewhere in Section 5.3. There is no particular order to the following comments and opinions.

The Monte Carlo simulations of Sections 5.3.2 and 5.3.4 might seem to violate our general philosophy that the actual data-generating model f would (should) in reality be more general than the global model G used as the basis for data analysis (an expanded vector of predictors $\underline{x}_T$ would apply under the true generating model; the global model does not use all of these predictors). This is only partly true. The part that is not true is thinking that because we generated the data under the global model, no more general model could actually apply. In fact, the residual variation of the global regression model, $\sigma^2_{y|x}$, is a confounding of average (with respect to f) model structural variations arising from the differences $E(y \mid \underline{x}, G) - E(y \mid \underline{x}_T, f)$, plus the "true" unexplained residual variation σ^2_ϵ (it might be 0) under f. Thus almost all aspects of a conceptually more general data-generating model are swept into $\sigma^2_{y|x}$ of the global model. Hence, it is more economical simply to generate data under an assumed global model.

One way in which this lacks generality is that the numerical values of the components of the true parameter vector β_T that are in the lower-dimensional global $\underline{\beta}$ may not exactly equal their counterpart component values in β_T. This would affect confidence interval coverage, which should be relative to the appropriate components of β_T, not to β. This seems like a small concern in initial Monte Carlo studies intended to explore basic model selection issues.

The more important lack of generality relates to how we conceive the asymptotic sequence of models as sample size increases. Classical theory holds the model or set of models fixed, independent of sample size. This is not in accord with reality, because as sample size grows we will include more structure in the data and in our models (e.g., in the GPA example, effects of year, high-school type, university attended, major, student age, and so forth). For that reason we should simply have a larger global generating model than the one we used in simulations here, and include more factors so that the size of the selected model can grow without the arbitrary bound of a global model with only four predictors. In a sense the issue becomes one of not strongly "bumping" up against a bound (i.e., large π_i for the generating model) as sample size increases, because this feature of data analysis is often unrealistic for observational studies. This problem is solved simply by having a sufficiently general global generating model, and it is then still acceptable (but not required) to have that generating model also as the global model for data analysis.

It is well known that selecting a best model from a set of regression models can lead to important biases in parameter estimates and associated estimated standard errors (see, e.g., Miller 1990, Hjorth 1994). The estimated residual variance $\hat{\sigma}^2_{y|x}$ is especially susceptible to being biased low due to the process of selecting a good model, because all selection criteria involve, to some extent, seeking a fitted model with a relatively small residual sum of squares. Use of AIC_c will not entirely protect one from this possible bias, but it helps (see, e.g., Table 5.18; the relative bias in $\hat{\sigma}^2_{y|x}$ is not high for the K-L good models, and this example is for a small sample size). The bias in $\hat{\sigma}^2_{y|x}$ is worse the more infrequently a model is selected because for such poor models they are selected only when they fit a sample unusually well. In the data analysis phase of a study (hence, sample size is then a given) the best way to avoid serious bias in $\hat{\sigma}^2_{y|x}$ is to keep the candidate set of models small.

When the predictors are intercorrelated and model selection is used, such selection tends to induce a bias in the estimators of regression coefficients of selected predictors. The less important a predictor x_i is, the less likely it is to be selected, and then when selection occurs, both of the associated estimators $\hat{\beta}_i$ or $\overline{\hat{\beta}}_i$, conditional on the model, tend to be biased away from zero. That is, let $E(\hat{\beta}_i \mid g_r$ always) denote the expected value of $\hat{\beta}_i$ under model g_r when that model is always fit to the data. Let $E(\hat{\beta}_i \mid g_r$ selected) denote the conditional expectation of $\hat{\beta}_i$ when model g_r is selected, as by AIC_c. If $E(\hat{\beta}_i \mid g_r$ always) > 0, then we usually find that $E(\hat{\beta}_i \mid g_r$ selected) $> E(\hat{\beta}_i \mid g_r$ always); whereas if $E(\hat{\beta}_i \mid g_r$ always) < 0, then we find that $E(\hat{\beta}_i \mid g_r$ selected) $< E(\hat{\beta}_i \mid g_r$ always). The strength of the bias depends mostly on the importance of the predictor, as measured by its overall selection probability (and that probability depends on sample size and goes to 1 as n goes to infinity if $|\beta_i| > 0$).

Consider Table 5.17, which gives Monte Carlo results for the simulation mimicking the GPA data. Using that information, and extending it to β_4, we computed the percent relative bias of $\hat{\beta}_i$, PRBias($\hat{\beta}_i$) below, attributed solely to model selection. The reference value for computing bias is the true value of β_i from the data-generating model, not the parameter value $\beta_{i,r}$ that applies conditionally to model g_r when model g_r is always fit to the data. The relative frequency of occurrence of the given parameter (i.e., predictor) in any selected model is denoted by Pr$\{x_i\}$:

i	Pr$\{x_i\}$	PRBias($\hat{\beta}_i$)
1	0.954	14%
2	0.751	25%
3	0.650	31%
4	0.162	267%

A 31% relative bias in conventional estimators due to model selection should be of concern (let alone 267%, but x_4 is not in the K-L best model).

We have looked at this issue for other models and sample sizes for all-subsets selection, and it is quite clear that this aspect of model selection bias in estimators is, as above, strongly related to the importance of the predictor: Model selection bias is less for a predictor always included in the selected model, but it can be very strong for predictors rarely selected. The more predictors one considers, the more likely it is that many are unimportant (especially in the presence of better predictors they correlate with) and the more likely it is that a few of those unimportant predictors will end up in the selected model. When that happens, all the model selection biases operate in a direction to make you think that the selected variables are important ("significant" in hypothesis-testing terms). The best way to reduce this risk of misleading results is to have a small list of carefully considered candidate variables. (To guarantee wrongly selecting one or more unimportant variables; just have a large list of poorly conceived variables and a small sample size; see, e.g., Freedman 1983, Rexstad et al. 1988, 1990).

Two undesirable, but mutually exclusive, properties of model selection strategies particularly relevant to all-subsets regression for observational data are worth noting here: overfitting the data or overfitting the model. If your strategy is to always fit and use the global model, you will probably overfit the model (i.e., include unnecessary variables). This approach to analysis will avoid subjectively tailoring the model to the data, but you probably will greatly inflate standard errors of all the $\hat{\beta}_i$. This loss of precision can be so bad that all the estimates are worthless. Thus, usually one is forced into some sort of model selection with multivariable observational data (it should be firstly by a priori considerations).

If you use a subjective selection procedure of first fitting a model and then examining the results (e.g., residual plots, r-squares, leverages, effect of trans-formations of variables) in search of a better model based on some vague synthetic criterion of your own choosing, you probably will overfit the data. Thus, you will include in model structure what are really stochastic aspects of the data, thereby possibly biasing $\hat{\sigma}^2_{y|x}$ quite low and as a result inferring "noise" as real structure. The resultant model may become more of a descrip-tion of the particular data at hand than a valid inference from those data. All samples have their nearly unique peculiarities as well as their main features that would show up in all, or most, samples you might get for the inference sit-uation at hand. Inference is about correctly identifying the repeatable features of samples. When you overfit the data, you mistakenly include in model struc-ture uncommon data features that would not be found in most such samples that might arise.

When you have a large number of models for a much smaller number of variables (like $R = 1,024$, $m = 10$, all-subsets) and all those models are fit and considered for selection, you run a high risk that some models will overfit the data. The use of AIC_c reduces this risk (because *heuristically*, it looks at model fit penalized by a function of model size K and sample size n; but there is no built-in "penalty" for having a huge set of models), but does not eliminate it for all-subsets selection: Some degree of selection bias remains. For this reason, and the instability of all-subsets selection, it is critical to properly evaluate model selection uncertainty under all-subsets regression and use inference rather than just use the selected best model.

For some cases where we applied the bootstrap to evaluate aspects of uncertainty for all-subsets model selection we found the results problematic. That is, either the bootstrap failed (the cement data, Section 4.3.1), or it produced some peculiar results that the theoretical approach did not produce (GPA example, Section 5.3.1). Two small studies that evaluated the bootstrap method of assessing aspects of model reliability after model selection in regression expressed pessimism that the bootstrap would always be a reliable method for the task (Freedman et al. 1988, Dijkstra and Veldkamp 1988). The only opinion we can now offer is that for even moderately high-dimensional problems (say $m \geq 7$, hence $R > 100$) one should not blithely think that the bootstrap will be a reliable way to assess model selection uncertainty for all-subsets selection; the method needs more study.

In fairness, it can also be said that AIC_c and associated methods presented here need more evaluation for their performance under all-subsets selection. However, a more basic issue is whether or not ever to do all-subsets selection (especially when the number of predictors is large) when this means selecting a single best model and ignoring all other models. The practical problems are instability of what model is selected (cf. Brieman 1996) and substantial model selection biases. Model instability arises when all model selection probabilities (i.e., the π_i) are low. For large R (hence if m is at all large, even $m = 7$) even the expected K-L best model might have selection probability less than 0.1, so the set of supposed important regressor variables, as judged by the selected best model, can vary dramatically over samples (An instructive example is given in Chapter 6 for $R = 8,191$ models).

At a fundamental level the question of variable selection ought really to be a question of the strength of evidence in the data for the importance of each predictor variable. If the problem is thought of this way, then strict model selection as such is an illegitimate discretization of what ought to be a problem of estimating continuous parameters (the regression coefficients of the global model). The flaw in using model selection is then just like the flaw of using hypothesis testing that makes a problem a reject-or-not dichotomy when it ought to be approached as an evaluation of strength of evidence (the use of P-values rather than strict reject-or-not procedures is also not acceptable; it is a flawed methodology; see, e.g., Harlow et al. 1997, Sellke et al. 2001). We strongly recommend against doing all-subsets selection when the only purpose is identifying a single "best" model: Promoting this practice is not good science and is a failing of statistical science.

We believe that the only defensible reason for fitting all-subsets of regression models should be to obtain the full set of Akaike weights, and then inferences are based on the full set of models as mediated by their associated Akaike weights (i.e., model averaging). The selected best model constitutes only one subset (of R) of the predictor variables. Unless the Akaike weight for that best model is very high (say $w_k \geq 0.9$), we maintain that it is totally misleading to infer that one has found *the* important predictors, and that the predictors not selected are unimportant. As noted above, it is not properly a yes-or-no issue as regards importance of a predictor variable; rather, it is a matter of quantifying predictor importance on a continuous scale, say 0 to 1. Also, this importance value is only with respect to prediction; reliable causal inferences cannot be made from just the data alone when those data are from an observational study.

We are led to believe that the only legitimate application of all-subsets model fitting with purely observational data (and then only after serious reduction of the number of predictors, as discussed in Section 5.3.5) is prediction. For prediction in this context we recommend model averaging. That is, a prediction is made with each model, and the Akaike weights are used to compute a weighted average of these predictions. We do not know who invented model averaging, but we have seen it only in the Bayesian literature, using of course Bayesian-based model weights (see, e.g., Madigan and Raftery 1994, Draper 1995, Hoeting et al. 1999, Raftery et al. 1997, Hoeting et al. 1999).

As noted by Breiman (1996), selection of a best model in all-subsets fitting is inherently unstable in its outcome. The solution proposed by Breiman to produce stabilized inferences is a type of model averaging: Generate many perturbed sets of the data (such as bootstrap samples create), select the best model in each case, and produce some sort of averaged inference. Our solution, for stabilized inference, is a sort of reverse strategy: Keep the one actual data set as is, but find for each fitted model its Akaike weight; then compute inferences as some form of weighted average over all the models.

Interest in regression parameter estimates in conjunction with large R and all-subsets model fitting will no doubt continue. Perhaps there is a legitimate need for this (we are not convinced). Motivated by this need and our recommendation to use model-averaged predictions, we decided to analytically relate such prediction to parameter estimation. This led to some surprising new ideas and issues that we will outline here. These are issues pursued further in Chapter 6, but still need additional research.

The model-averaged prediction (estimate) of $E(y \mid \underline{x})$ is

$$\hat{\bar{E}}(y \mid \underline{x}) = \sum_{r=1}^{R} w_r \hat{E}(y \mid \underline{x}, g_r).$$

We define an indicator function for when a predictor is in a model:

$$I_i(g_r) = \begin{cases} 1 & \text{if predictor } x_i \text{ is in model } g_r, \\ 0 & \text{otherwise.} \end{cases}$$

For model g_r the value of β_i is denoted here by $\beta_{i,r}$. One version of a model-averaged parameter estimator is

$$\hat{\bar{\beta}}_i = \frac{\sum_{r=1}^{R} w_r I_i(g_r) \hat{\beta}_{i,r}}{\sum_{r=1}^{R} w_r I_i(g_r)} = \frac{\sum_{r=1}^{R} w_r I_i(g_r) \hat{\beta}_{i,r}}{w_+(i)}, \qquad (5.7)$$

$$w_+(i) = \sum_{r=1}^{R} w_r I_i(g_r).$$

An alternative to the above conditional parameter estimator is the full model-averaged estimator over all models wherein if predictor x_i is not in model g_r, we simply set $\hat{\beta}_{i,r} = 0$. Thus a new estimator, denoted by $\bar{\hat{\beta}}_i$, is

$$\bar{\hat{\beta}}_i = w_+(i)\hat{\bar{\beta}}_i.$$

This is just $\hat{\bar{\beta}}_i$ shrunk toward zero by the amount $(1 - w_+(i))\hat{\bar{\beta}}_i$. Moreover, we found, based on empirical results, that we could also consider this shrinkage estimator as

$$\bar{\hat{\beta}}_i \equiv \hat{\bar{\beta}}_i - (1 - w_+(i))\hat{\bar{\beta}}_i = \hat{\bar{\beta}}_i - \text{model se\widehat{lecti}on bias;} \qquad (5.8)$$

that is, (5.8) is our original model-averaged estimator adjusted for (estimated) model selection bias. Certainly, the term $(1 - w_+(i))\hat{\bar{\beta}}_i$ is not an unbiased estimator of model selection bias, but it is a usable estimator of that bias.

Then we realized that $\bar{\hat{\beta}}_i$ is of fundamental importance because the model-averaged prediction can be expressed as

$$\hat{\bar{E}}(y \mid \underline{x}) = \hat{\bar{\beta}}_0 + \sum_{i=1}^{m} w_+(i)\hat{\bar{\beta}}_i x_i = \bar{\hat{\beta}}_0 + \sum_{i=1}^{m} \bar{\hat{\beta}}_i x_i.$$

If we accept $\hat{\bar{\beta}}_i$ as the appropriate naive estimate of β_i given multimodel inference, then heuristically, the above suggests that prediction is improved by shrinkage toward zero of each parameter's estimate by a measure of that parameter's unimportance ($= 1 - w_+(i)$). The value of shrinkage is well established in statistics (see, e.g., Copas 1983, Tibshirani 1996); hence this seems like a line of thought worth pursuing.

Thus we have compelling reasons to want to replace, at least in all-subsets regression, the conditional estimator of (5.7) by the unconditional estimator of (5.8). This would allow us to ignore the issue of what is a best model, and simply make inferences from the full set of models as regards any parameter or prediction. In our limited Monte Carlo evaluation of this idea we have found that $\bar{\hat{\beta}}_i$ is less biased by model selection than is $\hat{\bar{\beta}}_i$.

An unresolved matter is a simple, yet reliable, estimator for the sampling variance of $\bar{\hat{\beta}}_i$ and an associated confidence interval for $\bar{\beta}_i$. The derived theoretical sampling variance formula for a model average estimator applies here:

$$\text{var}\left(\bar{\hat{\beta}}_i\right) = \left[\sum_{r=1}^{R} \pi_r \sqrt{\text{var}\left(\hat{\beta}_{i,r} \mid g_r\right) + \left(\hat{\beta}_{i,r} - \bar{\hat{\beta}}_i\right)^2} \right]^2.$$

However, what we need is an estimator of $\text{var}(\bar{\hat{\beta}}_i)$. When the parameter of interest, say θ, appears in every model (hence, θ_r in model g_r) then we have found that a good estimator is

$$
\widehat{\mathrm{var}}\left(\hat{\hat{\theta}}\right) = \left[\sum_{r=1}^{R} w_r \sqrt{\widehat{\mathrm{var}}\left(\hat{\theta}_r|g_r\right) + \left(\hat{\theta}_r - \hat{\hat{\theta}}\right)^2}\right]^2.
$$

This variance estimator makes no allowance for the uncertainty that should be inherent in $\hat{\hat{\theta}}$ because the weights $w_r(= \hat{\pi}_r)$ are random variables. However, because these weights are positive and sum to 1, a decrease in one weight is compensated for by an increase in one or more other weights in such a way that the result is an acceptably stable variance estimator even ignoring inherent sampling variation of the weights.

Applying the above variance estimator to $\tilde{\bar{\beta}}_i$ the result, expressed for $\widehat{\mathrm{se}}(\tilde{\bar{\beta}}_i)$, is

$$
\widehat{\mathrm{se}}(\tilde{\bar{\beta}}_i) = \sum_{r=1}^{R} w_r \sqrt{\widehat{\mathrm{var}}\left(\hat{\beta}_{i,r}|g_r\right) + \left(\hat{\beta}_{i,r} - \tilde{\bar{\beta}}_i\right)^2} + |\tilde{\bar{\beta}}_i|(1 + w_+(i)).
$$

Now the sampling variation in the weights matters very much to the performance of this variance estimator. Another way to see this issue is to note that $\tilde{\bar{\beta}}_i = w_+(i)\hat{\bar{\beta}}_i$, and the variance of the conditional $\hat{\bar{\beta}}_i$ can be reliably estimated (because one renormalizes the Akaike weights over the subset of models that contain β), but now we need a formula like $\widehat{\mathrm{var}}(\tilde{\bar{\beta}}_i) = (w_+(i))^2 \widehat{\mathrm{var}}(\hat{\bar{\beta}}_i) + (\hat{\bar{\beta}}_i)^2 \widehat{\mathrm{var}}(w_+(i))$. We do not know $\widehat{\mathrm{var}}(w_{+}i)$); from many simulations we know that it is not trivial (i.e., cannot be ignored).

Despite seeking one, we do not yet know a reliable analytical estimator for $\widehat{\mathrm{se}}(\tilde{\bar{\beta}}_i)$. Moreover, if we had one, there is a second issue as regards a confidence interval on β_i under this unconditional model averaging. The interval given by $\tilde{\bar{\beta}}_i \pm 2\,\widehat{\mathrm{se}}(\tilde{\bar{\beta}}_i)$ is not justified, in general, because the sampling distribution of $\tilde{\bar{\beta}}_i$ can be *very* skewed (i.e., quite nonnormal) when $E(w_+(i))$ is not near 1.

While finding a reliable estimator of $\mathrm{var}(\tilde{\bar{\beta}}_i)$ is worthwhile, we would still have the issue that the sampling distribution of $\tilde{\bar{\beta}}_i$ can be far from a normal distribution. Hence, the only reliable approach to frequentist inference with $\tilde{\bar{\beta}}_i$ seems to be the bootstrap. To apply the bootstrap to this shrinkage estimator one must compute this estimator in each of the B bootstrap samples. Thus, one will obtain $\tilde{\bar{\beta}}^*_{i,b}$, $b = 1, \ldots, B$, and then determine the percentile confidence interval, and, if desired, the usual bootstrap-based estimate of $\widehat{\mathrm{se}}(\tilde{\bar{\beta}}_i)$.

We again consider the Monte Carlo evaluation of the GPA data example. Results in Table 5.17 were extended to β_4 and $\tilde{\bar{\beta}}_i$ to compare expected values. We need to be clear on what was done in this new Monte Carlo example. If predictor variable x_i was in the selected AIC_c best model, then we estimated

β_i by just $\hat{\beta}_i$ from that selected best model. Otherwise, no inference was made about β_i; this corresponds to classical model selection practice. The average of these estimates is $E(\hat{\beta}_i)$ under model selection; the expected relevant sample size is $10{,}000 \times \pi_+(i)$ for $\pi_+(i)$ the probability that x_i is in the selected best model. In contrast, $\tilde{\beta}_i$ was computed for every Monte Carlo sample, giving $E(\tilde{\beta}_i)$. The results are below; $\tilde{\beta}_i$ has the better performance:

i	$\pi_+(i)$	β_i	$E(\hat{\beta}_i)$	$E(\tilde{\beta}_i)$
1	0.95	0.00201	0.00229	0.00217
2	0.75	0.00125	0.00156	0.00112
3	0.65	0.18945	0.248	0.1577
4	0.16	0.08752	0.321	0.0588

5.4 Estimation of Density from Line Transect Sampling

5.4.1 Density Estimation Background

Animal inventory and monitoring programs often focus on the estimation of population density (i.e., number per unit area). Buckland et al. (1993, 2001) provide the theory and application for field sampling and analysis methods using line transects. We will illustrate several aspects of statistical inference in the face of model selection uncertainty using line transect data collected by Southwell (1994) on the eastern grey kangaroo (*Macropus giganteus*) at Wallaby Creek, in New South Wales, Australia. The program DISTANCE (Laake et al. 1994) was written for the analysis of line transect data, uses AIC in model selection, and has an option for bootstrapping the sample (see Buckland et al. 1997 for a similar example). Thus, line transect sampling and the program DISTANCE will be used to provide some deeper insights concerning model selection uncertainty and will serve as another comprehensive example.

In line transect sampling, the estimator of density (D) is

$$\hat{D} = \frac{n}{2wL\hat{P}},$$

where n (= 196 in this example) is the number of objects detected on r (= 78) transects of total length L (= 88.8 km) and width w (= 263 m). The unconditional probability of detection, within the strips examined, is P. The focus of the estimation of animal density is on the probability of detection, and this is defined as

$$P = \frac{\int_0^w g(x)dx}{w},$$

where $g(x)$ is the detection function (i.e., the probability of detection given that an animal is at perpendicular distance x from the line). The detection function $g(x)$ is a confounding of animal density, environmental factors affecting detection, differences in detectability by individual observers, and differences in detectability among animals being surveyed. The detection function can be estimated from perpendicular distances taken from the transect line to each object detected. Assumptions required in line transect sampling and other details and theory are given in Buckland et al. (2001).

Substituting the expression for P into the estimator of D, and canceling out the w and $1/w$, gives

$$\hat{D} = \frac{n}{2L \int_0^w \hat{g}(x)dx}.$$

Thus, the essence of data analysis here is to find a good approximating model for $g(x)$, the detection function. Buckland et al. (2001) recommend models of the general form

$$g(x) = key(x)[1 + series(x)].$$

The key function alone may be adequate for modeling $g(x)$, especially if sample size is small or the distance data are easily described by a simple model. Often, one or more adjustment terms must be added to achieve an acceptable model for the data. For the purposes of this example we chose to use four specific models of the above form. Each of these models provides a reasonable, but not unique, basis for data analysis in this example.

5.4.2   Line Transect Sampling of Kangaroos at Wallaby Creek

Eastern grey kangaroos often occur in family groups; thus an estimate of total animal density is the product of the estimated number of groups and the average group size. In this example we will focus only on estimating the number of groups of this species of kangaroo. We define the set of candidate models for this example in Table 5.24. The analysis theory of line transect sampling has been the subject of a great deal of work since about 1976; thus the set of candidate models is relatively well based in this example. The program DISTANCE (Laake et al. 1994) was used to compute MLEs of the model parameters in the key functions (σ, a, or b, in Table 5.24) and the series expansions (the a_j in Table 5.24), and choose, using AIC, the best model among the four.

5.4.3   Analysis of Wallaby Creek Data

The results of the initial analysis of these data are given in Table 5.25. Model 1 was selected using AIC and provides an estimated density of 9.88 groups per km^2 (conditional se $= 1.00$ and conditional cv $= 10.12\%$). All four models produce relatively similar point estimates of D for these data; the largest Δ_i value was 3.86 (model 3). In this example, the estimated log-likelihood values,

TABLE 5.24. Model set for the line transect example on kangaroo data; Hermite polynomials are special polynomials, defined recursively (see Buckland et al. 2001) for use with the half-normal key function.

Model	$g(x)$	K	Key function	Series expansion
1	$\left\{\dfrac{1}{w}\right\}\left\{1+\displaystyle\sum_{j=1}^{2}a_j\cos\left(\dfrac{j\pi x}{w}\right)\right\}$	2	uniform	cosine
2	$\left\{e^{-x^2/(2\sigma^2)}\right\}\left\{1+\displaystyle\sum_{j=2}^{3}a_j H_{2j}\left(\dfrac{x}{\sigma}\right)\right\}$	3	half-normal	Hermite polynomials
3	$\left\{1-e^{-(x/a)^{-b}}\right\}\left\{1+\displaystyle\sum_{j=2}^{3}a_j\left(\dfrac{x}{w}\right)^{2j}\right\}$	4	hazard	simple polynomials
4	$\left\{e^{-x^2/(2\sigma^2)}\right\}\left\{1+\displaystyle\sum_{j=2}^{3}a_j\cos\left(\dfrac{j\pi x}{w}\right)\right\}$	3	half-normal	cosine

TABLE 5.25. Summary statistics for the line transect data on eastern grey kangaroos at Wallaby Creek, New South Wales, Australia (from Southwell 1994). The AIC-selected model is shown in bold; $\hat{D}_i$ is used to clarify that $\hat{D}$ is based on model g_i.

Model	K	$\log\left(\mathcal{L}(\hat{\theta})\right)$	AIC	Δ_i	$\exp(-\Delta_i/2)$	w_i	$\hat{D}_i$	$\widehat{\mathrm{var}}(\hat{D}_i\,\vert\,g_i)$
1	**2**	**−1,021.725**	**2,047.449**	**0.000**	**1.000**	**0.499**	**9.88**	**0.999**
2	3	−1,021.546	2,049.092	1.643	0.440	0.220	10.43	1.613
3	4	−1,021.654	2,051.307	3.858	0.145	0.072	10.83	2.761
4	3	−1,021.600	2,049.192	1.743	0.418	0.209	10.46	1.935

number of model parameters, and estimated density are similar across models; however, the estimated conditional sampling variances differ by a factor of almost 3. In this case, all four models contain the same parameter (D, or equivalently, P); thus, model averaging (4.1) and (4.9) should be considered. In either analysis, estimates of unconditional variances and associated confidence intervals should be used in making inferences about population density.

Using the Akaike weights w_i and the conditional sampling variances $\widehat{\mathrm{var}}(\hat{D}_i\vert g_i)$ for each model, we computed the model-averaged estimate of density $\hat{\bar{D}} = 10.19$ and an estimate of its unconditional sampling variance $\widehat{\mathrm{var}}(\hat{\bar{D}}) = 1.51$ (4.9). Hence, the (unconditional) standard error of $\hat{\bar{D}}$ is 1.23, and its cv is 12.06%. This unconditional cv is slightly higher than the cv of 10.12% conditional on the AIC-selected model. Inferences would be essentially the same here whether based on the model-averaged results or based on the density estimate from the AIC-selected model but using for its variance the unconditional estimate of 1.51. In either case the achieved confidence interval coverage can be expected to be better than that based on the conditional standard error, and often very near the nominal level.

5.4.4 Bootstrap Analysis

The most obvious advantages of using Akaike weights as the basis to compute estimated unconditional sampling variances (and $\bar{\hat{D}}$) are simplicity and speed. However, the bootstrap method also can be used to make unconditional inferences; the bootstrap is especially useful in complex situations where theory for analytical variances, even given the models, is lacking. Here we do have such analytical theory to compare to the bootstrap results.

We used the program DISTANCE to draw and analyze bootstrap samples, based on transects as the sampling unit (thus, there were 78 sampling units for the bootstrap), and thereby compute lower and upper confidence limits on D as well as an estimated unconditional sampling variance for $\hat{D}$. We computed 10,000 bootstrap samples; we present first the results from all 10,000 samples. Then we examine the variability inherent here in a "mere" 1,000 bootstrap samples, based on the 10 sets of 1,000 samples each from cases 1–1,000, 1,001–2,000, and so forth.

The resultant density estimates, by model, and the model selection frequencies are shown in Table 5.26. The mean of the estimates from the 10,000 bootstrap samples, 10.39, is quite close to the estimate based on the Akaike weights and (4.1) and (4.9) (10.19, $\widehat{se} = 1.23$). Based on all 10,000 values of $\hat{D}^*$, the bootstrap estimate of the unconditional standard error of $\hat{D}$ (and of $\bar{\hat{D}}$, and $\bar{\hat{D}}^*$) is 1.48. The model selection relative frequencies from the bootstrap procedure are similar to, but do not exactly match, the Akaike weights (this is expected). However, the results are close for the favored model g_1: Akaike weight $w_1 = 0.50$ (Table 5.25) and from the bootstrap, $\hat{\pi}_1 = 0.45$ (Table 5.26).

5.4.5 Confidence Interval on D

There are several options for setting a confidence interval on D based on the estimated density and its estimated unconditional sampling variance. First, there is the usual procedure that assumes that the sampling distribution of the estimator is approximately normal. Hence, an approximate 95% confidence interval is based on

$$\hat{D} \pm 2\,\widehat{se}(\hat{D}),$$

where $\widehat{se}(\hat{D})$ ($= 1.23$ from Section 5.4.3) is the estimated (by theory) unconditional standard error. For this example, $9.88 \pm 2 \times 1.23$ gives the interval (7.42, 12.34).

The second method assumes (in this example) that the sampling distribution of $\hat{D}$ is log-normal. This is a plausibly better assumption than a normal sampling distribution for an estimator $\hat{\theta}$ in any context where the parameter θ is strictly positive, and for fixed sample size the $\text{cv}(\hat{\theta})$ tends to be independent of the actual value of θ. Then one computes lower and upper bounds as (from Burnham et al. 1987)

$$D_L = \hat{D}/C \quad \text{and} \quad D_U = \hat{D}C,$$

where

$$C = \exp\left[t_{\alpha/2,df}\sqrt{\log[1 + (\text{cv}(\hat{D}))^2]}\,\right].$$

The confidence level is $1 - \alpha$; $t_{\alpha/2,df}$ is the upper $1 - \alpha/2$ percentile point of the t-distribution on df degrees of freedom. The degrees of freedom are those of the estimated $\text{var}(\hat{D})$. For an approximate 95% interval, if df are 30 or more, it suffices to use 2 in place of $t_{0.025,df}$.

For this example $\hat{D} = 9.88$ (from the AIC-selected model), with unconditional $\text{cv}(\hat{D}) = 0.124$, and thus $C = 1.28$. Therefore, if we base inference on the AIC-selected model, the approximate 95% confidence interval is 7.72 to 12.65. If we base inference on the model-averaged estimate of density (which increasingly strikes us as the preferred approach), then the results are $\overline{\hat{D}} = 10.19$, again with standard error estimate 1.23, hence $C = 1.272$ and approximate 95% confidence interval 8.01 to 12.96. The bootstrap method would provide a point estimate of $\overline{\hat{D}}$, hence the corresponding confidence interval is more comparable to the analytical results for model averaging than to the results based on the selected single best model.

A third option is to use the bootstrap to produce a robust confidence interval, for example, based on the percentile method (Efron and Tibshirani 1993, Shao and Tu 1995). Here the 10,000 values of $\hat{D}_b^*$ generated in producing Table 5.26

TABLE 5.26. Summary of results from 10,000 bootstrap samples of the line transect data for eastern grey kangaroos at Wallaby Creek, New South Wales, Australia (from Southwell 1994): Empirical means of the $\hat{D}^*$ by selected model and overall, standard error estimates, and selection frequencies.

Model	$\overline{\hat{D}^*}$	Standard error estimate	Selection frequency
1	9.97	1.10	4,529
2	10.63	1.41	2,992
3	10.92	2.34	1,239
4	10.75	1.38	1,240
All	10.39	1.48	10,000

are sorted in ascending order. Thus we have $\hat{D}^*_{(1)}, \ldots, \hat{D}^*_{(10000)}$. The 2.5 and 97.5 percentiles of the bootstrap sampling distribution are used as the 95% confidence interval endpoints on D: $\hat{D}^*_{(250)} \leq D \leq \hat{D}^*_{(9750)}$. The results here were $\hat{D}^*_{(250)} = 7.88 \leq D \leq 13.78 = \hat{D}^*_{(9750)}$.

The interval lower bounds from the three methods are more alike than the upper bounds. Results from the bootstrap in this example estimate more model selection uncertainty than the results based on use of Akaike weights (Section 2.9); we rectify this matter in Section 5.4.6 below. In general, with either a good analytical approach or the bootstrap, achievement of nominal confidence interval coverage is likely if a good model is selected, if model selection uncertainty has been incorporated into an estimate of the unconditional standard error, and if nonnormality has been accounted for.

It can be problematic to identify a correct unit of data as the basis for bootstrap resampling. Aside from this fundamental issue, the bootstrap is conceptually simple and can effectively handle model selection uncertainty if computer software exists or can be written. The program DISTANCE allows bootstrapping in the context of distance sampling (Laake et al. 1994). In contrast, bootstrapping the experimental starling data (Section 3.4) would have been nearly impossible. Specialized software development for just this case would be prohibitive; and the computer time required might be measured in weeks. In these cases, we recommend use of Akaike weights to compute the estimate of an unconditional standard error, and then use of some suitable analytical confidence interval procedure.

5.4.6 Bootstrap Samples: 1,000 Versus 10,000

The $B = 10,000$ bootstrap samples were partitioned, in the order they were generated, into 10 sets of 1,000 samples per set, and estimates were computed on a per-set basis. The results are given in Table 5.27. Before discussing these results we need to establish our goals for precision of the bootstrap-based computation (estimate, actually) of quantities such as

$$\widehat{se}^*(\hat{D} \mid B) = \sqrt{\frac{\sum(\hat{D}^*_b - \bar{\hat{D}}^*)^2}{B - 1}}.$$

The true bootstrap estimate of the standard error of $\hat{D}$ (given the data) is actually the limit of $\widehat{se}^*(\hat{D} \mid B)$ as B goes to infinity. We denote that limit simply by $\widehat{se}(\hat{D})$; however, this bootstrap standard error need not be exactly the same number as the analytically computed standard error of $\hat{D}$ (for which we use the same notation). For any value of B we have $\widehat{se}^*(\hat{D} \mid B) = \widehat{se}(\hat{D}) + \epsilon$, where $E(\epsilon)$ goes to 0 quickly as B gets large and $var(\epsilon) = \phi/B$ (ϕ unknown, but estimable). The goal in selecting B should be to ensure that $\sqrt{\phi/B}$ is small relative to the value of $\widehat{se}(\hat{D})$. Our preference is to achieve a bootstrap uncertainty coefficient of variation of 0.005 or less; hence $\sqrt{\phi/B}/\widehat{\mathrm{se}}(\hat{D}) \leq 0.005$, because this means that we get our result for $\widehat{\mathrm{se}}^*(\hat{D} \mid B)$ (taken as $\widehat{\mathrm{se}}(\hat{D})$) reliable, essentially, to two significant digits. That is, we target a large enough B that the bootstrap result for $\widehat{\mathrm{se}}(\hat{D})$ (or whatever is being computed) is nearly stable in the first two significant digits over all bootstrap samples of size B. If the true result should be 100, we want to be assured that generally our bootstrap result will be between about 99 and 101. This does not seem like too much precision to ask for; yet even this precision may require in excess of 10,000 bootstrap samples; it is rarely achieved with $B = 1,000$.

Now consider the variation exhibited in Table 5.27 in bootstrap estimates of π_1, $\overline{D}$, unconditional standard error of $\hat{D}$, percentile confidence interval endpoints (95%), and the interval width, $\hat{D}_U - \hat{D}_L$. Only $\hat{D} \equiv \overline{D} = \overline{\hat{D}^*}$ satisfies our precision criterion for $B = 1,000$. However, we do not do bootstrapping to get $\overline{\hat{D}^*}$: We already have $\hat{D}$, from the best model and $\overline{D}$ from model averaging. It is the other quantities in Table 5.27 that we use the bootstrap method to compute. We find (empirically or theoretically) that $\hat{\pi}_1$ for $B = 1,000$ falls generally within 0.42 to 0.48; this does not meet our precision criterion. Similarly, none of $\widehat{\mathrm{se}}(\hat{D})$, the confidence interval bounds, or width, in Table 5.27 meet our (modest) precision criterion when $B = 1,000$. Based on the variation over the 10 sets of samples in Table 5.27 we estimate that for 10,000 samples the percent coefficients of variation on the bootstrap estimates are as follows: $\mathrm{cv}(\hat{\pi}_1) = 0.005$, $\mathrm{cv}(\widehat{\mathrm{se}}(\hat{D})) = 0.007$, $\mathrm{cv}(\hat{D}_L) = 0.004$, $\mathrm{cv}(\hat{D}_U) = 0.006$, and $\mathrm{cv}(\hat{D}_U - \hat{D}_L) = 0.01$. Thus here $B = 10,000$ is not too many samples to produce bootstrap-computed quantities reliable to (almost) 2 significant digits. When using the bootstrap, think in terms of $B = 10,000$.

5.4.7 Bootstrap Versus Akaike Weights: A Lesson on QAIC$_c$

The estimated unconditional standard error of $\hat{D}$ is 1.23 based on an analytical formula and use of the Akaike weights. However, based on the computer-intensive bootstrap method we obtained 1.48 for the estimated unconditional standard error of $\hat{D}$. The bootstrap method is telling us that there is more uncertainty in our density estimator than our analytical (i.e., theoretical) formula accounts for. We perceived a need to resolve this issue. Unfortunately, we took the wrong approach: We assumed that the bootstrap result might be wrong, and tried to find out why. It is not wrong, but we mention some of our thinking before giving the correct resolution of this matter.

The correct analytical variance of $\hat{D}$, given a model, is conditioned on total line length L (88.85 km) and has two parts: $\mathrm{var}(n/L) = \mathrm{var}(n)/L^2$ and $\mathrm{var}(\hat{P})$. The $\mathrm{var}(\hat{P})$ component is conditional on n (196). Because detections and kangaroo locations may not be independent within line segments, the units used here as the basis for the bootstrapping are the separate line seg-

TABLE 5.27. Some bootstrap estimates from 10 independent sets of bootstrap samples each of size 1,000 from the line transect data for eastern grey kangaroos at Wallaby Creek, New South Wales, Australia (from Southwell 1994); π_1 is the selection probability for model g_1; standard errors and (percentile) confidence intervals for D are unconditional, and hence include model selection uncertainty. Results for "All" are based on the full 10,000 samples.

Set	$\hat{\pi}_1$	$\hat{D}^*$	$\widehat{SE}(\hat{D})$	95% Conf. Int.		Width
1	0.478	10.39	1.47	7.77	13.64	5.86
2	0.412	10.41	1.52	7.71	13.84	6.13
3	0.473	10.43	1.42	7.98	13.56	5.59
4	0.418	10.37	1.49	7.84	13.73	5.89
5	0.442	10.40	1.49	8.00	13.80	5.80
6	0.410	10.37	1.48	8.03	13.82	5.79
7	0.461	10.40	1.50	7.73	13.90	6.17
8	0.447	10.40	1.53	7.92	13.82	5.89
9	0.448	10.39	1.48	7.84	13.75	5.90
10	0.540	10.30	1.44	7.92	13.47	5.55
All	0.453	10.39	1.48	7.88	13.78	5.90

ments (78 of them). The length of these segments varies from 0.5 to 1.6 km. In generating a bootstrap sample the value of g_b^* is not held fixed at 88.85 over bootstrap samples b. Instead, g_b^* varies considerably. Also, the value of n_b^* varies substantially over bootstrap samples. Might these aspects of variation incorporated into the bootstrap samples result in an inflated estimate of $se(\hat{D})$? We investigated this issue very intensively for this example and concluded that the bootstrap estimate of $se(\hat{D})$ was acceptable here. However, the simple theoretically computed unconditional $se(\hat{D})$ did not account for all uncertainty in $\hat{D}$ (even though it accounts for all model selection uncertainty).

The resolution of the matter also turned out to be simple: We had forgotten to consider the need for a variance inflation factor $\hat{c}$. The $var(\hat{P})$ component above was based on theoretical formulae under ML estimation given the model. However, this variance is underestimated if important assumptions fail: The assumption of independence of detections within a line segment may fail; there may be spatio-temporal variation in true detection probabilities by detection distance x; there may be errors in recording detection distances (there usually are). All these problems lead to more variance than theory accounts for. We can adjust the theoretical $\widehat{se}(\hat{D})$ to allow for these sources of variation (in a way analogous to what the bootstrap does). The simplest adjustment is to use $\sqrt{\bar{c}} \cdot \widehat{se}(\hat{D})$ as our theory-based unconditional standard error.

When all models considered are subsets of one global model, then $\hat{c}$ for QAIC, and variance inflation, comes from the goodness-of-fit of the global model: $\hat{c} = \chi^2/\text{df}$. However, here we have four models, but there is no global model, so the approach to obtaining $\hat{c}$ must consider the goodness-of-fit of all four models. Below we give the goodness-of-fit chi-square statistic, its degrees of freedom, and $\hat{c}$ for these four models, as well as the Akaike weights based on use of AIC:

Model	χ^2	df	$\hat{c}$	w_i
1	25.11	17	1.48	0.499
2	23.73	16	1.48	0.220
3	24.66	15	1.64	0.072
4	23.40	16	1.46	0.209

The weighted average of $\hat{c}$, weighted by w_i, is 1.49. Here we would use either $\hat{c}$ from the selected model or this weighted average. It makes no difference here; hopefully, this would be the usual situation in distance sampling. Hence, we use here $\hat{c} = 1.48$, df $= 17$ from model g_1.

We should, however, have been using QAIC rather than AIC because our Akaike weights might then change (along then with other results). From Table 5.25 we obtain $-2\log(\mathcal{L})$ for each model and thus compute QAIC $= (-2\log(\mathcal{L})/\hat{c}) + 2K$ and the associated weights w_i:

Model	QAIC	Δ_i	w_i
1	1,384.71	0.00	0.511
2	1,386.47	1.76	0.211
3	1,388.61	3.90	0.073
4	1,386.54	1.83	0.205

The differences between the Akaike weights based on AIC verses QAIC are here trivial (this is because of a large sample size here). Using the above weights with each $\hat{D}$ from Table 5.25 gives a model-averaged result of 10.18; the original result was 10.19. For an unconditional standard error based on the QAIC-derived w_i we get 1.23 (the same as with AIC-based weights). We will stay with the originally computed $\bar{\hat{D}} = 10.19$. In this example the only effect of using QAIC is to make us realize that we need to use a variance inflation factor with our theoretical standard errors.

The quick way to adjust the theoretical unconditional standard error is to compute $\sqrt{\hat{c}} \cdot \text{se}(\hat{D}) = \sqrt{1.48} \cdot 1.23 = 1.22 \cdot 1.23 = 1.50$; the bootstrap-based result for the unconditional standard error of $\hat{D}$ was 1.48. However, the use of $\sqrt{\hat{c}} \cdot \text{se}(\hat{D})$ is not the correct formula (we have used it here for its heuristic epistemological value). Rather, one should adjust each theoretical $\widehat{\text{var}}(\hat{D}_i \mid g_i)$ to be $\hat{c} \cdot \widehat{\text{var}}(\hat{D}_i \mid g_i)$ and then apply (4.9), which here becomes

$$\widehat{\text{se}}(\hat{D}) = \sum_{i=1}^{4} w_i \sqrt{\hat{c} \cdot \widehat{\text{var}}(\hat{D}_i \mid g_i) + (\hat{D}_i - \hat{D}_a)^2}. \tag{5.9}$$

The two approaches will give almost identical results when the values of $(\hat{D}_i - \bar{\hat{D}})^2$ are small relative to $\hat{c} \cdot \widehat{\text{var}}(\hat{D}_i \mid g_i)$, as they are here. Applying (5.9) using the quantities from Table 5.25 and $\hat{c} = 1.48$, we get as an analytical formula-based result $\widehat{se}(\hat{D}) = 1.48$. This is exactly the same result as generated by the bootstrap (this may be a coincidence).

The bootstrap method to obtain the unconditional standard error of a parameter estimator will, if done correctly, automatically include all sources of uncertainty in that standard error. Estimation based on theoretical–analytical formulae, for models that do not automatically estimate empirical residual variation, will not automatically include overdispersion variation that exceeds what theory assumes. Thus in these cases we must always consider the need to include an empirical variance inflation factor $\hat{c}$ in our calculations.

5.5   Summary

Model-based data analysis is very important, as illustrated by examples in this book, and as demonstrated by the much improved estimation results (better precision, less bias) for the chain binomial survival data examples of Section 5.2. For example, rather than try to separately estimate survival rate for every age one should produce smoothed estimates of these parameters by using suitable parametric models. For such observational data (this applies to the other examples here—GPA example, gas mileage data, Kangaroo data) we would rarely, if ever, know a priori the single best model to use for the analysis.

However, in all such cases the investigator can and should postulate a priori a small set of suitable candidate models for data analysis; this then entails creating a meaningful, reduced number of predictor variables. Then AIC_c- or $QAIC_c$-based model selection can be very effective at providing a ranking of the models based on Akaike weights. If it makes sense to select a best model (if the models mean something as alternative scientific or mechanistic explanations), one can use the expected K-L best model to draw inferences (bearing in mind that the selection of that model as best is itself an inference). Sampling standard errors of estimated parameters can and should include model selection uncertainty.

If the models are only a means to the end of "smoothing" the data, as is the case for prediction, then we recommend computing model-averaged parameter (prediction) estimators and their unconditional sampling standard errors based on the Akaike weights. Monte Carlo methods showed that this procedure worked well for the chain binomial models; unconditional confidence interval coverage is generally close to the nominal 95%, while traditional intervals conditioned on the selected best model may achieve only 70 to 80% coverage. Monte Carlo studies in this chapter also show that there is substantial model uncertainty but that the Akaike weights are effective at measuring this uncertainty. The sampling distribution of Δ_p was examined for many situations, and we found that generally, for a small number of candidate models, a value ≥ 10 corresponds to at least the 95th percentile and more often at least the 99th percentile. This supports our contention that an observed $\Delta_i \geq 10$ is strong evidence against model g_i.

For reliable results from simulation we recommend at least 10,000 Monte Carlo samples at each set of conditions used to generate data. This holds true for the bootstrap also: For the results to be stable to two significant digits one must often use at least 10,000 bootstrap samples. Too many applications of these simulation methods do not use enough replications.

We do not recommend the dimension-consistent criteria (e.g., BIC, HQ) for model selection in the biological sciences or medicine when there is an a priori set of well thought out candidate models. Such criteria are not estimates of K-L information, are based on poor assumptions, and perform poorly even when sample size is quite large. We do not recommend using any form of hypothesis testing for model selection.

The choice of models to examine is important. The chain binomial examples demonstrated that a class of logistic models produced better results than a model class that assumed constant survival rate after a given age. The GPA example demonstrates that in variable-selection problems, thoughtful considerations can lead to much better models than unthoughtful all-subsets selection.

In Section 5.3 we note that model selection bias occurs in variable selection: Regression coefficient estimators $\hat{\beta}_i$ are biased away from 0 because the variable x_i is included in the model only when that variable seems to be important (i.e., when $\hat{\beta}_i$ is sufficiently different from 0). The less important a variable, the more biasing effect model selection has on $\hat{\beta}_i$. Estimated error mean square $\hat{\sigma}^2_{y|x}$ is negatively biased by model selection. The use of AIC$_c$, more so than other methods, provides some protection against both model selection biases. **The best way to minimize model selection bias is to reduce the number of models fit to the data by thoughtful a priori model formulation.**

Usually, selection of a best model is needed if scientific understanding is the goal. However, often it is better to think in terms of multimodel inference using the full set of models, rather than selecting just one model and basing inferences on that single model. This is especially true in all-subsets variable selection as practiced in regression, because the selected best model is highly variable. Model averaging is then particularly useful, as is computing the relative importance of a variable as the sum of the Akaike weights over all models in which that variable appears and examination of the model-averaged regression parameters.

Erroneous results have stemmed from the frequent misuse of Monte Carlo simulation in judging various model selection approaches. In many cases, the generating model has had a few parameters (very often < 8 and often < 5) with no or few tapering effects, and the objective has been to see which selection method most often chooses the generating model. This conceptualization is in the spirit of BIC and is a sterile exercise as regards real-world applications; hence there is no reason why results would apply to real biological problems.

Robust confidence intervals can be established using (4.9) with either Akaike weights (w_i) or bootstrap estimated selection probabilities ($\hat{\pi}_i$). In all the examples we have examined, such intervals have excellent achieved coverage. There are surely cases where this simple approach does not perform well, but we have not found any during our investigations.

6

Advanced Issues and Deeper Insights

6.1 Introduction

Much of this chapter is new material not in the first edition. The rest is material moved from other chapters because we judged it to be more distracting than helpful on a first reading of introductory ideas. In either case we thought the material here did not fit well in a logical, linear progression of *introductory* ideas about K-L-based model selection and multimodel inference under a confirmatory orientation with a relatively small set of models. There is no natural ordering to the sections of this chapter; they can be read in any order.

We consider $R < 100$, or perhaps even $R < 200$, as relatively not large because many classical variable selection analyses, or all-subsets selection, consider thousands, tens of thousands, or even millions or models (hence have $R \gg n$). We consider the analysis as exploratory rather than confirmatory when the number of models exceeds the sample size, which usually means that no real thought has been expended on the issue of meaningful models for the data. The detailed properties of model selection, and subsequent inferences, when the number of models considered is huge are not well studied because of the gargantuan amount of computing required. To illustrate issues for this situation Section 6.2 looks in some detail at a published all-subsets (variable-selection) regression example with sample size 252 (n) and 13 predictors. We do not include the no-effects null model, hence $R = 8,191$ models ($= 2^{13} - 1$).

Another subject in this chapter is an overview of selection criteria and approaches, followed by a more detailed contrasting of BIC and AIC. Basically, all current model selection criteria fall into two classes, either efficient (includes AIC) or consistent (includes BIC). There has been much confusion because AIC and BIC have different bases, objectives, and performance. We hope to cast some light on this matter. Another issue is extension of AIC to random coefficient (effects) models. This is a rapidly developing area of much importance and promise. A partially related issue is determining the sample size for a data set. The issue of "the" sample size is often not clear because there is not a single sample size for complex data structures, and random effects correspond to noninteger effective sample size.

Also delved into here is goodness-of-fit for count data with multiple models; essentially, this is about estimating c for overdispersion and QAIC. More general handling of overdispersion is considered wherein more than one overdispersion parameter can be estimated and used. There is a brief look at formulas for Bayesian model averaging, for the interested reader. The importance of a small-sample version of AIC is discussed and the utility of AIC_c (as we have defined it). Another subject considered is comparison of models when the assumed probability distributions are different; in most applications there is a single "error" distribution (e.g., normal or multinomial) and only model structural aspects vary.

6.2   An Example with 13 Predictor Variables and 8,191 Models

6.2.1   Body Fat Data

In this example multiple regression is used to predict percent body fat based on predictors that are easily measured. The data are from a sample of 252 males, ages 21 to 81. A key reference is Johnson (1996), which is in a web journal (http://www.amstat.org/publications/jse/toc.html). The data are available on the web in conjunction with Johnson (1996). The web site states, "The data were generously supplied by Dr. A. Garth Fisher, Human Performance Research Center, Brigham Young University, Provo, Utah 84602, who gave permission to freely distribute the data and use them for noncommercial purposes." Reference to the data is also made in Penrose et al. (1985). These data have also been used in Hoeting et al. (1999), a seminal paper on Bayesian model averaging.

We take the response variable as $y = 1/D$; D is measured body density (observed sample minimum and maximum are 0.9950 and 1.1089). At a given weight, lower body density means more body fat because fat is not as dense as muscle and bone. The reciprocal of body density is regarded as linearly related to percent body fat; however, there is no agreement among medical experts on the parameters of that calibration, which is why we simply use $1/D$ as our response variable. Measuring body density requires an expensive, time-consuming underwater weighing method. For each subject 13 easy to measure potential predictors (x) were recorded. The goal is to predict y given $x_1, \ldots, x_{13}$ (age, weight, height, and 10 body circumference measurements). Sampling aspects for the study were not stated, we suspect that the 252 subjects were a self-selected sample (i.e., volunteers) from the Provo, Utah, area and that this sample was obtained and processed in a short time interval in the early 1980s.

We consider aspects of five possible approaches, ordered as least to most desirable, in our opinion:

1) fit the full (i.e., global) model only;
2) select one model by standard stepwise selection from all $2^{13} - 1 = 8{,}191$ possible simple regression models, then ignore selection uncertainty;
3) select the best model using AIC_c and consider selection uncertainty;
4) do full multimodel inference, such as model-averaged predictions, over all 8,191 models,
5) first reduce in number and refine the predictors based on theory, and/or logic, to a set of meaningful derived variables, then do step 4 (with far fewer models).

We then explore using the nonparametric and parametric bootstrap, mostly to compare aspects of selection under AIC and BIC. Our main objective in this example is to demonstrate how much model selection uncertainty there is when the model set is huge, and how this uncertainty is reduced if a better crafted set of models is used.

6.2.2 The Global Model

Table 6.1 shows basic results of fitting the global regression model $y|\underline{x} = \beta_0 + \sum \beta_i x_i + \epsilon, \epsilon \sim \text{normal}(0, \sigma^2)$. Hoeting et al. (1999) report that standard model checking showed this to be an acceptable model (we agree). We note also that correlations among the predictors are strong, but not extreme, almost entirely positive, and range from -0.245 (age & height) to 0.941 (weight & hips). The design matrix is of full rank.

The absolute value of the usual t-test statistic (Wald version) for a regression coefficient is $1/|\text{cv}|$. Hence, in Table 6.1 any parameter with $|\text{cv}| < 0.5$ would be considered "significant" at the $P = 0.05$ level. Inspection of results unambiguously suggests dropping knee (x_g), chest (x_5), and height (x_3) from the global model. The issue of other predictors that one might drop is obscured by the strong correlations among the predictors. However, because there is a strong suggestion that not all 13 predictors need to be in the best model, one is motivated to apply formal model selection.

6.2.3 Classical Stepwise Selection

We used SAS PROC REG, at its defaults, for stepwise variable selection. In only a few steps a model was selected, thus giving, perhaps, an erroneous im-

TABLE 6.1. Regression parameter estimates, standard errors, and absolute coefficients of variation for the full 13-predictor model for the body fat data, X4 to X13 are circumferences; $R^2 = 0.7420$.

Variable		$\hat{\beta}_i$	$\widehat{se}(\hat{\beta}_i \vert g)$	$\vert cv \vert$
INTERCEPT		0.873844	0.04594	0.053
X1	age	0.000109	0.00007	0.610
X2	weight	−0.000215	0.00013	0.596
X3	height	−0.000163	0.00037	2.273
X4	neck	−0.000971	0.00049	0.503
X5	chest	−0.000106	0.00021	2.024
X6	abdomen	0.002036	0.00019	0.092
X7	hips	−0.000432	0.00030	0.693
X8	thigh	0.000525	0.00030	0.577
X9	knee	0.000024	0.00051	21.739
X10	ankle	0.000571	0.00046	0.807
X11	biceps	0.000492	0.00036	0.725
X12	forearm	0.000923	0.00041	0.447
X13	wrist	−0.003649	0.00110	0.303

TABLE 6.2. Selected predictors, hence selected best model, for several model selection methods applied to the body fat data.

Selection	Indices of predictor variables selected										
Stepwise		2		4	6				11	12	13
Forward	1	2		4	6	7	8	10	11	12	13
Backward			3			7		9	10	11		
Mallows C_p	1	2		4	6		8		11	12	13
AIC	1	2		4	6	7	8			12	13
AIC_c		2		4	6				11	12	13
BIC		2			6					12	13

pression of confidence in the selected model. Common belief is that stepwise, rather than forward or backward, selection is the best of the testing-based selection methods. We give all three results in Table 6.2, plus results for Mallows C_p, AIC, AIC_c, and BIC.

There is substantial variation in the best model by method, especially for stepwise versus forward and backward methods. Even the best models under AIC and AIC_c differ by four predictors. However, this need not concern us, because we know that we should use AIC_c in deference to AIC here because $n/K = 252/13 \ll 40$. This example shows that using AIC_c, rather than AIC, makes a difference even with $n = 252$ and global $K = 15$ (14 structural parameters plus σ^2). Mallows's C_p does not select the same model as AIC. Some literature erroneously claims that these are identical procedures for linear models; in general they give similar, but not identical, results. Finally, as expected, BIC is more conservative than AIC.

An all too common inference procedure is to select a best model and then act as if the selected model was really the only model fit. In this situation, with 8,191 models, there will be considerable model selection uncertainty. It is unconscionable not to evaluate this uncertainty and use some sort of model-unconditional inference about importance of predictors. That is, one is not justified in saying that the selected predictors are important and those not selected are not important (this is the same false dichotomy that plagues null hypothesis testing). Even if we can agree on a selection method, there is still model selection uncertainty, and we consider this next.

6.2.4   Model Selection Uncertainty for AIC_c and BIC

We assume that most readers are now aware of the Bayesian Information Criterion (BIC, Schwarz 1978, Hoeting et al. 1999): BIC $= -2 \log \mathcal{L}(\hat{\theta}|data, g) + \log(n) \cdot K$ (whereas AIC $= -2 \log \mathcal{L}(\hat{\theta}|data, g) + 2 \cdot K$. Correspondingly, for model j, $\Delta BIC_j = BIC_j - BIC_{min}$ (BIC_{min} is the minimum BIC_j over the R models). Also, the same structural formula that gives the Akaike weights, from ΔAIC, is used with ΔBIC to give the (posterior) probabilities of models $g_1, \ldots, g_R$. More information about BIC appears in Sections 6.3 and 6.4).

For this data analysis we quantify model selection uncertainty partly by ΔAIC_c and ΔBIC, but mostly by Akaike weights and posterior model probabilities (for BIC). Table 6.3 shows Akaike weights, w_j, for the top six models and also shows weights and models at a few other ranks as determined by all 8,191 models ordered by largest to smallest w_j. The AIC_c best model ($r^2 = 0.733$) has a weight of only 0.010738; the other models all have smaller weights (the 8,191 weights sum to 1). The next best model has $w_2 = 0.010711$. The weight of evidence is essentially identical for both these models (evidence ratio is 1.0025). Table 6.3 shows a few other evidence ratios, in particular, the model ranked 52nd by AIC_c was ranked 1st by BIC. Plausible models here include those ranked first to 176th, or even to 642nd.

Another way to determine a confidence set on models is to include all models where the sum of these *ordered* Akaike weights is some value like 0.95, or 0.99 (this method is not the best one in general, but it is useful and convenient). All we care about here is how large such confidence sets are: How many models have some plausibility (as opposed to models we can discount with near certainty)? The number of models in such sets under AIC_c selection are shown below:

Σw_j	# of models	ER	
0.900	649	$w_1/w_{649} =$	34
0.950	876	$w_1/w_{876} =$	68
0.990	1449	$w_1/w_{1449} =$	413
0.999	2266	$w_1/w_{2266} =$	3579

TABLE 6.3. Akaike weights w_j and $\Delta_j = \Delta AIC_c$ values for some of the 8,191 models, ordered most- to least-supported by the data; also, a few evidence ratios, denoted by ER, are shown.

Model order j	w_j	Δ_j	
1	0.010738	0.00000	
2	0.010711	0.00496	
3	0.010333	0.07690	
4	0.009752	0.19273	
5	0.009459	0.25371	
6	0.009104	0.33008	ER: $w_1/w_6 = 1.2$
.	...	...	
36	0.003924	2.01	
52	0.003349	2.33	ER: $w_1/w_{52} = 3.2$
176	0.001450	4.00	
642	0.000321	7.02	ER: $w_1/w_{642} = 33.5$
1103	0.000071	10.00	ER: $w_1/w_{1103} = 151.3$
8191	1.3E-72	322.01	

There is no computational impediment here to using all the models for any model-averaged results, but if we were to use Occam's window (Madigan and Raftery 1994), we would want $\sum w_j$ at least 0.95, and preferably 0.999. Thus, model-averaged inferences require here using on the order of 1,000 of the possible models. This is not at all like using just the single (esti-mated as) best model and erroneously thinking that the model is a stable basis for reliable inferences in a repeated sampling (or Bayesian) framework (the Akaike weights approximate repeated sampling-based selection relative frequencies).

One type of model-averaged inference is the variable relative importance weight $w_+(i)$, which is the sum of the Akaike weights for predictor i over all models in which predictor i occurs. Table 6.4 shows these variable-importance weights and the variables included in the six top-ranked models under AIC_c (Table 6.3 shows w_j for these top six models).

Results for BIC, analogous to those for AIC_c in Table 6.3, are given in Table 6.5. The model ranked 12th by BIC is the model ranked first by AIC_c. The defining operational (frequentist) property of BIC is that as sample size goes to infinity, the posterior probability of a single model goes to 1 (this requires both the sampling context and model set to be fixed, independent of n). In this example the BIC best model has associated posterior probability of only 0.14; hence there is again substantial model selection uncertainty. Occam's window for $\Sigma \Pr(\text{model } j) = 0.999$ includes 1,611 models (Table 6.5). Thus, whether approached in a K-L or Bayesian context there is considerable model uncertainty here, and inferences, after selection, should reflect this uncertainty.

It is beyond our intended use of this example to use all variables and do model-averaged prediction under AIC (see Hoeting et al. 1999 for Bayesian

TABLE 6.4. Variable relative importance weights and predictors included in the top six models (1 if included, 0 otherwise) and K for these models.

| | | | AIC$_c$ top models | | | | | |
$w_+(i)$	variable	i	1	2	3	4	5	6
0.495	age	1	0	1	1	1	1	0
0.933	weight	2	1	1	1	1	1	1
0.314	height	3	0	0	0	0	0	0
0.652	neck	4	1	1	1	1	1	1
0.283	chest	5	0	0	0	0	0	0
1.000	abdomen	6	1	1	1	1	1	1
0.445	hips	7	0	1	0	0	1	0
0.588	thigh	8	0	1	1	1	1	0
0.293	knee	9	0	0	0	0	0	0
0.448	ankle	10	0	0	0	0	0	1
0.600	biceps	11	1	0	1	0	1	1
0.828	forearm	12	1	1	1	1	1	1
0.976	wrist	13	1	1	1	1	1	1
		$K =$	8	10	10	9	11	9

TABLE 6.5. Some posterior model probabilities for BIC, based on all 8,191 models ordered most to least probable, and the cumulative probabilities.

Model order j	Pr(model j)	ΣPr
1	0.13930	0.13930
2	0.08980	0.22911
3	0.05681	0.28591
4	0.03829	0.32420
5	0.03488	0.35908
6	0.03118	0.39027
.	...	...
12	0.01484	0.50689
158	0.00060	0.90010
292	0.00023	0.95001
757	0.00003	0.99001
1611	0.00000	0.99900
8191	9.4E-69	1.00000

model-averaged prediction with these data). Moreover, we recommend against simply accepting these 13 predictors and using either the global model or doing stepwise selection in the first place. Rather, we encourage the following approach (noted as #5 in Section 6.2.1): First reduce the number of variables, hence models, based on theory and/or logic to a set of meaningful derived variables related to y, then do multimodel inference.

6.2.5   An A Priori Approach

We obtained the data, and information about the data, from the web site. Initially, no analysis at all was done. Rather, one of us (KPB) thought about the matter intermittently over several weeks and decided to try the derived variables below. The actual data were not studied before this a priori thinking was done; i.e., none of the above model fitting was done until after the six variables below were decided on. A knowledgeable health-trained specialist should do even better at generating derived variables and suitable model forms. We did not consider improved model forms beyond linear regression.

Weight and height jointly ought to be very important for body fat prediction, but not as separate predictors in linear regression; they should be considered together. Allometric relationships are common in biology, so an ideal adult body might have a nearly constant ratio of some function of weight and height. In many animals mass tends to be proportional to the cube of height, and within species that proportionality is often very stable. Hence, we might expect that $z_1 = \log(\text{weight})/\log(\text{height})$ would be very stable for the biologically ideal body and thus variation in this derived variable would be positively correlated with variation in body fat (hence, 1/density).

Additional considerations led to five more derived variables thought to supply information about different dimensions of the prediction problem. The full set of six, ordered as considered most to least important, a comment on the rationale for each, and the predicted sign ($+$ or $-$) of each regression coefficient are given below:

$$z_1 = \frac{\log(\text{weight})}{\log(\text{height})} \qquad \text{based on ideas of allometry } (+)$$

$$z_2 = \frac{\text{abdomen}}{\text{chest}} \qquad \text{beer gut factor } (+)$$

$$z_3 = \frac{(\text{knee} * \text{wrist} * \text{ankle})^{\frac{1}{3}}}{\text{height}} \qquad \text{heavyset or light-boned } (-)$$

$$z_4 = \left[\frac{\text{biceps} * \text{thigh} * \text{forearm}}{\text{knee} * \text{wrist} * \text{ankle}}\right]^{\frac{1}{3}} \qquad \text{fleshiness index } (+)$$

$$z_5 = \text{age} \qquad \text{standardized by mean and standard deviation } (+)$$

$$z_6 = \text{age}^2 \qquad \text{based on standardized age } (+).$$

Part of the thinking here is that because the response variable is essentially percent body fat, only ratios of body measurements should be important. After z_1 the most important predictor seemed as if it ought to be abdomen size (gut), but only relative to some other body size metric. Based on years of observing shapes of men (and experience with his own measurements), KPB opted for z_2. Next there are issues of genetic variation in being slight or heavyset (hence z_3: For a given weight, more bony is less fat), and there is variation in fitness (hence z_4). Finally, it seemed reasonable that age might be predictive, even given z_1 to z_4. Percent body fat would tend to increase with age, beyond the twenties, but asymptotically, so not exactly linear on age. Now we have only 6 (derived) predictors, and therefore 63 possible models (we did not fit the null model of no predictors), not 8,191 models. Also, we have an a priori ordering on these predictors and a predicted sign of each "effect." Thus we can learn (via feedback) from the data analysis whether our reasoning is plausible or not. This feedback aspect of data analysis is critical in the scientific method.

Table 6.6 gives some basic results about model selection uncertainty under AIC_c, for the fat data with these six predictors. Noteworthy is that now the AIC_c best model ($r^2 = 0.659$) is also the best model under both Mallows C_p and BIC. The number of models in the confidence set with weights summing to different values is shown below:

$\sum w_j$	# of models	ER	
0.900	3	$w_1/w_3 =$	2.8
0.950	4	$w_1/w_4 =$	5.9
0.990	5	$w_1/w_5 =$	63.2
0.999	7	$w_1/w_7 =$	178.3

Thus, we need to consider only a few models here, not hundreds or a thousand. Table 6.7 shows the top six models and the variable-importance weights.

TABLE 6.6. Akaike weights w_j and $\Delta \, AIC_c$ values for some of the 63 models, ordered as most to least supported by the data, based on the z_i predictors.

Model order j	w_j	Δ_j
1	0.48867	0.000
2	0.23627	1.453
3	0.17745	2.026
4	0.08316	3.542
5	0.00773	8.294
6	0.00287	10.278
.	. . .	. . .
63	2.0E-58	264.258

TABLE 6.7. Variable relative-importance weights and predictors included in the top six models (1 if included, 0 otherwise) based on the z_i and K for that model.

z_i	description	$w_+(i)$	1	2	3	4	5	6
					AIC_c top models			
z1	wt/ht	1.000	1	1	1	1	1	1
z2	gut	1.000	1	1	1	1	1	1
z3	bony	0.323	0	1	0	1	0	1
z4	fleshy	0.986	1	1	1	1	0	0
z5	age	1.000	1	1	1	1	1	1
z6	age*age	0.264	0	0	1	1	0	0
		$K =$	6	7	7	8	5	6

The regression model using all six z_i produced the results below ($r^2 = 0.660$):

	$\hat{\beta}_i$	$\widehat{se}(\hat{\beta}_i|g)$	$|cv|$
z_1	0.18693	0.03714	0.199
z_2	0.14404	0.01717	0.119
z_3	0.04520	0.05828	1.290
z_4	0.00554	0.00168	0.303
z_5	0.00310	0.00070	0.227
z_6	0.00011	0.00053	4.785

We did not intuit the correct sign for predictor z_3, but it is not a useful predictor. The other signs we predicted correctly, but we can discount this for predictor z_6 as it is relatively unimportant here ($w_+(6) = 0.264$). The estimated order of importance of the z_i is 2, 1, 5, 4, 3, 6, with z_3 and z_6 having negligible effects.

Overall we think that the a priori considerations here are more important and influential than just the purely statistical model selection aspects. And it is always possible to do exploratory analysis after any a priori thinking (but not vice-versa). In fact, we did some final exploratory analysis, with the goal of having only one or two simple predictors and a high model r^2. After a good deal of probing we ended up with a very competitive 1-predictor model based on $z_e =$ abdomen/height ($r^2 = 0.682$), which gave the results below:

parameter	estimate	$\widehat{se}(\hat{\beta}|g)$	$|cv|$
β_0	0.8259	0.00528	0.0064
β_e	0.0924	0.00398	0.0432

This type of *post hoc* hypothesis-generating analysis is acceptable as long as it is reported for what it is: strictly exploratory, hypothesis generating, not confirmatory.

6.2.6   Bootstrap Evaluation of Model Uncertainty

We now return to the original 13 predictors problem and our purpose for having this extended example: to illustrate the extent of model uncertainty when there is a huge number of models ($R \gg n$). We explore the bootstrap to estimate model selection probabilities (π) for these selection methods. For AIC_c we want to know how well the estimated selection probabilities match the Akaike weights (or for BIC, the posterior model probabilities; however, for a Bayesian this is not a fair question).

For each of several model selection methods we created $B = 10,000$ bootstrap samples, all of size 252, from the data. In addition to getting selection relative frequencies we want to know how many different models, of 8,191 possible, ever get selected (this number depends weakly on B, for large B). For AIC_c and BIC we also looked at the sampling distributions of ΔAIC_c and ΔBIC.

Applying stepwise model selection (SAS PROC REG defaults) to each bootstrap sample resulted in 1,206 distinct models being selected at least once. These 1,206 models were listed in rank order, most to least frequently selected. Table 6.8 shows a few records from this list. Models are denoted by whether predictor x_1 to x_{13} (in that order) is in or out of the model. For example, 0101010000111 denotes the model with predictors 2, 4, 6, 11, 12, and 13. The models selected as best, by method, with the actual data are indicated in Table 6.8.

From Table 6.8 we see that even the most commonly selected model under stepwise selection ($\hat{\pi} = 0.0215$) has a very low selection probability. The model selected by the stepwise method from the actual data has $\hat{\pi} = 0.0184$. This bootstrap assessment corroborates that there is considerable model selection uncertainty. Corresponding bootstrap-based assessments for AIC_c and BIC are in Tables 6.9 and 6.10. Applying AIC_c selection to 10,000 bootstrap samples, 1,233 distinct models were selected with $\hat{\pi} = 0.013$ for the AIC_c best model for the actual data. This $\hat{\pi}$ compares well to the Akalke weight of 0.01. Applying BIC selection to 10,000 bootstrap samples, there were 562 distinct models selected with $\hat{\pi} = 0.0891$ for the BIC best model for the actual data. From the data the posterior model probability for the BIC best model is 0.14. Whatever one thinks of 0.09 versus 0.14, the comparison is not fair, because posterior probabilities, which are conditional on the data, are not required to be comparable to predata random variable frequencies.

For AIC_c model selection the bootstrap assessment of model uncertainty matches well to the Akaike weights. However, when we looked at the bootstrap-based estimate of the sampling distribution of the ΔAIC_c the results were different from our earlier assessments of this distribution. Average ΔAIC_c was 9.1; maximum was 49.3. This maximum depends weakly on B. This mo-

TABLE 6.8. Model ranks and selection frequencies for a few of the 1,206 distinct models selected by the stepwise method applied to 10,000 bootstrap samples from the body fat data; the models selected by different methods with the actual data are also indicated.

Rank	Model	Frequency	
1	1101011100011	215	AIC
2	0101010000111	184	AIC_c & stepwise
3	0101010001111	141	
4	1101011101011	141	
.	. . .	.	
16	1100010100011	89	
17	0100010000011	87	BIC
18	1101011100111	87	
.	. . .	.	
21	0101011100111	70	
22	1101010100111	63	C_p
23	0100010100011	62	

TABLE 6.9. Model ranks and selection frequencies for a few of the 1,233 distinct models selected by AIC_c applied to 10,000 bootstrap samples from the body fat data; the models selected by AIC, AIC_c, and BIC with the actual data are also indicated.

Rank	Model	Frequency	
1	1101011100011	218	AIC
2	1001011100011	151	
3	1011011100011	149	
4	1101011101011	148	
5	1101010100011	145	
6	0101010000111	130	AIC_c
7	1101010101111	125	
8	0101011100011	112	
.	. . .	.	
53	1011110000111	36	
54	0100010000011	34	BIC
55	1001011101011	34	

TABLE 6.10. Model ranks and selection frequencies for a few of the 562 distinct models selected by BIC applied to 10,000 bootstrap samples from the body fat data; the models selected by AIC_c and AIC with the actual data are also indicated.

Rank	Model	Frequency	
1	0100010000011	891	BIC
2	0100010000101	689	
3	0100010001101	470	
4	0010010000001	388	
5	0100010000001	359	
.	. . .	.	
23	0100010100001	90	
24	0101010000111	87	AIC_c
25	1000010000001	87	
.	. . .	.	
142	1101011100011	10	AIC

tivated looking at the same sampling distribution information for BIC because the guidelines for interpreting Δ are the same for K-L criteria and BIC: average $\Delta BIC = 7.3$, maximum was 49.2. Bootstrap-based estimated sampling percentiles:

Percentile	ΔAIC_c	ΔBIC
0.50	8.0	6.2
0.90	17.5	15.3
0.95	20.8	18.6
0.99	27.5	25.3

These sampling results are not consistent with what we have seen when the number of models is small, such as $R < 100$ (and certainly $R \ll n$). We wondered whether bootstrap performance was breaking down, so we decided to do some Monte Carlo simulations mimicking these body fat data (i.e., parametric bootstrap).

6.2.7 Monte Carlo Simulations

We assumed that the measurements y and $\underline{x} = (x_1, \ldots, x_{13})'$ on a subject could be suitably modeled as multivariate normal with a variance-covariance matrix taken to be the observed variance-covariance matrix (this method is also called the parametric bootstrap). This full variance-covariance matrix is partitioned as below; $\underline{c}$ is 13×1 and Σ_x is 13×13:

$$\Sigma = \begin{bmatrix} \sigma_y^2 & \underline{c}' \\ \underline{c} & \Sigma_x \end{bmatrix}.$$

The global model is now also the generating model under which the vector of true regression parameters is given by $\underline{\beta}' = \underline{c}'(\Sigma_x)^{-1}$. It suffices, for our limited purposes, to set $\beta_0 = 0$ and generate the data in two steps. First, an observation $\underline{x}$ is generated from the marginal $MVN(\underline{0}, \Sigma_x)$ then $E(y|\underline{x}) = \underline{x}'\underline{\beta}$ and $y = E(y|\underline{x}) + \epsilon$, where ϵ is a normal random variable with mean $\overline{0}$ and variance $\sigma_{y|x}^2 = \sigma_y^2 - \underline{c}'(\Sigma_x)^{-1}\underline{c}$. More details, and philosophy about this approach are given in Sections 4.3.6, 5.3.2, and 5.3.4.

The approximate theoretical standard error for each $\hat{\beta}_i$ can be determined (see Section 5.3.2). Hence, we computed the "effect sizes" for $n = 252$ as $\lambda_i = \beta_i/\text{se}(\hat{\beta}_i|\text{global } g)$; this is essentially the mean for $\hat{\beta}_i$ standardized to be a normal$(\lambda_i, 1)$ random variable. These λ_i values are

i	λ_i	i	λ_i	i	λ_i
1	1.687	6	11.123	10	1.274
2	−1.727	7	−1.484	11	1.419
3	−0.453	8	1.783	12	2.303
4	−2.048	9	0.048	13	−3.401
5	−0.509				

There clearly are tapering effects, and the only trivial predictor is x_9, knee circumference. The actual average ordering of predictors by their variable importances may not match the ordering by $|\lambda_i|$ because of the correlated nature of the predictors.

We generated 10,000 independent samples of size 252 and applied AIC_c, and BIC model selection. Our interest is in regard to, first, the frequency distribution of models selected: Do those relative frequencies match the Akaike weights and results from the (nonparametric) bootstrap? Second, do the Monte Carlo based sampling distributions of ΔAIC_c and ΔBIC match results from the bootstrap. The answers are yes; there was no substantial discrepancy between the bootstrap and Monte Carlo approaches. Some summary results for selection frequencies are given in Tables 6.11 (AIC_c) and 6.12 (for BIC).

TABLE 6.11. Model ranks and selection frequencies for a few of the 1,137 distinct models selected by AIC_c applied to 10,000 Monte Carlo samples that mimic the essential properties of the body fat data; the models selected by AIC_c and BIC with the actual data are also indicated.

Rank	Model	Frequency	
1	1101010100011	197	
2	1101011100011	177	
3	1101011101011	165	
4	1101010101011	152	
5	0101010000111	149	AIC_c
.	...	.	
9	1101010100111	132	
10	1101010101111	119	
.	...	.	
21	1100011101011	72	
22	0100010000011	70	BIC
23	0100010000111	70	

TABLE 6.12. Model ranks and selection frequencies for a few of the 532 distinct models selected by BIC applied to 10,000 Monte Carlo samples that mimic the essential properties of the body fat data; the models selected by AIC_c and BIC with the actual data are also indicated.

Rank	Model	Frequency	
1	0100010000011	1063	BIC
2	0100010000101	852	
3	0100010000001	371	
4	0100010001011	349	
5	010100000100	331	
.	...	.	
9	0101010000010	221	
10	0100010100011	212	
.	...	.	
19	0100010001001	112	
20	0101010000111	111	AIC_c
21	1001011100011	109	

The sampling distribution percentiles for the 10,000 Δ_p values obtained from the Monte Carlo samples are below:

Percentile	ΔAIC_c	ΔBIC
0.50	5.6	5.5
0.90	12.8	14.9
0.95	15.3	18.6
0.99	21.2	26.6

The mean and maximum Δ values were 6.5 and 38.6 for AIC_c, and 6.8 and 40.4 for BIC. These results, while similar to the bootstrap estimates of percentiles, are generally a little smaller than those from the bootstrap (but for such an inference we only have a sample of size 1). However, both sets of distributional results show larger percentiles of Δ_p than what we have seen when the number of models is very much smaller than 8,191. We believe that the results obtained here generally apply when R is so large.

The bootstrap simulation relative frequencies of model selection match well to the Akaike weights. However, we noticed that the sampling distribution of Δ_p was stretched to the right. We wondered whether this result was an artifact of the bootstrap in this case. Therefore, we then did the Monte Carlo simulations to verify the bootstrap; both approaches gave about the same results. Now we had to reconsider the distribution of Δ_p because the guidelines we gave about interpreting Δ as regards inferential evidence strength about models could be questioned, at least to the extent those guidelines were partly supported with sampling distribution ideas. Our inferential guidelines are essentially the same as those for ΔBIC (Raftery 1996a), and therefore the sampling distribution of ΔBIC is here also out of line with those guidelines.

The resolution of this concern is that we need to realize fully that the deeper basis for inference about model selection uncertainty under the information theoretic approach is the model likelihood $\mathcal{L}(g_i|data)$ and what follows from it (evidence ratios and Akaike weights). This is analogous to inference being based on posterior model probabilities for BIC in a Bayesian approach. Under both of these approaches inference is conditional on the data through the likelihood, rather than being justified by ideas of sampling uncertainty. Thus, we are justified in retaining our guidelines as being useful, but they must not be interpreted strictly in a sampling distributional framework.

6.2.8   Summary Messages

The first general point illustrated by this example is that substantial model selection uncertainty should be expected when the number of models is quite large, such as under many instances of all subsets (i.e., variables) selection. This example has a good sample size ($n = 252$) relative to a moderate number of predictor variables (13) for such applications, and still $R = 8,191$ is a lot of models. With so many models we find here that the selected-as-best model has a very small Akaike weight (0.010738), and is essentially tied with the second- and third-place models (Table 6.3). Moreover, a confidence set of models here easily includes over 100 models. Any all-subsets application of model selection with R far exceeding n can be expected to have such extreme model selection uncertainty wherein even the best model has a very small Akaike weight.

When all the models have very low weights, such as here, there is no inferential credibility for any single model regarding what are the "important" predictor variables. It is foolish to think that the variables included in the best model are "the" important ones and the excluded variables are not important.

This all-or-nothing (i.e., important or not) thinking in the context of variables selection is not in the spirit of statistics, and it should be banished; measures of variable importance are needed (Brieman 2001). The summed weights for variable i, $w_+(i)$ (Table 6.4), provide a model-averaged measure of the relative importance of each predictor variable. They are relative, not absolute, because the baseline value that corresponds to no predictive value of variable i occurs not at $w_+(i) = 0$, but at some value > 0 (randomization methods can be used to estimate this baseline value, see Section 6.9.8).

Even if prediction is the goal, it is foolish to think that the selected-best model has any special credibility when its Akaike weight is low, as here. Rather, model-averaged prediction should be used (this is being realized in the literature, see e.g., Brieman 1996, 2001, regarding "bagging" and " random forests"). For a vector of predictors each fitted model yields a prediction, $\hat{y}_j$, and the model-averaged prediction is $\hat{\bar{y}} = \sum w_j \hat{y}_j$. For linear models this implies the best measure of the absolute importance for a variable should be the model-averaged partial regression coefficient for that variable, $\bar{\hat{\beta}}$ (Section 5.3.6).

If there is so little inferential weight for the best model, why has model selection been considered to be so useful? Because the best (by whatever criterion) model gives good in-sample prediction, relative to the global model, as measured by the coefficient of determination, r^2. However, the same, or virtually the same, r^2 is achieved by many competitor models. The AIC_c best model here has $r^2 = 0.733$. However, in the confidence set of 876 models determined by the sum of the ordered (large to small) w_j being $= 0.95$, the minimum and maximum r^2 are 0.718 and 0.742. Any of these 876 models provides essentially the same average in-sample predictability, but each one uses a different subset of predictor variables. This phenomenon of many near-equivalent models as judged by r^2 values is acute when there are many predictors and they are strongly intercorrelated, which is the usual case in variables selection.

A second point we make here is that stepwise model selection should not be used. Almost any thoughtful model selection will find model here a that has an r^2 above 0.7. Even stepwise selection, ad hoc though it is, will usually lead to a model with decent r^2, relative to what is possible, and may give the same model as AIC. So why not use stepwise selection? Because (1) there is no theoretical basis for stepwise selection, as regards any optimality criterion. (2) there is no simple way to compute model (inference) weights in the context of stepwise selection and, as practiced, no such model inferential weights are provided by stepwise selection (unless one resorts to the bootstrap, which never seems to be done); this is a major failing. (3) stepwise selection gives the subjective appearance of much less model uncertainty than exists because only a small number of all possible models are fit, and much of the software for stepwise selection lists only a few (perhaps < 10) models even of those that were fit to the data. As a result, the user is mislead about how much model selection uncertainty exists. (4) as practiced, stepwise selection cannot lead to model-averaged inference, nor reliable inference about importance of predictors, nor unconditional measures of uncertainty. Rather, one pretends that the selected model was the one and only a priori model considered. (5) the results of stepwise selection depend nontrivially on the choice of α levels to enter and drop predictors; there is no theory for these choices. (See also McQarrie and Tsai 1998, 427–429 about stepwise selection).

A third point follows from the results of the bootstrap and Monte Carlo evaluation of model selection uncertainty and the sampling distribution of the $\triangle$AIC values. This is one of the few cases where we have looked at this issue for the number of models, R, over several hundred, as opposed to when R is a several dozen or fewer. The guidelines we have given for interpreting a large $\triangle$ did not hold up with $R = 8{,}191$; the same guidelines have been used for BIC differences and they also did not hold up here. Rather than $\triangle = 10$ being big here, hence discounting the model with the bigger AIC, it was more like $\triangle = 20$ is "big." However, the model selection relative frequencies from these simulations were very consistent with Akaike weights from the actual data analysis of all 8,191 models. Thus, using these weights (and things like evidence ratios) as the basis for inference about model selection was supported by the simulations.

We are quite convinced now, from all of our research and thinking on the matter, that the w_j are valid and useful inferential statistics in model selection. Conversely, rigorous inference should not be based on the sampling distribution of the $\triangle$AIC, even though rough guidelines on this matter do seem useful when R is small. It is a principle that sampling variation across replicate data sets is not the same as inferential uncertainty (as reflected in the likelihood) about models, or parameters in models, given the single data set at hand. The two types of "variation" are often similar, but when they differ, inference should be based on the likelihood.

A fourth point is the advantage of reducing a priori the number of models to consider, especially by reducing the number of predictor variables. This can best be done by thoughtful creation of meaningful derived predictors and dropping meaningless predictors (or ones whose usefulness is hard to measure). This idea was illustrated in Section 6.2.5. Instead of 8,191 models we ended up with only 63 models; of these 63 only 4 had substantial weight (Table 6.6), and the best model had $w = 0.489$, with $r^2 = 0.682$. It is hoped that the greater interpretability and logic underlying these models would render their out-of-sample predictions better than those from the brute-force all-subsets approach (alas, we have no other data with which to test this hope).

A fifth point is that after the a priori analyses one is free to do exploratory, more judgment-based analyses and model selection, as long as one is honest about the inferences one makes: a priori versus ranging from careful exploratory to reckless data dredging. For example, it is clear in Section 6.2.5 (see Table 6.7) that variables z_3 and z_6 are useless predictors (our judgment, based on the "objective" methodology and our understanding of that methodology). So one might chose to drop them and just use the AIC$_c$ selected-best model as the only model. This ignores model selection uncertainty, but that might be one's professional judgment if the purpose is to suggest a single model that health professionals can use to quickly predict percentage of body fat. Or you can do the sort of uninhibited exploration we did here, after all the other analyses, of really simple models to end up with a linear model based on only the one derived predictor, $z_e = $ abdomen/height ($r^2 = 0.682$). A number of models were considered in arriving at this one, so it should not be accepted without testing it by application to a new set of data. However, we do not consider this model as being the result of reckless data dredging because we restricted ourselves to single predictor models.

6.3   Overview of Model Selection Criteria

There is a variety of model selection methods. However, from the point of view of statistical performance of a method, and intended context of its use, there are only two distinct classes of methods: These have been labeled *efficient* and *consistent*. We will characterize these two classes in Section 6.3.4 after introducing other model selection criteria.

Under the frequentist paradigm for model selection one generally has three main approaches: (I) optimization of some selection criteria, (II) tests of hypotheses, and (III) ad hoc methods. One has a further classification under (I): (1) criteria based on some form of mean squared error (e.g., Mallows's C_p, Mallows 1973) or mean squared prediction error (e.g., PRESS, Allen 1970), (2) criteria that are estimates of K-L information or distance (e.g., TIC and the special cases AIC, AIC_c, and $QAIC_c$), and (3) criteria that are consistent estimators of K, the dimension of the "true model" (e.g., BIC). We will explore (2) and (3) in the following material.

6.3.1   Criteria That Are Estimates of K-L Information

AIC, AIC_c, and $QAIC_c$ are estimates of the relative K-L distance between truth $f(x)$ and the approximating model $g(x)$. These criteria were motivated by the concept that truth is very complex and that no "true model" exists (or at least that it was immaterial to the argument). Thus, one could only *approximate* truth with a model, say $g(x)$. Given a good set of candidate models for the data, one could estimate which approximating model was best (among those candidates considered, given the data and their sample size). Linhart and Zucchini (1986) speak of "approximating families" of models. Hurvich and Tsai (1994) explain that these criteria select the best finite-dimensional approximating model in large samples when truth is infinite-dimensional. The basis for these criteria seems reasonable in the biological sciences.

When sample sizes are quite large, there are other criteria derived that might offer advantages in model selection and inference (e.g., TIC in Chapter 7). These criteria specifically allow for "misspecification" of the approximating models: the fact that the set of candidate models does not include $f(x)$ or any model very similar to $f(x)$. Here we will note four criteria, even though their operating properties have received little attention in the published statistical literature (but see Konishi and Kitagawa, 1996).

Takeuchi (1976) provides a general derivation from K-L information to AIC. An intermediate result indicated that a selection criterion was useful when the candidate models were not particularly close approximations to f. He derived TIC (Takeuchi's information criterion) for model selection that has a more general bias adjustment term to allow $-2\log_e(\mathcal{L})$ to be adjusted to be an asymptotically unbiased estimate of relative K-L,

$$\text{TIC} = -2\log(\mathcal{L}) + 2 \cdot \text{tr}\big(J(\theta)I(\theta)^{-1}\big).$$

The matrices $J(\theta)$ and $I(\theta)$ involve first and second mixed partial derivatives of the log-likelihood function, and "tr" denotes the matrix trace function. AIC is an approximation to TIC, where $\text{tr}\big(J(\theta)I(\theta)^{-1}\big) \approx K$. The approximation is excellent when the approximating model is quite "good" and can become poor when the approximating model is poor. One might consider always using TIC and worry less about the adequacy of the models in the set of candidates. This consideration involves two issues that are problematic. First, one must always worry about the quality of the set of approximating models being considered; this is not something to shortcut. Second, using the expanded bias adjustment term in TIC involves estimation of the elements of the matrices $J(\theta)$ and $I(\theta)$ (details provided in Chapter 7). Shibata (1989) notes that estimation error of these two matrices can cause instability of the results of model selection (note that the matrices are of dimension $K \times K$). If overdispersion is found in count data, then the log-likelihood could be divided by an estimated variance inflation factor, given QTIC. In most practical situations, AIC and AIC_c are very useful approximations to relative K-L information.

Linhart and Zucchini (1986) proposed a further generalization, and Amari (1993) proposed a network information criterion (NIC) potentially useful in training samples in neural network models. Konishi and Kitagawa (1996) suggest even more general criteria for model selection and provide further insights into AIC and TIC and their derivation. Shibata (1989) developed a complicated criterion, based on the theory of penalized likelihoods. His method has been called RIC for "regularized information criterion." We will not explore these methods, since they would take us too far afield from our stated objectives and they do not have the direct link with information theory and the estimation of relative K-L distance. However, we note that almost no work has been done to evaluate the utility of these extensions in applied problems. Surely, the use of these criteria must be reserved for problems where the sample size is quite large and good estimates of the elements of the matrices ($I(\theta)$ and $J(\theta)$) in the bias adjustment term are available.

Mallows's C_p (Mallows 1973, 1995) statistic is well known for variable selection, but limited to LS regression problems with normal errors. However, C_p lacks any direct link to K-L information. Atilgan (1996) provides a relationship between AIC and Mallows's C_p, shows that under some conditions

AIC selection behaves like minimum mean squared error selection, and notes that AIC and C_p are somewhat equivalent criteria. When the usual multiple linear regression assumptions hold, the two criteria seem to often select the same model and rank the contending models in the same order, but they are not equivalent. We have not found a small-sample version of C_p that would be useful when the sample size is small compared to the number of regressor variables (like AIC_c) (see Fujikoshi and Satoh 1997). Ronchetti and Staudte (1994) provide a robust version of C_p (also see Sommer and Huggins 1996). Of course, adjusted R^2 has been used in classical multiple linear regression analysis; however, it has very poor performance (see e.g., McQuerrie and Tsai, 1998).

6.3.2   Criteria That Are Consistent for K

This section deals with a class of criteria used in model selection that are "consistent" or "dimension-consistent" and with how these criteria differ from those that are estimates of Kullback–Leibler information. Several criteria have been developed, based on the assumptions that an exactly "true model" exists, that it is one of the candidate models being considered, and that the model selection goal is to select the *true* model. Implicit is the assumption that truth is of fairly low dimension (i.e., $K = 1$–5 or so) and that K, and the data-generating (true) model, is fixed as sample size increases. Here, the criteria are derived to provide a consistent estimator of the order or dimension (K) of this "true model," and the probability of selecting this "true model" approaches 1 as sample size increases. Bozdogan (1987) provides a nice review of many of the "dimension-consistent" criteria. The best known of the "dimension-consistent criteria" was derived by Schwarz (1978) in a Bayesian context and is termed BIC for Bayesian information criterion (or occasionally SIC for Schwarz's information criterion); it is simply

$$\text{BIC} = -2\log(\mathcal{L}) + K \cdot \log(n).$$

BIC arises from a Bayesian viewpoint with equal prior probability on each model and very vague priors on the parameters, given the model. The assumed purpose of the BIC-selected model was often simple prediction; as opposed to scientific understanding of the process or system under study. BIC is not an estimator of relative K-L.

Rissanen (1989) proposed a criterion that he called minimum description length (MDL), based on coding theory, another branch of information theory (see also Yu 1996, Bryant and Cordero-Braña 2000). While the derivation and its justification are difficult to follow without a strong background in coding theory, his result is equivalent to BIC. Hannan and Quinn (1979) derived a criterion (HQ) for model selection whereby the penalty term was

$$c \cdot \log(\log(n)),$$

where n is sample size and c is a constant greater than 2 (see Bozdogan 1987:359). This criterion, while often cited, seems to have seen little use in practice. Bozdogan (1987) proposed a criterion he called CAICF (C denoting "consistent" and F denoting the use of the Fisher information matrix),

$$\text{CAICF} = -2\log(\mathcal{L}) + K\{\log(n) + 2\} + \log|I(\hat{\theta})|,$$

where $\log|I(\hat{\theta})|$ is the natural logarithm of the determinant of the estimated Fisher information matrix. He has recently advanced a somewhat similar criterion based on a notion of complexity (ICOMP, Bozdogan 1988). Neither CAICF nor ICOMP is invariant to 1-to-1 transformations of the parameters, and this feature would seem to limit their application. AIC, AIC_c, QAIC, and TIC are invariant to 1-to-1 transformations.

We question (deny, actually) the concept of a simple "true model" in the biological sciences (see the Preface) and would surely think it unlikely that even if a "true model" existed, it might be included in the set of candidate models! If an investigator knew that a true model existed and that it was in the set of candidate models, would she not know which one it was? We see little philosphical justification for these criteria in the biological, social, or medical sciences, although they have seen frequent application. Relatively few people seem to be aware of the differences in the basis and assumptions for these dimension-consistent criteria relative to criteria that are estimates of K-L information. The dimension-consistent criteria are directed at a different objective than those addressed by criteria that are estimates of K-L.

People have often (mis) used Monte Carlo methods to study the various criteria, and this has been the source of confusion in some cases (such as in Rosenblum 1994). In Monte Carlo studies, one *knows* the generating model and often considers it to be "truth." The generating model is often quite simple, and it is included in the set of candidate models. In the analysis of the simulated data, attention is (mistakenly) focused on what criterion most often finds this true model (e.g., Bozdogan 1987, Fujikoshi and Satoh 1997, Ibrahim and Chen 1997). Under this objective, we would suggest the use of the dimension-consistent criteria in this artificial situation, especially if the order of the true model was quite low (e.g., $K = 3$–5), or the residual variation (σ^2) was quite small, or the sample size was quite large. However, this contrived situation is far from that confronted in the analysis of empirical data in the biological sciences. Monte Carlo studies to evaluate model selection approaches to the analysis of real data must employ generating models with a range of tapering effect sizes and substantial complexity. Such evaluations should then focus on selection of a best approximating model and ranking of the candidate models; the notion that the true (in this case, the generating) model is in the set should be discarded.

Research into the dimension-consistent criteria has often used a generating model with only a few large effects. More realistic models employing a range of tapering effects have been avoided. In addition, the basis for the dimension-consistent criteria assumes that the true model remains fixed as sample size approaches infinity. In biological systems increased sample size stems from the addition of new geographic field sites or laboratories, the inclusion of additional years, and the inclusion of new animals with genetic variation over individuals. Thus, as substantial increases in sample size are achieved, the number of factors in the model also increases. The data-generating model does not remain fixed as $n \to \infty$. We have found that the dimension-consistent criteria perform poorly in open population capture–recapture models even in the case where K is small, but the parameters reflect a range of effect sizes (Anderson et al. 1998).

Notwithstanding our objections above, the sample sizes required to achieve the benefits of dimension-consistent estimation of model order (K) are often very, very large by any usual standard. In the examples we have studied (that have substantial residual variances) we have seen the need for sample sizes in the thousands or much more before the consistent criteria begin to point to the "true model" with a high probability. In cases where the sample size was very large, say 100,000, one might merely examine the ratios $\hat{\theta} / \widehat{\text{se}}(\hat{\theta})$ to decide on the parametrization, with little regard for the principle of parsimony (given the assumption that the true model is being sought, and it *is* in the set of candidates). It should be emphasized that these dimension-consistent criteria are not linked directly to K-L information and are "information-theoretic" only in the weakest sense. Instead, their motivation veered to consistent estimation of the order (K) of the supposed "true model" by employing alternative penalty terms (but see Section 2.12.2).

When sample size is less than very large for realistic sorts of biological data, these dimension-consistent criteria tend to select underfitted models with the attendant large bias, overestimated precision, and associated problems in inference. Umbach and Wilcox (1996:1341) present the results of Monte Carlo simulations conducted under the BIC-type assumptions. For sample size up to 100,000, AIC performed better than BIC in terms of the selected set coinciding with the "correct" set. The two criteria were tied at sample size 125,000. However, even at that large sample size, BIC selected the "correct" set in only 79% of the cases; this is still far from selecting the correct set with probability 1. While these criteria might be useful in some of the physical sciences and engineering, we suspect that they have relatively little utility in the biological and social sciences or medicine. Findley (1985) notes that ". . . consistency can be an undesirable property in the context of selecting a model."

6.3.3 Contrasts

As Reschenhofer (1996) notes, regarding criteria that are estimates of relative K-L information vs. criteria that are dimension consistent, they ". . . are often employed in the same situations, which is in contrast to the fact that they have been designed to answer different questions" (also see Pötscher 1991, Hurvich and Tsai 1995a and 1996, and Anderson and Burnham 1999b). In the biological and social sciences and medicine, we argue that the AIC-type criteria (e.g., AIC, AIC_c, QAIC, $QAIC_c$, and TIC) are reasonable for the analysis of empirical data. The dimension-consistent criteria (e.g., BIC, MDL, HQ, CAICF, and ICOMP) might find use in some physical sciences where a simple true model might exist and where sample size is quite large (perhaps thousands or tens of thousands, or more). Still, we question whether this true model would be in the set of candidate models. Even in cases where a simple true model exists and it is contained in the set of candidates, AIC might frequently have better inferential properties than the dimension-consistent criteria.

Still other, somewhat similar criteria have been derived (see Sclove 1987, 1994a, b, and Stoica et al. 1986 for recent reviews). A large number of other methods have appeared, including the lasso (Tibshirani 1996), the little bootstrap (Breiman 1992), the nonnegative garrote (Breiman 1995), predictive least quasi-deviance (Qian et al. 1996), various Bayesian methods (e.g., Ibrahim and Chen 1997) including the use of Gibbs sampling (George and McCulloch 1993). Some of these approaches seem somewhat ad hoc, while others are difficult to understand, interpret, or compute. Often the methods lack generality; for example, several are applicable only to regression-type models. We will not pursue these methods here, since they take us too far from our objectives.

In summary, we recommend the class of information-theoretic criteria that are estimates of relative K-L information such as AIC, AIC_c for general use in the selection of a parsimonious approximating model for statistical inference for sample sizes that occur in practice. If count data are found to be overdispersed, then QAIC and $QAIC_c$ are useful. If large samples are available, then TIC might offer an improvement over AIC or AIC_c. However, our limited investigations suggest that the simpler criteria perform as well as TIC in cases we examined (Chapter 7).

6.3.4   Consistent Selection in Practice: Quasi-true Models

The original motivation for a consistent model selection criterion is based on the idea that the true (i.e., data-generating) model is in the set of models and is, or may be, nested within some overly general models and that as sample size goes to infinity we want to select that true model with probability 1. This scenario is also based on the condition that one can increase n to be arbitrarily large while keeping the data generating context fixed: No additional factors may enter as sample size n increases. This sort of sample size augmentation can be done in Monte Carlo computer simulation, but not in real world studies. More formally, the frequentist motivation for BIC is idealized as below.

Assume that we have a nested sequence of models, g_1 to g_R and that the true model, g_t, is neither the first nor last model. The additional parameters nominally in models $g_i, t < i \leq R$ are actually not needed. The simplest example is regression based on predictors x_1 to x_R where $y = \beta_0 + \beta_1 x_1 + \cdots + \beta_t x_t + \epsilon$, and x_{t+1} to x_R have zero correlation with y and with all of x_1 to x_t. Thus, in models more than g_t we have $\beta_{t+1} = \cdots = \beta_R \equiv 0$. Hence, these models are also true, as theoretical models, but as fitted models they are over-parametrized. Therefore, the unique, lowest dimension true ("the" true) model is g_t. This is the model that consistent criteria must select with probability 1 as n gets large. BIC will do this, and the inferred posterior probability of g_t from BIC will also go to 1 as n gets large (this holds for any consistent criterion, but it suffices to restrict ourselves to BIC).

The inferential model "weights" from BIC selection have the same formula as the Akaike weights, but may be interpreted as probabilities of the model, given the data, the model set, and the prior model probabilities $(1/R)$ on each model. Define BIC differences as $\Delta \text{BIC}_i = \text{BIC}_i - \text{BIC}_{min}$, where in context, BIC_{min} is the minimum BIC value over all models and it occurs at model g_{min}. By context we mean that the index min may differ for AIC versus BIC, but we do not complicate the notation to distinguish these two possible values of min.

Under BIC the posterior model probabilities are given by

$$\Pr\{g_i\} = \frac{\exp\left(-\frac{1}{2}\Delta \text{BIC}_i\right)}{\sum_{r=1}^{R} \exp\left(-\frac{1}{2}\Delta \text{BIC}_r\right)}.$$

If there is a true model, g_t, in the set then $\Pr\{g_t\}$ goes to 1 as n goes to infinity; and of course $\Pr\{g_i\}$ goes to 0 for all other models (very large sample sizes may be required). For model g_t the Kullback-Leibler distance is 0, i.e., $I(f, g_t) = 0$. However, when model, g_t is nested in any more general model structures, in the model set considered, then for those other models we also have $I(f, g) = 0$. Hence, from the standpoint of its selection, "true model" must mean the smallest dimension representation of this true model: the model with smallest K that has $I(f, g) = 0$. For the nested models case $\Pr\{g_i\}$ goes to zero for all $i > t$ as well as for all $i < t$, while $\Pr\{g_t\}$ goes to 1. This is the large-sample behavior of BIC that we are to look for if there is a nonunique true model in our set, and it is this pattern of posterior probabilities that might be taken as evidence for the fitted model, $\hat{g}_t$, as being true. But there is a logical fallacy here that we need to make very clear.

Whereas this is the asymptotic behavior BIC and $\Pr\{g_i\}$ will have if the true model is in the set and is nested in (unnecessarily) more general models, and sample size is quite large, it is also what will happen if $I(f, g_t) > 0$. For any model g_j that g_t is nested in (i.e., the added parameters have values equal to 0), $I(f, g_t) = I(f, g_j)$. However, asymptotically BIC actually selects based on relative distances $I(f, g_t) - I(f, g_j)$, not absolute. Only these differences in K-L distances are estimable. Thus, even if we have the needed large sample size, we cannot infer that model g_{min} selected by BIC is truth just because $\Pr\{g_{min}\}$ is 1, or nearly 1. This is a type of nonidentifiability, but for models, rather than parameters in models. Model g_{min} may be very far from truth, and it may not even have the correct form or the correct predictors.

The best that BIC can do asymptotically is identify unnecessarily more general versions of the apparent true model in which g_{min} is nested. Tapering effects would mean that the sequence $I(f, g_j)$ is strictly decreasing, which means that the global model would be the closest to truth and would be selected asymptotically by both BIC and AIC. Thus, the concerns motivating BIC are theoretically based on the idea that for a subset of the models, no tapering effects exist: All models in that subset have the identical value of $I(f, g)$, and this is the minimum of the K-L information loss over all models in the set. Then BIC is a criterion that selects from this subset of models the model with smallest dimension K.

We need a name for this property of a selected model that can appear to be "true" (as explained above) even though it is not the true model. We will call it a quasi-true model. In practice, consistent model selection allows, at best, an inference that a quasi-true model has been found if the selected model has $\Pr\{g_{min}\}$ virtually 1 and that model is nested in more general models in the set. We do not need this concept of a quasi-true model if the global model is selected; we would not be inclined to think that it is truth. The information-theoretic approach also does not need this concept, especially given that in practice we expect tapering effects in our model set rather than any ties in the K-L information loss values.

To make these concepts less abstract we give a simple example. Let $x_1, \ldots, x_6$ be independent normal(0, 1) random variables and let $z = x_4 x_5 x_6$. Given the x_i let $y = 100 + 15x_1 + 10x_2 + 5x_3 + 3z + \epsilon$, for ϵ an independent normal(0, 1) random variable. Let the model set considered be the nested six regression models for response variable y:

model	predictors
1	x_1
2	x_1, x_2
3	x_1, x_2, x_3
4	x_1, x_2, x_3, x_4
5	x_1, x_2, x_3, x_4, x_5
6	$x_1, x_2, x_3, x_4, x_5, x_6$

Although y depends on z, y is uncorrelated with x_4, x_5 and x_6; in models 4, 5, and 6, $\beta_4 = \beta_5 = \beta_6 = 0$. In this model set, model 3 is a quasi-true model: The inferential properties of BIC, as regards model 3, are here the same as if model 3 was the true model.

Another set of six models was also considered: The 6 models above but with x_4 replaced by z. Denote these as models $1z$ to $6z$. Models 1, 2, 3 are identical to models $1z$, $2z$, $3z$. Now the true model ($4z$) is in this second model set, but is nested in models $5z$ and $6z$. Table 6.13 presents Akaike weights (AIC$_c$ was used) and BIC model probabilities from one random sample of this example for each power of 10, $n = 10$ to 1 million.

In Table 6.13 consider results in the first model set for AIC$_c$ at $n = 10$ and BIC at $n = 100,000$ (also perhaps at $n = 10,000$ and one million): $w_3 = 1$, $\Pr(g_3) = 1$. It would be careless, and wrong, to infer from these samples that model 3 is true. But this pattern of inferential statistics is exactly what will

TABLE 6.13. Akaike weights (w) from AIC_c and model probabilities (Pr) from BIC, for the two model sets, for one random sample at each sample size (see text for more details).

					Inference by sample size						
Model	10		100		1,000		10,000		100,000		1,000,000	
set	w	Pr	w	Pr	w	Pr	w	Pr	w	Pr	w	Pr
1	0	0	0	0	0	0	0	0	0	0	0	0
2	0	0	0	0	0	0	0	0	0	0	0	0
3	100	49	66	89	27	87	57	99	63	100	19	99
4	0	35	23	10	44	12	26	1	24	0	42	1
5	0	12	8	1	16	1	12	0	9	0	25	0
6	0	4	3	0	12	0	5	0	4	0	14	0
$1z$	0	0	0	0	0	0	0	0	0	0	0	0
$2z$	0	0	0	0	0	0	0	0	0	0	0	0
$3z$	77	1	0	0	0	0	0	0	0	0	0	0
$4z$	23	51	57	83	55	96	41	97	29	99	61	100
$5z$	0	23	32	15	22	4	42	3	41	1	28	0
$6z$	0	25	11	2	23	0	17	0	30	0	11	0

occur if model 3 is true. In applying model selection with these same samples but using the second model set the inferences change dramatically. For BIC for $n \geq 100$ model 3z ($\equiv$ model 3) is ruled out; it gets an inferential probability of 0. This is a very different inference than for the first model set. The point is that statistically we can infer only that a best model (by some criterion) has been selected, never that it is the true model. Yet the initial frequentist thinking underlying BIC was that it would select the true model, or its true dimension, given a large enough sample size (it seemed to be implicitly assumed that of course the true model was in the set—where else would it be?). We will pursue in Section 6.4.2 what criterion BIC is optimizing and how we think the prior and posterior probabilities for BIC should be interpreted.

A few more comments on the example used here. We did look at replicated Monte Carlo simulations, and results were as expected. However, such results are both extensive and mostly irrelevant to our purpose in this section. We do note below the average value of $Pr\{g_3|$ models 1 to 6$\}$ and $Pr\{g_4|$ models 1z to 6z$\}$. The number of Monte Carlo samples used is also shown:

# M.C. samples	n	models 1 to 6 $\bar{P}r\{g_3\}$	models 1z to 6z $\bar{P}r\{g_{4z}\}$
10,000	10	0.2664	0.3041
1,000	100	0.8067	0.8147
1,000	1,000	0.9294	0.9297
100	10,000	0.9694	0.9670
10	100,000	0.9947	0.9835
1	1,000,000	0.9989	0.9902

Under the conditions of this example, ruling out uninformative predictors with near certainty takes, on average, n at least on the order of 1,000, but more like 10,000 (or better yet 100,000).

The summary messages of this subsection: **Truth and true models are not statistically identifiable from data**. BIC selection producing $\Pr\{g_{min}\} \approx 1$ justifies only an inference that we have the quasi-true model of a model set, and this strained concept requires that there be a subset of the models that have identical K-L distances (an unlikely event, it seems). Convergence may require very large sample sizes. So we think that in practice BIC cannot really do what frequentists want it to do in the unrealistic, idealized context wherein AIC is not consistent (an asymptotic property), which is the justification sometimes given for recommending against AIC. The Bayesian perspective, Section 6.4, is more general. Finally, to argue, after selection, that you have selected the true model (an oxymoron) you must argue a priori that the true model is in the model set; true models are not statistically identifiable.

6.4 Contrasting AIC and BIC

6.4.1 A Heuristic Derivation of BIC

The derivation of BIC holds both the model set and the data-generating (i.e., true) model fixed as sample size goes to infinity. It is also clear that if the model set contains the true (generating) model, then BIC selection converges with probability 1 to that generating model as $n \to \infty$ (and the posterior probability of that model goes to 1), even if the generating model is nested in some too-general set of models. The literature has not been clear on whether the derivation of BIC requires the true model to be in the set. For example, in his derivation Schwarz (1978) interprets the prior probability for model g_j as being the probability that model g_j is the true model (hence, for him posterior probabilities are to be interpreted this same way). However, Cavanaugh and Neath (1999) make it clear that the derivation of BIC does not require any assumption about the true model being in the set of models. Yet the difference between AIC and BIC is the $\log(n)$ in BIC (and not in AIC), and this $\log(n)$ is needed for idealized asymptotic consistency.

So a question is, why does the $\log(n)$ arise in deriving BIC? We set out to understand the answer to this question, thinking it would shed light on the issue of the role of the "true model." It was evident that assumptions and interpretations about prior probabilities are irrelevant in deriving the basic BIC result. As used, BIC assumes equal prior probability for each model, but it is easily adapted to allow any model priors. Because the derivation of BIC is free of any aspect of the priors on the models, its derivation and mathematics tell us nothing about how we should interpret model prior and posterior probabilities. The Bayesian literature we have seen simply refers to "the probability of model g_j," without clarifying what these probabilities mean. We will give a plausible interpretation below, after a simple heuristic derivation of BIC. (See McQuarrie and Tsai 1998, pages 22–23 and 50–63 for further insights).

The critical quantity to be approximated is the marginal probability of the data:

$$\int \left[\prod_{i=1}^{n} g(x_i|\theta) \right] \pi(\theta)d\theta.$$

(Section 6.4.4 gives basic formulas for the Bayesian approach, hence puts BIC in context.) The parameter θ has dimension K, as does the integral. As a function of θ, the product in $g(\cdot)$ under the integral is the likelihood. Hence, we can write it symbolically as

$$\int [\mathcal{L}(\theta|x, g)]\pi(\theta)d\theta,$$

where x represents the data. Under general regularity conditions, as sample size increases the likelihood function "near" the MLE, $\hat{\theta}$ (near is in terms of the probability distribution of $\hat{\theta}$), can be well approximated as

$$\mathcal{L}(\theta|x, g) = \mathcal{L}(\hat{\theta}|x, g)e^{-\frac{1}{2}(\theta-\hat{\theta})'V(\hat{\theta})^{-1}(\theta-\hat{\theta})}.$$

Here, $V(\hat{\theta})$ is the (estimated) $K \times K$ variance–covariance matrix of the MLE. This form of the likelihood is related to the fact that the sampling distribution of the MLE becomes multivariate normal as sample size goes to infinity, with $\hat{\theta}$ converging to a fixed value θ_0 (see Section 7.1). As regards these formulas there is no requirement that g be the true model. It suffices to take $V(\hat{\theta})^{-1} = I(\hat{\theta})J(\hat{\theta})^{-1}I(\hat{\theta})$ (same I, J as used in TIC, Section 7.3.1). If g is the true model, $I \equiv J$ and $V^{-1} = I$. Nevertheless, for a random sample we have $V(\hat{\theta})^{-1} = nV_1(\hat{\theta})^{-1}$ where the matrix $V_1(\cdot)$ is independent of sample size and $V_1(\hat{\theta})^{-1}$ converges to $V_1(\theta_0)^{-1}$.

Now we consider the needed integral, which is approximately

$$\mathcal{L}(\hat{\theta}|x, g) \int e^{-\frac{1}{2}(\theta-\hat{\theta})'V(\hat{\theta})^{-1}(\theta-\hat{\theta})}\pi(\theta)\,d\theta.$$

As n goes to infinity the approximation becomes exact, the likelihood concentrates near $\hat{\theta}$ (which is converging to θ_0) and the prior is effectively uniform (over the space where $\hat{\theta}$ has any substantial probability of being), so we can treat $\pi(\theta)$ as a constant. Alternatively, in the spirit of having a vague prior we can just directly use the improper prior $d\theta$. The needed integral is directly related to the underlying multivariate normal distribution and can be evaluated because we know the needed normalizing constant:

$$\int (2\pi)^{-K/2}\|V(\hat{\theta})^{-1}\|^{1/2}e^{-\frac{1}{2}(\theta-\hat{\theta})'V(\hat{\theta})^{-1}(\theta-\hat{\theta})}\,d\theta = 1,$$

where $\| \cdot \|$ denotes the determinant of a matrix. Therefore, we get

$$\int \left[\prod_{i=1}^{n} g(x_i|\theta) \right] \pi(\theta) \, d\theta \approx \mathcal{L}(\hat{\theta}|x, g) \left[(2\pi)^{K/2} \| V(\hat{\theta})^{-1} \|^{-1/2} \right]$$

$$= \mathcal{L}(\hat{\theta}|x, g) \left[(2\pi)^{K/2} \| n V_1(\hat{\theta})^{-1} \|^{-1/2} \right],$$

and by a property of the determinant, $\| n V_1(\hat{\theta})^{-1} \| \equiv n^K \| V_1(\hat{\theta})^{-1} \|$. So we have the approximation

$$\int \left[\prod_{i=1}^{n} g(x_i|\theta) \right] \pi(\theta) \, d\theta \approx \mathcal{L}(\hat{\theta}|x, g) \left[(2\pi)^{K/2} n^{-K/2} \| V_1(\hat{\theta})^{-1} \|^{1/2} \right].$$

Taking -2 times the log of the right hand side above, we have essentially the BIC criterion:

$$-2 \log(\mathcal{L}(\hat{\theta}|x, g)) + K \log(n) - K \log(2\pi) - \log(\| V_1(\hat{\theta})^{-1} \|).$$

The literature drops the last two terms of the expression above presumably because, asymptotically, they are dominated by the term of order $\log(n)$ as well as by the log-likelihood term (which is of order n).

We now see that the $\log(n)$ term arises because of the quintessential Bayesian feature of marginalization over θ (i.e., integrating out θ). There is no mathematical requirement in the derivation of BIC that the model g be true; hence, the model set does not need to contain the true model. However, there is also nothing in the foundation or derivation of BIC that addresses a bias-variance tradeoff, and hence addresses parsimony as a feature of BIC model selection. This is not a strike against BIC because this tradeoff is a frequentist concept not explicitly invoked in Bayesian statistics. But we are left with no theoretical basis to know what sort of parsimony the BIC model selection procedure has. Simulation studies of this question have been done, but the results when comparing AIC and BIC performance depend on the nature of the data generating model (such as having many tapering effects or not), on whether the model set contains the generating model, on the sample sizes considered, and on the objective: select the true model or select the K-L best approximating model. One can simulate situations where either BIC or AIC is the clear "winner." Thus, it is the unknown context and intent (i.e., true model or best model) of their use that is critical for deciding which method is "correct."

6.4.2   A K-L-Based Conceptual Comparison of AIC and BIC

The motivation for this section was to clarify what objectively ought to be meant, in the Bayesian-oriented literature about BIC, by the "probability of the model," or similar such vague phrases. We precede by relating BIC to objective K-L discrepancy, as opposed to allowing "model probabilities" to be subjective, hence meaningless (to us). Here are two examples of Bayesian usage. From Morgan (2000:96) "The Bayesian framework for modelling attributes probabilities to models . . . ." Leonard and Hsu (1999:82) say ". . . ϕ_j denotes your probability that the jth model is the most appropriate." Most commonly the literature simply refers to the probability of the model, with no clarity about what this means as regards a model being true, quasi-true, or "appropriate" by some unspecified criterion.

From its operating characteristics we know that BIC's main extolled feature is that it asymptotically will select, with probability 1, the true model—if that true model is in the set. However, such convergence in a sampling probability sense to a single model does not, and cannot, logically mean that model is truth (Section 6.3.4). In fact, as sample size $n \to \infty$, the model selected by BIC is consistent for the quasi-true model in the model set. We formally define a quasi-true model below.

For a set of R models the Kullback-Leibler "distance" of model g_r from truth is denoted $I(f, g_r)$ (Section 2.1.3). If $g_r \equiv g_r(x|\underline{\theta})$ nominally would denote a parametric family of models with $\underline{\theta} \in \Theta$, Θ being a K_r dimensional space, then g_r is the family member for the unique $\underline{\theta}_0 \in \Theta$ which makes g_r closest to truth in K-L distance (see Section 7.2). For our purposes here we also assume the models are indexed worst (g_1) to best, i.e., so that $I(f, g_1) \geq I(f, g_2) \geq \cdots \geq I(f, g_R)$. Let Q be the tail-end subset of the models defined by $\{g_r, r \geq t, 1 \leq t \leq R | I(f, g_{t-1}) > I(f, g_t) = \cdots = I(f, g_R)\}$. Set Q exits because $t = R$ is allowed, in which case the K-L best model (of the R models) is unique. For the case when Q contains more than one model (i.e., $1 \leq t < R$) we assume the models g_t to g_R are ordered such that $K_t < K_{t+1} \leq \cdots \leq K_R$ (in principle $K_t = K_{t+1}$ could occur).

The set Q contains models that are all equally good approximations, by K-L distance, to truth f. However, we can further distinguish them by their parameter space dimension, and we must prefer the smallest dimension model. If $t < R$, and $K_t < K_{t+1}$ holds, then model g_t is the unique quasi-true model of the R models. As a matter of inference from data, BIC model selection is consistent for this quasi-true model, which is not absolute truth unless $I(f, g_t) = 0$. (In principle, there might not be a unique quasi-true model).

Both AIC and BIC model selection actually depend on the K-L differences, $I(f, g_i) - I(f, g_j)$, not on absolute K-L values. Only these differences are estimable. For a random sample we can write $I(f, g_i) = nI_1(f, g_i)$, where $I_1(f, g_i)$ being for $n = 1$ is a constant as regards sample size. Hence, $I(f, g_i) - I(f, g_j) = n(I_1(f, g_i) - I_1(f, g_j))$. Because the MLE $\hat{\underline{\theta}}$ converges to $\underline{\theta}_0$ the basic convergence properties of AIC and BIC for large n can be deduced from approximations such as

$$\text{AIC}_i - \text{AIC}_j \approx 2n[I_1(f, g_i) - I_1(f, g_j)] + (K_i - K_j)2,$$

$$\text{BIC}_i - \text{BIC}_j \approx 2n[I_1(f, g_i) - I_1(f, g_j)] + (K_i - K_j)\log(n).$$

This level of approximation will not lead to reliable distributional results but does show the heuristics of large sample convergence regarding the selected model.

In the case of tapering effects, so $t = R$,

$$2n(I_1(f, g_i) - I_1(f, g_R)) > 0, \quad i < R.$$

Hence, as $n \to \infty$ all these differences diverge to ∞ at a rate proportional to n. Also, the magnitude of these differences dominates the "penalty" terms, which at best only grow at a rate proportional to $\log(n)$. Therefore, both the AIC- and BIC-selected model will converge to model g_R with certainty as $n \to \infty$.

The case of a nontrivial quasi-true model (i.e., $t < R$) primarily (not exclusively) corresponds to model g_t nested in models g_i, $i > t$. We assume such a case here. The relevant differences are

$$\begin{aligned}
\text{AIC}_i - \text{AIC}_t &\approx 2n[I_1(f, g_i) - I_1(f, g_t)] + (K_i - K_t)2, & i < t, \\
\text{AIC}_i - \text{AIC}_t &\approx -\chi_i^2 + (K_i - K_t)2, & i > t, \\
\text{BIC}_i - \text{BIC}_t &\approx 2n[I_1(f, g_i) - I_1(f, g_t)] + (K_i - K_t)\log(n), & i < t, \\
\text{BIC}_i - \text{BIC}_t &\approx -\chi_i^2 + (K_i - K_t)\log(n), & i > t.
\end{aligned}$$

Here, χ_i^2 is a central chi-square random variable on $K_i - K_t$ degrees of freedom. For all $i < t$ the differenes $\text{AIC}_i - \text{AIC}_t$ and $\text{BIC}_i - \text{BIC}_t$ become infinite as $n \to \infty$, with probability 1, hence model g_t is always selected over models g_1 to g_{t-1}. For all $i > t$ the differences $\text{BIC}_i - \text{BIC}_t$ become infinite as $n \to \infty$, with probability 1, as long as $K_i > K_t$ because then $\log(n)$ diverges to ∞. Hence, if there is a nontrivial quasi-true model the probability (*sensu* frequentist sampling theory) of the model being the one selected by BIC goes to 1 for a big enough sample size.

By contrast, for $i > t$ we only have $\text{E}(\text{AIC}_i - \text{AIC}_t) = K_i - K_t$, which is > 0 but is independent of n. While these expected AIC differences are positive, the actual AIC values are random variables with enough variability that AIC does not select model g_t with certainty in this hypothetical situation of $t < R$. However, even if this were the situation, Shibata (1983, 1986, 1989) shows that there is a sense in which use of AIC leads to optimal parameter estimates and predictions as $n \to \infty$.

It is clear to us, as argued for above, that the Bayesian "probability of model g_i" used in conjunction with BIC can, and must, be interpreted more precisely. Mathematically it is

$$\Pr\{g_i|\, data\} = \frac{\exp(-\frac{1}{2}\Delta\text{BIC}_i)}{\sum_{r=1}^{R} \exp(-\frac{1}{2}\Delta\text{BIC}_r)}.$$

This posterior probability assumes equal prior probabilities (i.e., $1/R$) on the models and is conditional on that set of R models. Conceptually, $\Pr\{g_i|\, data\}$ must be interpreted as the probability that model g_i is the quasi-true model in the set of R models. This model will generally be unique, but need not be; its dimension, K_t, is unique. Whether one wants to interpret probability in a subjective or frequentist sense does not change the necessity of this interpretation if inference from data is to have credibility.

Finally, given the necessity of this interpretation on the posterior probability of model g_i, then logically one must interpret the model prior probabilities in the same way, as the prior probability (it can be degree of belief) that model g_i is the quasi-true model in the set of R models. If one wants to go further and argue that model g_i might be the true model, then one must believe (or know) that the true model is already in the model set (see e.g., Wasserman 2000). Because of possible one-to-one transformations of models the focus is sometimes on model dimension, K_t, which is unique, hence the dimensional-consistent idea.

In summary, K-L distance is fundamental to understanding the properties of both AIC and BIC model selection. Both selection criteria can be derived, and applied, without assuming the true model is in the model set. However, the defining characteristic of BIC (i.e., what is it trying to do) is only evident asymptotically in relation to the concept of a quasi-true model. In contrast AIC seeks to select only a best model at a given sample size; "best" is in relation to an expected estimated K-L criterion which serves to recognizes a bias-variance trade-off in model selection. For AIC, "best" varies with n. For BIC, its "best," i.e., the quasi-true model, does not depend on n. However, on average the BIC-selected model approaches its target "best" model from below in terms of the model ordering imposed here by the $I(f, g_r)$. How researchers assess AIC and BIC performance depends on the performance criteria they adopt (true model or best model), the assumptions they make (usually only implicitly, as in simulation studies) about the underlying K-L values, $I(f, g_r), r = 1, \ldots, R$, and the sample sizes considered. Failure to properly recognize all of these factors and issues has led to much confusion in the model selection literature about AIC versus BIC.

6.4.3 Performance Comparison

A generally accepted measure of model and model selection performance is predictive mean square error (MSE). We evaluated the predictive MSE for AIC_c and BIC using Monte Carlo simulated data that mimics the body fat data used in Section 6.2. Thus, the simulated variable y has the properties of the body fat data example (such as tapering effects) and is generated by a linear model using 13 predictors (the $\underline{x}$) as described in Section 6.2.7. This performance comparsion involved generating a random sample of size 253 observations for $\underline{x}$, generating $E(y)$ ($\equiv$ here to $E(y|\underline{x})$) for each given $\underline{x}$, and finally generating $y = E(y) + \epsilon$ for the first 252 observations. Model selection used the sample of size $n = 252$; $\underline{x}_{253}$, and $E(y_{253})$ were set aside.

Initially, classical model selection was done; hence we obtained just the one best model selected under AIC_c and under BIC for each simulated sample. Then using these selected best models and $\underline{x}_{253}$, $E(y)$ ($\equiv E(y|\underline{x}_{253})$) was predicted.

Hence, we get $\hat{E}(y_{AIC_c})$ and $\hat{E}(y_{BIC})$ to compare to $E(y)$. This was done for 10,000 simulated samples. Predictive MSE was estimated as the average of the 10,000 values of $(\hat{E}(y_{AIC_c}) - E(y))^2$ and $(\hat{E}(y_{BIC}) - E(y))^2$ for AIC_c and BIC, respectively.

The model selection procedure with the smallest MSE is the better procedure. The results of this simulation were a MSE of 5.6849×10^{-6} for AIC_c and 7.6590×10^{-6} for BIC. Thus the ratio of predictive MSE of AIC_c to BIC is estimated as 0.74. Both MSEs have a coefficient of variation of 1.5%. Rigorous statistical comparison is based on the paired nature of the comparisons: we look at the 10,000 values of $diff = (\hat{E}(y_{BIC}) - E(y))^2 - (\hat{E}(y_{AIC_c}) - E(y))^2$. The mean of $diff$ was 1.9741×10^{-6} with a standard error of $0.0908 \times 10^{-6}(cv = 4.6\%)$; hence, a 95% confidence interval on $E(diff)$ of 1.8×10^{-6} to 2.2×10^{-6}. Some results about the sample distribution of the 10,000 values of $(\hat{E}(y_{BIC}) - E(y))^2$ and $(\hat{E}(y_{AIC_c}) - E(y))^2$ are listed below:

| performance | sample percentiles $\times 10^6$ | | | | | |
measure	5	10	50	90	95	maximum
$(\hat{E}(y_{AIC_c}) - E(y))^2$	0.019	0.089	2.5	14.9	22.2	117
$(\hat{E}(y_{BIC}) - E(y))^2$	0.029	0.110	3.3	20.5	29.5	159

This shows that the predictions based on the AIC_c-selected model are stochastically closer to the true $E(y)$ values than are the predictions from the BIC-selected model (as opposed to the result for MSE being due to a few BIC-produced outliers).

In 5.7% of the simulated sample $diff = 0$ occurred; almost surely because AIC_c and BIC selected the same model. In the other 94.3% of samples $diff > 0$ occurred in 57.3% of cases. Thus, overall by our performance measure, AIC_c-model selection performed as well as or better than BIC in 60% of the samples.

Simple linear regression of $\hat{E}(y)$ on $E(y)$ was also done, to get an estimated intercept (β_0) and slope (β_1) parameter by each model selection procedure. The results are listed below, with standard errors in parentheses:

procedure	$\hat{\beta}_0 \times 10^6$	$\hat{\beta}_1$	r^2
AIC_c	$-8.83(23.8)$	$0.9936(0.0016)$	0.9739
BIC	$12.80(27.6)$	$0.9818(0.0019)$	0.9646

These results suggests that prediction based on the AIC_c-selected model is much closer to being unbiased than prediction based on the BIC-selected model. It is interesting that the smaller prediction MSE of AIC_c did not come at the expense of greater prediction bias.

A second set of 10,000 simulations was done wherein $\hat{E}(y)$ for $\underline{x}_{253}$ was based on model averaging the $\hat{E}_r(y)$, $r = 1, \ldots, R = 8,191$ different predictions, one for each fitted model. Under such a model-averaged prediction only the model weights vary by selection (i.e., weight generation) method, not the set of $\hat{E}_r(y)$. The results of this simulation example of model averaging were a MSE of 4.8534×10^{-6} for AIC_c and 5.8819×10^{-6} for BIC. Thus for model averaging the ratio of predictive MSE of AIC_c to BIC was 0.83. Both MSEs have a coefficient of variation of 1.5%. Other aspects of comparisons for prediction-based model averaging are about the same as those for the traditional best-model strategy.

A final noteworthy comparison is that the MSE values are significantly smaller under model averaging, as shown below, for MSEs $\times 10^6$:

method	model averaged	best model	ratio
AIC_c	4.8534	5.6849	0.85
BIC	5.8819	7.6590	0.77

For this simulation scenario (which mimics real data) model-averaged prediction beats the traditional best model approach, and AIC_c beats BIC.

Producing the above simulation results for the best-model approach took 9 hours of CPU time on a dedicated 1.9 GHz speed computer with 512M RAM and 80G hard drive (and about 14 hours clock time). The model averaged example took 12 hours of CPU time and 19 hours of clock time on the same computer. Thus, whereas, extensive simulation study along these lines is needed, it will be very computer intensive.

We also examined a few aspects of AIC_c versus BIC model selection under the simulation scenario of the example in Section 3.4. The simulation is based on a real capture–recapture experiment, and the simulated data mimic the real data. The reader would benefit from scanning parts of Section 3.4, especially Section 3.4.6, before reading the next two paragraphs. In particular, Table 3.7 shows the results of AIC model selection on a set of 14 nested models, based on 50,000 Monte Carlo data sets. BIC model selection was also applied to these 14 models. The key point we make here is that there was little overlap between the models selected by AIC and BIC.

BIC selection in the face of tapering treatment effect size, a sample size of "only" 2,500 (but still quite less than ∞), and a generating model with 34 parameters that was not (quite) in the set of candidate models performed poorly, as theory would suggest. BIC selection most frequently chose model $g_{2\phi}$ (26.8%), followed by model $g_{1\phi}$ (24.5%), and model g_{2p} (19.4%). BIC selection frequencies fell rapidly for models g_{3p}, $g_{3\phi}$, g_{4p}, and $g_{4\phi}$ (13.0, 9.8, 3.1, and 1.3, respectively). The BIC selected models are substantially underfit and would have poor confidence interval coverage. If the set of candidate models were to be expanded to include the generating model ($g_{9\phi}$), then as sample size increased, BIC should select model $g_{9\phi}$ with probability one. The initial number of nestlings required for BIC to select the generating model ($K = 34$) with probability approaching one in this moderately complex example is approximately 108,000 birds (instead of the 600 used in this example).

Clearly, it would often be impossible to find and band 108,000 birds on one small island for a particular year; it might be quite unusual to have such a large number of nestlings present! Thus, to realize substantial increases in initial sample size, one must include other islands and conduct the study over several years. However, in so doing, other factors become important, and the conceptualization of truth must include obvious factors such as island and year, in addition to slightly less obvious factors such as technicians with differing resighting probabilities and islands with differing vegetation that also affects resighting probabilities. The "year" effect is not so much the actual calendar year, but a host of covariates (most unmeasured) that affect both survival and resighting probabilities in complex, nonlinear ways across time. Of course, there is individual heterogeneity that is substantial (e.g., weight, hatching data, growth rate, dispersal distance). Thus, the concept of truth, or full reality, is very complex. To think that such reality exists as an exactly true model is not useful; to think that such a true model is included in the set of candidate models seems absurd.

The primary foundations of the BIC criteria do not apply in the biological sciences and medicine and the other "noisy" sciences. Reality is not fixed as sample size is increased by orders of magnitude in biological systems; rather, the target "true model" sought by BIC increases in size as n increases. This simple fact is a violation of the assumptions that form the basis of BIC; however, this is allowed under the AIC-type criteria.

Many other published works have compared AIC or AIC_c and BIC model selection. Some, but not all, of these studies note the realities of the situation. For example Hjorth (1994:46) says "... the asymptotic consistency of BIC-measures is theoretically correct, but in the model selection field we have to be skeptical against asymptotic results when we are analyzing data sets of say 50 or even 500 observations. The asymptotics may require several powers of ten more data before a reasonable accuracy is achieved, if many models are possible." McQuarrie and Tsai (1998) report on extensive Monte Carlo studies of model selection methods. A parsimonious summary of their conclusions is on pages 410–411 of that book.

6.4.4 Exact Bayesian Model Selection Formulas

The analytical formulae for Bayesian model selection are given here, partly to help put BIC in context. Implementation of these formulae is usually by computer-intensive methods such as Markov chain Monte Carlo.

For structural model g_i the likelihood of θ_i (a $K_i \times 1$ vector) given the data is denoted as $g_i(x|\theta_i)$ $(\equiv \mathcal{L}(\theta|x, g)$ as used in Section 6.4.1). The prior probability for θ_i is denoted $\pi_i(\theta_i)$. A key quantity needed is the marginal likelihood,

$$g_i(x, \pi_i) = \int g_i(x|\theta_i)\pi_i(\theta_i)\, d\theta_i.$$

This is taken (by us) as the likelihood of model structure g_i given the data and the prior on θ_i. In essence this quantity is what was approximated in Section 6.4.1 to obtain the BIC criterion. The posterior distribution for θ_i assuming model g_i is

$$h_i(\theta_i | x, \pi_i) = \frac{g_i(x|\theta_i)\pi_i(\theta_i)}{g_i(x, \pi_i)}.$$

Let the prior probability for model structure g_i be denoted by p_i. This prior relates only to the structure of model g_i, not its parameters. We think that p_i should be interpreted as the prior probability that model g_i is the quasi-true model in the set of R models. The posterior probability that model g_i is the quasi-true model in the set of R models is given by

$$m_i(x, g_i, \pi_i) = \frac{g_i(x, \pi_i)p_i}{\sum_{r=1}^{R} g_r(x, \pi_r)p_r}.$$

From data analysis alone one cannot conclude that a model is the true data generating model even if it gets $m_i = 1$. There is an identifiability problem here. All we can know from the data is that a model is the quasi-true model; i.e., for a large enough sample size we can infer that there is no other model in the model set that has a smaller K-L discrepancy, even when the model in question is nested in some other models. In order to infer truth from the data we need to believe (i.e., *know*) a priori that the true model is in the model set (we just do not know a priori which model it is). This seems implausible. But it is justifiable to believe that there is always a single best model in the model set, if we have a suitable criterion for what is a best model.

The posterior odds ratio is informative:

$$\frac{m_i(x, g_i, \pi_i)}{m_j(x, g_j, \pi_j)} = \left[\frac{g_i(x, \pi_i)}{g_j(x, \pi_j)} \right] \left[\frac{p_i}{p_j} \right].$$

The ratio $g_i(x, \pi_j)/g_i(x, \pi_j)$ is called the Bayes factor; it is analogous to the information-theoretic evidence ratio. Often, BIC is motivated by the desire for a simple approximation to the Bayes factor, inasmuch as the Bayes factor can be difficult to compute exactly. The prior odds ratio is modified to give the posterior odds ratio solely by the Bayes factor.

6.4.5   Akaike Weights as Bayesian Posterior Model Probabilities[1]

For a large sample size a good approximation to the Bayesian posterior model probability can be based on BIC, provided one is willing to assume equal prior model probabilities (useful background Sections here are 6.3.4, 6.4.1, 6.4.2, and 6.4.4). However, the expression BIC can be used more generally with any model priors. Let p_i be the prior probability placed on model g_i. Then the

---

[1]A note to the reader: At the time we were checking the second set of page proofs for this book (late March 2002) we found relationships of such importance that we felt we had to include them. The results are this section.

Bayesian posterior model probability is

$$\Pr\{g_i|data\} = \frac{\exp\left(-\frac{1}{2}\Delta\text{BIC}_i\right)p_i}{\sum_{r=1}^{R}\exp\left(-\frac{1}{2}\Delta\text{BIC}_r\right)p_r}.$$

To get Akaike weights we use the model prior

$$p_i = B \cdot \exp\left(\tfrac{1}{2}\Delta\text{BIC}_i\right) \cdot \exp\left(-\tfrac{1}{2}\Delta\,\text{AIC}_i\right).$$

B is a normalizing constant (p_i simplifies, as will be shown below). Clearly,

$$\exp\left(-\tfrac{1}{2}\Delta\text{BIC}_i\right)\cdot\exp\left(\tfrac{1}{2}\Delta\text{BIC}_i\right)\cdot\exp\left(-\tfrac{1}{2}\Delta\,\text{AIC}_i\right) = \exp\left(-\tfrac{1}{2}\Delta\,\text{AIC}_i\right);$$

hence, with this prior probability distribution on models we get

$$\Pr\{g_i|data\} = \frac{\exp\left(-\frac{1}{2}\Delta\text{BIC}_i\right)p_i}{\sum_{r=1}^{R}\exp\left(-\frac{1}{2}\Delta\text{BIC}_r\right)p_r} = \frac{\exp\left(-\frac{1}{2}\Delta\,\text{AIC}_i\right)}{\sum_{r=1}^{R}\exp\left(-\frac{1}{2}\Delta\,\text{AIC}_r\right)} = w_i,$$

which is the Akaike weight for model g_i.

Moreover, this model prior is actually simple and not dependent on the data as shown by simplifying how it is expressed (there is an associated change in the formula for the normalizing constant):

$$\begin{aligned} p_i &= B \cdot \exp\left(\tfrac{1}{2}\Delta\text{BIC}_i\right)\cdot\exp\left(-\tfrac{1}{2}\Delta\,\text{AIC}_i\right)\\ &= B \cdot \exp\left(\tfrac{1}{2}[\Delta\text{BIC}_i - \Delta\,\text{AIC}_i]\right)\\ &= C \cdot \exp\left(\tfrac{1}{2}[\text{BIC}_i - \text{AIC}_i]\right)\\ &= C \cdot \exp\left(\tfrac{1}{2}K_i\log(n) - K_i\right) \end{aligned}$$

and

$$C = \frac{1}{\sum_{r=1}^{R}\exp\left(\frac{1}{2}K_r\log(n) - K_r\right)}.$$

The result easily generalizes to AIC_c and formally to QAIC_c; however, a Bayesian would not handle overdispersion by simply using $\hat{c}$ in "QBIC." Formally, the Akaike weights from AIC_c are Bayesian posterior model probabilities for the model prior

$$p_i = C_c \cdot \exp\left[\tfrac{1}{2}K_i\log(n) - \frac{nK_i}{n - K_i - 1}\right],$$

$$C_c = \frac{1}{\sum_{r=1}^{R}\exp\left[\frac{1}{2}K_r\log(n) - \frac{nK_r}{n-K_r-1}\right]}.$$

We will call this the K-L model prior. Because BIC is for large n, the applicability of the result (i.e., AIC_c as Bayesian) can be questioned for small n, but it does apply for large n and small or large K_i.

Do not be confused by the $\log(n)$ appearing in both BIC and in the K-L model prior. BIC actually arises in the context of obtaining a large sample approximation to the Bayes factor, a quantity that is unrelated to the model priors. Thus BIC is not in any way intrinsically encumbered by any assumed model priors. The $\log(n)$ in BIC has nothing to do with priors on models. Rather, the $\log(n)$ in BIC arises owing to the renormalization of the likelihood, assuming vague priors on the parameters in model g_i (Section 6.4.1). Also, if the BIC formula is used with any prior p_i that is not a function of sample size, then the posterior will converge (as $n \to \infty$) to the quasi-true model in the set of models.

When the K-L model prior is used with BIC so that we get AIC as a Bayesian result, the interpretation of the model probabilities is not the same as for BIC. Whereas the quasi-true model g_i is the "target" of BIC (a target unrelated to sample size), the target model of AIC is the model that minimizes expected estimated K-L information loss. From Section 7.3 (7.18), that model is the one that minimizes

$$E_{\underline{y}}\left[I(f, g(\cdot|\hat{\underline{\theta}}(\underline{y})))\right] = \text{constant} - E_{\underline{y}}E_{\underline{x}}\left[\log[g(\underline{x}|\hat{\underline{\theta}}(\underline{y}))]\right].$$

For simplicity we call this target model the K-L best model. It depends on sample size and it is essentially a fitted model as opposed to the quasi-true $g_t(x|\underline{\theta}_o)$ for BIC, which is the model of smallest dimension that produces minimum $I(f, g_i)$. Thus as a Bayesian result we must interpret an Akaike weight w_i and a model prior p_i as the probability that model g_i is the K-L best model.

The target K-L best model has a variance-bias trade-off as a fitted-to-data model. Such a trade-off depends on both sample size and number of parameters to be estimated. Still, we did not expect that the model prior needed to get AIC in a Bayesian context had to depend on n and K: That prior on models cannot be independent of these values (which are known prior to data analysis). It is easy to numerically explore the K-L model prior so we do not provide any numerical examples.

There are now two ways to compare AIC and BIC. One is to use the frequentist framework of looking at sampling measures of performance, for example, predictive mean square error and confidence interval coverage. The other way is to consider AIC as Bayesian and think about and compare the BIC model prior to the K-L model prior in conjunction with knowing that the targeted models for selection are different, and the interpretation of model probabilities is different, for AIC versus BIC. Such comparisons must consider the context of what we are assuming about the information in the data regarding parameter estimation and the models as approximations to some conceptual underlying generating distribution ("truth"). It is useful to think in terms of effects, as $|\theta|/\text{se}(\hat{\theta})$. We would assume few or no effects are truly zero. Thus, we assume meaningful, informative data and thoughtfully selected predictors and models. We assume tapering effects: some may be big (values like 10 or 5), but some are only 2, 1 or 0.5, or less. We assume we can only estimate, say, n/m parameters reliably; m might be 20 or as small as 10. These ideas lead us to a concept of savvy model priors, with properties like the K-L prior, which depend on n and K. The K-L prior is a particularly important savvy model prior.

In summary, we have shown that AIC can be justified in a Bayesian context and an Akaike weight is a valid posterior, i.e., data-dependent, model probability. However, the interpretation of what the probability of model g_i means is different for AIC versus BIC. Also, to use the Bayes factor approximation provided by BIC in the context required by AIC implies that we must have the prior probability on model g_i be an increasing function of n and a decreasing function of K (i.e., a savvy prior). The implicit BIC prior of $1/R$ is not sensible in the information-theoretic context.

6.5   Goodness-of-Fit and Overdispersion Revisited

Overdispersion of count data, relative to a theoretical model, must be dealt with to obtain valid inferences. A more sophisticated approach than we have advocated here is to incorporate one or more variance parameters directly into the parametric model, hence into the likelihood. As noted by Lindsey (1999a) this is a desirable approach that can be quite flexible and effective. But it is not the simple omnibus approach that QAIC and QAIC$_c$ are; to develop Lindsey's ideas here would take us too far afield from our model selection objective. Instead, we present a generalization of QAIC to allow more than one overdispersion parameter. This, however, assumes that the data are structured by some factor, or factors, that allow partitioning the data. Then the degree of overdispersion, c, can vary by data subset. Before dealing with these issues, we consider a strategy for obtaining $\hat{c}$ when there is not a single global model.

6.5.1   Overdispersion $\hat{c}$ and Goodness-of-Fit: A General Strategy

When there is a global model we can usually compute from it an unambiguous $\hat{c}$. The logic is that the global model is theoretically the best-fitting model, because all other models are special cases of the global model. The special cases cannot, on average, fit the data better than the global model. If problematic overdispersion exists in the data, then the goodness-of-fit statistic will on average exceed its degrees of freedom (as shown by a small P-value). This will be true even if the global model is structurally adequate. However, if we find $\hat{c}$ meaningfully > 1 we do not know whether this is a result of overdispersion or inadequate model structure. In the end this distinction does not matter operationally if we cannot generalize the global model. Rather, if the global model is not adequate, this must be accounted for, and it can be by use of $\hat{c}$. We prefer to say that the lack of fit problem is due to overdispersion because then the global model is structurally adequate.

However, the use of $\hat{c}$ (when $c > 1$ is clearly indicated) in all aspects of inference for count data (such as inflated standard errors as well as QAIC) means that the inferences are based on empirical residuals, rather than theoretical variances. This situation is analogous to most models for continuous data wherein variance estimates are based on empirical residuals, which automatically reflect both lack of model structural adequacy as well as "true" stochastic variation about "truth." Thus using $\hat{c}$ we are conservative in our inferences because importance of a structural data feature (i.e., predictability) is judged against the totality of unexplained stochastic variations and structural lack of fit, just as with models for continuous data.

With no global model, there is the question of how to obtain a defensible $\hat{c}$. For example, in the dose-response example of Section 4.10 there are three distinct a priori models, all on an equal footing. None is nested in another one; all have the same number of parameters. We must compute goodness-of-fit (and $\hat{c}$) for each model, because there is no theoretical basis to know which is the best-fitting model. If the best-fitting model has an acceptable fit, we use $c = 1$. If even the best-fitting model is a poor fit (say $P < 0.15$), we use $\hat{c}$ from that model. The logic here is that overdispersion, if present, will show up in each goodness-of-fit test, but so will inadequate model structure. If at least one model is adequate, then on average, its goodness-of-fit reflects only overdispersion, and that would, on average, be the smallest $\hat{c}$ value. If more than one model fits the data (no overdispersion, $c = 1$) we might by chance get $\hat{c} < 1$. Fortunately, this is not a problem, because we then set $c = 1$.

The general strategy to obtain $\hat{c}$ is as follows. Partition the set of R models into s subsets, where each subset (size R_i, $i = 1, \ldots, s$) has its own (sub-) global model. Thus, one gets s ($\leq R$) subglobal models, none of which are subsets of each other. Compute $\hat{c}_i$ for each subglobal model; note that all the data are used in fitting each subglobal model. Even if one or more of these subglobal models structurally fits the data, if there is problematic overdispersion, it will stochastically inflate all of the goodness-of-fit tests. On average, if a model structurally fits the data, its $\hat{c}$ value estimates actual overdispersion. Use the smallest $\hat{c}$ as the estimate: $c = 1$ if for that goodness-of-fit test $P > 0.15$ (as a guideline). Otherwise, use the computed $\hat{c}$.

We do not see a simple alternative to this strategy, although it is not without potential biases arising from taking the minimum of a set of statistics. The smaller is s, the better, as this minimizes potential selection bias. However, such bias is not a big concern here, for the reason that the same data are used in each calculation of goodness-of-fit. As a result, the $\hat{c}$ values are positively pairwise correlated and this reduces selection bias (which is usually thought of for selection over independent random variables). Of more concern should be small degrees of freedom for a goodness-of-fit statistic, especially if some other subglobal models have much larger degrees of freedom. Small degrees of freedom for goodness-of-fit leads to less reliable $\hat{c}$. However, if each subglobal model allows ample, or similar degrees of freedom for $\hat{c}$ and if s is small (2, 3, or 4) this strategy should work. Assessing goodness-of-fit (in general), hence $\hat{c}$ (in particular), requires some judgment. If nothing else it is a judgment call as to when we use $c = 1$, i.e., when we judge that the model "fits."

6.5.2 Overdispersion Modeling: More Than One $\hat{c}$

Having only a single overdispersion adjustment parameter, $\hat{c}$, may be too restrictive for many instances of count data. Such data can have a built-in structure due to being collected across factors such as gender, age, areas, years, treatments, and so forth. The degree of overdispersion may vary by factor-levels. For example, it is easy to imagine that in a survival study, c is enough different for males versus females that a common $\hat{c}$ should not be used.

Assume that the data are naturally partitioned into V independent subsets, hence symbolically $data = \cup_{v=1}^{V} data_v$. Then the nominally correct log-likelihood is a simple sum of separate log-likelihoods:

$$\log \mathcal{L}(\theta | data, g) = \sum_{v=1}^{V} \log \mathcal{L}(\theta | data_v, g).$$

We let θ represent the full vector of structural parameters, components of which may be in common over different parts of the likelihood. The vth subset of the data has likelihood component denoted by $\mathcal{L}_v(\theta | data_v, g)$ for any model g. Let the dispersion parameter for $data_v$ be c_v. The appropriate quasi log-likelihood is actually

$$\sum_{v=1}^{V} \frac{\log \mathcal{L}_v(\theta | data_v, g)}{c_v}.$$

The overdispersion parameters must be estimated under the global model. The same approaches apply here as are discussed in Section 6.5.1, but they now apply for each subset of the data. The overall goodness-of-fit test statistic is the sum of the separate chi-square statistics on their summed degrees of freedom. However, if even by this overall result we judge the global model to fit, we should check whether any separate component, $data_v$, clearly fails to fit, as we can use some $c_v = 1$ if warranted, while other $\hat{c}_v > 1$ are used.

There are some caveats. In theory, the V data subsets should be independent, hence without correlations across subsets. However, one of the main causes of overdispersion is correlation structure in the data, which presumably could be both within and between subsets. Hence, it would be best if the data arise so that subsets are independent. Still, even if there are weak correlations across data subsets, if the c_v are quite variable, this generalized quasi-likelihood is to be preferred to the simple case of only a single $\hat{c}$. We also emphasize that the data partitioning is a priori to data analysis. In a sense, it is a priori to any model, even though it is based on factors that may also be used in model construction.

Determination of the data partition, hence also the value of V, should be based on subject-matter knowledge. Strive to have V small. For biological data, partition on factors such a sex (i.e., males vs. females) and age (juveniles vs. adults), especially if the sexes and ages have different behaviors. Finally, be sure to use the total number of parameters as $K = p+V$, where p is the number of structural parameters, hence the dimension of θ. For each data subset there is a corresponding sample size n_v; total sample size is $n = \sum_{v=1}^{V} n_v$.

The QAIC formula is not changed by this data partitioning:

$$\text{QAIC} = \sum_{v=1}^{V} \frac{-2 \log \mathcal{L}_v(\hat{\theta}|data_v, g)}{\hat{c}_v} + 2K,$$

However, its small sample version, QAIC_c is problematic. A sum of terms involving reciprocals in n_v should be used, but these terms also involve a partitioning of K over v. We considered the issue of a tradeoff of a better, but more complex, formula versus the value of just using the existing simple formula, hence minimizing the number of formulae one needs to know. For now we suggest using just

$$\text{QAIC}_c = \sum_{v=1}^{V} \frac{\log \mathcal{L}_v(\hat{\theta}|data_v, g)}{\hat{c}_v} + 2K + \frac{2K(K+1)}{n - K - 1}.$$

When $V = 1$ we can ignore the estimate $\hat{c}$ while finding the MLE. That is, we can work directly with $\log \mathcal{L}(\theta|data, g)$. Moreover, we can compute the nominal empirical variance-covariance matrix $\hat{\Sigma}$ (as the inverse of the Hessian) directly from $\log \mathcal{L}(\theta|data, g)$, and then the appropriate variance-covariance matrix of $\hat{\theta}$ is taken as $\hat{c}\hat{\Sigma}$. Also, the degrees of freedom (df) to associate with variance estimates is the df of $\hat{c}$. However, for $V > 1$ we must work directly with the quasi log-likelihood. To find the MLE, we must directly maximize

$$\sum_{v=1}^{V} \frac{\log \mathcal{L}_v(\theta|data_v, g)}{\hat{c}_v}$$

over θ. The likelihood equations to solve are

$$\sum_{v=1}^{V} \left[\frac{1}{\hat{c}_v} \right] \left[\frac{\partial \log \mathcal{L}_v(\theta|data_v, g)}{\partial \theta_i} \right] = 0, \qquad i = 1, \ldots, K.$$

The elements of the $K \times K$ appropriate Hessian are

$$\sum_{v=1}^{V} \left[\frac{1}{\hat{c}_v} \right] \left[\frac{\partial^2 \log \mathcal{L}_v(\theta|data_v, g)}{\partial \theta_i \partial \theta_j} \right] = 0, \qquad i, j = 1, \ldots, K.$$

Each $\hat{c}_v$ has associated degrees of freedom $\text{df}_v = n_v - K_v - 1$, where K_v is the number of structural parameters in the global model for data subset v. For this global model there should be no parameters in common over data subsets. Hence, another complication is determing the appropriate df for the Hessian, hence the variance of any component of $\hat{\theta}$. As an exact result the df_v do not simply add as $\text{df} = \Sigma \text{df}_v$. However, to keep it simple we suggest just using this summed df for the applicable degrees of freedom.

A single $\hat{c}$ should often suffice, and should not be quickly abandoned for more complicated approaches: Favor parsimony even in this variance modeling. The type of variation in c we have in mind is not small-scale such as 1.5 versus 1.6; for this case use a single $\hat{c}$. Even bigger differences than this can be ignored (e.g., 1.3 versus 1.7) given that the c_v are only estimated. Given sufficient df_v the levels of differences to be concerned about are such as 1 versus 2, or 1.4 versus 2.8. Bear in mind that exact modeling of variation in overdispersion is not as important as having at least a basic adjustment (i.e., $V = 1$) for overdispersion, as by quasi log-likelihood.

6.5.3 Model Goodness-of-Fit After Selection

Often the set of models under consideration contains a most general model (the global model), in which case we recommend assessing the fit of that global model to the data (preferably before commencing with model selection). If the global model fits, as by some standard goodness-of-fit test, then the AIC-selected model will fit the data. We think that this is true also for AIC_c model selection (but we are not sure). If the global model does not fit statistically, one might decide that the lack of fit is not of concern and then resort to QAIC or $QAIC_c$. (In fact, if the global model does not fit, but you proceed with K-L–based model selection, you must use QAIC or $QAIC_c$ selection.)

There is a philosophy under which one would want to use BIC; and people are using BIC, even when the context is such that AIC should be used. It is clear that BIC selects more parsimonious models than AIC. What does this do to model fit; if the AIC selected model fits, will the BIC selected model for the same data also fit? This is a question we have never seen addressed in the literature, and we do not know the answer.

The paper by Leroux (1992) motivated our interest in this question. Leroux (1992) reports the observed versus expected count frequencies for some automobile accident data ($n = 9{,}461$). A pure Poisson model ($K = 1$) and two mixture models are fit to the data. BIC selects the two-component mixture model ($K = 3$), while AIC selects the three-component mixture model ($K = 5$). Model selection tends to lead to overly optimistic assessments of model fit. Hence, model selection may result in optimistic indications of statistical model fit for the selected model. However, a goodness-of-fit test applied to the global model will not be biased, because no selection process has first occurred. The usual chi-square goodness-of-fit procedure applied to the models selected in Leroux (1992) (this entails some pooling of sparse cells) produces $\chi^2 = 1.11$ (1 df, $P = 0.2921$) for the AIC-selected model and $\chi^2 = 11.53$ (2 df, $P = 0.0031$) for the BIC-selected model.

Whereas the goodness-of-fit $P = 0.0031$ is small, there is a large sample size here, and perhaps therefore it is acceptable to use for inference a model that statistically is not a good fit to the data. We think that this practice can be acceptable, but it must be argued for on a case-by-case basis. However, statisticians have consistently cautioned about drawing inferences from a model that does not fit the data. We should not ignore the issue of whether model selection procedures systematically select models that do, or do not, fit. There is good reason to think that AIC selects models that do fit, especially if the global model fits. It is an open question whether BIC-selected models fit the data at the nominal α-levels used in goodness-of-fit tests. Research is needed to understand this general issue of the fit, and assessing the fit, of models to data after model selection.

6.6 AIC and Random Coefficient Models

6.6.1 Basic Concepts and Marginal Likelihood Approach

Parameters are sometimes considered as "random effects" (or more generally as random coefficients; see Longford 1993). In the simplest case $\theta_1, \ldots, \theta_K$ are parameters all of the same type (e.g., survival rates in K years), and we consider the K elements of $\underline{\theta}$ as random variables. Thus, in the simplest case, we conceptualize $\theta_1, \ldots, \theta_K$ as independent random variables with mean μ and variance σ^2. Now the inference problem could be entirely about the underlying fixed population-level parameters μ and σ^2. However, the likelihood we can directly write down is for $\underline{\theta}$ as if the elements of $\underline{\theta}$ were the fixed parameters of direct inference interest. In using the likelihood $\mathcal{L}(\underline{\theta})$ we are ignoring issues of how $\theta_1, \ldots, \theta_K$ may have arisen from some process or some real or conceptual population. The likelihood $\mathcal{L}(\underline{\theta})$ is appropriate for when the parameters are "fixed effects." This is a valid approach if we interpret the parameters $\theta_1, \ldots, \theta_K$ as deterministic. It is then possible to fit this global model by standard likelihood methods and also fit simpler models based on deterministic constraints on $\underline{\theta}$, such as $\theta_i \equiv \mu$, where the likelihood is $\mathcal{L}(\mu)$.

However, we may also want to consider an intermediate model based on only the two parameters μ and σ^2, where we regard θ as a random variable with mean μ and variance σ. Thus while we directly have likelihoods $\mathcal{L}(\underline{\theta})$ and $\mathcal{L}(\mu)$ (hence models g_K and g_1), we also want the likelihood, say $\mathcal{L}(\mu, \sigma^2)$, for the two-parameter model g_2. The parameter σ^2 in model g_2 serves to fit the possible stochastic nature of the θ_i. Model g_K allows arbitrary variation in the θ_i, but this freedom costs us $K - 2$ extra parameters compared to model g_2. If the unexplained variation in these K parameters is substantial and consistent with them being considered as exchangeable random variables, we should select model g_2 rather than model g_K. Model g_2 with only two parameters, μ and σ^2, parsimoniously allows for variation in the θ_i (something model g_K does not do parsimoniously and model g_1 does not do at all).

Whereas the likelihood $\mathcal{L}(\mu)$ is a special case of either $\mathcal{L}(\mu, \sigma^2)$ or $\mathcal{L}(\underline{\theta})$, the conceptually intermediate model g_2 is not mathematically an intermediate model between models g_K and g_1 in the simple sense of being just a deterministically constrained version of the global model g_K. Therefore, model g_2 cannot be fit by standard likelihood methods based only on the global model likelihood $\mathcal{L}(\underline{\theta})$ and deterministic constraints. These random-coefficient (or random-effects) models are different from other models that arise as just deterministic constraints on the parameters of some more general model because of the reliance on an assumed probability distribution for $\underline{\theta}$.

Our focus here is on how we can compute a valid AIC for random-coefficient models, such as this case of having only the parameters μ and σ^2, even though our only obvious starting point is model g_K and its likelihood. The standard likelihood approach to stochastic parameters is to postulate a probability distribution, hence a model, for the random variable $\underline{\theta}$, say $h(\underline{\theta} \mid \mu, \sigma^2)$, and obtain the needed (proper) likelihood $\mathcal{L}(\mu, \sigma^2)$ based on the marginal distribution

$$g(\underline{x} \mid \mu, \sigma^2) = \int g(\underline{x} \mid \underline{\theta}) h(\underline{\theta} \mid \mu, \sigma^2) d\underline{\theta}; \qquad (6.1)$$

thus $\underline{\theta}$ has been integrated out. Considering $g(\underline{x} \mid \mu, \sigma^2)$ as a function of the parameters given the data, we have $\mathcal{L}(\mu, \sigma^2) = g(\underline{x} \mid \mu, \sigma^2)$. Thus the AIC for model g_2 is computed based on $g(\underline{x} \mid \mu, \sigma^2)$, which we can get by computing the integral in (6.1).

A more informative way to think about random-coefficient models is that sometimes a few parameters, defined by deterministic constraints on $\underline{\theta}$, cannot explain all the variation in the much larger set $\theta_1, \ldots, \theta_K$, when K is not small (like 2 or 3). We might have $K = 10, 15,$ or 20 (or more). If we do have 20 values of the same type of parameter (perhaps for 20 years or sites), it is likely that there is some "explainable" (i.e., consistent, simple, and understandable) smooth, low-level pattern, such as a linear trend, to the variation in these parameters. However, to be consistent with our philosophy of models, we must admit that the actual values of the 20 parameters will not perfectly fit such a simple model (a two-parameter linear trend). There will be unignorable yet unexplainable residual variation in the parameters; the modeling issue is that of how much of this residual variation we can detect with the data. If the (unknown) residuals behave like *iid* random variables, then random-coefficient models can be very effective tools for data analysis when there are large numbers of the same types of parameters. The explainable variation is fit by smooth, parsimonious structural models, and the unexplainable (not smooth) variation in $\theta_1, \ldots, \theta_K$ is swept into σ^2.

Conceptually, we still think of a parsimonious structural model imposed on $\underline{\theta}$, but one allowing homoskedastic residuals. For example, we may have reason to try the constrained model structure $\underline{\theta} = X\underline{\psi}$ (for known covariate regressors in matrix X), but we think that this model would not exactly fit the θ_i even if we could apply it to those exact θ_i. If the model did have σ nearly 0 (relative to the size of components of $\underline{\psi}$), we could safely use the deterministic interpretation of $\underline{\theta}$ and define any new structural model by simple constraints and directly get the likelihood for the new parameters $\underline{\psi}$ as

$$\mathcal{L}(\underline{\psi}) = \mathcal{L}(\underline{\theta} \mid \underline{\theta} = X\underline{\psi});$$

$\underline{\psi}$ would have only, say, 1 to 4 components.

It is often no more reasonable to assume that $\underline{\theta} = X\underline{\psi}$ is exact for unobservable $\underline{\theta}$ than it is to assume that $Y = \underline{Z}'\underline{\theta}$ is exact for the observable random variable Y. In both cases we must allow that the structural model may not fit exactly. For the case of random-coefficient models we must now have a second model $h(\underline{\theta} \mid \underline{\psi})$ imposed on $\underline{\theta}$ in terms of fixed parameters $\underline{\psi}$. We might conceptualize $\underline{\theta}$ as a normal random variable with $\underline{\theta} = X\underline{\psi} + \underline{\delta}$, $\bar{E}(\underline{\delta}) = \underline{0}$, and variance–covariance matrix of $\underline{\delta}$ as $\sigma^2 I$. This serves to define $h(\underline{\theta} \mid \underline{\psi})$, and then we compute the actual parsimonious reduced model for the fixed parameters as

$$g(\underline{x} \mid \underline{\psi}, \sigma^2) = \int g(\underline{x} \mid \underline{\theta}) h(\underline{\theta} \mid \underline{\psi}, \sigma^2) d\underline{\theta}.$$

It is more likely that we will be faced with mixed models: some fixed and some random parameters in one or more of our models. To make this idea explicit we extend our notation by partitioning the generic parameter vector $\underline{\theta}$ into two parts: $\underline{\theta} = (\underline{\alpha}', \underline{\beta}')'$ with $\underline{\alpha}$ fixed and $\underline{\beta} = (\beta_1, \ldots, \beta_{K_\beta})'$ random (we of course may choose to consider $\underline{\beta}$ as fixed for some models). As above we will have a model imposed on $\underline{\beta}$ in terms of a distribution for β_i as a random variable, hence $h(\underline{\beta} \mid \underline{\psi}, \sigma^2)$. The needed marginal distribution (model) is

$$g(\underline{x} \mid \underline{\alpha}, \underline{\psi}, \sigma^2) = \int g(\underline{x} \mid \underline{\alpha}, \underline{\beta}) h(\underline{\beta} \mid \underline{\psi}, \sigma^2) d\underline{\beta}.$$

Further generalizations are possible, but for our purposes here we will stick to the simple case, hence (6.1) and issues of likelihoods for fixed-effects reduced models,

$$\mathcal{L}(\underline{\psi}) = \mathcal{L}(\underline{\theta} \mid \underline{\theta} = X\underline{\psi}), \tag{6.2}$$

versus random-effects reduced models, hence

$$\mathcal{L}(\underline{\psi}, \sigma^2) = g(\underline{x} \mid \underline{\psi}, \sigma^2), \tag{6.3}$$

where computing (6.3) requires the multidimensional integration illustrated by (6.1). The MLE $\hat{\underline{\psi}}$ from these two models (6.2 vs. 6.3) will be essentially the same value even if $\hat{\sigma}^2 > 0$ occurs ($\hat{\sigma}^2$ might be zero). However, the two likelihood functions will differ. Therefore, AIC values for these two models will be different, and it is important to consider the random-effects model as well as its more restrictive fixed-effects version (wherein $\sigma^2 = 0$ is assumed) in sets of models fit to data. Also, the variation represented by σ^2 may be of interest in its own right. This is the case in Section 3.5.6, where the random coefficient model is indirectly fit to real sage grouse survival data to estimate the process variation, σ^2, from a set of annual survival rates.

One classical basis for the concept and use of random effects is from designed experiments where levels of some factor are selected randomly from a defined population, for example animals from a herd, farms in a country, or corn cultivars from a population of cultivars (see, e.g., Cox and Reid 2000). There is then a well-defined inference to a real population. However, random coefficient models need not have this "random selection" feature, especially when time is involved. The set of K annual survival probabilities, $S_1, \ldots, S_K$, for consecutive yearly time intervals 1 to K does not correspond to a randomly selected set of years, and they never could. In this case the "random" in random effects modeling is conceptual and refers to the envisioned residuals for a suitable low-level smoothing model imposed on the parameters, such as $S_i = \alpha_1 + i\alpha_2 + \epsilon_i$ or even just $S_i = \alpha + \epsilon_i$. If the imposed smoothing (model) captures the explainable variation in the survival parameters over time, then we expect the residuals $\epsilon_i, \ldots, \epsilon_K$ to have the properties of exchangeable random variables. Hence, we can treat them as independent, identically distributed with $E(\epsilon) = 0$ and $var(\epsilon) = \sigma^2$. It is these residuals, relative to some low-level smoothing, that are the random effects. Therefore, there is no concept, or requirement, that the years (hence the S_i) be in any way selected at random from a defined population of years.

In collapsing the problem to $g(\underline{x} \mid \psi, \sigma^2)$ we are restricting our inference to ψ and σ^2, hence ignoring the original individual $\theta_1, \ldots, \theta_K$. We may want to "have our cake and eat it too," that is, get estimators of $\underline{\theta}$ as well as ψ and σ^2. This can be accomplished using shrinkage estimators. Shrinkage estimators $\tilde{\underline{\theta}}$ arise in both Bayesian and frequentist theories (see, e.g., Efron and Morris 1975, Morris 1983, Longford 1993, Casella 1995, and Carlin and Louis 1996). Shrinkage estimators, $\tilde{\underline{\theta}}$, for random-coefficient models such as $\underline{\theta} = X\psi + \underline{\delta}$, $E(\underline{\delta}) = \underline{0}$, $E(\underline{\delta}\underline{\delta}') = \sigma^2 I$ can be based on the MLE of $\hat{\underline{\theta}}$ under model g_K in such a way that the residuals from direct simple linear regression of $\tilde{\underline{\theta}}$ on $X\psi$ reproduce the estimate of σ^2 computed in obtaining $\tilde{\underline{\theta}}$. This can be interpreted as saying that we may be able to find a suitable proxy for the maximized likelihood of the fitted model in (6.3), $\mathcal{L}(\hat{\psi}, \hat{\sigma}^2)$, by use of the original $\mathcal{L}(\underline{\theta})$ evaluated at such a shrinkage estimator, hence via $\mathcal{L}(\tilde{\underline{\theta}})$.

It would be a considerable advantage if such random-coefficient models could be fit without ever computing the integral in (6.1). This would allow a practical approach to getting a nearly correct AIC value for the model $g(\underline{x} \mid \psi, \sigma^2)$ yet based on $\log(\mathcal{L}(\tilde{\underline{\theta}}))$. As it is, the shrinkage approach is a pragmatic way to fit what amounts to model $g(\underline{x} \mid \psi, \sigma^2)$, thus getting an estimate of $\underline{\theta}$ subject to the stochastic "constraint" inherent in the random coefficients model, without making distributional assumptions.

If we do have to compute the integrals as in (6.1), it is certainly possible using the recent developments from Bayesian methods; see, e.g., Gelfand and Smith (1990), Zeger and Karim (1991), and Carlin and Chib (1995).

6.6.2 A Shrinkage Approach to AIC and Random Effects

Making inferences about all the random and fixed effects parameters in a random effects model can be accomplished by Bayesian methods (Gelman et al. 1995), empirical Bayes methods (Carlin and Louis 1996), or frequentist shrinkage methods (Tibshirani 1996, Royle and Link 2002). However, only recently has formal model selection been considered in conjunction with Bayesian

MCMC methods for models that include random effects (Spigelhalter et al., 2002) or for AIC-model selection in conjunction with frequentist shrinkage methods (Burnham in review, Burnham and White 2002). The purpose of this section is to give an overview of one way AIC can be extended to models with simple random effects based only on the likelihood $\mathcal{L}(\underline{\alpha}, \underline{\theta})$. This extension is not straightforward because we need to consider the θ_i as random, even though in this likelihood they are technically to be considered as fixed effects. It is not proposed that this is the ultimate methodology we should use for AIC-type model selection with random effects. But it is a practical method when only one factor is a random effect and sample size is not small.

Fixed effects inference based on the likelihood $\mathcal{L}(\underline{\alpha}, \underline{\theta}) \equiv \mathcal{L}(\underline{\alpha}, \underline{\theta} | data, g)$ focuses on the MLE $\hat{\underline{\theta}}$, which is considered conditional on $\underline{\theta} = (\theta_i, \ldots, \theta_k)'$, a k dimensional vector. This MLE has conditional sampling variance–covariance matrix W (it may depend on $\underline{\theta}$) which applies to $\epsilon = \hat{\underline{\theta}} - \underline{\theta}$. We augment model g, and hence $\mathcal{L}(\underline{\alpha}, \underline{\theta})$, with a random-effects model wherein $\underline{\theta} = X\beta + \underline{\delta}$, $E(\underline{\delta}) = \underline{0}$, $VC(\underline{\delta}) = \sigma^2 I$; $\underline{\beta}$ is an r-dimensional vector ($r < k$). It then follows that an unconditional structural model applicable to the (otherwise conditional) MLE is

$$\hat{\underline{\theta}} = X\underline{\beta} + \underline{\delta} + \underline{\epsilon}, \qquad VC(\underline{\delta} + \underline{\epsilon}) = D = \sigma^2 I + E_{\underline{\theta}}(W);$$

VC means variance-covariance matrix. Parameters $\underline{\beta}$, as well as $\underline{\alpha}$, are fixed effects. In practice, $\hat{E}_{\underline{\theta}}(W) = \hat{W}$ from standard likelihood inference methods.

From generalized least square theory, for σ^2 known, the best linear unbiased estimator of $\underline{\beta}$ is

$$\hat{\underline{\beta}} = (X'D^{-1}X)^{-1}X'D^{-1}\hat{\underline{\theta}}.$$

Assuming normality of $\hat{\underline{\theta}}$ (approximate normality suffices), the weighted residual sum of squares $(\hat{\underline{\theta}} - X\hat{\underline{\beta}})'D^{-1}(\hat{\underline{\theta}} - X\hat{\underline{\beta}})$ has a central chi-squared distribution on $k - r$ degrees of freedom. Therefore, a method of moments estimator of σ^2 is obtained by solving the equation

$$k - r = (\hat{\underline{\theta}} - X\hat{\underline{\beta}})'D^{-1}(\hat{\underline{\theta}} - X\hat{\underline{\beta}}).$$

Under random effects for inference about $\underline{\theta}$ we use shrinkage estimates $\tilde{\underline{\theta}}$, not the MLE (see e.g., Burnham in review). Shrinkage is a type of generalized smoothing. Computing the shrinkage estimator requires the matrix

$$H = \sigma D^{-1/2} = \sigma \left(\sigma^2 I + \hat{E}_{\underline{\theta}}(W)\right)^{-1/2} = \left(I + \frac{1}{\sigma^2}\hat{E}_{\underline{\theta}}(W)\right)^{-1/2},$$

evaluated at $\hat{\sigma}$. Then $\tilde{\underline{\theta}} = H(\hat{\underline{\theta}} - X\hat{\underline{\beta}}) + X\hat{\underline{\beta}}$. Move informatively, let $G = H + (I - H)AD^{-1}$, where $A = X(X'D^{-1}X)^{-1}X'$. Then G is a projection matrix such that $\tilde{\underline{\theta}} = G\hat{\underline{\theta}}$. This shrinkage estimator is such that the sum of squares of the shrunk residuals (i.e., $\tilde{\underline{\theta}} - X\hat{\underline{\beta}}$), divided by $k - r$, equals $\hat{\sigma}^2$.

Because of how it is computed $\tilde{\theta}$ essentially "contains" $\hat{\beta}$ and $\hat{\sigma}^2$, and this is the key to being able to compute a likelihood value for the fitted random effects model from $\tilde{\theta}$ using only the fixed effects likelihood evaluated at $\tilde{\theta}$.

Let ℓ be the dimension of $\underline{\alpha}$. Then as a fixed effects model $K = k + \ell$ and AIC $= -2 \log \mathcal{L}(\hat{\underline{\alpha}}, \hat{\underline{\theta}}) + 2K$. The random effects log-likelihood value is taken as

$$\log \mathcal{L}(\hat{\underline{\alpha}}, \tilde{\underline{\theta}}) \equiv \log \mathcal{L}(\hat{\underline{\alpha}}(\tilde{\underline{\theta}}), \tilde{\underline{\theta}}) = \max_{\underline{\alpha}}[\log \mathcal{L}(\underline{\alpha}, \tilde{\underline{\theta}})].$$

Reoptimizing over α at fixed $\tilde{\underline{\theta}}$ is necessary. The dimension of the parameter space to associate with this random effects model is K_{re}, when

$$K_{re} = \text{tr}(G) + \ell.$$

We note that $1 \leq \text{tr}(G) \leq k$. This corresponds to the fact that the random-effects model for $\theta_1, \ldots, \theta_k$ is intermediate between a model in which the variation among $\theta_1 = \cdots = \theta_k$ is unstructured and unrestricted and the no-effects model wherein $\theta_1 = \cdots = \theta_k$.

AIC for the random effects model on $\underline{\theta}$ is $-2 \log \mathcal{L}(\hat{\underline{\alpha}}, \tilde{\underline{\theta}}) + 2K_{re}$. The general small sample version is

$$\text{QAIC}_c = \frac{-2 \log \mathcal{L}(\hat{\underline{\alpha}}, \tilde{\underline{\theta}})}{\hat{c}} + 2K_{re} + 2\frac{K_{re}(K_{re} + 1)}{n + K_{re} - 1}.$$

If it is justified to take $\hat{c} = 1$, then the above becomes AIC$_c$. Some evaluation of this methodology is given in Burnham and White (2002), including many inference formulae not given here.

Examples are given in Burnham (in review). In particular there is a band recovery example for which $k = 41$ (units are years); θ is the annual survival probability. The fixed effects model means fitting 40 more parameters than just a mean (μ) of the θ_i merely to "capture" average annual variation that might be better represented by a single parameter σ^2, which is of interest in its own right. The full model also requires 42-band recovery rate parameters, r_i. Results for three models (the r_i are unrestricted) are below ($\hat{c} = 1.195$, small but justified because sample size was 42,015 birds banded):

Model	K	ΔQAIC	Akaike weight	survival model
$\theta_{\mu,\sigma}$	73.26	0.00	0.9984	random time effects
θ_t	83	12.87	0.0016	fixed-time effects
θ	43	100.11	0.0000	time-constant θ

Summary points we want to make from this section: First, AIC can be generalized to random-effects models, ultimately probably in way better than given here. Second, the correct measure of parameter dimension under random effects is the trace of the assoicated projection (smoothing) matrix, G (this way of computing K_{re} is not unique). Extensions of AIC to other nonstandard (i.e., not simple, fixed effects) models often require computing K from the trace of a matrix.

6.6.3   On Extensions

Kullback–Leibler-based model selection is well developed, studied, and understood for models that correspond to fixed-effects likelihoods. Also, as shown in Section 6.6.2 it is applicable to, and has been developed for, simple random effects models. Extensions of AIC-type model selection (i.e., model weights are with respect to a best approximating model, not a true model) and multimodel inference to other contexts are mostly state-of-the-art, but are ongoing successfully, Our point here is that AIC does not dead end at simple fixed-effects models. It has a wider spectrum of applications, some of which we will mention here, but without details.

Data smoothing by semi- or non-parametric methods, is an active subject area in statistics. Hurvitch et al. (1998) provide versions of AIC_c for smoothing parameter selection on nonparametric regression (see also Naik and Tsai 2001). For data vector $\underline{y}$ Hurvitch et al. (1998) note that the smoothed data correspond to $\hat{\underline{y}} = H\underline{y}$ for a smoothing matrix H that must be determined as a function of a smoothing constant; this constant is not analogous to a single traditional parameter. Moreover, for several AIC-type smoothing-constant selectors "Each of these selectors depends on H through its trace, which can be interpreted as the effective number of parameters used in the smoothing fit . . . " (Hurvitch et al. 1998:273). Indeed, the role of the number of structual parameters is taken here by $tr(H)$ and then, generally, $K = tr(H) + 1$ because there is also one variance parameter, σ^2. There are links between AIC for generalized smoothing and generalized cross-validation as well as generalized additive models (Hastie and Tibshirani 1990:49, 158), and in each case the trace of a smoother matrix takes the role of the number of fixed-effects structural parameters.

Other general statistical-modeling methodologies to which K-L-based model selection has been extended include generalized estimating equations (Pan 2001a,b). The method of Pan also serves to extend K-L to quasi-likelihood modeling. Robust regression is sometimes approached using least-absolute deviations (also called L1 regression, a type of quantile regression). Hurvich and Tsai (1990a) give the small-sample AIC for L1 regression; they denote it as L1cAIC. The formula for L1cAIC is very different from AIC_c. However, one of their conclusions was that both AIC_c and L1cAIC (Hurvich and Tsai 1990a:263) ". . . provide good model selections in small samples from a linear regression model with double exponential errors." This supports our recommendation that AIC_c is useful in general. Recent work on AIC-type model selection for robust and nonparametric regression has been done by Shi and Tsai (1998, 1999), Hurvich and Tsai (1998), and Simonoff and Tsai (1999).

A general approach to K-L model selection when the models include random effects remains elusive. However, the recent deviance information criterion (DIC) approach within a Bayesian framwork may provide one solution (Spigelhalter et al. 2002). As with AIC the intent of DIC is to select a best model, not the true model. DIC does not use Bayes factors and behaves like AIC rather than like BIC. The main disadvantage of DIC is that it requires an MCMC approach to model fitting. Heuristically, the basic idea seems to be to minimize the posterior expected value of (relative) Kullback–Leibler information loss (see also Thabane and Haq 1999, regarding this idea). Fincally, a general likelihood approach to fitting and selecting among models that include random effects (without integrating out the random effects) may be possible within the framework of h-likelihood (Lee and Nelder 1996, Ha 2001). Research along this line seems especially worthwhile. Some aspects of the results in Section 6.6.2 combined with ideas from the DIC approach suggest to us that a likelihood solution may be possible (i.e., AIC for general random-effects models).

6.7 Selection When Probability Distributions Differ by Model

6.7.1 Keep All the Parts

Most model selection focuses on questions about model structure, i.e., explainable variation in data, within the context of a single assumed probability distribution. For example, if y is the response variable and we have potential predictors x_1 to x_6, we may assume that the structural aspect of any model is some functional form for $E(y|\text{predictors}$ based on the $x_i)$; the x_i may be transformed with impunity. The (so-called) error distribution is placed on the residual $\epsilon = y - E(y|\text{predictors})$, i.e., unexplained variation. Kullback–Leibler-based model selection allows the error distribution to vary over models. For example, we can compare models wherein ϵ has a normal distribution to models wherein ϵ has a lognormal distribution. However, care must be taken when so doing: No component part of either probability density function can safely be dropped in forming the likelihoods. Also, the comparison cannot be based on y for some models and $\log(y)$ for other models (see also Section 2.11.3). We can contrast models for y as normal versus log-normal, but this must be done in a correct way.

Denote the model structural aspects by $\mu_i = E(y_i|\text{predictors})$. The μ_i will depend functionally on some smaller number of parameters. Assuming a normal probability distribution, variance homogeneity, and independence, the likelihood, as $\prod g_1(y_i|\mu_i, \sigma)$, is

$$\mathcal{L}_1 = \left[\frac{1}{\sqrt{2\pi}}\right]^n \left[\frac{1}{\sigma}\right]^n \exp\left[-\frac{1}{2}\sum_{i=1}^{n}\frac{(y_i - \mu_i)^2}{\sigma^2}\right]. \tag{6.4}$$

As long as every model considered is concerned just with modeling the μ_i given the assumed normality of y, then we can drop from $\mathcal{L}_1$ the constant involving 2π. This is a general principle: If the likelihood for each model arises as a special case of a global likelihood (hence global model), then we can drop (but do not have to) any component term not involving the parameters and comparisons based on relative expected $\hat{K}$-L distance (i.e., AIC and its siblings) remain valid.

As another example assume that you think that the probability distribution for y should be a special case of the gamma: $g_2(y|\mu) = (y/\mu^2)\exp(-y/\mu)$. Then,

$$\mathcal{L}_2 = \left[\prod_{i=1}^{n} y_i\right]\left[\prod_{i=1}^{n} \frac{1}{\mu_i}\right]^2 \exp\left[-\sum_{i=1}^{n} \frac{y_i}{\mu_i}\right], \tag{6.5}$$

and as long as this is the global likelihood, the leading term in the y_i's can be dropped. However, to compare g_1 to g_2 no parts of either can be dropped.

Keep All the Parts to Compare Different Distributions

To compare two models g_1 and g_2 that are based on different probability distributions (those models may have the same or different structures on the μ_i) we must keep in $\mathcal{L}_1$ and $\mathcal{L}_2$ all component parts arising from the underlying probability distributions.

If in both (6.4) and (6.5) we dropped the leading term, then comparison of resultant AIC_1 to AIC_2 is invalid: it gives meaningless results. Heuristically, this is because we would have confounded real model "effects" with differences arising just because $\log\left[\frac{1}{\sqrt{2\pi}}\right]^n$ and $\log\left[\prod_{i=1}^{n} y_i\right]$ are different.

There is a link here to comparing models on data transformations, which also cannot be done directly. For example, rather than compare assumed normal distribution models for y versus $\log(y)$, we must compare models for y wherein y can have either a normal or log-normal distribution. The easy way to do this is use software for generalized linear models, and that is what we recommend. Under GLM (McCullagh and Nelder 1989) it is not the data that are transformed, rather it is the parametric part of the model that is transformed and linked with different assumed "error" distributions.

6.7.2   A Normal Versus Log-Normal Example

To compare a model based on the normal distribution (g_1) to one based on the log-normal distribution (g_2) we have to use the log-normal form below:

$$g_2(y|\theta, \sigma) = \frac{1}{y\sigma\sqrt{2\pi}} \exp\left[-\frac{1}{2}\frac{[\log(y) - \theta]^2}{\sigma^2}\right].$$

(Again, a GLM approach would be doing this for you without you having to know the correct probability density form for y being log-normal). The likelihood, in general, is now

$$\mathcal{L}_2 = \left[\prod_{i=1}^{n} \frac{1}{y_i}\right]\left[\frac{1}{\sqrt{2\pi}}\right]^n\left[\frac{1}{\sigma}\right]^n \exp\left[-\frac{1}{2}\frac{\sum_{i=1}^{n}[\log(y_i) - \theta_i]^2}{\sigma^2}\right]. \tag{6.6}$$

To have the parameterization $\mu = E(y)$ with the log-normal distribution we must set $\theta_i = \left[\log(\mu_i) - \sigma^2/2 \right]$ in (6.6). Then any two models for the μ_i can be compared based on likelihoods (6.4) and (6.6) for normal and log-normal distributions, respectively. It will generally make more sense, we think, to compare the same structural models; hence the comparison is concerned just with a plausible probability model for the data. (Note, however, that if the primary focus of data analysis is the question of a suitable structural model, then the issue of what to use for the error distribution part of the model is not a major matter, as long as the distribution chosen is not a terrible approximation).

We provide results of a small Monte Carlo study to select either the normal or log-normal model, based on likelihoods in (6.4) and (6.6), under the structural model wherein the μ_i are constant, and hence $\mu_i \equiv \mu$. The data were generated from either the normal or log-normal model, with the scale parameter σ fixed at 1 (results are scale invariant). Because both alternative models have $K = 2$, issues of using AIC or AIC$_c$ (or even BIC) are moot. Table 6.14 shows results in terms of selection relative frequency of the normal distribution model and expected (i.e., average) value of the Akaike weight for the normal model. Results in Table 6.14 are accurate to the two decimal places shown.

This likelihood-based discrimination between the normal and half-normal distributions performs well (Table 6.14). The two distributions differ more at small values of $E(y)$ (which actually should be interpreted here as being $E(y/\sigma)$). However, these distributions are increasingly similar as $E(y)$ increases. Correspondingly, the two distributions can be well distinguished, given sufficient sample size, if they are meaningfully different. Discrimination

TABLE 6.14. Results from 10,000 Monte Carlo trials to select between the simple normal and log-normal models, variance$(y) = 1$ for both distributions; π is the percentage of cases wherein the normal model was selected, $E(w)$ is the average of the Akaike weight for the normal model, given as a percentage.

| | | | | E(y), generating model: normal | | | | |
| | 5 | | 10 | | 25 | | 50 | |
n	π	$E(w)$	π	$E(w)$	π	$E(w)$	π	$E(w)$
10	58	54	53	51	52	50	51	50
50	77	71	65	57	56	51	53	50
100	88	83	72	63	59	53	54	51
500	100	100	92	88	71	62	61	54

| | | | | E(y), generating model: log-normal | | | | |
| | 5 | | 10 | | 25 | | 50 | |
n	π	$E(w)$	π	$E(w)$	π	$E(w)$	π	$E(w)$
10	38	44	44	49	48	50	48	50
50	21	29	34	43	43	49	47	50
100	12	17	27	36	41	47	45	49
500	0	1	9	13	29	38	39	47

is difficult at $E(y) = 50$, but it also does not then matter which probability model is assumed as the basis of data analysis (because they are then nearly identical distributions).

The π reflect sampling variation, whereas the $E(w)$ reflect inferential uncertainty. Sampling variation and inferential uncertainty are conceptually very different in general, and numerically not identical in Table 6.14, except for some cases. Notice that the average inferential uncertainty actually exceeds sampling variation. That is, $E(w)$ is always intermediate between 50% and $100\pi\%$ in Table 6.14.

It is also worth noting again that these results are equivalently considered as arising from AIC or BIC (because $K = 2$ for both models). Therefore, we feel entitled (pragmatically) to interpret w_i, for large samples, as the probability that model g_i is the K-L best model, but not that it is the true model. To further illustrate this last point about inferring truth, we generated samples from the negative exponential distribution $g_3 = \exp(-y/\lambda)/\lambda$ and did selection between the normal and log-normal models. The parameter $\lambda = E(y)$ is a scale parameter so results are invariant to its value. Hence, we need only present model selection results for different sample sizes. As with Table 6.14, results are for selection of the normal model as the best approximating model:

n	π	$E(w)$
5	24	32
10	17	22
20	10	12
50	2	3
100	0	0

For n greater than about 100 the result is selection of the log-normal model, essentially with no inferential uncertainty. Of course, that model did not generate the data: Selecting a model with (inferential) certainty does not mean that the model is truth.

6.7.3   Comparing Across Several Distributions: An Example

Lindsey and Jones (1998) gives an example based on observed T_4 cell counts per cubic millimeter of blood. This is a type of leukocyte cell that is part of the immune system. The data are from 20 patients in remission from Hodgkin's disease (considered "treatment" here) and 20 patients as unmatched controls (their Table I, shown here in Table 6.15). The interest is in the average difference of this white cell count between the two groups of patients. Means and standard deviations are $\bar{y}_T = 823$, $\bar{y}_C = 522$, $s_T = 566$, and $s_C = 293$. The distribution to use as the basis of parametric inference is problematic and might be ignored in favor of distribution-free inference, except that the sampling distribution across patients for this type of count is of interest in its own right.

Standard practice has been to assume a normal distribution and compute a t-test, or preferably, a point estimate and a confidence interval, possibly based on

TABLE 6.15. The data on T_4 blood cell counts per mm^3, from Lindsey and Jones (1998); treatment data are from patients in remission from Hodgkin's disease, controls are patients in remission from a non-Hodgkin's disease.

Treatment				Control			
171	397	795	1212	116	375	440	736
257	431	902	1283	151	375	503	752
288	435	958	1378	192	377	675	771
295	554	1004	1621	208	410	688	979
396	568	1104	2415	315	426	700	1252

TABLE 6.16. AIC results for the T_4 blood cell count data (Table 6.15), from Lindsey and Jones (1998, Table II), "difference" means a treatment effect, hence $\mu_T \neq \mu_C$; see text for more explanation.

Model	K	No difference AIC	Δ	K	Difference AIC	Δ
normal	2	608.8	22.8	3	606.4	20.4
log-normal	2	590.1	4.1	3	588.6	2.6
gamma	2	591.3	5.3	3	588.0	2.0
inverse Gaussian	2	590.0	4.0	3	588.2	2.2
Poisson	1	11652.0	11066.0	2	10294.0	9708.0
negative binomial	2	589.2	3.2	3	586.0	0.0

the t-distribution. However, one might postulate several possible distributions as the basis of the model and then use AIC to compute a weight of evidence for the suitability of each model. As long as we compute each log-likelihood based on the complete probability distribution (i.e., no dropped parts) this approach is valid, as noted in Lindsey and Jones (1998). The AIC results given by Lindsey and Jones (their Table II) for models he considered are shown here in Table 6.16. The parameters of these models are either for expected values, $E(y) = \mu$, or are a dispersion parameter such as for the normal distribution. Within a distribution there is a pair of models: either $\mu_T = \mu_C$ (no difference) or $\mu_T \neq \mu_C$. We will show aspects of calculation of three AIC values in Table 6.16.

For the Poisson model when $\mu_T = \mu_C = \mu$ the probability distribution is given by

$$g(y|\mu) = \frac{e^{-\mu}\mu^y}{y!}.$$

Hence the likelihood is to be taken here as (T is the sum of all 40 counts)

$$\mathcal{L}(\mu) = \prod_{i=1}^{40} \frac{e^{-\mu}\mu^{y_i}}{y_i!} = \left[\prod_{i=1}^{40} \frac{1}{y_i!}\right]\left[e^{-40\mu}\mu^T\right]$$

$$= \exp\left[-\sum_{i=1}^{40} \log(y_i!)\right]\left[e^{-40\mu}\mu^T\right]$$

$$= \exp(-195150.88)\left[e^{-40\mu}\mu^{26905}\right].$$

The huge term involving factorials could be ignored (dropped) if this were our only model and inference were just about μ (given this model), because we would care only about the ratios $\mathcal{L}(\mu)/\mathcal{L}(\hat{\mu})$. But we must keep all the parts to make a comparison of this model (i.e., probability distribution) to a different probability distribution. The MLE is $\hat{\mu} = 26905/40 = 672.625$; $\log\mathcal{L}(\hat{\mu}) = -5824.9478$ and AIC $= 11651.896$, rounded to 11652 in Table 6.16.

Consider the normal model for the no-difference case. The likelihood to use is given by (6.4). The MLEs are $\hat{\mu} = 672.625$ and $\hat{\sigma}^2 = 215824.28 = (464.57)^2$; $\log\mathcal{L}$ at the MLEs is given by

$$\log\mathcal{L} = -\frac{n}{2}\log(2\pi) - \frac{n}{2}\log(\hat{\sigma}^2) - \frac{n}{2} = -302.4019.$$

Hence, here AIC $= -2\log\mathcal{L} + 4 = 608.8$.

For the normal model with a difference by treatment group, but common dispersion parameter σ^2, the likelihood is

$$\mathcal{L} = \left[\frac{1}{\sqrt{2\pi}}\right]^{40}\left[\frac{1}{\sigma}\right]^{40}\exp\left[-\frac{1}{2}\frac{\sum_{i=1}^{20}(y_{T_i} - \mu_T)^2 + \sum_{i=1}^{20}(y_{C_i} - \mu_C)^2}{\sigma^2}\right].$$

The MLEs are $\hat{\mu}_T = \bar{y}_T = 823$, $\hat{\mu}_C = \bar{y}_C = 522$ and $\hat{\sigma}^2 = (SS_T + SS_C)/40 = 193151.45 = (439.49)^2$ (SS denotes sum of squares). The log-likelihood for this model, evaluated at the MLEs, reduces to the same form as for the no-difference model case, but $\hat{\sigma}^2$ is different:

$$\log\mathcal{L} = -\frac{n}{2}\log(2\pi) - \frac{n}{2}\log(\hat{\sigma}^2) - \frac{n}{2} = -300.182,$$

AIC $= 606.364$.

To consider the question of the best approximating model we should not make comparisons that confound structural and stochastic model components. So in Table 6.16 we must compare within model structures (within columns). The correct comparison is automatic when we use an evidence ratio by structural model pairs. For example, to compare the normal and log-normal distributions the two evidence ratios are

$$11499 = \frac{e^{-4.1/2}}{e^{-22.8/2}}, \quad \text{forcing } \mu_T = \mu_C,$$

$$7332 = \frac{e^{-2.6/2}}{e^{-20.4/2}}, \quad \text{allowing } \mu_T \text{ and } \mu_C \text{ to differ.}$$

Either one of these ratios is overwhelming evidence against the normal distribution, relative to the log-normal, as a best approximating distribution. While this says that we can discard the normal distribution here, we do not know how good the log-normal is.

In practice, to address the distribution issue we would have computed only the difference-allowed cases. There it is clear that the best model is negative binomial, but the log-normal (evidence ratio in favor of the negative binomial is 3.7), gamma (2.7), and inverse Gaussian (3.0) are plausible competitors. Akaike weights for this subset of four models (ordered as in Table 6.16) are 0.14, 0.18, 0.17, 0.51.

A final observation here is that inference about the "treatment" effect is not much affected by choice of model, for models that fit well (the Poisson model is a terrible fit). For the other model pairs, evidence ratios in favor of an effect are

model	ER for an effect
normal	3.3
log-normal	2.1
gamma	2.7
inverse Gaussian	2.5
negative binomial	5.0

How would we analyze these data for an effect? First, we would not include the no-effect models. Second, we would reparametrize each model structure from μ_T and μ_C to $\delta = \mu_T - \mu_C$ and $\gamma = \mu_T + \mu_C$. Third, fitting the data to each reparametrized model leads to $\hat{\delta}$, $\widehat{se}(\hat{\delta}|g)$, and Akaike weights for each of the six models (those w_i are 0.000, 0.138, 0.186, 0.169, 0.000, 0.507). Fourth, we would base inference about δ on model averaging (note: serious inference requires substantially more data).

6.8   Lessons from the Literature and Other Matters

6.8.1   Use AIC$_c$, Not AIC, with Small Sample Sizes

It is far too common that papers examining AIC, by itself or compared to BIC, fail to use AIC$_c$ when the latter must be used because the number of parameters, at least for some models considered, is not small relative to sample size. For example, Chatfield (1996) considered model selection issues and used a time series example with $n = 132$ and $R = 12$ a priori designated models wherein K ranged from 6 to 61. Overall we commend the paper; however, in this particular example the conclusion that AIC performed poorly is misleading. AIC did do poorly; but it is well known, documented, and commented on in the literature on K-L–based model selection that in such an example it is imperative to use AIC$_c$, not AIC (e.g., Sakamota et al. 1986, Bozdogan 1987, Hurvich and Tsai 1989, Hurvich et al. 1990).

Recall that

$$\text{AIC}_c = \text{AIC} + 2\frac{K(K+1)}{n-K-1} = -2\log(\mathcal{L}) + 2K + 2\frac{K(K+1)}{n-K-1}.$$

Consider for $n = 132$ the effect of the bias-correction term for $K = 6$ and 61 for these two models with likelihoods denoted by $\mathcal{L}_6$ and $\mathcal{L}_{61}$:

K	AIC	AIC_c
6	$-2\log(\mathcal{L}_6) + 12$	$-2\log(\mathcal{L}_6) + 12.672$
61	$-2\log(\mathcal{L}_{61}) + 122$	$-2\log(\mathcal{L}_{61}) + 230.057$

The difference here between AIC and AIC_c is huge for $K = 61$, and this will greatly affect which model is selected.

We present in Table 6.17 the results that Chatfield (1996) should have presented as regards K-L–based model selection versus BIC. In so doing we also use Δ_i values, not absolute values of these model selection criteria. We do show for comparison the ΔAIC values implicitly used by Chatfield. The results in Table 6.17 are based on the results in Table 1 of Chatfield (1996). The nature of the models need not concern us, so we label them just 1 to 12, but keep them in the same order as used in Table 1 of Chatfield. The Δ_i values in Table 6.17 for the AIC_c criterion do have here the interpretations and uses described in Section 2.6. Those interpretations are not true in this example for the Δ_i derived here from AIC, because for large K relative to n, AIC is too biased an estimator of the expected K-L distance.

As noted by Chatfield, in this example AIC and BIC lead to very different selected models. However, AIC_c (which must be used here) gives (seemingly) acceptable results. In fact, the Akaike weights here (see Table 6.17) show that only four fitted models have any plausibility in this set of 12 fitted models.

TABLE 6.17. The ΔAIC_c that must be used for K-L model selection on the 12 models considered in Table 1 of Chatfield (1996), and corresponding ΔAIC and ΔBIC values; also, the Akaike weights based on AIC_c.

Model	K	ΔAIC_c	ΔAIC	ΔBIC	$w_i\,\text{AIC}_c$
1	6	4.5	68.0	**0.0**	0.048
2	11	**0.0**	62.0	13.0	0.459
3	21	2.5	58.3	47.0	0.132
4	9	86.1	148.8	92.2	0.000
5	17	98.6	157.4	130.9	0.000
6	41	155.5	181.4	246.0	0.000
7	11	83.7	145.7	96.7	0.000
8	21	94.7	150.5	139.3	0.000
9	13	0.5	61.6	20.0	0.358
10	25	10.1	62.0	65.7	0.003
11	31	15.9	60.3	86.8	0.000
12	61	43.9	**0.0**	139.8	0.000

Those models are (in order of their likelihood) g_2 ($K = 11$), g_9 ($K = 13$), g_3 ($K = 21$), and g_1 ($K = 6$). Thus the evidence, as interpreted with AIC_c, definitely eliminates 8 of the 12 models, because relative to other models in the set, they are extremely implausible. Also, most of the weight of evidence is put on models with low K. Based on comparing AIC to BIC, Chatfield (1996) concluded that BIC was a better criterion than AIC. That conclusion is not justified in that here one must compare AIC_c to BIC for a proper comparison of K-L information-theoretic model selection versus BIC.

6.8.2   Use AIC_c, Not AIC, When K Is Large

Leirs et al. (1997) report the analysis of an extensive set of capture–recapture data from Tanzania on the rat *Mastomys natalensis*. The objective of their data analysis was to examine factors potentially important in the population survival dynamics of the species. The data were collected between October 1986 and February 1989 by live trapping on a 1 ha grid of 100 live trapping positions (several traps per position). There were three consecutive nights of trapping each month (hence 29 primary trapping periods). There were a total of 6,728 captures of 2,481 individual animals. We take the relevant sample size to be the latter, i.e., $n = 2,481$. Leirs et al. (1997) carefully formulate six a priori models to represent how environmental (rainfall) and population density factors might affect survival probabilities (S) and the probability of subadults maturing to adults (ψ). Capture probabilities (p) are another subset of parameters in these models. Data analysis was by ML methods for multistate capture–recapture models (see, e.g., Brownie et al. 1993, Nichols and Kendall 1995) with incorporation of covariates for rainfall and population density. The goodness-of-fit of the global model was quite acceptable ($P > 0.9$).

The global model used by Leirs et al. (1997) allows full (unexplained) temporal variation in all model parameters (hence S and p vary by time and age, and ψ varies by time). In their Table 1, this is model g_1 with $K = 113$ parameters. Their model g_2 is the most restricted model: no temporal variation in the parameters (S and p vary by age only, subadult versus adult, and there is only one maturation probability parameter, ψ). Model 3 allows capture probabilities to vary by time, but S and ψ are not time-varying. Models 4, 5, and 6 allow structured time variation in the three classes of parameters. These latter, quite complex, models are based on population dynamics models melded with the general capture–recapture model. Model 4 has temporal parameter variation as functions only of population density (internal factors only for population regulation); Model 5 has functions only of rainfall (i.e., external factors for population regulation); Model 6 has temporal parameter variation as functions of both population density and rainfall. Leirs et al. (1997) used AIC (not AIC_c) for model selection. In Table 6.18 we present the ΔAIC values from their analyses, as well as results for AIC_c, which we computed, and Akaike weights.

TABLE 6.18. Summary model selection results from, or based on, Table 1 of Leirs et al. (1997); AIC_c should be used here for model selection, not AIC.

Model	K	AIC results		AIC_c results	
		Δ_i	w_i	Δ_i	w_i
1 Global	113	0.0	0.99	0.0	0.76
2 No effects	5	540.7	0.00	529.8	0.00
3 No dynamics	49	207.4	0.00	198.5	0.00
4 Density effects	52	205.8	0.00	197.2	0.00
5 Rainfall effects	55	25.9	0.00	17.6	0.00
6 Rainfall and Density	64	9.7	0.01	2.3	0.24

Even though the sample size is large here (2,481), the fact of having a model with 113 parameters means that AIC_c should be used (and it then must be used for all models). The term added to AIC to get AIC_c for model g_1, $(2 \times 113 \times 114)/2367 = 10.9$, is not trivial. Clearly, using AIC_c here results in a different interpretation of the relative evidence for model g_1 versus g_6. Despite using AIC, Leirs et al. (1997) opted to select model g_6 ($K = 64$) as a useful model and therefore to infer that there were population dynamics occurring that could be substantially explained only by both external (rainfall) and internal (population density) factors. They worried some about this selection (J. D. Nichols, personal communication); they did not need to. Using AIC_c, which should be done here, model g_6 is a tenable model.

6.8.3   When Is AIC_c Suitable: A Gamma Distribution Example

The K-L approach to model selection is exact, philosophically, for any realized sample size (i.e., it is not intrinsically asymptotic). However, its implementation in the face of truth being unknown means some degree of approximation to the target model selection criterion, usually large-sample, must be made to get a practical estimator of this criterion. The simplest solution is to derive asymptotic results, which produces TIC; AIC is a practical and parsimonious implementation of TIC. Simulation studies and experience demonstrate that these "large sample" formulas will perform very poorly as K approaches n, or when n is small. Useful insights to small sample versions of K-L based selection are obtained by deriving exact versions of the target criterion under various assumed-true models (Section 7.4). The results have always been expressible as AIC+ one or more terms of the form $h(K)/(n - m(K))$, for simple functions $h(\cdot)$ and $m(\cdot)$.

The term added to AIC to get AIC_c is just $2K(K + 1)/(n - K - 1)$. While not unique, AIC_c is especially compelling as an omnibus small-sample form of AIC because essentially it requires only that the likelihood function be proportional to a normal distribution. This will be a good approximation even at quite small sample sizes if the sample elements are (nearly) independent and the underlying distribution is unimodal, and neither badly skewed nor heavy tailed. We will present below nearly exact small sample results for the needed adjustment term to AIC for the gamma distribution and compare it to $2K(K+1)/(n-K-1)$. We strongly recommend using a "corrected" version of AIC when K is not large relative to n; use AIC_c unless a better form is known.

Theoretically, no small sample adjustment term is needed in some situations, for example regression when the residual variation, σ^2, is known (hence not estimated). However, performance of model selection will be improved even in this case by use of AIC_c. The added term basically prevents model size K from reaching and exceeding n, which must somehow be enforced.

The theoretical bias of AIC (Section 7.2) is given by

$$2\left(\text{E}_x \log[\mathcal{L}(\hat{\theta}(x)|x)] - \text{E}_x\text{E}_y \log[\mathcal{L}(\hat{\theta}(x)|y)]\right) - 2K. \tag{6.7}$$

For the normal distribution model as truth, and with $g = f$, the result of (6.7) is

$$\frac{2K(K+1)}{n-K-1}. \tag{6.8}$$

We evaluated a simpler form of (6.7) for $f = g$ under the gamma model by Monte Carlo methods to compare those results to (6.8). Table 6.19 gives results of these Monte Carlo evaluations; however, we first need to give the technical details of what was done.

The gamma model probability density function is

$$g(x|\alpha, \beta) = \frac{x^{\alpha-1}e^{-x/\beta}}{\Gamma(\alpha)\beta^\alpha}, \qquad 0 < x, 0 < \alpha, 0 < \beta.$$

Here $\text{E}(x) = \alpha\beta$ and $\text{var}(x) = \alpha\beta^2$. Because β is a scale parameter results are invariant to the value of β. Thus it sufficed to just set $\beta = 1$ in the simulations (we still must estimate β from the data).

The gamma distribution is in the exponential family. Therefore, from Section 7.5 a simpler alternative to (6.7) is

$$2\text{tr}[\text{COV}(\hat{\theta}, S)] - 2K, \tag{6.9}$$

where $\text{COV}(\hat{\theta}, S)$ is here a 2×2 matrix and θ and S are both 2×1 vectors. The canonical form for the gamma distribution is

$$g(x) = \exp\left[x\left[-\frac{1}{\beta}\right] + \log(x)(\alpha) - \log(x) - \log(\Gamma(\alpha)) - \alpha\log(\beta)\right].$$

Hence, $\theta_1 = -1/\beta$ and $\theta_2 = \alpha$ are a 1-1 transformation. The minimal sufficient statistic, S (in 6.9), has the components $\sum x$ and $\sum \log(x)$.

For given α, a random sample of size n was generated from the gamma distribution, and then S was computed and $\hat{\theta}$ found by standard numerical methods. This was repeated for 100,000 independent trials. Then $\widehat{\text{COV}}(\hat{\theta}, S)$

TABLE 6.19. Exact small-sample bias of AIC (6.7) for the gamma distribution ($K = 2$) contrasted to the bias correction term (6.8) used in AIC_c.

	Sample Size, n			
α	10	20	50	100
0.25	2.78	1.06	0.34	0.16
0.50	2.03	0.80	0.28	0.14
0.75	1.93	0.77	0.28	0.13
1.00	1.83	0.76	0.25	0.13
5.00	1.74	0.66	0.26	0.14
10.00	1.73	0.71	0.27	0.13
25.00	1.68	0.69	0.28	0.14
50.00	1.71	0.69	0.24	0.13
100.00	1.75	0.71	0.27	0.13
From (6.8)	1.71	0.71	0.26	0.12

was found, and finally (6.9) was computed. Results, and levels used for n and α, are shown in Table 6.19, along with the value of (6.8) for $K = 2$. Each tabled value based on (6.9) is reliable to (almost) two decimal places. We can deduce from Table 6.19 that AIC_c is very adequate for use with the gamma distribution (at $K = 2$), except for when both α and n are both small. The worst case of $\alpha = 0.25$ and $n = 10$ is quite extreme in that the underlying gamma(0.25, 1) is very skewed, hence the likelihood is not near to having a normal distribution form at sample size 10. However, even for $n = 10$ the approximation is quite good at $\alpha = 1$, which corresponds to the negative exponential distribution.

More research is desirable on the issue of small-sample versions of AIC, and on the general suitability of AIC_c. However, this example and other results we have examined support AIC_c as generally suitable unless the underlying probability distribution (for a single sample) is extremely nonnormal, especially in terms of being strongly skewed.

6.8.4   Inference from a Less Than Best Model

We continue with some ideas, exemplified by the example in Section 6.8.2 of Leirs et al. (1997), about inference from other than the K-L best model. In some circumstances this is justified, especially if (1) the model, say $g_{(2)}$ (as generic notation for the second-best AIC model), used for inference is nested within the best model, $g_{(1)}$; and (2) the unexplained "effects" in the data represented by the additional parameters added to model $g_{(2)}$ to generate model $g_{(1)}$ are small relative to the explained effects represented by model $g_{(2)}$. Conceptually, this assumes a parametrization of the models as $g(x \mid \theta_2, \theta_1)$ for model $g_{(1)}$ with model $g_{(2)}$ arising under the imposed constraint $\theta_1 = 0$.

We elaborate these ideas further using the Leirs et al. (1997) example. Their AIC_c best model (g_1) was not interpretable in its entirety, but their second-best model (g_6) was interpretable, and because $\Delta_6 = 2.3$, that model is a plausible model for the data. Moreover, that second-best model is nested within the best model here. In principle here, the best model, $g_{(1)}$, could be parametrized as being the model $g_{(2)}$ structure plus an additional 49 parameters structurally additive to (and preferably orthogonal to) the 64 parameters of that second-best fitted model. Therefore, their results provide overwhelming support (in the set of models used) for the joint importance of rainfall and population density as at least good predictors of the observed population variation in survival and capture probabilities (if not outright support for a causal link to those variables).

In choosing to make inferences based on model $g_{(2)}$ (their model 6) and ignoring model $g_{(1)}$ (their model 1), Leirs et al. (1997) are in effect saying that they cannot interpret the meaning of the additional 49 parameters that constitute the difference between their best and second-best models. This does not in any way invalidate inference from the second-best model in this situation where $g_{(2)}$ is nested within $g_{(1)}$. This sort of argument holds in general if the models are nested.

The only pressing concern here, in ignoring the best model, i.e., ignoring the 49 "effects" defining the difference here between the best and second-best models, is the issue of the relative magnitude of the two sets of effects. In analysis of variance terms this issue is about the partition of the total variation of effects represented by the difference in their fitted model g_1 versus g_2 into a sum of squares for effects of g_6 versus g_2 plus a sum of squares for effects of g_1 versus g_6. Analogous to ANOVA, we can use here analysis of deviance (ANODEV) (see, e.g., McCullagh and Nelder 1989, Skalski et al. 1993) to accomplish a useful partition.

In this example, ANODEV proceeds as follows to measure the relative importance of the ignored effects left unexplained in model g_1 beyond the explained effects in model g_6. First, some baseline "no effects" model is needed; here that baseline is model g_2 of Leirs et al. (1997). Note the nesting $g_2 \subset g_6 \subset g_1$ and corresponding values of K: 5, 64, and 113. The ANODEV proceeds by obtaining the log-likelihood values and computing the partition of total deviance of model g_2 versus g_1 as

$$\left[2\log(\mathcal{L}(\hat{\underline{\theta}} \mid g_1)) - 2\log(\mathcal{L}(\hat{\underline{\theta}} \mid g_2))\right] = \left[2\log(\mathcal{L}(\hat{\underline{\theta}} \mid g_6)) - 2\log(\mathcal{L}(\hat{\underline{\theta}} \mid g_2))\right]$$
$$+ \left[2\log(\mathcal{L}(\hat{\underline{\theta}} \mid g_1)) - 2\log(\mathcal{L}(\hat{\underline{\theta}} \mid g_6))\right].$$

The result here is $756.8 = 649.0 + 107.8$. The above three bracketed differences are also interpretable as likelihood ratio test statistics on 108, 59, and 49 df.

The above partitions a measure of the magnitude of the total effects (756.8, on 108 df) represented by fitted model g_1 into a measure of the effects explained by model g_6 alone (649.0, on 59 df), plus the additional measure of effects (107.8, on 49 df) explained by the added 49 parameters that "create" model g_1 from model g_6. Based on this partition we can define a type of multiple coefficient of determination, R^2, as (here)

$$R^2 = \frac{2\log(\mathcal{L}(\hat{\underline{\theta}} \mid g_6)) - 2\log(\mathcal{L}(\hat{\underline{\theta}} \mid g_2))}{2\log(\mathcal{L}(\hat{\underline{\theta}} \mid g_1)) - 2\log(\mathcal{L}(\hat{\underline{\theta}} \mid g_2))} = \frac{649.0}{756.8} = 0.858.$$

The interpretation is that 86% of the total structural information about parameter variation in model g_1 is contained in model g_6. Thus, in some sense 14% of potentially interpretable effects has been lost by making inferences based only on model g_6 (i.e., the second-best AIC_c model), rather than based on model g_1. However, that other 14% of information was left as not interpretable. It was judged to be real information, as evidenced by AIC_c selection of model g_1 as the best model, but ignoring it does not invalidate the inferences made from model g_6.

Clearly, the addition to model g_6 of all the structure represented by the additional 49 parameters (to get model g_1) does, for the data at hand, lead to the K-L best-fitted model. However, in principle there is some intermediate model, between models g_6 and g_1, that adds far fewer than 49 parameters and would produce an even smaller AIC_c than model g_1. Such an additional model would extract additional useful information from the data; it might be some form of random-effects model, or some interaction effect of rainfall and population density. The situation faced here is, essentially, considered in Sections 3.5.5, 3.5.6, and 6.9.3, where we point out that if there are two models, one nested in the other and differing by a large number of parameters (say 10 or more), then anomalies can arise in data analysis based on K-L model selection.

In general, there are situations where choosing to make inferences based on other than the AIC_c best model can be justified. However, this situation is not satisfied if the AIC_c best model has many additional parameters compared to the model one uses for the basis of inference. If we find ourselves in this situation, it suggests that we did not think hard enough a priori about our set of models, because we probably left out at least one good model. Now some a posteriori (to the initial data analysis) model building and fitting could be done; just admit, then, which models were a priori and which were motivated by initial data analyses.

6.8.5 Are Parameters Real?

Consideration of what is a parameter seems important, inasmuch as we are focused entirely on parametric models. With only one class of exceptions we regard a parameter as a hypothetical construct. Hence, a parameter is usually the embodiment of a concept and does not have the reality of a directly recordable variable. As such, a parameter in a statistical setting is (usually) just a useful, virtually essential, conceptual abstraction based on the fundamental concept of the expected value of a measurable variable that is not fully predictable. There also needs to be a large number of actual occurrences possible for this measurable variable, or at least a well-defined conceptually possible large number of occurrences. Then the concept of an average of observed values converging to some stable number is at the heart of the concept of a statistical parameter. As such, a statistical parameter cannot be determined exactly by one, or a few, simple measurements. There is no instrument, or simple protocol, to record the exact value of a parameter used in a statistical model. (The exception occurs in measurement error models where the quantity measured is real but becomes the parameter of interest because each recorded measurement is recognized to be imprecise at a nonignorable level of imprecision).

We go a step further and recognize two classes of parameters in statistical models: (1) parameters that appear in the log-likelihood; these may or may not have any associated physical or biological reality; and (2) parameters as noted above that are directly related to expectations of measurable, hence predictable, variables. The second class of parameters are tied to measurable reality, but need not appear in the likelihood (they often do appear).

As an example, consider the analysis of cohort survival data, such as represented by examples in Section 5.2. The age-specific survival probability parameters S_r cannot be directly measured (such as the weight of an animal can be). However, the concept represented by S_r has clear and obvious ties to a measurable event: survival of an animal over a defined time interval. The event can be repeated based on a sample of animals (from a large, if not conceptually infinite, population of animals). These survival probability parameters are in the second class of parameters above. To provide both a useful representation of a set of age-specific survival probabilities, $\{S_r\}$, and provide the basis for parsimonious estimation of this set of parameters from limited data, statistical science adopts smooth, deterministic parametric mathematical functions ("models" for short) such as

$$S_r = \frac{1}{1 + \exp[-(\theta_1 + \theta_2 \cdot r + \theta_3 \cdot r^2)]}$$

(as emphasized in this book, we should not pretend that exact equality really holds). The parameters θ_1, θ_2, and θ_3 appear in the likelihood function $\mathcal{L}(\underline{\theta})$. These parameters are in our first class of parameters above, and they need not have any direct physical or biological reality. In this context the θ_i are very useful in making parsimonious predictions of the S_r, which now become derived parameters based on the interpretable and parsimonious parametric model. Often, interpretability is as important as parsimony, and it is fortuitous that the two criteria of model usefulness are complementary, rather than in conflict. (Interpretability is a subject-matter criterion, not a statistical one, so we have not focused on it here).

The relationship of a parameter to prediction and expectation (which are themselves concepts) is straightforward in a simple linear model like

$$E(y \mid x) = \beta_0 + \beta_1 x.$$

If we can measure the values of y when separately x and $x + 1$ occur (we may be able to control x), then

$$\beta_1 = E(y \mid x + 1) - E(y \mid x).$$

Hence, measurements directly relatable to the parameter β_1 can be made. However, β_1 remains as the embodiment of a concept, whereas specific instances of y can be discovered by direct measurement. As Mayr (1997) notes, concepts are often the driving force in science, much more so than specific discoveries. The concept of parametric models in statistical science is, and remains, a powerful force.

6.8.6   *Sample Size Is Often Not a Simple Issue*

Students are introduced to statistical data using the concept of a sample of size n of a single response variable, y, and possibly some explanatory predictors, x. This simple data structure fails to convey the possible complexity of data, especially when that structure is not describable by a single sample size. For example, there may be a sample n_s of subjects, and on each subject repeated measurements are taken at n_t time points. To then claim that total sample size is $n = n_s \times n_t$ can be misleading; it is only defensible under a fully fixed-effects inference model. However, if subjects are treated as random effects, then there is one sample size for subjects and a different sample size for measurements within subjects. This latter case creates difficulties in the proper assessment of sample size and thus the computation of AIC_c.

A related example is the Durban storm data (Section 4.10). We therein used sample size as all 2,474 weeks in which a storm event could occur. However, one might argue that the sample size should be the number of years (about 47), or maybe n should be taken as 52, the number of weeks. The latter number makes some sense because if we knew the weekly storm probabilities p_1 to p_{52}, then $n = 52$ would be correct (we would still use a model to smooth the p_i). Results of the Durban storm analysis are slightly different if we take $n = 52$ and hence use QAIC_c. There is no definitive basis to know from mathematics the "correct" sample size.

For some data structures there may be two (or more) distinct sample sizes. In line transect sampling (simple designs) there will be a sample size k of lines (often k ranges 10 to 30) and a sample of n detected objects from these lines, with total length L. For modeling the detection function n is the relevant sample size, whereas k is the relevant sample size for estimating the spatial variation of encounter rate, n/L. For capture–recapture there is sample size n_s for the number of distinct animals captured once or more, and n_r for the potential number of recapture events. There are logical arguments in favor of either of these as the sample size to use for AIC_c (n_s seems the better choice now). Further thought suggests that the sample size to associate with the survival parameters (under fixed-effects modeling) in capture-recapture could be n_s, while for the capture probability parameters sample size could be n_r. In general, one can envision different sample sizes as appropriate for different subsets of the parameters in models. Certainly this is true if some parameters are treated as random effects.

General hierarchical models (Carlin and Louis 2000, Lee and Nelder 1996) pose a problem for K-L-based model selection, as regards this sample size issue. Under a Bayesian approach Spigelhalter et al. (2002) have developed a model selection statistic called DIC that seems to behave like AIC and copes with complex sample-size structure. A similar approach might be possible under likelihood-based inference.

The issue of sample size can be complex and has implications for what to use as AIC_c and $QAIC_c$. We do not pursue solutions here. We just raise the issue as a future research area.

6.8.7 Judgment Has a Role

We have seen published statements to the effect that formal (presumably meaning automated or mathematical) model selection methods should not be used, presumably because they do not allow for judgment and they force one to make an inference based on a single model. Apparently, a proponent of such thinking would want to select a model by some ill-defined application of judgment and possibly do some form of ill-defined multimodel inference. We disagree with such thinking, even though we value and respect the role of judgment in science and data analyses. To paraphrase the first sentence of the abstract in Stewart-Oaten (1995), statistical analyses are based on a mixture of rigorous, formal mathematical-statistical methods and judgments based on subject matter knowledge and a (hopefully) deep understanding of strengths and the limitations of the formal data analysis methods used. The emphasis is on mathematical ("objective") methods with the admission that judgment is required for data analysis and interpretation of results.

Judgment does play an important role in data analysis. It is instrumental in the decisions on the general approach and the specific analysis methods one will use. Also, a great deal of judgment may be required in formalizing the a priori set of models to be considered. However, once this a priori model set is established, then we maintain that a formal and well-grounded (philosophically and mathematically) model selection methodology must be applied to these R models. The result will be at least an objectively (given the method) selected best model, and preferably, model credibility weights that facilitate full multimodel inference. This level of inferential model selection can only be achieved by objective, criterion-based methods such as AIC. For point and interval estimation of a parameter given a model surely no statistician would argue that inference should be just a matter of judgment.

Given a model, the statisticians insist on exactly this paradigm of a well-founded, objective criterion, or approach, that is precisely describable by mathematics and that, for given data, gives the same numerical results for all who do the analysis. After various such formal analyses of the data, final conclusions and recommendations may indeed be the result of applying judgment to the situation. But at least the formal results can be said to have a type of objectivity. Just as we insist on a well-founded objective method (often likelihood-based) for a parameter point estimate, we should insist on the use of an objective well-founded method for model selection (by which we mean multimodel inference). There is room for subsequent professional judgment about a final best model or inference but only after the results of using formal selection methods have been presented and defended as relevant.

The other aspect of the critique of formal-objective (i.e., automated) model selection that is sometimes heard is that it forces one to select a single model. This is a valid concern; it is also now mute. Formal multimodel inference methods will greatly reduce the need for what are difficult judgments as a means to cope with the obvious inadequacies of methods that produce only a single best model.

The real issue about the use of judgment is not whether is it used (it is), but rather who is qualified to exercise good judgment, and how do we (or they) know they are qualified? Good judgment can be taught, and should be. However, it takes a lot of training and experience to achieve, and in statistics (as opposed to say medicine), we do not get frequent and reliable feedback about whether our judgments (decisions) are good ones.

6.9   Tidbits About AIC

The section contains miscellaneous ideas and results that do not fit well elsewhere, but are worth understanding.

6.9.1   Irrelevance of Between-Sample Variation of AIC

Likelihood-based inference, including AIC-based model selection, is only concerned with relative evidence about alternatives, conditional on the data at hand. For AIC the technical nature of the inference is about comparing estimates of relative K-L information loss, as bias-adjusted log-likelihood values, over different models; all such comparisons must be based on a given data set. It is neither meaningful nor valid to compare individual AIC, or log-likelihood, values between different samples. However, people have sometimes simulated data and noted the sample-to-sample variation in the value of AIC for a given fitted model and concluded that our guidelines about AIC differences, the Δ, could not be correct. This "insight" is wrong because such between-sample variation for a single model (rather than for Δ) is totally irrelevant to the issue of inference about alternative models given the data. The situation is analogous to a randomized complete block (RCB) experiment wherein treatment effects are estimated only from within-block differences.

This error is common enough that we illustrate it here. Consider the simple linear model $y_i - x_i'\beta + \epsilon_i, \epsilon \sim$ normal$(0, \sigma^2)$ where β has p component parameters. For a random sample under this model, conditional on the predictors and σ^2, we have for the maximinzed likelihood,

$$-2\log(\mathcal{L}) = n\log\left[2\pi\sigma^2\left(\frac{n-p}{n}\right)\frac{\text{RSS}}{(n-p)\sigma^2}\right] + n,$$

where RSS is the residual sum of squares. As a random variable the quantity RSS $/(n-p)\sigma^2$ is central chi-square on $\nu = n - p$ degrees, χ_ν^2. Hence, as a random variable

$$-2\log(\mathcal{L}) = n\log(\chi_\nu^2) + n\log\left[2\pi\sigma^2\left(\frac{n-p}{n}\right)\right] + n.$$

Thus, ignoring the additive constant, the sample-to-sample variation of the maximized -2 log-likelihood is easy to simulate in this common model. That variation depends strongly on sample size. For example, for $p = 10$, $n = 100$ or 500, and 10,000 Monte Carlo repetitions of this model we got the results below for $-2\log(\mathcal{L})$:

sample	n	
%-tile	100	500
1	412.6	3,021.2
10	429.3	3,056.1
25	438.9	3,075.3
50	449.0	3,096.5
75	458.7	3,117.7
90	467.8	3,136.9
99	482.5	3,168.5

The "absolute" variation over samples for a given model is here much greater than our guidelines, such as a $\Delta \geq 10$ being generally strong evidence against the model with the bigger AIC. But this sampling variation of AIC for a single model is meaningless and misleading. The only relevant quantities for inference about alternative models are differences over models, such as Δ and Δ_p. The variation of such a difference is not related to, and cannot be inferred from, the (irrelevant) sampling variation of a single AIC value anymore than treatment effect in an RCB experiment can be inferred from block-to-block variation of a given treatment.

A related misconception arises because AIC values can be quite large, as above. Sometimes authors and analysts have AIC values such as 5000, 5010, and 5020 for three models under consideration and conclude that the models are a short distance apart and "one model is nearly as good as the other two." This is a poor interpretation and is probably influenced by the large sample size that contributes to the fact that AIC values in this case are in the 5,000 range. The focus of attention must always be on the differences in AIC values, the Δ_j and the associated Akaike weights, w_j, and the ranking and scaling of the models based on these w_j.

6.9.2   The G-Statistic and K-L Information

For discrete count data for k mutually exclusive categories there is a close relationship between the G-statistic for goodness-of-fit testing and the K-L distance. The G-statistic is usually written as

$$G = 2 \sum_{j=1}^{k} O_j \log \left(\frac{O_j}{E_j} \right),$$

where O_j is the observed count and E_j is the expectation under some fitted model. Under mild conditions, G is asymptotically distributed as chi-squared under the null hypothesis that the model is an adequate fit to the discrete data. Such G-statistics are additive, whereas the more traditional Pearson's goodness-of-fit test statistic

$$\text{Pearson} = \sum_{j=1}^{k} ((O_j - E_j)^2 / E_j)$$

is not. The K-L distance for discrete data is written as

$$I(f, g) = \sum_{i=1}^{k} p_i \log \left(\frac{p_i}{\pi_i} \right)$$

and is almost identical in form to the G-statistic.

Given a sample of count data $n_1, \ldots, n_k$ ($n = \sum n_i$), let $p_j = n_j/n$ correspond to the observed relative frequencies. Denote the estimated expected probabilities under the approximating model by $\hat{\pi}_j(\theta)$; thus $n\hat{\pi}_j(\theta) = E_j$. In the discrete case, we have $0 < p_i < 1$, $0 < \pi_i < 1$, and these quantities each sum to 1, as do their estimators. Then $I(\hat{f}, \hat{g})$ can be rewritten as

$$\sum_{j=1}^{k} (n_j/n) \log \left(\frac{n_j/n}{E_j/n} \right).$$

Now K-L distance between these (estimated) distributions can be written as

$$\frac{1}{n} \sum_{j=1}^{k} (n_j) \log \left(\frac{n_j}{E_j} \right),$$

or

$$\frac{1}{n} \sum_{j=1}^{k} O_j \log \left(\frac{O_j}{E_j} \right).$$

Thus, the G-statistic and K-L information differ by a constant multiplier of $2n$, i.e., in this context, $G = 2n \cdot I(\hat{f}, \hat{g})$. Similar relationships exist between K-L information expectations of likelihood ratio statistics for continuous data (G is a likelihood ratio test (LRT) for discrete data). Thus, the LRT is fundamentally related to the K-L distance.

6.9.3 AIC Versus Hypothesis Testing: Results Can Be Very Different

The use of the information-theoretic criteria in model selection can be quite different from that of hypothesis testing, and this is an important issue to understand. These differences can be illustrated by considering a set of nested candidate models, each successive model differing by one parameter. Model g_i is the null model with i parameters, and model g_{i+j} is the alternative with $i + j$ parameters. Model i is nested within model $i + j$; thus likelihood ratio tests (LRT) can be used to compare the null model with any of the alternative models g_{i+j}, where $j \geq 1$. Thus, if model g_i has 12 parameters, then model g_{i+1} has 13, model g_{i+2} has 14, and so on.

This concept of a set of nested models is useful in illustrating some differences between AIC versus LRT for model selection. First, assume that the AIC value for each of the models is exactly the same; thus no model in the set has more support than any other model. Second, in each case we let the null hypothesis be model g_i and assume that it is an adequate model for the data. Then, we entertain a set of alternative hypotheses, models g_{i+j}; these are each hypothesized to offer a "significantly" better explanation of the data. That is, g_i (the null) is tested individually against the $j \geq 1$ alternative models in the set. The first test statistic (g_i versus g_{i+1}) here is assumed to be distributed as χ^2 with 1 df, while the second test statistic (g_i versus g_{i+2}) has an assumed χ^2 distribution with 2 df, and so on. The following relations will be useful:

$$\text{AIC}_i = -2\log(\mathcal{L}_i) + 2i,$$
$$\text{AIC}_{i+j} = -2\log(\mathcal{L}_{i+j}) + 2(i + j),$$
$$\text{LRT} = -2\big(\log(\mathcal{L}_i) - \log(\mathcal{L}_{i+j})\big) \text{ with } j \text{ df.}$$

Then, in general,

$$\text{LRT} = \text{AIC}_i - \text{AIC}_{i+j} + 2j.$$

Now, for illustration of a point about the difference between LRTs and AIC in model selection, assume

$$\text{AIC}_i \equiv \text{AIC}_{i+j}.$$

If this boundary condition were to occur (where K-L–based selection is indifferent to the model), then we would have,

$$\text{LRT} = 2j \text{ on } j \text{ degrees of freedom.}$$

Now, a difference of 1 df between g_i and g_{i+1} corresponds to a χ^2 value of 2 with 1 df, and a P-value of 0.157 (Table 6.20). Similarly, a difference of 4 df ($j = 4$) between g_i and g_{i+4} corresponds to a χ^2 value of 8 and a P value of 0.092. If the degrees of freedom is less than about 7 (assuming $\alpha = 0.05$), then hypothesis-testing methods support the null model (g_i) over any of the alternative models ($g_{i+1}, g_{i+2}, g_{i+3}, \ldots$) (Table 6.20). This result is in contrast with AIC selection, where in this example all the models are supported equally.

TABLE 6.20. Summary of P-values (i.e., $\text{Prob}\{\chi^2 \geq 2\text{df} = 2j\}$) for likelihood ratio tests between two nested models where the two corresponding AIC values are equal, but the number of estimable parameters differs by j (after Sakamoto et al. 1986).

j	χ^2	P
1	2	0.157
2	4	0.135
3	6	0.112
4	8	0.092
5	10	0.075
6	12	0.062
7	14	0.051
8	16	0.042
9	18	0.035
10	20	0.029
15	30	0.012
20	40	0.005
25	50	0.005
30	60	0.001

Test results change in this scenario when there are more than $j = 8$ additional parameters in the alternative model (Table 6.20). Here, the null model (g_i) is rejected with increasing strength since the alternative model has an increasing number of parameters. For example, the likelihood ratio test of g_i versus g_{i+10} has 10 df, $\chi^2 = 20$, and $P = 0.029$. More striking is the test of g_i versus g_{i+30}, which has 30 df, $\chi^2 = 60$, and $P = 0.001$, even though the AIC value is the same for all the models (the null and the various alternatives). In these cases (i.e., > 8 parameters difference between the null and alternative model), the testing method indicates increasingly strong support of the models with many parameters and strong rejection of the simple null model g_i (see Sakamoto 1991 and Sakamoto and Akaike 1978:196 for additional insights on this issue).

More extreme differences between the two approaches can be shown by letting $\text{AIC}_i = \text{AIC}_{i+j} - x$ for x in the range of about 0 to 4. It is convenient to work with the Δ_{i+j} values; then relative to the selected model, Δ for model g_{i+j} is x. If $x = 4$, the choice of model g_i is compelling in the context of nested models, as judged by AIC. For comparison, the LRT statistic is $2j - x$. Let $x = 4$ and $j = 20$; then the LRT statistic is 36 on 20 df and $P = 0.0154$. Most would take this P-value as compelling evidence for the use of model g_{i+j}. Thus, AIC can clearly support the simple model g_i, while LRT can clearly support model g_{i+j} with 20 additional parameters. The solution to this dilemma is entirely a matter of which the model selection approach has a sound theoretical basis: Information criteria based on K-L information does; likelihood ratio testing does not.

Those individuals holding the belief that the results of hypothesis tests represent a "gold standard" will be surprised at the information in Table 6.20 and may even believe that AIC "loses power" as the difference in parameters between models increases beyond about 7. [Note: The concept of "power" has no utility in the information-theoretic approach because it is not a "test" in any way.] Akaike (1974) noted, "The use of a fixed level of significance for the comparison of models with various numbers of parameters is wrong, because it does not take into account the increase of the variability of the estimates when the number of parameters is increased." The α-level should be related to sample size and the degrees of freedom if hypothesis testing is to be somehow used as a basis for model selection (see Akaike 1974; Lindsey 1999b). However, the α-level is usually kept fixed, regardless of sample size or degrees of freedom, in the hypothesis testing approach. This practice of keeping the α-level constant corresponds to asymptotically inconsistent results from hypothesis testing. For example, if the null hypothesis is true and α is fixed (at, say, 0.05), then even as the degrees of freedom approach ∞ we still have a 0.05 probability of rejecting the null hypothesis, even with near infinite sample size. The inconsistency is that statistical procedures in this simple context should converge on truth with probability 1 as $n \to \infty$.

6.9.4 A Subtle Model Selection Bias Issue

Consider having 10 independent one degree-of-freedom central chi-square random variables, denoted as x_1 to x_{10}. Let z be the minimum over the set $\{x_i, i = 1, \ldots, 10\}$. Because of the selection of z as a minimum of *iid* random variables, z is not distributed as central chi-square on 1 df. Rather, z is stochastically smaller ("biased") compared to a central chi-square random variable on 1 df. The selection process induces what may be considered a type of bias. A few authors have expressed concern for a similar sort of selection bias on AIC as a random variable; they are both right, yet mostly wrong as regards model selection as such. By right, we mean that if one focuses on the underlying K-L-based criterion, T (Section 7.2), to be estimated, then for a single specified (good) model we can indeed have $E(\hat{T}) = T$. Thus a single AIC may be nearly an unbiased estimator of T (which is related to relative K-L information loss) if only one model is considered. However, if we have 10 models and we compute $\hat{T}$ (i.e., AIC) for each model and then select the smallest $\hat{T}$, we induce a bias by this selection process as regards $\hat{T}$ for the selected model.

Technically, just assume $E(\hat{T}_j) = T_j$, where the expectation is over all possible samples. If model j produces the minimum $\hat{T}_r$ value, then $min = j$ and we can define $E(\hat{T}_j|min = j)$. The selection bias that some people might mistakenly worry about occurs because $E(\hat{T}_j|min = j) < E(\hat{T}_j) = T_j$.

However, the situation of AIC is not really analogous to the above chi-square example for two reasons. First, the same data are used to compute each AIC (i.e., each $\hat{T}$); this induces a strong positive correlation in the set of AIC values over the models considered. It would be like having the 10 chi-square random variables very positively correlated. It those correlations were all actually 1, then the 10 x_i would all be equal, so selecting their minimum would not induce a bias; the selected variable would always be just the original central chi-square random variable.

Second, we do not care about a single AIC; we only care about the differences ΔAIC. Most of any bias induced by selection (i.e., ordering the AIC values) will drop out of these differences. In particular, let the ordered AIC (i.e., $\hat{T}$ values) smallest to largest be AIC_1 to AIC_{10}. Then we care most about being correct (not biased) for models close to each other in K-L measure, hence producing on average small Δ, such as expected for $\Delta_2 = AIC_2 - AIC_1$. Here, one model was selected as best and the other as second best, so we expect that the selection biases are very similar, and hence essentially drop out of the difference. Moreover, it is highly relevant that the same data are used for fitting both (all) models; only the models differ, not the data and the models. Again, this means potential selection bias in an individual AIC is an irrelevant concept. Rather, the possible selection bias noted herein strongly tends to drop out of the differences, ΔAIC. Because only these differences are relevant, the possible biasing of a single AIC by selection of the minimum over a model set is an almost irrelevant issue. Possible selection bias of the Δ as estimators of K-L differences is relevant, but should be quite small for models close to each other in K-L information loss, especially if R is not large.

6.9.5 The Dimensional Unit of AIC

The dimensional unit of AIC is the unit of $-\log(g(x|\theta))$, where g is a probability distribution, not a pdf (which is for a continuous random variable), but a proper probability distribution. Thus, g is strictly for a discrete random variable. From a philosophical point of view all random variables are discrete: their possible values increment by some minimal step size, δ. Also, they take only a countable number of possible values. For example, we might think weight is continuous, but we can only measure it to some number of places, perhaps four digits (and at best to the nearest discrete atom of mass). So weight can properly be considered a discrete random variable. For convenience we model it as continuous, which allows both the use of models based on the normal probability and all the convenience and power of calculus.

The reason for this observation is simply that likelihood, properly, derives from probability, not pdf's (see Lindsey 1999b). Thus we should write $g(x|\theta) \equiv \Pr\{\tilde{X} = x|\theta\}$, where $\tilde{X}$ is the random variable. While technically not correct, we prefer to use the briefer form. However, that form obscures issues about the dimensional unit of $-\log(\mathcal{L}(\theta|x)) = -\log(\Pr\{\tilde{X} = x|\theta\})$. In particular, the dimensional unit of x is irrelevant to the dimensional unit of $\Pr\{\tilde{X} = x|\theta\}$, which is *always* probability. The probability of the event $\tilde{X} = x$ is invariant to the measurement units used for x. The probability of an event does not depend on the unit of measurement. Therefore, for $-\log \mathcal{L}(\theta|x) = -\log(\Pr\{\tilde{X} = x|\theta\}) = -\log(g)$ we can think in terms of units being negative log-probability. But Claude Shannon (Shannon 1948), the founder of information theory, established that $-\log(\text{probability})$ is the unique mathematical representation of information (and K-L is information loss for g as a model of f). The intrinsic unit for $-\log(\mathcal{L})$ is information; hence, the units of AIC are always information.

There does not seem to be an established name for a unit of information. Claude Shannon died in February 2001. It has now been proposed that the unit of information be called the "Shannon." Regardless of the name, the key point here is that the units of the data are irrelevant when we compute probabilities (which then become the units); and likelihood, properly, is based on probability of data.

More clarification is needed. A continuous random variable is to be viewed as a convenient way of dealing with the situation when δ would be taken as very small relative to the range of $\tilde{X}$, for example, a range of 0 to 1 with $\delta = 0.0001$. The range could be 0 to infinity, as long as the probability of big values of $\tilde{X}$ drops off sufficiently fast for large x. If $g(x|\theta)$ represents the pdf for a (conceptual) continuous random variable, then we can use the approximation $\Pr\{x|\theta\} \approx g(x|\theta)\delta$. For sufficiently small δ this approximation is excellent and justifies using $g(x|\theta)\delta$ for $\mathcal{L}(\theta|x) = \Pr\{x|\theta\}$. The constant δ drops out of all uses of the likelihood as a measure of relative evidence about parameter values given the model, or about models given the data.

Thus, for models based on the presumption of continuous data, $-\log(\mathcal{L}) = -\log(\Pr\{x|\theta\}) = -\log(g(x|\theta)) - \log(\delta)$ is technically required. However, it suffices to use $-\log(\mathcal{L}) = -\log(g(x|\theta))$ for purposes of inference. Properly computed, negative log-likelihood has information as its unit of dimension. However, $-\log(g(x|\theta))$ does not have this unit of dimension because of the missing $\log(\delta)$. This carries over to AIC. Because of such multiplicative constants possibly left out of the likelihood (i.e., dropped from g) there can be confusion over the inferential dimensional unit of AIC: it is information. This is justified because the only inferential way we use AIC is in a comparative manner, as Δ AIC. The dimensional unit of Δ is information, independent of any additive constants common to, but left out of, each underlying log-likelihood.

It is easy to lose sight of these deep matters when considering, for example, $-\log \mathcal{L}$ for normal distribution-based models. We should have

$$-\log(\mathcal{L}(\hat{\theta}|x, g)) = \frac{n}{2} \log\left(\frac{\text{RSS}}{n}\right) + C(g),$$

where RSS is the residual sum of squares for the fitted model and the constant C depends only on the assumed normality of "errors," and includes $\log(\delta)$. If all models considered assume normality and variance homogeneity, then $C(g)$ is identical over all R models, and thus it drops out of all inferential comparisons of models. This is pragmatic justification for then simply using $\log(\mathcal{L}(\hat{\theta}|x, g)) = -\frac{n}{2} \log\left(\frac{\text{RSS}}{n}\right)$.

When we bypass all of this important information about what a likelihood really is, we risk confusion about what is the dimensional unit of log $\mathcal{L}$, and hence AIC. For example, for the normal distribution one might erroneously think the dimensional unit is log(dimension of x^2). In fact this is totally off the mark because we are now dealing with units in $-$ log-probability, or "information."

6.9.6 AIC and Finite Mixture Models

Finite mixture models (see e.g., McLachlan and Peel 2000) are a useful class of statistical models, however, they are nonstandard in many respects. In particular, the usual likelihood ratio test statistic is not applicable because of issues about parameter values under the null hypothesis being on the boundary of the parameter space (McLachlan and Peel 2000: 185–186). This irregularity has caused concerns about the use of AIC for model selection when finite mixture models are in the model set. People wonder, should the "$2K$" in AIC be modified, just as the likelihood ratio test for mixtures must be modified? We have considered the issue enough to provide a preliminary opinion: The formula for AIC does not need to be modified for mixture models. Heuristically (i.e., at a shallow level), this is because AIC is not a test. At a deeper level, in applying AIC to mixture models we encounter two issues: (1) using the correct count for K, in light of failure of parameter estimability when the MLE estimates are on a parameter space boundary, and (2) model redundancy occurs, when such estimability failure occurs, and must be dealt with (redundant models must be dropped from he model set). Thus, it is an aspect of how AIC is used that must be modified.

Consider an example of a strictly positive response variable wherein a two-component mixture negative exponential model be useful (see e.g., Burnham 1988). The mixture model is

$$g_3(x) = \pi \left[\frac{1}{\lambda_1} e^{-x/\lambda_1}\right] + (1 - \pi)\left[\frac{1}{\lambda_2} e^{-x/\lambda_2}\right], \tag{6.10}$$

$0 < \pi < 1, 0 < \lambda_i$, and $\lambda_1 \neq \lambda_2$. The model of (6.10) can be represented as

$$g_3(x) = \pi g_1(x) + (1 - \pi)g_2(x),$$

where

$$g_i(x) = \frac{1}{\lambda_i} e^{-x/\lambda_i}, \qquad i = 1, 2.$$

The nominal number of parameters in model g_3 is $K = 3$, whereas for model g_1, or g_2, $K = 1$. The problem with fitting mixture models such as (6.10) is that the MLE will be on a boundary if either $\hat{\pi} = 0$ (or 1), or $\hat{\lambda}_1 = \hat{\lambda}_2$, occurs and then the model is no longer a mixture. In the first case only a single λ is estimable. In the second case π is not estimable; $\hat{\pi}$ can take an arbitrary value as π actually drops out of the likelihood. In either case, when this 2-component mixture model is not estimable the proper count for K is 1.

At a minimum, model selection here would have two models, g_3 and $g(x)$, equivalent to either g_1 or g_2, i.e., no subscript is needed for λ. The simple model always has $K = 1$. When the likelihood based on model g_3 is properly maximized with $\hat{\pi}$, $\hat{\lambda}_1$, and $\hat{\lambda}_2$ not on boundaries, only then do we use $K = 3$ and compute the usual AIC or AIC_c for the mixture model. If the MLE is on a boundary, then the fitted mixture model actually collapses to model g. Hence, in that case fitted models g_3 and g are redundant (Section 4.6), and no model selection occurs because the only fitted model one has is g.

If model g_3 is not estimable, for the data at hand, it is a mistake to compute an AIC for it as if $K = 3$. Such an AIC would actually have its underlying log-likelihood identical to that for model g but would be four units larger than the AIC for model g. However, model selection or multimodel inference is meaningless in this situation where the mixture model in fact cannot be fit and $\hat{g}_3$ becomes $\hat{g}$.

The ideas here generalize if the mixing models are of different types or if the mixture model has more than two components. In the first case we might have $g_3 = \pi g_1 + (1 - \pi)g_2$, with models g_1 and g_2 being of different forms (e.g., negative exponential and half-normal) with K_1 and K_2 parameters. Now model g_3 will collapse to either model g_1 or g_2 if $\hat{\pi}$ is on a boundary. In this case model redundancy occurs, and there really are only models g_1 and g_2. Only if the mixture model is estimable does one have three models to consider, with $K_3 = 1 + K_1 + K_2$ for model g_3.

If the same type of basic model is used in a three (or more) component mixture, the possible complications increase. But the principle is simple: If the full mixture model cannot be fit because the MLE of its vector parameter is on a boundary, then the model set must be adjusted for any resultant model redundancy. Also, the true nature of the reduced-component fitted model must be recognized and its correct K determined. For example, a 3-component mixture, when fit to the data, might collapse to a 2-component mixture (which may have already been a model in the model set). Thus the use of mixture models and AIC-model selection entails some traps that must be avoided.

The above does not address AIC and mixture models at a theoretical level. One way to examine theory is to numerically compare the expected value of AIC_c, as properly used for mixture models, to the theoretical target value it is estimating (Section 7.2), e.g., $target = -2E_{\hat{\theta}}E_x[\log(g_3(x|\hat{\theta}))]$. Expectations are with respect to the actual data-generating distribution. To make sure we were "on track" we did some Monte Carlo evaluations wherein the model g_3 is (6.10) and data are generated under either (6.10) or g_1. Sample sizes, n, were $25, 50, 100$, and 200, with $25,000$ or $50,000$ Monte Carlo replications. It sufficed to fix $\lambda_1 = 1$ and vary λ_2 over $1, 5, 10, 15, 20$. The values of π ranged over 0 to 0.5 by 0.1. The resulting comparisons, as $\delta = |E(AIC_c) - target|$ were quite good for $n = 100$ and 200, and (to us) acceptable at the smaller values of

n. However, there is clearly room for improvement regarding a small-sample version of AIC for mixture models. Basic results are given below, by whether the generating distribution was a mixture or a simple negative exponential (mean and maximum δ are over the set of values used for λ_2 and π):

generating distribution	n	δ	
		mean	maximum
simple	25	1.63	1.79
	50	1.10	1.31
	100	0.82	0.97
	200	0.55	0.90
mixture	25	0.66	1.32
	50	0.29	0.66
	100	0.24	0.53
	200	0.23	0.57

6.9.7   Unconditional Variance

The formula for estimating the unconditional variance of a model-averaged parameter estimate is a derived result (Section 4.3.2, denoted there as formula 4.9):

$$\widehat{\text{var}}\left(\hat{\bar{\theta}}\right) = \left[\sum_{i=1}^{R} w_i \sqrt{\widehat{\text{var}}(\hat{\theta}_i | g_i) + \left(\hat{\theta}_i - \hat{\bar{\theta}}\right)^2}\right]^2. \qquad (6.11)$$

A corresponding formula is given in Section 4.3.2 for an unconditional covariance, $\widehat{\text{cov}}(\hat{\bar{\theta}}, \hat{\bar{\tau}})$. However, in obtaining (6.11) we assume perfect pairwise conditional (on the models) correlation of estimators of θ from different models. Thus, just on that ground there might be weak motivation for more theoretical exploration of an alternative to (6.11).

A better, but not overwhelming in practive, motivation is an inconsistency underlying (6.11), as for example using linear models and all-subsets models with predictors x_1 to x_p. Thus, model g_i uses some subset of the p predictors, and $\theta_i = E(y|\underline{x}, g_i)$ is the appropriate linear combination of the β_j regression parameters under model g_i. Because all the models considered are linear we have (Section 5.3.6)

$$\hat{\bar{\theta}} = \tilde{\beta}_0 + x_1 \tilde{\beta}_1 + \cdots + x_p \tilde{\beta}_p = \underline{x}' \tilde{\underline{\beta}}$$

(see also Section 4.2.2 regarding $\tilde{\beta}_i$). Using (6.11) and the corresponding formula for unconditional covariances we can obtain an unconditional variance–covariance matrix for $\tilde{\underline{\beta}}$, say $\hat{\Sigma}$; $\hat{\Sigma}$ is very complicated, whereas $\widehat{\text{var}}\left(\hat{\theta}_i | g_i\right)$, hence (6.11), is simple. An alternative for $\widehat{\text{var}}\left(\hat{\bar{\theta}}\right)$ should be $\underline{x}' \hat{\Sigma} \underline{x}$; however, because of the nonlinearities involving the weights, $\underline{x}' \hat{\Sigma} \underline{x} \neq \widehat{\text{var}}\left(\hat{\bar{\theta}}\right)$

from (6.11). This inequality is suffcient to motivate interest in either a different covariance formula to use with (6.11), or an alternative to (6.11).

One heuristic approach is to note what is used with BIC, wherein the model-averaged posterior is a mixture distribution. Similarly, a model-averaged pseudo-likelihood can be defined in the K-L framework, which is a mixture of each model-specific likelihood. Either motivation leads one to postulate a possible alternative to (6.11) as

$$\widehat{\widehat{\text{var}}}(\hat{\theta}) = \sum_{i=1}^{R} w_i \left[\widehat{\text{var}}(\hat{\theta}_i | g_i) + (\hat{\theta}_i - \bar{\hat{\theta}})^2 \right]. \tag{6.12}$$

This formula is linear in the weights. Using the Cauchy–Schwarz inequality we can show that $\widehat{\text{var}}(\hat{\theta}) \leq \widehat{\widehat{\text{var}}}(\hat{\theta})$, hence (6.12) actually yields a bigger variance. We emphasize that (6.12) is not a derived result in the K-L model selection framework. We have not studied this matter further; however, the issue of the unconditional variance and covariance for AIC are subject areas worthy of more research.

6.9.8   A Baseline for $w_+(i)$

The measure $w_+(i)$ of the relative importance of variables was introduced in Section 4.2.2. We envision it as applied mostly when there are p predictor variables (thus, $w_+(i)$ for $i = 1, \ldots, p$), and the R models considered are all subsets of variables, such as just main-effect terms like $x_i \beta_i$ ($R = 2^p$ possible models), or these terms plus interaction-type terms, such as $x_i x_j \beta_{ij}$ (for when p is small). One important point is that the interpretation of $w_+(i)$ is only within both the context of the set of models and predictors used. Change either the model set or the set of predictors, and $w_+(i)$ can change. A second point is that $w_+(i)$ is not expected to be 0, even if x_i has no contextual predictive value at all. Rather, in this case $E(w_+(i)) > 0$, regardless of sample size. This is the reason these summed weights give only a relative importance of variables.

A randomization method can be used to estimate the baseline value for $w_+(i)$ if x_i has no predictive value. We denote this unknown baseline value as $w_{0+}(i)$. The data structure is an n by 1 response-variable column vector y, and the full design matrix X, which is n by $p + 1$ if the models include an intercept (they usually do). Based on this data structure all R models are fit. For each model AIC is obtained; Akaike weights are obtained, and then the $w_+(i)$ are computed.

To estimate $w_{0+}(i)$ there is one variation on this scenario. First, randomly permute the n values of x_i that are in column $i + 1$ of matrix X; leave the other columns unaltered. Then proceed in the usual way with model fitting and compute what is nominally $w_+(i)$, except what you get is one value of $\hat{w}_{0+}(i)$. This computation is quite easy, especially if one already has coded the bootstrap for the data. A slight alteration in code changes the bootstrap sample generator to generate a sample in terms of the needed randomly permuted values of

TABLE 6.21. Values of baseline $\hat{w}_{0+}(i)$ to compare with $w_+(i)$, for the body fat data example (Section 6.2); each $\hat{w}_{0+}(i)$ is the median of 100 independent values of $w_{0+}(i)$ from random permutations of predictor variable i; see text for details.

i	variable	$w_+(i)$	$\hat{w}_{0+}(i)$
1	age	0.50	0.31
2	weight	0.93	0.31
3	height	0.31	0.29
4	neck	0.65	0.29
5	chest	0.28	0.29
6	abdomen	1.00	0.30
7	hips	0.45	0.31
8	thigh	0.59	0.31
9	knee	0.29	0.30
10	ankle	0.45	0.31
11	biceps	0.60	0.31
12	forearm	0.83	0.29
13	wrist	0.98	0.31

$x_{i1}, \ldots, x_{in}$. Obviously, the random permutation renders y_i and x_i uncorrelated on average; the permuted x_i and all x_j, $j \neq i$, are also uncorrelated.

We have tried this methodology; it seems it could be useful, except for one drawback. A single permutation sample is not enough; $\hat{w}_{0+}(i)$ is quite variable from permutation sample to sample. We suggest doing at least 100 samples; more might be needed. Also the random variable $\hat{w}_{0+}(i)$ can have a very skewed distribution. Hence, rather than average the resultant sample of $\hat{w}_{0+}(i)$ values we suggest using the sample median as the single best $\hat{w}_{0+}(i)$.

An example of estimating baseline values for the $w_+(i)$ is given in Table 6.21. One hundred independent permutation samples were used for each predictor variable i, as described above, and the sample median was used for $\hat{w}_{0+}(i)$. Because of the correlations among these predictors we thought $\hat{w}_{0+}(i)$ might distinctly vary, but they did not do so here. Perhaps the symmetry of the model set combined with the randomization process will always mean the value of $w_{0+}(i)$ is the same for all x_i. From Table 6.21 we judge that for these data and in this context of usage the predictors height, chest circumference, and knee circumference have essentially zero importance. There is potential to define a measure of absolute variable importance based on something like $w_+(i) - \hat{w}_{0+}(i)$. Similar ideas appear in Breiman (2001). Clearly, more research of these methods and ideas is possible and worthwhile.

Another informative idea for a baseline here is to look simultaneously at the full set of $w_+(1), \ldots, w_+(p)$ by leaving the X matrix alone and randomly permuting the elements of vector y. Then fit all models, get AICs, and so forth (the reader might think formal null hypothesis test here, but that is not our intention). If each $w_+(i)$ and $\hat{w}_{0+}(i)$ pair are about the same there is little or no predictability of y by the entire set of x_i. We have seen this occur. This sort procedure is especially informative to guard against spurious results (Anderson et. al. 2001a) when there are lots of predictors and a relatively small sample, e.g., $p = 30$ and $n = 100$.

For the body fat data example we also randomly permuted the elements of the y vector to estimate the $w_{0+}(i)$. This is a much faster appraoch, if valid. Again, the median of the randomization sample was used as the estimator. For 100 samples the results varied from 0.30 to 0.34 (mean of 0.32); for 500 estimates the results varied from 0.31 to 0.34 (mean of 0.32). For either approach, by x_i or y, we note that $w_{0+}(i)$ will depend upon p, and in general on R; there is nothing special about 0.31 or 0.32.

We conclude that more research on these ideas seems warranted.

6.10   Summary

This chapter is a more in-depth examination of some aspects of K-L based model selection; included are some comparisons to other model selection methods. In particular, AIC is contrasted to BIC to better understand the nature of both methods. Those results are mostly in Sections 6.3 and 6.4. Because of their importance we will start this summary with reference to those sections and compare AIC and BIC.

The derivation of BIC (Section 6.4.1) can be done without any assumption that the set of models contains the true model. Thus, neither K-L nor BIC (or Bayesian, in general) model selection methods require for their derivation, validity, or use that the true data-generating model is in the set of models under consideration. Moreover, in the commonly assumed BIC framework (a fixed generating model and a fixed model set) as sample size gets arbitrarily large so that selection converges with probability 1 to a single model it is not logically valid to infer that the selected model is truth (see e.g., Section 6.3.4).

Rather, the model selected by BIC converges to the model with the smallest dimension (i.e., the minimum K) in the subset (of size ≥ 1) of models that all have the identical minimum Kullback–Leibler distance from truth (Section 6.4.2). Denote this model, which BIC selection converges to, as $g_b(\underline{x}|\underline{\theta}_o)$; $\underline{\theta}_o$ is the value of $\underline{\theta}$ that minimizes K-L information loss $I(f, g_b(\cdot|\underline{\theta}))$ for the family g_b of models. Model $g_b(\underline{x}|\underline{\theta}_o)$ is the model with both the smallest parameter dimension K_i and for which $I(f, g_i(\cdot|\underline{\theta}_o))$ is minimized over $i = 1, \ldots, R$.

In reality with real data we expect the model that minimizes K-L distance to be unique in the model set. But in many, if not most, simulation evaluations of model selection the set of models used includes the data-generating model and has it nested in some overly parametrized models. This results in there being a subset of the R models that all have the same K-L distance ($= 0$) from the generating ("true") model. BIC is then consistent for the dimension of that generating model, which has the smallest dimension in this set of models. Thus the dimension-consistent property of BIC is motivated by an unrealistic context. Given that these sorts of simulations are not realistic, conclusions drawn from them may not apply to real data analysis. In particular, for real data analysis the BIC target model cannot validly be inferred to be truth. Rather, it is merely the model with minimum dimension that is nearest to truth as measured by K-L distance.

Whereas both the context and the target model for BIC are independent of sample size, the context for AIC is sample-size specific. Although truth f is unchanging, under the information–theoretic approach the set of candidate models is allowed (in fact, assumed) to grow if n increases substantially (e.g., an order of magnitude). It is not realistic to let n go to infinity while holding the set of candidate models fixed (as in BIC) because substantially more data means both more information and more factor levels, factors, or both, in the study. It then follows that more parameters need to be, and can be, reliably estimated. In this spirit the AIC target model g_a depends on sample size. It is the model for which $E_{\hat{\theta}}[I(f, g_i(\cdot|\hat{\theta}))]$ (expectation is with respect to f) is minimized over $i = 1, \ldots, R$.

Thus the AIC target model (the K-L best model) is generally different from the BIC target model. Because both g_a and the model set are sample-size specific it is not logical to compare g_b to the model that g_a converges to when $n \to \infty$. Furthermore, the model that is g_a at huge sample size is generally different from the realized model g_a at actual n.

Section 6.4.5 provides a milestone result: It shows that AIC can be justified as a Bayesian model selection criterion. The key is to use the BIC formula with a quite different prior probability distribution on the model set; used in a Bayesian context BIC assumes a uniform prior distribution on the model set. The K-L model prior that yields AIC is proportional to $\exp(\frac{1}{2}K \log(n) - K)$; it can be generalized for QAIC$_c$. One consequence of this result is a justification for interpreting the Akaike weights as a posterior probability distribution over the model set. Hence, $w_i = $ the probability that model g_i is the K-L best model.

Another result of Sections 6.3 and 6.4 is a clear understanding of the scientific meaning we must associate with prior and posterior model probabilities. Saying "p_i is the probability of model g_i" we must be referring to the probability that this model is the target model of the model selection procedure. These target models, g_a and g_b noted above, are different for AIC versus BIC (and neither target is necessarily truth f). Now a Bayesian basis for comparison of AIC and BIC is to argue for or against their respective model priors and in general to understand the implications of those priors. The prior for BIC is $p_i = 1/R$. The prior for AIC makes p_i an increasing function of sample size and a decreasing function of the number of estimable parameters in the model. Such a prior corresponds to the idea that we expect the number of parameters we can reliably estimate to depend on n and K_i, and that there is substantial information in the data.

The alternative way to understand and compare selection procedures is in frequentist terms of their actual performance and expected operating characteristics. Some such comparisons are the focus of Sections 6.2 and 6.4.3. The body fat data (Section 6.2) employs multiple regression with 13 predictors and $n = 252$ ($R = 8,191$); the AIC_c-selected model has an Akaike weight of 0.01. A 95% confidence set on the K-L best model includes nearly 900 models. Thus it is totally unacceptable to say the best model has any unique meaning. In particular, it is absurd to interpret the included variables as *the* important ones. This example illustrates the point that with high dimensional data it will typically be the case that even the best model will have very small evidential support. The extensive body fat example of Section 6.2 has its own summary, Section 6.2.8; we direct the reader there rather than repeat that material here.

A few theoretical simulations were done based on the body fat example in Section 6.2 to examine predictive mean square error for the AIC_c and BIC best-model selection strategy and for model averaging (Section 6.4.3). The predictive mean square error (MSE) for AIC_c was 26% lower than for BIC for the best model strategy and 17% lower under model averaging. Also, model averaging was superior to the traditional best-model strategy for both AIC_c (MSE 15% lower) and BIC (23% lower). Ongoing research suggests that model-averaged inferences are generally superior in all subsets selection.

The remainder of Chapter 6 (Sections 6.5 to 6.9) has a variety of topics. Section 6.5 shows that overdispersion adjustment can be generalized to allow far more than one $\hat{c}$. The key idea is to partition the data and apply different overdispersion estimates by these data subsets. Partitioning might be on sex, year, area, treatment, and so forth. When this is done, the count K must include the number of different $\hat{c}$ values used. Issues of goodness-of-fit are interwoven with estimation of overdispersion factors. Therefore, a general strategy for these issues is given in Section 6.5.1, including suggestions for when there is no global model. We also note the matter of goodness-of-fit of the selected model, assuming there is a global model and that it fits the data. The issue is, does the selected model then also fit the data? This is an area needing research. Ongoing work (not included here) has shown that in this context (global model fits) BIC can select a model that is in fact a poor fit to the data, but AIC virtually never does so. It is not clear it either, both, or neither of these operating characteristics should be of any concern.

Almost all model selection literature is only really applicable to the case of parameters as fixed effects. However, the range of application of AIC can be expanded to random effects. It can also be expanded to other nonstandard situations, such as generalized estimation equations. Section 6.6 gives information about these matters, especially application of AIC to models that include simple random effects. These extensions are in an early stage of development. What is important is that AIC can be extended beyond the simple fixed-effects ML approaches of this book. An AIC-like Bayesian procedure (DIC) can be applied in general to Bayesian hierarchical models.

Another seeming nonstandard application occurs when there are models in the set based on different probability distributions. For example, one might want to compare models for the data based on a gamma versus a half-normal distribution. As shown in Section 6.7 this only requires that we keep all terms of the exact probability distributions as part of our likelihoods and then simply use the standard formula for AIC. That section gives some informative examples and details.

In Section 6.8.3 we use numerical methods to evaluate exactly the small sample bias of AIC when the data are from a gamma distribution. We compare this result to the AIC bias-correction term used to get AIC_c (a result derived by assuming a normal distribution). The result is basically that AIC_c is generally an adequate small sample version of K-L model selection for the gamma distribution for the situation studied ($n \geq 10$). This is in line with other results confirming the general usefulness of AIC_c even when the data do not follow a normal distribution.

This book focuses on the formal (i.e., objective) aspects of model selection and multimodel inference. Whereas we do not offer much advice about how to apply professional judgments in the course of data analysis, we recognize the substantial role that judgment plays. We therefore recommend that people try to be clear in their work about what supports the conclusions they draw from data. What is the quantifiable evidence; upon what judgments does this assessment of evidence rely; are the persons making the judgments qualified to be making these judgments?

In Section 6.9.5 we consider the dimensional unit of AIC: It is information, in the Shannon sense that negative $\log(probability)$ mathematically characterizes information. Although the data have associated units of measurement, those units are lost, in a sense, as soon as one interprets the information in the data by using a likelihood, which properly has units of probability; hence, the units for $-\log(\mathcal{L})$ are information, regardless of the units of the original data. It also follows that the dimensional interpretation of Δ_i is information loss when using model $\hat{g}_i$ to approximate model $\hat{g}_{min}$ (the estimated K-L best model).

Model selection has been applied to finite mixture models, but this type of model is nonstandard and the likelihood ratio test must be modified for use with mixture models. This motivates a concern that theoretically the formula for AIC might need changing to apply to selection of finite mixture models. Our thought (Section 6.9.6) at this time is that the formula for AIC does not need to be modified for use with mixture models (heuristically because AIC is not a test, such as a LRT is). Instead, the big issue with finite mixture models is model redundancy that arises when a mixture model cannot be fit to the data, i.e., the fitted mixture model actually collapses to a simpler model. With mixture models it is critical to properly deal with this model redundancy (see Section 6.9.6).

We did some theoretical evaluation of whether AIC_c achieves its nominal target for a two-component mixture negative exponential model when properly accounting for model redundancy. It did quite well at $n \geq 100$ and seemed acceptable even for $n \geq 50$. Research on an improved small-sample version of

AIC (i.e., an alternative to AIC_c) is worth pursuing for use with finite mixture models.

The relative importance of a predictor variable x_i can be quantified by $w_+(i)$. This number is relative, not absolute, because it will be greater than 0 even if predictor x_i has no predictive value at all in the given context. A baseline value for $w_+(i)$ can be estimated by computer-intensive data permutation methods. One can then better judge what predictors or factors are really irrelevant in the data at hand. The method is simple, but computer intensive; details are in Section 6.9.8.

There is a lot of material in this chapter; some of it we have not summarized here. Many sections probe issues about AIC without fully resolving them and thereby suggest additional research areas. A couple of big-picture messages follow: The body fat example shows that one can expect substantial model selection uncertainty with all subsets selection applied to over 8,000 models and illustrates ways to deal with this uncertainty. Those ways are primarily types of multimodel inference. A seminal result is that AIC is Bayesian to the same extent as BIC and shows that the difference is all in the prior distribution over the model set (i.e., model probabilities). Finally, we show exactly how Bayesian model probabilities must be interpreted for BIC and AIC; these interpretations are different.

7
Statistical Theory and Numerical Results

This chapter contains theory and derivations relevant to Kullback–Leibler information-theory–based model selection. We have tried to make the other chapters of this book readable by a general audience, especially graduate students in various fields. Hence, we have reserved this chapter for the theoretical material we believe should be made available to statisticians and quantitative biologists. For many, it will suffice to know that this theory exists. However, we encourage researchers, especially if they have some mathematical–statistical training, to read and try to understand the theory given here, because that understanding provides a much deeper knowledge of many facets of K-L–based model selection in particular, and of some general model selection issues also.

The material given here is a combination of our distillation and interpretation of the existing literature and what we feel are clarifications and extensions of the existing theory. In the former case we have not drawn heavily or directly from any one source; hence there is no particular reference we could cite for these derivations. We have not indicated what results might be truly new to the literature about the estimation of expected K-L information, partly because this is sometimes not clear even to us.

7.1   Useful Preliminaries

The sole purpose of this section is to provide a summary of the basic notation, concepts, and mathematical background needed to produce and understand the derivation of AIC and related issues that follow this section. Even researchers who totally understand the mathematics involved will benefit from this section in that it establishes much of the notation and conventions to be used in Section 7.2 and beyond in this chapter.

As a model selection criterion, it is clear what AIC is: $-2\log(\mathcal{L}(\hat{\theta})) + 2K$ for a model with K estimated parameters, $\hat{\theta}$ being the MLE of those parameters, computed from the data $\underline{x}$, under an assumed model (i.e., pdf) $g(\underline{x}\,|\,\underline{\theta})$. However, we need more detailed notation than just $\mathcal{L}(\hat{\theta})$, and in the derivations we need to alternate between the likelihood and the pdf interpretations of the model. Therefore, without loss of generality we take the likelihood of $\underline{\theta}$ as $\mathcal{L}(\underline{\theta}\,|\,\underline{x}) = g(\underline{x}\,|\,\underline{\theta})$ by simply then interpreting g as a function of $\underline{\theta}$ given $\underline{x}$. If instead of using this convention we had constantly switched notation between $g(\underline{x}\,|\,\underline{\theta})$ and $\mathcal{L}(\underline{\theta}\,|\,\underline{x})$, that would be more confusing than simply staying with the single notation $g(\underline{x}\,|\,\underline{\theta})$. This dual usage of the notation $g(\underline{x}\,|\,\underline{\theta})$ is thus noted; the reader must follow the mathematics with an eye to which usage is being made at any point.

A second dual usage of notation for the random variable $\underline{x}$ arises: Sometimes $\underline{x}$ denotes the data (as a random variable), and sometimes $\underline{x}$ denotes the variable of integration, always with respect to $f(\underline{x})$, under an integral sign (over an n-dimensional space). Because we are dealing with random variables, integration is usually denoted in terms of the statistical expectation operator, but that operator is just an integral. At times we must have both an integral (hence $\underline{x}$) and, separately, data, say $\underline{y}$. But the notation for data versus integrand variable is arbitrary and sometimes must be switched back and forth in the derivations. It becomes impossible always to use $\underline{x}$ for a variable of integration and $\underline{y}$ for data; hence, we do not try to do so, and instead we often use $\underline{x}$ to denote data even though at other times $\underline{x}$ is an integrand variable and $\underline{y}$ are the data. *Always*, however, the data, no matter how denoted ($\underline{x}$ or $\underline{y}$, or otherwise), actually arise from truth $f(\cdot)$, not from $g(\cdot\,|\,\underline{\theta})$ (when $f \neq g$); this is a critically important point.

AIC has been motivated, justified, and derived in a variety of ways (see, for example, Akaike 1973, Sawa 1978, Sugiura 1978, Chow 1981, Stone 1982, Shibata 1989, Bozdogan 1987), but these derivations are often cryptic and thus difficult to follow. Here we give a general derivation in some detail, but without being rigorous about all required conditions (they are not very restrictive). We do note where approximations are made. The data have some sample size n, and the general result is justified for "large" n. That is, the result is justified asymptotically as $n \to \infty$. Also, the integrals and expectations shown are over an n-dimensional sample space, although that fact is not fully indicated by the notation used.

The most general approach to deriving AIC uses the Taylor series expansion to second order. An elementary introduction to the Taylor series is given in Peterson (1960) (or any introductory calculus book); a more rigorous treatment, including results for real-valued multivariable functions, is given by Apostol

(1957) (or any rigorous book on real analysis). If $h(\underline{\theta})$ is a real-valued function on K dimensions, then the Taylor series expansion about some value $\underline{\theta}_o$ near to $\underline{\theta}$ is given below:

$$h(\underline{\theta}) = h(\underline{\theta}_o) + \left[\frac{\partial h(\underline{\theta}_o)}{\partial\underline{\theta}}\right]'[\underline{\theta}-\underline{\theta}_o] + \frac{1}{2}[\underline{\theta}-\underline{\theta}_o]'\left[\frac{\partial^2 h(\underline{\theta}_o)}{\partial\underline{\theta}^2}\right][\underline{\theta}-\underline{\theta}_o] + Re \quad (7.1)$$

($\underline{\theta}$ and $\underline{\theta}_o$ are just two different points in the space over which $h(\cdot)$ is defined). Here, Re represents the exact remainder term for the quadratic Taylor series expansion; the exact nature of Re is known (see Apostol 1957). Various approximations for the error that results from ignoring Re can be given. For its heuristic value only, we can claim that an approximation to this error, Re, is of order

$$O(\|\underline{\theta} - \underline{\theta}_o\|^3).$$

Here, for any vector argument $\underline{z} - \underline{w}$,

$$\|\underline{z} - \underline{w}\| = \sqrt{\sum_{i=1}^{K}(z_i - w_i)^2}$$

denotes the Euclidean distance between the two points in the K-dimensional space. Thus, the order of the approximation error is the cube of the Euclidean distance between $\underline{\theta}$ and $\underline{\theta}_o$. This is quite a simplification of what Re is, but it makes the point that the error of approximation is quite small if this distance is small.

The notation $O(x)$ denotes an unspecified (but possibly complicated) function of the scalar argument x that satisfies the condition that $O(x)$ is approximately equal to cx for small x, where c is a constant. Hence, $O(x)$ goes to 0 at least at a linear rate in x as x gets near 0. In the case of (7.1) the quadratic approximation to $h(\underline{\theta})$ "near" $\underline{\theta}_o$ is arbitrarily good, as $\underline{\theta}$ becomes nearer to $\underline{\theta}_o$ for $h(\cdot)$ a suitably smooth and bounded function.

In (7.1) the notation

$$\left[\frac{\partial h(\underline{\theta}_o)}{\partial\underline{\theta}}\right]$$

denotes a $K \times 1$ column vector of the first partial derivatives of $h(\underline{\theta})$ with respect to $\theta_1, \ldots, \theta_K$, evaluated at $\underline{\theta} = \underline{\theta}_o$; hence,

$$\left[\frac{\partial h(\underline{\theta}_o)}{\partial\underline{\theta}}\right] = \begin{bmatrix} \dfrac{\partial h(\underline{\theta})}{\partial\theta_1} \\ \vdots \\ \dfrac{\partial h(\underline{\theta})}{\partial\theta_K} \end{bmatrix}_{|\underline{\theta}=\underline{\theta}_o}.$$

The notation

$$\left[\frac{\partial^2 h(\underline{\theta}_o)}{\partial \underline{\theta}^2}\right] = \left\{\frac{\partial^2 h(\underline{\theta})}{\partial \theta_i \partial \theta_j}\right\}_{|\underline{\theta}=\underline{\theta}_o}, \qquad i = 1, \ldots, K, \quad j = 1, \ldots, K, \quad (7.2)$$

denotes the $K \times K$ matrix of second mixed partial derivatives of $h(\underline{\theta})$ with respect to $\theta_1, \ldots, \theta_K$, evaluated at $\underline{\theta} = \underline{\theta}_o$. This matrix is often called the Hessian of $h(\underline{\theta})$.

The expansion in (7.1) when terminated at the quadratic term is only an approximation to $h(\underline{\theta})$. In this deterministic case, as indicated above, the error of approximation is related roughly to the cube of the Euclidean distance between $\underline{\theta}$ and $\underline{\theta}_o$. For a sufficiently small distance, this is a good order of approximation. For the cases of interest, $h(\cdot)$ will be a log-likelihood function based on a probability distribution. One special value of $\underline{\theta}$, denoted by $\underline{\theta}_o$, needed in these expansions is the large-sample (hence approximate) expected value of the MLE $\hat{\underline{\theta}}$; that is, $E(\hat{\underline{\theta}}) \approx \underline{\theta}_o$ for large n (the exact nature of $\underline{\theta}_o$ in relation to K-L information will be given below). The approximation here is often of order $1/n$, denoted by $O(1/n)$. This notation means that the error of approximation in $E(\hat{\underline{\theta}}) \approx \underline{\theta}_o$ is less than or equal to a constant divided by the sample size for large sample sizes (the constant might even be 0).

Stronger statements about large-sample limits are possible. In particular, as sample size $n \to \infty$, $\hat{\underline{\theta}} \to \underline{\theta}_o$ with probability 1, and the Taylor series approximation given by (7.1) is quite good. In this case (7.1) becomes

$$h(\hat{\underline{\theta}}) = h(\underline{\theta}_o) + \left[\frac{\partial h(\underline{\theta}_o)}{\partial \underline{\theta}}\right]'[\hat{\underline{\theta}} - \underline{\theta}_o] + \frac{1}{2}[\hat{\underline{\theta}} - \underline{\theta}_o]'\left[\frac{\partial^2 h(\underline{\theta}_o)}{\partial \underline{\theta}^2}\right]'[\hat{\underline{\theta}} - \underline{\theta}_o] + O_p(1/n).$$
$$(7.3)$$

Now the error of approximation in (7.3) is stochastic, but its expectation is generally on the order of $1/n$ with probability going to 1 as $n \to \infty$, hence the added "p" notation of the form $O_p(\cdot)$. The exact size of the expected error of approximation in expansions like (7.3) is not known (in general), but it is negligible for large sample sizes, subject to mild regularity conditions of the same type needed to ensure that the MLE is well behaved (see, for example, Lehmann 1983).

In the context of parametric MLE the standard approach is to assume that the data are generated by one specific member of a family of models. That family of models, denoted here by $g(\underline{x} \mid \underline{\theta})$, is a set of probability distributions indexed by an unknown parameter that may be estimated by any value in the parameter space Θ. By assumption, truth corresponds to one specific (but unknown) value of $\underline{\theta}$, which we could for clarity denote by $\underline{\theta}_o$. One would not ask where $\underline{\theta}_o$ comes from; it simply exists as (unknown) truth. Thus even when we assume that the known model structure of g is true, there is still a fundamental concept of an underlying unknown truth to the problem of inference from data (and we cannot know, metaphysically, where this truth $\underline{\theta}_o$ comes from).

When we acknowledge that g is just a model of truth, hence must be mis-specified, the issue arises as to what unique parameter in Θ, hence what unique distribution in the class g, we are estimating. In fact, there is a unique parameter in Θ that the MLE $\hat{\underline{\theta}}$ is estimating given the parametric class of models and given the concept of a fixed underlying unknown truth, as a pdf $f(\underline{x})$. As part of this essential conceptualization of the inference problem we must assume that the data arose from some deep truth, denoted without loss of generality by f. Now, one cannot usefully ask where truth f comes from, in the same metaphysical sense that one cannot ask where $\underline{\theta}_o$ comes from under the assumption that $g(\underline{x} \mid \underline{\theta}_o)$ is truth, but we just do not happen to know true $\underline{\theta}_o$.

Given this essential framework of f as truth (rather than any model structure as truth) we can, and must, ask whether there is a unique model $g(\cdot \mid \underline{\theta}_o)$ in the class of models $g(\cdot \mid \underline{\theta})$ that best describes the data. Hence, given the set of models $g(\cdot \mid \underline{\theta})$, $\underline{\theta} \in \Theta$, is there a unique $\underline{\theta}_o$ that the MLE is estimating, and is this $g(\cdot \mid \underline{\theta}_o)$ a best model in some sense? In fact, the MLE is (for large samples) estimating a unique parameter value that we will denote by $\underline{\theta}_o$; it is this parameter value that indexes our target model under likelihood inference (we will say more on this below).

Approached theoretically, ignoring issues of data and estimation, the best approximating model g in the class of models considered, under the (compelling) K-L information measure, is simply the model that produces the minimum K-L value over Θ. Hence we look for a unique value of $\underline{\theta} \in \Theta$, which we will denote by $\underline{\theta}_o$, that provides the K-L best approximating model. Therefore, $\underline{\theta}_o$ is the solution to the optimization problem

$$\min_{\underline{\theta} \in \Theta} [I(f, g)] = \int f(\underline{x}) \log \left(\frac{f(\underline{x})}{g(\underline{x} \mid \underline{\theta}_o)} \right) d\underline{x}.$$

Clearly, $g(\underline{x} \mid \underline{\theta}_o)$ is the best model here, and this serves, in fact, to define truth as a target $\underline{\theta}_o$ given f and given the class of models g. As we will discuss below, the MLE of $\underline{\theta}$ under model g is estimating $\underline{\theta}_o$.

Given the assumed regularity conditions on the model, $\underline{\theta}_o$ satisfies the vector equations

$$\frac{\partial}{\partial \underline{\theta}} \int f(\underline{x}) \log \left(\frac{f(\underline{x})}{g(\underline{x} \mid \underline{\theta}_o)} \right) d\underline{x} = \underline{0}. \tag{7.4}$$

Rewriting (7.4) using that $\log(a/b) = \log(a) - \log(b)$, we have

$$\frac{\partial}{\partial \underline{\theta}} \int f(\underline{x}) \log(f(\underline{x})) - \frac{\partial}{\partial \underline{\theta}} \int f(\underline{x}) \log(g(\underline{x} \mid \underline{\theta})) d\underline{x} = \underline{0}.$$

Because $\underline{\theta}$ is not involved in $f(\cdot)$, the first term of the above is $\underline{0}$. The second term (ignoring the minus sign) can be written as

$$\int f(\underline{x}) \left[\frac{\partial}{\partial \underline{\theta}} \log(g(\underline{x} \mid \underline{\theta})) \right]_{|\underline{\theta}=\underline{\theta}_o} d\underline{x} = \mathrm{E}_f \left[\left[\frac{\partial}{\partial \underline{\theta}} \log(g(\underline{x} \mid \underline{\theta})) \right]_{|\underline{\theta}=\underline{\theta}_o} \right] = \underline{0}.$$

A more compact way to denote this result is

$$E_f \left[\frac{\partial}{\partial \underline{\theta}} \log(g(\underline{x} \mid \underline{\theta}_o)) \right] = \underline{0}. \tag{7.5}$$

The well-known asymptotic consistency property of MLEs and strong convergence of means of *iid* random variables allow, in conjunction with (7.5), another interpretation of $\underline{\theta}_o$. If $\underline{x}$ represents an *iid* sample of size n from pdf $f(\underline{x}) \equiv \prod_{i=1}^{n} f(x_i)$ and we consider the MLE under model $g(\underline{x} \mid \underline{\theta}) \equiv \prod_{i=1}^{n} g(x_i \mid \underline{\theta}_o)$, then for every n we have the K likelihood equations (expressed as a mean, without loss of generality)

$$\frac{1}{n} \left[\sum_{i=1}^{n} \frac{\partial}{\partial \underline{\theta}} \log(g(x_i \mid \hat{\underline{\theta}})) \right] = \underline{0}.$$

As $n \to \infty$ two limits are approached with probability one (almost sure convergence). The sequence of MLEs $\hat{\underline{\theta}}(n)$ (adding notation to denote the MLE as a function of sample size), converges to something. In fact, $\hat{\underline{\theta}}(n)$ has to converge to $\underline{\theta}_o$, because the means on the left-hand sides of the above likelihood equations converge (as n gets large) to their expected values, which must all equal 0. Under suitable regularity conditions, (7.5) is satisfied only for the unique value of $\underline{\theta} = \underline{\theta}_o$. Hence, the sequence $\hat{\underline{\theta}}(n)$ must converge almost surely to $\underline{\theta}_o$, which is the K-L minimizer (see, e.g., White 1994).

Some deep ideas and philosophy are involved in the above results. In particular, we have the distinction that unknown truth $f(\underline{x})$ implicitly incorporates the numerical values on the (often only) conceptual, but unknown, parameter $\underline{\theta}$ of interest to us. Yet $f(\underline{x})$ is not a mathematical function of $\underline{\theta}$. Only our model $g(\underline{x} \mid \underline{\theta})$ is a mathematical function of $\underline{\theta}$, because therein $\underline{\theta}$ is unknown, but interpretable, hence useful to consider, and varies over a defined parameter space. Even if we think that $g(\cdot)$ represents truth, this is only the case at the single point $\underline{\theta}_o$ in the parameter space (in a frequentist philosophy of statistics). Hence, in this case we would be saying that $f(\underline{x}) \equiv g(\underline{x} \mid \underline{\theta}_o)$, where $\underline{\theta}_o$ is a single fixed point; thus even in this context $\underline{\theta}$ is not a variable in $f(\underline{x})$. Therefore, in this or any case, $f(\underline{x})$ is not a function of $\underline{\theta}$, and therefore

$$\frac{\partial}{\partial \underline{\theta}} \int f(\underline{x}) \log(f(\underline{x})) d\underline{x} = \underline{0}.$$

The derivation of AIC occurs in the context of probability distributions and expectations of functions of random variables. Such expectations are just types of integrals, but the notation and "machinery" of statistical expectations are more convenient to use here than the explicit notation of integration. One particular aspect of this matter that needs to be noted is the validity of interchanging the order of taking two expectations of the form $E_{\underline{x}} E_{\underline{y}} [h(\underline{x}, \underline{y})]$ when $\underline{x}$ and $\underline{y}$ denote random variables. The function $h(\cdot, \cdot)$ is arbitrary. From basic calculus of integrals as linear operators, $E_{\underline{x}} E_{\underline{y}} [h(\underline{x}, \underline{y})] = E_{\underline{y}} E_{\underline{x}} [h(\underline{x}, \underline{y})]$. This interchange equivalence is true for the case of $\underline{x}$ and $\underline{y}$ having the same or different probability distributions and whether or not $\underline{x}$ and $\underline{y}$ are independent.

Another aspect of preliminaries concerns (7.2). If $h(\,\cdot\,)$ is the log-likelihood, $\log(g(\underline{x}\,|\,\underline{\theta}))$, then (7.2) is

$$\left\{\frac{\partial^2 \log(g(\underline{x}\,|\,\underline{\theta}))}{\partial\theta_i\,\partial\theta_j}\right\}_{|\underline{\theta}=\underline{\theta}_o},$$

which is related to the Fisher information matrix

$$\mathcal{I}(\underline{\theta}_o) = \mathrm{E}_g\left\{-\frac{\partial^2 \log(g(\underline{x}\,|\,\underline{\theta}))}{\partial\theta_i\,\partial\theta_j}\right\}_{|\underline{\theta}=\underline{\theta}_o} \qquad (7.6)$$

(expectation here is with respect to $g(\,\cdot\,)$). If $g(\,\cdot\,)$ is the true model form (which it is if f is a special case of g, or if $g = f$), then the sampling variance–covariance matrix Σ of the MLE is (for large samples) $\Sigma = [\mathcal{I}(\underline{\theta}_o)]^{-1}$. That is, $\Sigma = \mathrm{E}(\hat{\underline{\theta}} - \underline{\theta}_o)(\hat{\underline{\theta}} - \underline{\theta}_o)'$ is $[\mathcal{I}(\underline{\theta}_o)]^{-1}$. If g is not the true model for $\underline{x}$ (it may be less general than the true model, or otherwise different from f), then in general we must expect that $\Sigma \neq [\mathcal{I}(\underline{\theta}_o)]^{-1}$. In fact, in deriving AIC, we take expectations with respect to f, not g. Hence, we define

$$I(\underline{\theta}_o) = \mathrm{E}_f\left\{-\frac{\partial^2 \log(g(\underline{x}\,|\,\underline{\theta}))}{\partial\theta_i\,\partial\theta_j}\right\}_{|\underline{\theta}=\underline{\theta}_o}. \qquad (7.7)$$

In the case that $f = g$ or f is a special case of g, then and only then do we have $\mathcal{I}(\underline{\theta}_o) = I(\underline{\theta}_o)$. We will not generally make this distinction in our notation as to whether the situation allows $\mathcal{I}(\underline{\theta}_o) = I(\underline{\theta}_o)$ or not. It is an important matter, however, to be always cognizant of whether the expectation defining any given $I(\underline{\theta}_o)$ is with respect to f or g.

Additional notation useful here is the empirical, but unknown, matrix

$$\hat{I}(\underline{\theta}_o) = \left\{-\frac{\partial^2 \log(g(\underline{x}\,|\,\underline{\theta}))}{\partial\theta_i\,\partial\theta_j}\right\}_{|\underline{\theta}=\underline{\theta}_o}.$$

For simpler notation we will use

$$I(\underline{\theta}_o) = \mathrm{E}_f\left[-\frac{\partial^2 \log(g(\underline{x}\,|\,\underline{\theta}_o))}{\partial\theta^2}\right],$$

which means exactly the same as (7.7), and hence simpler notation for the $\hat{I}(\underline{\theta}_o)$ is

$$\hat{I}(\underline{\theta}_o) = -\frac{\partial^2 \log(g(\underline{x}\,|\,\underline{\theta}_o))}{\partial\theta^2}.$$

It is obvious that $\mathrm{E}_f[\hat{I}(\underline{\theta}_o)] = I(\underline{\theta}_o)$. When $\underline{x}$ is a random sample from $f(\,\cdot\,)$, $\hat{I}(\underline{\theta}_o)$ converges to $I(\underline{\theta}_o)$ as $n \to \infty$. We can express this alternatively as

$$\hat{I}(\underline{\theta}_o) = I(\underline{\theta}_o) + Re, \text{ and usually } Re \text{ is } O(1/n).$$

An actual estimator of $I(\underline{\theta}_o)$ is $\hat{I}(\underline{\hat{\theta}})$ (the negative Hessian of the log-likelihood equations):

$$\hat{I}(\underline{\hat{\theta}}) = -\frac{\partial^2 \log(g(\underline{x} \mid \underline{\hat{\theta}}))}{\partial \underline{\theta}^2}. \tag{7.8}$$

Because $\underline{\hat{\theta}}$ is the MLE under the model $g(\underline{x} \mid \underline{\theta})$, $\underline{\hat{\theta}}$ converges to $\underline{\theta}_o$ as $n \to \infty$, and hence $\hat{I}(\underline{\hat{\theta}})$ converges to $I(\underline{\theta}_o)$. Thus, $\hat{I}(\underline{\hat{\theta}}) \approx I(\underline{\theta}_o)$; the order of this approximation is at worst $O(1/\sqrt{n})$, and in most common applications it will be $O(1/n)$. If we could determine the analytical form (under actual analysis of data) of $I(\underline{\theta}_o)$, an alternative estimator would be $I(\underline{\hat{\theta}})$, i.e., (7.7) evaluated at the MLE; $I(\underline{\hat{\theta}})$ is often not the same as $\hat{I}(\underline{\hat{\theta}})$. Note also that the commonly used estimator $\mathcal{I}(\underline{\hat{\theta}})$ (i.e., (7.6) evaluated at the MLE) is not always the same as either $I(\underline{\hat{\theta}})$ or $\hat{I}(\underline{\hat{\theta}})$ and may not converge to $I(\underline{\theta}_o)$.

There are two ways to compute the Fisher information matrix, $\mathcal{I}(\underline{\theta})$, of (7.6) when $f = g$. This additional material, and more, is needed below, hence is given here. Because the model is a probability distribution,

$$\int g(\underline{x} \mid \underline{\theta}) d\underline{x} = 1,$$

and therefore (under the same mild regularity conditions already assumed)

$$\int \frac{\partial g(\underline{x} \mid \underline{\theta})}{\partial \underline{\theta}} d\underline{x} = \underline{0}.$$

Next, we use in the above the result

$$\frac{\partial \log(g(\underline{x} \mid \underline{\theta}))}{\partial \underline{\theta}} = \frac{1}{g(\underline{x} \mid \underline{\theta})} \left[\frac{\partial g(\underline{x} \mid \underline{\theta})}{\partial \underline{\theta}} \right],$$

and hence we get

$$\int g(\underline{x} \mid \underline{\theta}) \left[\frac{\partial}{\partial \underline{\theta}} \log(g(\underline{x} \mid \underline{\theta})) \right] d\underline{x} = \underline{0}. \tag{7.9}$$

Now take the partial derivative vector of (7.9) with respect to $\underline{\theta}$ to get (7.10); this derivation uses the chain rule of differentiation and some of the above algebraic results:

$$\int g(\underline{x} \mid \underline{\theta}) \left[\frac{\partial}{\partial \underline{\theta}} \log(g(\underline{x} \mid \underline{\theta})) \right] \left[\frac{\partial}{\partial \underline{\theta}} \log(g(\underline{x} \mid \underline{\theta})) \right]' d\underline{x}$$
$$+ \int g(\underline{x} \mid \underline{\theta}) \frac{\partial^2 \log(g(\underline{x} \mid \underline{\theta}))}{\partial \underline{\theta}^2} d\underline{x} = O \quad (7.10)$$

(O is a $K \times K$ matrix of zero elements). We can rewrite (7.10) as

$$E_g \left[\left[\frac{\partial}{\partial \underline{\theta}} \log(g(\underline{x} \mid \underline{\theta})) \right] \left[\frac{\partial}{\partial \underline{\theta}} \log(g(\underline{x} \mid \underline{\theta})) \right]' \right] = E_g \left[-\frac{\partial^2 \log(g(\underline{x} \mid \underline{\theta}))}{\partial \underline{\theta}^2} \right],$$

or

$$E_g\left[\left[\frac{\partial}{\partial\underline{\theta}}\log(g(\underline{x}\mid\underline{\theta}))\right]\left[\frac{\partial}{\partial\underline{\theta}}\log(g(\underline{x}\mid\underline{\theta}))\right]'\right]=\mathcal{I}(\underline{\theta}).$$

We will denote the left-hand side of the above by $\mathcal{J}(\underline{\theta})$; hence define

$$\mathcal{J}(\underline{\theta})=E_g\left[\left[\frac{\partial}{\partial\underline{\theta}}\log(g(\underline{x}\mid\underline{\theta}))\right]\left[\frac{\partial}{\partial\underline{\theta}}\log(g(\underline{x}\mid\underline{\theta}))\right]'\right]. \qquad (7.11)$$

Thus, $\mathcal{I}(\underline{\theta}) = \mathcal{J}(\underline{\theta})$, but the expectations underlying this result are taken with respect to $g(\underline{x}\mid\underline{\theta})$, not with respect to to $f(\underline{x})$. One implication of this is that the inverse Fisher information matrix may not be the theoretically correct conditional variance–covariance matrix of the MLE if the model is misspecified.

What we need more than (7.11) is

$$J(\underline{\theta})=E_f\left[\left[\frac{\partial}{\partial\underline{\theta}}\log(g(\underline{x}\mid\underline{\theta}))\right]\left[\frac{\partial}{\partial\underline{\theta}}\log(g(\underline{x}\mid\underline{\theta}))\right]'\right]. \qquad (7.12)$$

We can expect $J(\underline{\theta}) = \mathcal{J}(\underline{\theta})$ only when $f = g$, or f is a special case of g. Although $\mathcal{I}(\underline{\theta}) = \mathcal{J}(\underline{\theta})$, there is no such general equality between $I(\underline{\theta})$ and $J(\underline{\theta})$ when g is only an approximation to f, hence when the K-L discrepancy between f and g, $I(f, g)$, is > 0. Heuristically, however, we can expect near equalities of the sort $I(\underline{\theta}_o) \approx J(\underline{\theta}_o)$, $\mathcal{I}(\underline{\theta}_o) \approx I(\underline{\theta}_o)$, and $\mathcal{J}(\underline{\theta}_o) \approx J(\underline{\theta}_o)$ when $I(f, g) \approx 0$, hence when a good approximating model is used.

There is a large-sample relationship among $I(\underline{\theta}_o)$, $J(\underline{\theta}_o)$, and Σ that is worth knowing, and perhaps should be used more:

$$I(\underline{\theta}_o)\Sigma = J(\underline{\theta}_o)[I(\underline{\theta}_o)]^{-1}, \qquad (7.13)$$

and hence

$$\Sigma = [I(\underline{\theta}_o)]^{-1}J(\underline{\theta}_o)[I(\underline{\theta}_o)]^{-1}, \qquad (7.14)$$

where Σ is the true large-sample variance–covariance matrix of the MLE of $\underline{\theta}$ derived from model g when f is truth. It suffices to derive (7.14), although it is (7.13) that we will use more directly in deriving AIC.

Expanding the likelihood equations evaluated at $\underline{\theta}_o$ as a first-order Taylor series about the MLE, we have

$$\frac{\partial}{\partial\underline{\theta}}\log(g(\underline{x}\mid\underline{\theta}_o)) \approx \frac{\partial}{\partial\underline{\theta}}\log(g(\underline{x}\mid\hat{\underline{\theta}})) + \left[\frac{\partial^2\log(g(\underline{x}\mid\hat{\underline{\theta}}))}{\partial\underline{\theta}^2}\right](\underline{\theta}_o - \hat{\underline{\theta}}).$$

The MLE satisfies

$$\frac{\partial}{\partial\underline{\theta}}\log(g(\underline{x}\mid\hat{\underline{\theta}})) = \underline{0};$$

hence we have

$$\frac{\partial}{\partial \underline{\theta}} \log(g(\underline{x} \mid \theta_o)) \approx \left[-\frac{\partial^2 \log(g(\underline{x} \mid \hat{\theta}))}{\partial \theta^2} \right] (\hat{\theta} - \theta_o)$$

$$= \hat{I}(\hat{\theta})(\hat{\theta} - \theta_o) \approx I(\theta_o)(\hat{\theta} - \theta_o).$$

From the above we get

$$[I(\theta_o)]^{-1} \left[\frac{\partial}{\partial \underline{\theta}} \log(g(\underline{x} \mid \theta_o)) \right] \approx (\hat{\theta} - \theta_o). \tag{7.15}$$

Transpose (7.15) and use that transposed result along with, again, (7.15) to derive

$$[I(\theta_o)]^{-1} \left[\frac{\partial}{\partial \underline{\theta}} \log(g(\underline{x} \mid \theta_o)) \right] \left[\frac{\partial}{\partial \underline{\theta}} \log(g(\underline{x} \mid \theta_o)) \right]' [I(\theta_o)]^{-1} \approx (\hat{\theta} - \theta_o)(\hat{\theta} - \theta_o)'.$$

Now take the expectation of the above with respect to $f(\underline{x})$ to get (see (7.13))

$$[I(\theta_o)]^{-1} J(\theta_o)[I(\theta_o)]^{-1} \approx E_f(\hat{\theta} - \theta_o)(\hat{\theta} - \theta_o)' = \Sigma;$$

hence, we have (7.14) as a large-sample result.

The above likelihood-based results under either a true data-generating model or under model misspecification (i.e., truth is f, the model used is g) are all in the statistical literature. For very rigorous derivations see White (1994).

To take expectations of the quadratic forms that are in expansions like (7.3) we will need to use an equivalent expression of that form:

$$\underline{z}' A \underline{z} = \text{tr} \left[A \underline{z} \underline{z}' \right].$$

Here "tr" stands for the matrix trace function, the sum of the diagonal elements of a square matrix. The trace function is a linear operator; therefore, when the quadratic is a stochastic variable in $\underline{z}$, its expectation can be written as

$$E_{\underline{z}} \left[\underline{z}' A \underline{z} \right] = \text{tr} \left[E_{\underline{z}} \left[A \underline{z} \underline{z}' \right] \right].$$

If A is fixed (or stochastic but independent of $\underline{z}$), then $E_{\underline{z}} \left[A \underline{z} \underline{z}' \right] = A E_{\underline{z}}[\underline{z} \underline{z}']$. If $\underline{z}$ has mean $\underline{0}$ (such as $\underline{z} = \hat{\theta} - E(\hat{\theta})$), then $E_{\underline{z}}[\underline{z} \underline{z}'] = \Sigma$ is the variance–covariance matrix of $\underline{z}$; hence then

$$E_{\underline{z}} \left[\underline{z}' A \underline{z} \right] = \text{tr} \left[A \Sigma \right].$$

If A is stochastic but independent of $\underline{z}$, then we can use

$$E_A E_{\underline{z}} \left[\underline{z}' A \underline{z} \right] = \text{tr} \left[E_A E_{\underline{z}} \left[A \underline{z} \underline{z}' \right] \right] = \text{tr} \left[E_A(A) E_{\underline{z}}(\underline{z} \underline{z}') \right].$$

A final aspect of notation, and of concepts, reemphasizes some ideas at the start of this section: The notation for a random variable (i.e., data point) is arbitrary in taking expectations over the sample space. What is not arbitrary for such an expectation (i.e., integration) is the model used, which refers to its form, its assumptions, its parameters, and the distribution of unexplained residuals (i.e., "errors"). Moreover, it is just a convenience to switch thinking modes between integrals and expectations in this probabilistic modeling and data-analysis framework. Because an expectation is a type of integral over a defined space, the result of the integration is not dependent on the notation used in the integrand. Thus

$$
\begin{aligned}
E_f \left[\log(g(\underline{x} \mid \hat{\underline{\theta}}(\underline{x}))) \right] &= \int f(\underline{x}) \log(g(\underline{x} \mid \hat{\underline{\theta}}(\underline{x}))) d\underline{x} \\
&\equiv \int f(\underline{y}) \log(g(\underline{y} \mid \hat{\underline{\theta}}(\underline{y}))) d\underline{y} \\
&= E_f \left[\log(g(\underline{y} \mid \hat{\underline{\theta}}(\underline{y}))) \right].
\end{aligned}
$$

Changing notation for the integrand (i.e., $\underline{x}$ to $\underline{y}$) has no effect on the result; this type of useful notation change is required in derivations below, because in places, at a conceptual level, we recognize two independent samples, hence have two notations, $\underline{x}$ and $\underline{y}$. In fact, these derivations are about average frequentist properties of data-analysis methods, but there is no real data literally being used in these derivations. Rather, in these theoretical derivations the possible "data" are just points in an n-dimensional sample space that arise in accordance with some true probability distribution $f(\cdot)$.

7.2   A General Derivation of AIC

We now give a general conceptual and then mathematical derivation of AIC starting from K-L information for the best approximating model in the class of models $g(\underline{x} \mid \theta)$:

$$
I(f, g(\cdot \mid \underline{\theta}_o)) = \int f(\underline{x}) \log \left(\frac{f(\underline{x})}{g(\underline{x} \mid \underline{\theta}_o)} \right) d\underline{x}. \tag{7.16}
$$

Note that while for the model we do not know $\underline{\theta}$, the target K-L information value for the class of models is appropriately taken as $I(f, g)$ evaluated at $\underline{\theta}_o$ (i.e., 7.16), because the parameter value we will be estimating is $\underline{\theta}_o$. Also, note the expanded notation in (7.16), so we can represent $I(f, g)$ as dependent, in general, on the unknown parameter value, given the model form. However, $I(f, g)$ does not involve any data, nor any value of $\underline{x}$, since $\underline{x}$ has been integrated out.

Given that we have data $\underline{y}$ as a sample from $f(\cdot)$, the logical step would be to find the MLE $\hat{\underline{\theta}} = \hat{\underline{\theta}}(\underline{y})$ and compute an estimate of $I(f, g(\cdot \mid \underline{\theta}_o))$ as

$$
I(f, g(\cdot \mid \hat{\underline{\theta}}(\underline{y}))) = \int f(\underline{x}) \log \left(\frac{f(\underline{x})}{g(\underline{x} \mid \hat{\underline{\theta}}(\underline{y}))} \right) d\underline{x}.
$$

This $I(f, g(\cdot \,|\, \hat{\underline{\theta}}(\underline{y})))$ remains conceptual, since we do not know f. Still, it is useful to push ahead, and we shall do so with two different conceptual approaches, both will lead to the same basis for AIC (there is no unique path from K-L to AIC).

If we could find the $\underline{\theta}_o$ that minimizes K-L (for a given g), we would know that our target for a perfect model would be $I(f, g) = 0$. We could then judge how good any model is relative to this absolute value of zero. But matters change when we have only an estimate of $\underline{\theta}$. Even if our model structure was (miraculously) truth, hence $g(\underline{x} \,|\, \underline{\theta}_o) = f(\underline{x})$, our estimator $\hat{\underline{\theta}}(\underline{y})$ would not equal $\underline{\theta}_o$ almost surely for continuous parameters and distributions, and at best for some discrete distributions the equality would be with probability $\ll 1$. Any value of $\hat{\underline{\theta}}(\underline{y})$ other than $\underline{\theta}_o$ results in $I(f, g(\cdot \,|\, \hat{\underline{\theta}}(\underline{y}))) > I(f, g(\cdot \,|\, \underline{\theta}_o))$. Thus, even if we had the correct model structure, because we must estimate $\underline{\theta}$ we should think in terms of the (essentially estimated) K-L as taking, on average, a value > 0. This motivates us to revise our idea of what our target must be as a measure of perfect agreement of fitted model with truth f.

In the context of repeated sampling properties as a guide to inference we would expect our estimated K-L to have on average the positive value $\mathrm{E}_{\underline{y}}\left[I(f, g(\cdot \,|\, \hat{\underline{\theta}}(\underline{y})))\right]$. We should therefore readjust our idea of perfection of the model to be not the minimizing of $I(f, g(\cdot \,|\, \underline{\theta}_o))$ (given g), but the slightly larger value, on average, given by

$$\mathrm{E}_{\underline{y}}\left[I(f, g(\cdot \,|\, \hat{\underline{\theta}}(\underline{y})))\right] > I(f, g(\cdot \,|\, \underline{\theta}_o))$$

(and repeating ourselves because it is an important point: All expectations here are with respect to f regardless of the notation for random variables involved, such as $\underline{x}$, $\underline{y}$, or $\hat{\underline{\theta}}$). Thus, given the reality that we must *estimate* $\underline{\theta}$, we must adopt the criterion

$$\text{"select the model } g \text{ to minimize } \mathrm{E}_{\underline{y}}\left[I(f, g(\cdot \,|\, \hat{\underline{\theta}}(\underline{y})))\right]." \qquad (7.17)$$

Hence our goal must be to minimize the expected value of this (conceptually) estimated K-L information value. (If we could compute the value of $\underline{\theta}_o$ for each model, we could stay with the goal of minimizing K-L itself. For the curious we note here that the large-sample difference is $\mathrm{E}_{\underline{y}}\left[I(f, g(\cdot \,|\, \hat{\underline{\theta}}(\underline{y})))\right] - I(f, g(\cdot \,|\, \underline{\theta}_o)) = \frac{1}{2}\,\mathrm{tr}\left[J(\underline{\theta}_o)I(\underline{\theta}_o)^{-1}\right]$, which does not depend on sample size n.)

Rewriting the basis of this new target to be minimized, (7.17), we have

$$\mathrm{E}_{\underline{y}}\left[I(f, g(\cdot \,|\, \hat{\underline{\theta}}(\underline{y})))\right] = \int f(\underline{x}) \log(f(\underline{x})) d\underline{x} - \mathrm{E}_{\underline{y}}\left[\int f(\underline{x}) \log[g(\underline{x} \,|\, \hat{\underline{\theta}}(\underline{y}))] d\underline{x}\right];$$

hence

$$\mathrm{E}_{\underline{y}}\left[I(f, g(\cdot \,|\, \hat{\underline{\theta}}(\underline{y})))\right] = \text{constant} - \mathrm{E}_{\underline{y}}\mathrm{E}_{\underline{x}}\left[\log[g(\underline{x} \,|\, \hat{\underline{\theta}}(\underline{y}))]\right]. \qquad (7.18)$$

It turns out that we can estimate $E_y E_x \left[\log[g(\underline{x} \mid \hat{\underline{\theta}}(\underline{y}))] \right]$, and therefore we can select a model to minimize the expected estimated relative K-L information value given by (7.18). In most of our writing here about this matter we find it much simpler just to say that we are selecting an estimated relative K-L best model by use of AIC.

There is a second, less compelling, approach that we can take in going from K-L to AIC: Start with

$$I(f, g(\cdot \mid \underline{\theta}_o)) = \text{constant} - E_x \left[\log(g(\underline{x} \mid \underline{\theta}_o)) \right]$$

and see whether we can compute (or estimate) $E_x \left[\log(g(\underline{x} \mid \hat{\underline{\theta}}(\underline{y}))) \right]$ based on Taylor series expansions. As will be made evident below, we can derive the result

$$E_x \left[\log(g(\underline{x} \mid \hat{\underline{\theta}}(\underline{y}))) \right] \approx E_x \left[\log(g(\underline{x} \mid \hat{\underline{\theta}}(\underline{x}))) \right] - \frac{1}{2} \text{tr} \left[J(\underline{\theta}_o) I(\underline{\theta}_o)^{-1} \right]$$
$$- \frac{1}{2} (\hat{\underline{\theta}}(\underline{y}) - \underline{\theta}_o)' I(\underline{\theta}_o)(\hat{\underline{\theta}}(\underline{y}) - \underline{\theta}_o).$$

On the right-hand side above, the only component that absolutely cannot be estimated or computed (in any useful way) is the quadratic term involving $(\hat{\underline{\theta}}(\underline{y}) - \underline{\theta}_o)$ (and it is pointless therein to use $\hat{\underline{\theta}}_o = \hat{\underline{\theta}}(\underline{y})$). But if we take the expectation of both sides above with respect to $\underline{y}$, we get a quantity we can estimate:

$$E_y E_x \left[\log(g(\underline{x} \mid \hat{\underline{\theta}}(\underline{y}))) \right] \approx E_x \left[\log(g(\underline{x} \mid \hat{\underline{\theta}}(\underline{x}))) \right] - \text{tr} \left[J(\underline{\theta}_o) I(\underline{\theta}_o)^{-1} \right].$$

Thus, either line of derivation demonstrates that we have to change our objective from model selection based on minimum K-L with known $\underline{\theta}_o$ given g, to selecting the model with estimated $\underline{\theta}$ based on minimizing an expected K-L information measure. It is still the case that only a relative minimum can be found based on $E_y E_x \left[\log(g(\underline{x} \mid \hat{\underline{\theta}}(\underline{y}))) \right]$ as the target objective function to be maximized; the constant $E_x[f(\underline{x}) \log(f(\underline{x}))]$ cannot be computed or estimated.

Only some of the literature is clear that AIC model selection is based on the concept of minimizing the expected K-L criterion $E_y \left[I(f, g(\cdot \mid \hat{\underline{\theta}}(\underline{y}))) \right]$ (see, e.g., Sawa 1978, Sugiura 1978, Bozdogan 1987 (page 351), Bonneu and Milhaud 1994). It is the relative value of this criterion that is estimated over the set of models. That is, we want to estimate without bias, as our model selection criterion (denoted below by T for target) for each approximating model, the value of

$$T = \int f(\underline{y}) \left[\int f(\underline{x}) \log(g(\underline{x} \mid \hat{\underline{\theta}}(\underline{y}))) d\underline{x} \right] d\underline{y}. \qquad (7.19)$$

The change from conceptual model selection based on minimum K-L to actual model selection based on maximizing an estimate of T in (7.19) is forced on us because we must estimate the parameters in g based on a finite amount of data.

Sometimes the criterion given by (7.19), hence AIC, is motivated by the concept of Akaike's predicative likelihood $E_p[\log(\mathcal{L}(\hat{\theta}))] = E_y E_x[\log(\mathcal{L}(\hat{\theta}(y)) \,|\, x)] \equiv T$, which has a heuristic interpretation in terms of cross-validation and independent random variables $\underline{x}$ and $\underline{y}$. However, the quantity T, and selection by maximizing $\hat{T}$ (or minimizing $-2\hat{T}$), does arise from a pure K-L approach to the problem of model selection without ever invoking the idea of cross-validation.

In a slightly simplified, but obvious, notation, the K-L–based model selection problem is now to find a useful expression for, and estimator of, the target

$$T = E_{\hat{\underline{\theta}}} E_{\underline{x}} \left[\log(g(\underline{x} \,|\, \hat{\underline{\theta}})) \right], \tag{7.20}$$

where it is understood that the MLE $\hat{\underline{\theta}}$ is based on sample $\underline{y}$, and the two expectations are for $\underline{x}$ and $\underline{y}$ (hence $\hat{\underline{\theta}}$) both with respect to truth f. It is because T is also a double expectation based, conceptually, on two independent samples that AIC-based model selection is asymptotically equivalent to cross-validation (see, e.g., Stone 1977); cross-validation is a well-accepted basis of model selection.

Step 1 is an expansion of the form (7.3) applied to $\log(g(\underline{x} \,|\, \hat{\underline{\theta}}))$ around $\underline{\theta}_o$ for any given $\underline{x}$:

$$\log(g(\underline{x} \,|\, \hat{\underline{\theta}})) \approx \log(g(\underline{x} \,|\, \underline{\theta}_o)) + \left[\frac{\partial \log(g(\underline{x} \,|\, \underline{\theta}_o))}{\partial \underline{\theta}} \right]' [\hat{\underline{\theta}} - \underline{\theta}_o]$$
$$+ \frac{1}{2} [\hat{\underline{\theta}} - \underline{\theta}_o]' \left[\frac{\partial^2 \log(g(\underline{x} \,|\, \underline{\theta}_o))}{\partial \underline{\theta}^2} \right] [\hat{\underline{\theta}} - \underline{\theta}_o]. \tag{7.21}$$

Truncation at the quadratic term entails an unknown degree of approximation (but it is an error of approximation that goes to zero as $n \to \infty$). To relate (7.21) to (7.20) we first take the expected value of (7.21) with respect to $\underline{x}$:

$$E_{\underline{x}} \left[\log(g(\underline{x} \,|\, \hat{\underline{\theta}})) \right] \approx E_{\underline{x}} \left[\log(g(\underline{x} \,|\, \underline{\theta}_o)) \right] + E_{\underline{x}} \left[\frac{\partial \log(g(\underline{x} \,|\, \underline{\theta}_o))}{\partial \underline{\theta}} \right]' [\hat{\underline{\theta}} - \underline{\theta}_o]$$
$$+ \frac{1}{2} [\hat{\underline{\theta}} - \underline{\theta}_o]' \left[E_{\underline{x}} \frac{\partial^2 \log(g(\underline{x} \,|\, \underline{\theta}_o))}{\partial \underline{\theta}^2} \right] [\hat{\underline{\theta}} - \underline{\theta}_o]. \tag{7.22}$$

The vector multiplier of $[\hat{\underline{\theta}} - \underline{\theta}_o]$ in the linear term above is exactly the same as (7.5). It is just that for clarification $E_{\underline{x}}$ is used to mean E_f over the function of the random variable $\underline{x}$ (and keep remembering that $\hat{\underline{\theta}} \equiv \hat{\underline{\theta}}(\underline{y})$ is independent of $\underline{x}$). Therefore, upon taking this expectation, the linear term vanishes; that is, (7.5) applies:

$$E_{\underline{x}} \left[\frac{\partial \log(g(\underline{x} \,|\, \underline{\theta}_o))}{\partial \underline{\theta}} \right] = \underline{0}.$$

Also, for the quadratic term in (7.22), definition (7.7) applies; hence we can write

$$E_{\underline{x}}\left[\log(g(\underline{x}\,|\,\hat{\underline{\theta}}))\right] \approx E_{\underline{x}}\left[\log(g(\underline{x}\,|\,\underline{\theta}_o))\right] - \frac{1}{2}[\hat{\underline{\theta}} - \underline{\theta}_o]'I(\underline{\theta}_o)[\hat{\underline{\theta}} - \underline{\theta}_o]. \quad (7.23)$$

Now we can take the expectation of (7.23) with respect to $\hat{\underline{\theta}}$ (i.e., $\underline{y}$). Here is where the trace function is used, yielding

$$E_{\hat{\underline{\theta}}}E_{\underline{x}}\left[\log(g(\underline{x}\,|\,\hat{\underline{\theta}}))\right] \approx E_{\underline{x}}\left[\log(g(\underline{x}\,|\,\underline{\theta}_o))\right] - \frac{1}{2}\,\text{tr}\left[I(\underline{\theta}_o)E_{\hat{\underline{\theta}}}\left[[\hat{\underline{\theta}} - \underline{\theta}_o][\hat{\underline{\theta}} - \underline{\theta}_o]'\right]\right].$$

The left-hand side above is T from (7.20), and $E_{\hat{\underline{\theta}}}\left[[\hat{\underline{\theta}} - \underline{\theta}_o][\hat{\underline{\theta}} - \underline{\theta}_o]'\right] = \Sigma$ is the correct large-sample theoretical sampling variance of the MLE, because the expectation herein is taken with respect to truth f, not with respect to g. Thus we have

$$T \approx E_{\underline{x}}\left[\log(g(\underline{x}\,|\,\underline{\theta}_o))\right] - \frac{1}{2}\,\text{tr}\left[I(\underline{\theta}_o)\Sigma\right]. \quad (7.24)$$

Step 2 starts with the realization that we have not yet derived what we need: a relationship between T and $E_{\underline{x}}\left[\log[g(\underline{x}\,|\,\hat{\underline{\theta}}(x))]\right]$, which is the expectation of the actual log-likelihood at the MLE. We now do a second expansion, this time of $\log(g(\underline{x}\,|\,\underline{\theta}_o)$ about $\hat{\underline{\theta}}(x)$, treating $\underline{x}$ as the sample data, hence getting the MLE of $\underline{\theta}$ for this $\underline{x}$. This procedure is acceptable, because all we are after is an expected value, which means taking an integral over all possible points in the sample space. Therefore, it does not matter what notation we use for these sample points: $\underline{x}$ or $\underline{y}$. Applying the Taylor series approximation (7.3) (but with the roles of $\hat{\underline{\theta}}$ and $\underline{\theta}_o$ switched; also note well that here, $\hat{\underline{\theta}} \equiv \hat{\underline{\theta}}(x)$), we obtain

$$\log(g(\underline{x}\,|\,\underline{\theta}_o)) \approx \log(g(\underline{x}\,|\,\hat{\underline{\theta}})) + \left[\frac{\partial \log(g(\underline{x}\,|\,\hat{\underline{\theta}}))}{\partial \underline{\theta}}\right]'[\underline{\theta}_o - \hat{\underline{\theta}}]$$

$$+ \frac{1}{2}[\underline{\theta}_o - \hat{\underline{\theta}}]'\left[\frac{\partial^2 \log(g(\underline{x}\,|\,\hat{\underline{\theta}}))}{\partial \underline{\theta}^2}\right][\underline{\theta}_o - \hat{\underline{\theta}}]. \quad (7.25)$$

The MLE $\hat{\underline{\theta}}$ is the solution of, hence satisfies, the equations

$$\frac{\partial \log(g(\underline{x}\,|\,\hat{\underline{\theta}}))}{\partial \underline{\theta}} = \underline{0}.$$

Therefore, the linear term in (7.25) vanishes. Taking the needed expectation we can write

$$E_{\underline{x}}\left[\log(g(\underline{x}\,|\,\underline{\theta}_o))\right] \approx E_{\underline{x}}\left[\log(g(\underline{x}\,|\,\hat{\underline{\theta}}))\right] - \frac{1}{2}\,\text{tr}\left[E_{\underline{x}}\left[\hat{I}(\hat{\underline{\theta}})\right][\underline{\theta}_o - \hat{\underline{\theta}}][\underline{\theta}_o - \hat{\underline{\theta}}]'\right].$$
$$(7.26)$$

See (7.8) for $\hat{I}(\hat{\underline{\theta}})$, the Hessian of the log-likelihood evaluated at the MLE.

To make analytical progress with (7.26) we use the approximation $\hat{I}(\hat{\theta}) \approx I(\underline{\theta}_o)$; hence we obtain

$$\mathrm{E}_{\underline{x}}\left[\hat{I}(\hat{\theta})\right][\underline{\theta}_o - \hat{\underline{\theta}}][\underline{\theta}_o - \hat{\underline{\theta}}]' \approx \left[I(\underline{\theta}_o)\right]\left[\mathrm{E}_{\underline{x}}[\underline{\theta}_o - \hat{\underline{\theta}}][\underline{\theta}_o - \hat{\underline{\theta}}]'\right]$$
$$= \left[I(\underline{\theta}_o)\right]\left[\mathrm{E}_{\underline{x}}[\hat{\underline{\theta}} - \underline{\theta}_o][\hat{\underline{\theta}} - \underline{\theta}_o]'\right]$$
$$= \left[I(\underline{\theta}_o)\right]\Sigma. \tag{7.27}$$

The approximation made in (7.27) is often good to $O(1/n)$, hence is justified. However, there are circumstances where the approximation may not be this good, and the overall approximation in (7.27) is equivalent to using $\hat{I}(\hat{\theta}) \approx I(\underline{\theta}_o)$ after first writing the approximation

$$\mathrm{E}_{\underline{x}}\left[\hat{I}(\hat{\theta})\right][\underline{\theta}_o - \hat{\underline{\theta}}][\underline{\theta}_o - \hat{\underline{\theta}}]' \approx \left[\mathrm{E}_{\underline{x}}\left[\hat{I}(\hat{\theta})\right]\right]\left[\mathrm{E}_{\underline{x}}[\underline{\theta}_o - \hat{\underline{\theta}}][\underline{\theta}_o - \hat{\underline{\theta}}]'\right]$$
$$= \left[\left[I(\underline{\theta}_o)\right]\right]\Sigma \tag{7.28}$$

to arrive at the same result as (7.27). In any case, (7.28) does improve with sample size, but the overall error involved in this approximation to $\mathrm{E}_{\underline{x}}\left[\hat{I}(\hat{\theta})\right][\underline{\theta}_o - \hat{\underline{\theta}}][\underline{\theta}_o - \hat{\underline{\theta}}]'$ is hard to assess, in general. (The matter is revisited below for the exponential family of distributions, and (7.28) is found to be there a good approximation to $O(1/n)$.)

Using either (7.27) or (7.28), along with (7.26), we have

$$\mathrm{E}_{\underline{x}}\left[\log(g(\underline{x} \mid \underline{\theta}_o))\right] \approx \mathrm{E}_{\underline{x}}\left[\log(g(\underline{x} \mid \hat{\underline{\theta}}(\underline{x})))\right] - \frac{1}{2}\,\mathrm{tr}\left[I(\underline{\theta}_o)\Sigma\right]. \tag{7.29}$$

Recall (7.24):

$$T \approx \mathrm{E}_{\underline{x}}\left[\log(g(\underline{x} \mid \underline{\theta}_o))\right] - \frac{1}{2}\,\mathrm{tr}\left[I(\underline{\theta}_o)\Sigma\right].$$

Substituting (7.29) into (7.24) we have a key result that is known in the literature:

$$T \approx \mathrm{E}_{\underline{x}}\left[\log(g(\underline{x} \mid \hat{\underline{\theta}}(\underline{x})))\right] - \mathrm{tr}\left[I(\underline{\theta}_o)\Sigma\right]. \tag{7.30}$$

The literature usually presents not (7.30), but rather an alternative equivalent form based on (7.13):

$$T \approx \mathrm{E}_{\underline{x}}\left[\log(g(\underline{x} \mid \hat{\underline{\theta}}(\underline{x})))\right] - \mathrm{tr}\left[J(\underline{\theta}_o)[I(\underline{\theta}_o)]^{-1}\right]. \tag{7.31}$$

The notation $\hat{\underline{\theta}}(\underline{x})$ rather than just $\hat{\underline{\theta}}$ is used above only to emphasize that on the right-hand side of (7.31) only one random variable $\underline{x}$ appears, and it can be taken to refer to the actual data. From (7.30) or (7.31), we can infer that a criterion for model selection (i.e., a nearly unbiased estimator of T) is structurally of the form

$$\hat{T} \approx \log(g(\underline{x} \mid \hat{\underline{\theta}})) - \widehat{\mathrm{tr}}\left[I(\underline{\theta}_o)\Sigma\right], \tag{7.32}$$

or

$$\hat{T} \approx \log(g(\underline{x} \mid \hat{\underline{\theta}})) - \widehat{\text{tr}}\left[J(\underline{\theta}_o)[I(\underline{\theta}_o)]^{-1}\right]. \tag{7.33}$$

Simple, direct estimation of Σ from one sample is not possible, because there is only one $\hat{\underline{\theta}}$ available (a bootstrap estimator of Σ is possible), whereas both $J(\underline{\theta}_o)$ and $I(\underline{\theta}_o)$ are directly estimable from the single sample. We note that (7.33), but not (7.32), requires a parametrization wherein $I(\underline{\theta}_o)$ is of full rank, whence its inverse exists. There is no loss in generality if we assume that all the probability distribution models have fully identifiable parameters, and hence are of full rank.

The maximized log-likelihood $\log(g(\underline{x} \mid \hat{\underline{\theta}}))$ in (7.31) is an unbiased estimator of its own expectation $E_{\underline{x}}[\log(g(\underline{x} \mid \hat{\underline{\theta}}))]$ (but is biased as an estimator of T). Hence, the only problem left is to get a reliable (low, or no, bias) estimator of the trace term, or at least an estimator with small mean square error. Then the best model to use is the one with the largest value of $\hat{T}$, because this would produce a model with the smallest estimated expected K-L distance. As a matter of convention the criterion is often stated as that of minimizing

$$-2\log(g(\underline{x} \mid \hat{\underline{\theta}})) + 2\widehat{\text{tr}}\left[J(\underline{\theta}_o)[I(\underline{\theta}_o)]^{-1}\right]. \tag{7.34}$$

If f is a subset of g (i.e., if $g = f$ or f is contained within g in the sense of nested models), then $I(\underline{\theta}_o) \equiv \mathcal{I}(\underline{\theta}_o) = \mathcal{J}(\underline{\theta}_o) = J(\underline{\theta}_o) = \Sigma^{-1}$, and hence $\text{tr}\left[I(\underline{\theta}_o)\Sigma\right] = K$. Even if g is just a good model (i.e., a good approximation) for f, the literature supports the idea that our best estimator is probably to use $\widehat{\text{tr}}\left[I(\underline{\theta}_o)\Sigma\right] = K$ (Shibata 1989).

When the model is too restrictive to be good, the term $-2\log(g(\underline{x} \mid \hat{\underline{\theta}}))$ will be much inflated (compared to this same term for a "good" model), and we will not select that model. In this case having a good estimate of the trace term should not matter. The practical key to making AIC (wherein we have assumed $\widehat{\text{tr}}\left[I(\underline{\theta}_o)\Sigma\right] = K$) work is then to have some good models in the set considered, but not too many good, but over parametrized, models. By a "good" model we mean one that is close to f in the sense of having a small K-L value, in which case such "closeness" also means that the use of $\widehat{\text{tr}}\left[I(\underline{\theta}_o)\Sigma\right] = K$ is itself a parsimonious estimator. This matter of estimation of the trace term and closeness of g to f is explored further in Section 7.6. It is those Section 7.6 derivations, and the above ideas in this paragraph, that to us justifies AIC, which is seen as a special case of (7.34):

$$\text{AIC} = -2\log(g(\underline{x} \mid \hat{\underline{\theta}})) + 2K.$$

The generalization given by (7.34) leads to Takeuchi's (1976) information criterion (TIC) for model selection (Shibata 1989). The result (7.32) suggests that we might use the bootstrap to compute $\widehat{\text{tr}}\left[I(\underline{\theta}_o)\Sigma\right]$ and hence implement the TIC criterion via

$$-2\log(g(\underline{x} \mid \hat{\underline{\theta}})) + 2\widehat{\text{tr}}\left[I(\underline{\theta}_o)\Sigma\right] \tag{7.35}$$

Kei Takeuchi was born in 1933 in Tokyo, Japan, and graduated in 1956 from the University of Tokyo. He received a Ph.D. in economics in 1966 (Keizaigaku Hakushi), and his research interests include mathematical statistics, econometrics, global environmental problems, history of civilization, and Japanese economy. He is the author of many books on mathematics, statistics, and the impacts of science and technology on society. He is currently a professor on the Faculty of International Studies at Meiji Gakuin University and Professor Emeritus, University of Tokyo (recent photograph).

or even use more exact forms for the trace term. These ideas are pursued a bit in the next section.

First, however, there is one more crucial point on which the reader must be clear: It is not required that truth f be in the set of models to which we apply AIC model selection. Many derivations of AIC are quite misleading by making the assumption (often implicitly, hence without realizing it) that $f \equiv g$ (or $f \subset g$). Such derivations lead directly to AIC, hence bypass the completely general result of (7.33), which does not require $f \subset g$. Once one has (7.33), then it is possible to see how a proper philosophy of having a set of good approximating models to complex truth in conjunction with the parsimonious choice of $\widehat{\text{tr}}\left[I(\underline{\theta}_o)\Sigma\right] = K$ justifies use of AIC.

There are a few odds and ends worth considering at this point. First, we state the result

$$\text{E}_{\underline{y}}\left[I(f, g(\cdot \mid \hat{\underline{\theta}}(\underline{y})))\right] - I(f, g(\cdot \mid \underline{\theta}_o)) = \frac{1}{2}\text{tr}\left[J(\underline{\theta}_o)I(\underline{\theta}_o)^{-1}\right].$$

The proof is simple, because the left-hand side of the above reduces to

$$E_y\left[E_x[\log(g(\underline{x}\,|\,\underline{\theta}_o))] - E_x[\log(g(\underline{x}\,|\,\hat{\underline{\theta}}(\underline{y})))]\right].$$

Now substitute (7.23) for $E_x[\log(g(\underline{x}\,|\,\hat{\underline{\theta}}(\underline{y})))]$ in the above to get the result

$$\frac{1}{2}E_y\left[\hat{\underline{\theta}}(\underline{y}) - \underline{\theta}_o]'I(\underline{\theta}_o)[\hat{\underline{\theta}}(\underline{y}) - \underline{\theta}_o\right],$$

which becomes $\frac{1}{2}\,\mathrm{tr}\left[I(\underline{\theta}_o)\Sigma\right] = \frac{1}{2}\,\mathrm{tr}\left[J(\underline{\theta}_o)I(\underline{\theta}_o)^{-1}\right]$.

It should be almost obvious (and it is true) that this trace term, $\mathrm{tr}\left[J(\underline{\theta}_o)I(\underline{\theta}_o)^{-1}\right]$, does not depend upon sample size. Rather, for good models it is about equal to K (these matters are explored in other Chapter 7 sections below). In stark contrast, quantities such as the log-likelihood, expected log-likelihood, and both of K-L $I(f, g(\cdot\,|\,\underline{\theta}_o))$ and the expected K-L $E_y\left[I(f, g(\cdot\,|\,\hat{\underline{\theta}}(\underline{y})))\right]$ increase linearly in sample size n. As a result, for large sample sizes, and K/n small, the ratio

$$\frac{E_y\left[I(f, g(\cdot\,|\,\hat{\underline{\theta}}(\underline{y})))\right]}{I(f, g(\cdot\,|\,\underline{\theta}_o))}$$

is essentially 1 even though the difference between expected and actual K-L is > 0. Thus, on an absolute scale TIC and AIC (when thoughtfully applied) model selection are producing the model estimated to provide the minimum K-L model from the set of models considered if sample size is large and K/n is small.

The reason that the criterion for practical model selection gets changed from minimum K-L to minimum expected K-L as $E_y[I(f, g(\cdot\,|\,\hat{\underline{\theta}}(\underline{y})))]$ is because we must estimate θ by the model-based MLE. This seemingly innocent little fact has deep ramifications. It is why the K-L–based conceptual motivation (at the start of this section) virtually forces us to adopt $E_y[I(f, g(\cdot\,|\,\hat{\underline{\theta}}(\underline{y})))]$ to be minimized, hence T, i.e., (7.20), to be maximized.

In this regard there is a nominally puzzling result: If we just start with K-L as

$$I(f, g(\cdot\,|\,\underline{\theta}_o)) = \text{constant} - E_x\left[\log(g(\underline{x}\,|\,\underline{\theta}_o))\right],$$

and no actual data in hand, hence no estimate of θ, we might notice a direct Taylor series expansion of $\log(g(\underline{x}\,|\,\underline{\theta}_o))$ about what would be the MLE given any value of the variable of integration $\underline{x}$ (which is *not* data). After taking the expectation over the sample space of the random variable $\underline{x}$, the result is

$$E_x\left[\log(g(\underline{x}\,|\,\underline{\theta}_o))\right] = E_x\left[\log(g(\underline{x}\,|\,\hat{\underline{\theta}}(\underline{x})))\right] - \frac{1}{2}\,\mathrm{tr}\left[J(\underline{\theta}_o)I(\underline{\theta}_o)^{-1}\right].$$

The above would suggest that K-L model selection could be based on maximizing $\log(g(\underline{x}\,|\,\hat{\underline{\theta}}(\underline{x}))) - \frac{1}{2}\,\mathrm{tr}[\hat{J}(\underline{\theta}_o)\hat{I}(\underline{\theta}_o)^{-1}]$; it cannot be so based.

This conclusion is not valid, because there are no data lurking anywhere. There never was a valid MLE of $\hat{\theta}$ injected into the process. The K-L criterion has already been integrated over the sample space, and properly there is no x and no data involved in K-L. Data cannot be manufactured by a Taylor series expansion on a random variable. Thus, the intriguing result is mathematically correct, but conceptually wrong for what we are trying to do, and hence misleading.

7.3   General K-L–Based Model Selection: TIC

7.3.1   Analytical Computation of TIC

There are other alternatives to estimation of relative K-L (not much used) that try to provide a data-based estimator of the trace term. These methods are computationally much more intense, and the resultant estimator of the trace term can be so variable, and may have its own biases, that it is questionable whether such approaches are worth applying (unless perhaps n is huge). Takeuchi (1976) proposed TIC (see also Shibata 1989, and Konishi and Kitagawa 1996): Select the model that minimizes (7.34) for specific estimators of $J(\underline{\theta}_o)$ and $I(\underline{\theta}_o)$, hence getting an estimator of tr $\left[J(\underline{\theta}_o)[I(\underline{\theta}_o)]^{-1} \right]$. The estimator of $I(\underline{\theta}_o)$ is (7.8), the empirical Hessian:

$$\hat{I}(\underline{\theta}_o) = \hat{I}(\hat{\theta}) = -\frac{\partial^2 \log(g(\underline{x} \,|\, \hat{\theta}))}{\partial \theta^2}. \tag{7.36}$$

General estimation of $J(\underline{\theta}_o)$ relies on recognizing the sample as structured on n independent units of information. In the simplest case we would have x as an *iid* sample, $x_1, \ldots, x_n$. It is required only that the sample be recognized as having n conditionally independent components so that the log-likelihood can be computed as the sum of n terms; hence we have

$$\log(g(\underline{x} \,|\, \hat{\theta})) = \sum_{i=1}^{n} \log(g_i(x_i \,|\, \hat{\theta})).$$

For the *iid* sample case, $g_i(x_i \,|\, \hat{\theta}) \equiv g(x_i \,|\, \hat{\theta})$. Using here $g(\cdot \,|\, \theta)$ for both the basic sample-size one pdf and for the probability distribution function of the full sample of size n is a minor abuse of notation. However, we think that the reader will understand the meaning of the formulas and that it is better to minimize notation to facilitate comprehension of concepts.

A general estimator of $J(\underline{\theta}_o)$ for TIC can be derived from (7.12):

$$J(\underline{\theta}_o) = E_f \left[\left[\frac{\partial}{\partial \underline{\theta}} \log(g(\underline{x} \,|\, \underline{\theta}_o)) \right] \left[\frac{\partial}{\partial \underline{\theta}} \log(g(\underline{x} \,|\, \underline{\theta}_o)) \right]' \right].$$

For the case of a random sample,

$$J(\underline{\theta}_o) = E_f \left[\left[\sum_{i=1}^{K} \frac{\partial}{\partial \underline{\theta}} \log(g(x_i \mid \underline{\theta}_o)) \right] \left[\sum_{i=1}^{K} \frac{\partial}{\partial \underline{\theta}} \log(g(x_i \mid \underline{\theta}_o)) \right]' \right]$$

$$= \sum_{i=1}^{K} E_f \left[\frac{\partial}{\partial \underline{\theta}} \log(g(x_i \mid \underline{\theta}_o)) \right] \left[\frac{\partial}{\partial \underline{\theta}} \log(g(x_i \mid \underline{\theta}_o)) \right]'.$$

Therefore, we are led to use

$$\hat{J}(\underline{\theta}_o) = \sum_{i=1}^{K} \left[\frac{\partial}{\partial \underline{\theta}} \log(g(x_i \mid \hat{\underline{\theta}})) \right] \left[\frac{\partial}{\partial \underline{\theta}} \log(g(x_i \mid \hat{\underline{\theta}})) \right]'. \qquad (7.37)$$

A general version of TIC can be defined based on (7.36) and (7.37) (see, e.g., Shibata 1989:222):

$$\text{TIC} = -2 \log(g(\underline{x} \mid \hat{\underline{\theta}})) + 2 \operatorname{tr} \left[\hat{J}(\underline{\theta}_o)[\hat{I}(\underline{\theta}_o)]^{-1} \right]. \qquad (7.38)$$

One selects the model that produces the smallest TIC. Because $-\text{TIC}/2 = \hat{T}$ is for each model an asymptotically unbiased estimator of $E_y \left[I(f, g(\cdot \mid \hat{\underline{\theta}}(\underline{y}))) \right]$ $-$ constant, the underlying optimization criterion is that we select the model that on average (over the set of models) minimizes this expected K-L information loss. For large n this expected criterion is almost the same as minimizing the criterion $I(f, g)$ $-$ constant; thus using (7.38), we are essentially targeting selecting the K-L best model of the set of models, and this is regardless of whether or not f is in the model set.

The estimator $\hat{J}(\underline{\theta}_o)$ converges to $J(\underline{\theta}_o)$, and $\hat{I}(\underline{\theta}_o)$ converges to $I(\underline{\theta}_o)$, so TIC is asymptotically unbiased (i.e., consistent) as a selection criterion for the minimum expected K̂-L model. In practice this estimator of the trace term is so variable (and is not unbiased), even for large n, that it seems better to just use the parsimonious "estimator" $\widehat{\operatorname{tr}} \left[J(\underline{\theta}_o)[I(\underline{\theta}_o)]^{-1} \right] = K$ (cf. Shibata 1989) (we will consider the matter further in later sections). This seems especially appropriate if we have done a good job of specifying our set of models from which to select a best-fitting model.

7.3.2 Bootstrap Estimation of TIC

The primary value of the bootstrap method herein is to assess model selection uncertainty based on applying an analytical model selection criterion (e.g., AIC, AIC_c, QAIC_c, or TIC based on formulas (7.33), (7.34), and (7.35)). However, a second and quite different use of the bootstrap can be made: Use some bootstrap method to estimate directly the quantity $T = E_x E_{\hat{\theta}} \left[\log(g(\underline{x} \mid \hat{\underline{\theta}})) \right]$; the K-L best model is the one that maximizes $\hat{T}$. Variations on this theme involve more direct bootstrap estimation of the key quantity $\operatorname{tr} \left[I(\underline{\theta}_o)\Sigma \right]$ (or equivalently, $\mathrm{tr}\left[J(\underline{\theta}_o)[I(\underline{\theta}_o)]^{-1}\right]$). We will describe a method designed to minimize the impact of approximations made in deriving (7.35).

From (7.24) and (7.26) (wherein $\hat{\underline{\theta}}$ denotes $\hat{\underline{\theta}}(\underline{x})$) we derive

$$T \approx \mathrm{E}_{\underline{x}}\left[\log(g(\underline{x} \mid \hat{\underline{\theta}}))\right] - \frac{1}{2}\,\mathrm{tr}\left[I(\underline{\theta}_o)\Sigma\right] - \frac{1}{2}\,\mathrm{tr}\left[\mathrm{E}_{\underline{x}}\left[\hat{I}(\hat{\underline{\theta}})\right][\underline{\theta}_o - \hat{\underline{\theta}}][\underline{\theta}_o - \hat{\underline{\theta}}]'\right].$$

Hence, a model selection criterion can be based on

$$\hat{T} = \log(g(\underline{x} \mid \hat{\underline{\theta}})) - \frac{1}{2}\,\mathrm{tr}\left[\hat{I}(\underline{\theta}_o)\hat{\Sigma}\right] - \frac{1}{2}\,\mathrm{tr}\left[\hat{\mathrm{E}}_{\underline{x}}\left[\hat{I}(\hat{\underline{\theta}})\right][\underline{\theta}_o - \hat{\underline{\theta}}][\underline{\theta}_o - \hat{\underline{\theta}}]'\right].$$
(7.39)

Additional approximations applied to (7.39), or to the basic derivations, lead to

$$\hat{T} = \log(g(\underline{x} \mid \hat{\underline{\theta}})) - \mathrm{tr}\left[\hat{I}(\underline{\theta}_o)\hat{\Sigma}\right],$$

which could also be the basis for a bootstrap estimator (as could (7.35)).

We assume that the sample structure allows a meaningful bootstrap sampling procedure (easily done in the *iid* sample case). Let a bootstrap sample be denoted by $\underline{x}^*$ with corresponding bootstrap MLE $\hat{\underline{\theta}}^*$. The needed likelihood second partial derivatives will have to be determined either analytically or numerically. To avoid more notation, we do not index the bootstrap samples, but rather just note that needed summations are over B bootstrap samples.

In the bootstrap estimators, the MLE $\hat{\underline{\theta}}$ plays the role of $\underline{\theta}_o$. Hence bootstrap estimators of $I(\underline{\theta}_o)$, Σ, and $\mathrm{E}_{\underline{x}}\left[\hat{I}(\hat{\underline{\theta}})\right][\underline{\theta}_o - \hat{\underline{\theta}}][\underline{\theta}_o - \hat{\underline{\theta}}]'$ are

$$\hat{I}(\underline{\theta}_o) = -\frac{1}{B}\left[\sum_B \frac{\partial^2 \log(g(\underline{x}^* \mid \hat{\underline{\theta}}))}{\partial \underline{\theta}^2}\right], \tag{7.40}$$

$$\hat{\Sigma} = \frac{1}{B}\left[\sum_B [\hat{\underline{\theta}}^* - \hat{\underline{\theta}}][\hat{\underline{\theta}}^* - \hat{\underline{\theta}}]'\right], \tag{7.41}$$

$$\hat{\mathrm{E}}_{\underline{x}}\left[\left[\hat{I}(\hat{\underline{\theta}})\right][\underline{\theta}_o - \hat{\underline{\theta}}][\underline{\theta}_o - \hat{\underline{\theta}}]'\right] \tag{7.42}$$

$$= \frac{1}{B}\left[\sum_B \left[-\frac{\partial^2 \log(g(\underline{x}^* \mid \hat{\underline{\theta}}^*))}{\partial \underline{\theta}^2}\right][\hat{\underline{\theta}}^* - \hat{\underline{\theta}}][\hat{\underline{\theta}}^* - \hat{\underline{\theta}}]'\right].$$

These estimators mimic the expectation over f, because the sample arises from f, the bootstrap resamples the sample, and under any model our best estimator of $\underline{\theta}_o$ is the MLE $\hat{\underline{\theta}}$ (note that $\underline{\theta}_o$ varies by model g). One should use the same B bootstrap samples with every model in the set of models over which selection is made.

The above suffices to compute TIC as

$$\mathrm{TIC} = -2\log(g(\underline{x} \mid \hat{\underline{\theta}})) + 2\,\mathrm{tr}\left[\hat{I}(\underline{\theta}_o)\hat{\Sigma}\right], \tag{7.43}$$

using (7.40) and (7.41) (this is estimating the same quantity as TIC, so we call it TIC here, because $J(\underline{\theta}_o)[I(\underline{\theta}_o)]^{-1} = I(\underline{\theta}_o)\Sigma$). To use (7.39) for bootstrap-based model selection, base the estimation of its second and third components on (7.40), (7.41), and (7.42); or in a form analogous to AIC and TIC, the model selection criterion to minimize is

$$- 2\log(g(\underline{x} \mid \hat{\underline{\theta}})) + \operatorname{tr}\left[\hat{I}(\underline{\theta}_o)\hat{\Sigma}\right] + \operatorname{tr}\hat{E}_{\underline{x}}\left[\left[\hat{I}(\hat{\underline{\theta}})\right][\underline{\theta}_o - \hat{\underline{\theta}}][\underline{\theta}_o - \hat{\underline{\theta}}]'\right]. \quad (7.44)$$

It may well be that (7.43), i.e., TIC, would suffice and (7.44) is not a better estimator of $-2T$.

Recent work on this use of the bootstrap to find $\hat{T}$ for K-L–based model selection is found in Ishiguro, et al. (1997), Cavanaugh and Shumway (1997), Shao (1996) and Chung et al. (1996). Shibata (1997a) has considered, in a general context, theoretical properties of many alternative implementations of the bootstrap to estimate the needed model selection criterion T. He notes that there is no unique way to do this bootstrapping to estimate the relative K-L model selection criterion, but that all reasonable bootstrap implementations are asymptotically equivalent to TIC. This use of the bootstrap has the advantage of bypassing concerns about all approximations used to get TIC or AIC. Despite this apparent advantage, Shibata (1997a, page 393) concludes that there is no reason to use the bootstrap this way to compute $\hat{T}$. It probably suffices to use a simple nonbootstrap computation of $\hat{T}$ (in particular, AIC$_c$).

It should thus be clear that there are two very different ways to use the bootstrap in model selection. Not much used is the case of getting a single estimate of T for each model based on the full set of bootstrap samples. The more common (and more useful) use of the bootstrap in model selection is first to accept some easily computable model selection criterion, such as AIC, and then to apply that criterion to all models considered for all the bootstrap samples created (and tabulate results like frequency of selection of each model). This use of the bootstrap leads to information about inference uncertainties after model selection. [There is also a large literature on use of the bootstrap under non–K-L–based model selection; see, e.g., Breiman 1992; Efron 1983, 1986; Hjorth 1994; Linhart and Zucchini 1986; and Shao 1996.]

7.4   AIC$_c$: A Second-Order Improvement

7.4.1   Derivation of AIC$_c$

The results above are completely general, and as such do not lead to some of the more specific results in the literature. In particular, if we assume a univariate linear structural model with homogeneous, normally distributed errors, conditional on any regressor variables, we can get the results of Hurvich and

Tsai (1989, 1995b) (see also Sugiura 1978). We let the model structure be

$$\mu_i = \mathrm{E}(x_i \mid \underline{z}) = \sum_{j=1}^{K-1} z_{ij}\beta_j, \qquad i = 1, \ldots, n.$$

More specifically (but without explicitly denoting the conditioning on "regressors" $\underline{z}_i$),

$$x_i = \sum_{j=1}^{K-1} z_{ij}\beta_j + \epsilon_i, \qquad i = 1, \ldots, n,$$

where the ϵ_i are *iid* normal$(0, \sigma^2)$. There are thus K parameters making up $\underline{\theta}$ (σ^2 is the Kth one), and $g(\underline{x} \mid \underline{\theta})$ is given by the multivariate–normal$(\mu, \sigma^2 I)$ distribution (MVN); I is the $n \times n$ identity matrix. If we let $f \equiv g$ or $\overline{f} \subset g$, then we can derive the AIC$_c$ results of Hurvich and Tsai. This last notation means that either g is the true data-generating "model," or f is actually the same distribution and structural form as *model* g but with one or more elements of $\underline{\theta}$ set to 0 (hence there are superfluous parameters). The superfluous parameters serve only to increase K; hence the simplest way to get AIC$_c$ is to assume this regression model g and assume that $f \equiv g$. The derivation is given below in some detail because of the importance of AIC$_c$.

Matrix notation is simpler to use, and hence $\underline{X} = Z\underline{\beta} + \underline{\epsilon}$ and $\mathrm{E}(\underline{X}) = \underline{\mu}$. Without loss of generality we assume that Z (n by $K-1$) is of full rank. The likelihood is

$$g(\underline{x} \mid \underline{\theta}) = \left[\frac{1}{\sqrt{2\pi}}\right]^n \left[\frac{1}{\sigma^2}\right]^{n/2} \exp\left[-\frac{1}{2}\frac{(\underline{X} - Z\underline{\beta})'(\underline{X} - Z\underline{\beta})}{\sigma^2}\right],$$

and we are here taking $f \equiv g$. Ignoring additive constants and simplifying, the log-likelihood can be taken as

$$\log(g(\underline{x} \mid \underline{\theta})) = -\frac{n}{2}\log(\sigma^2) - \frac{1}{2}\frac{(\underline{X} - Z\underline{\beta})'(\underline{X} - Z\underline{\beta})}{\sigma^2}.$$

The MLEs are well known here:

$$\underline{\hat{\beta}} = (Z'Z)^{-1}Z'\underline{X},$$

$$\hat{\sigma}^2 = \frac{(\underline{X} - Z\underline{\hat{\beta}})'(\underline{X} - Z\underline{\hat{\beta}})}{n}.$$

Therefore,

$$\log(g(\underline{x} \mid \underline{\hat{\theta}}(\underline{x}))) = -\frac{n}{2}\log(\hat{\sigma}^2) - \frac{1}{2}\frac{(\underline{X} - Z\underline{\hat{\beta}})'(\underline{X} - Z\underline{\hat{\beta}})}{\hat{\sigma}^2};$$

hence, the maximized log-likelihood is

$$\log(g(\underline{x} \mid \underline{\hat{\theta}}(\underline{x}))) = -\frac{n}{2}\log(\hat{\sigma}^2) - \frac{n}{2}$$

(the constant $-n/2$ can be dropped in practice).

We want to determine the bias if we use $\log(g(\underline{x} \mid \hat{\theta}(\underline{x})))$ as an estimator of our target

$$T = E_{\underline{x}}E_{\hat{\theta}(\underline{y})}\left[\log(g(\underline{x} \mid \hat{\theta}(\underline{y})))\right],$$

where $\underline{x}$ and $\underline{y}$ are two independent random samples of size n. To make the evaluation here we actually use the specified form of the model (and of course take expectations with respect to $f \equiv g$). Hence, we want (a simplified notation is used here)

$$T = E_{\underline{x}}E_{\hat{\theta}(\underline{y})}\left[\log(g(\underline{x} \mid \hat{\theta}(\underline{y})))\right]$$

$$= E_{\hat{\theta}(\underline{y})}E_{\underline{x}}\left[-\frac{n}{2}\log(\hat{\sigma}_y^2) - \frac{1}{2}\frac{(\underline{X} - Z\hat{\underline{\beta}}_y)'(\underline{X} - Z\hat{\underline{\beta}}_y)}{\hat{\sigma}_y^2}\right].$$

The order of integration was reversed for the right-hand side above. Thus our first task is to evaluate

$$E_{\underline{x}}\left[(\underline{X} - Z\hat{\underline{\beta}}_y)'(\underline{X} - Z\hat{\underline{\beta}}_y)\right]$$

$$= E_{\underline{x}}\left[((\underline{X} - Z\underline{\beta}) + (Z\underline{\beta} - Z\hat{\underline{\beta}}_y))'((\underline{X} - Z\underline{\beta}) + (Z\underline{\beta} - Z\hat{\underline{\beta}}_y))\right]$$

$$= E_{\underline{x}}\left[(\underline{X} - Z\underline{\beta})'(\underline{X} - Z\underline{\beta})\right] + E_{\underline{x}}\left[2(Z\underline{\beta} - Z\hat{\underline{\beta}}_y)'(\underline{X} - Z\underline{\beta})\right]$$

$$+ E_{\underline{x}}\left[(Z\underline{\beta} - Z\hat{\underline{\beta}}_y)'(Z\underline{\beta} - Z\hat{\underline{\beta}}_y)\right]$$

$$= E_{\underline{x}}\left[(\underline{X} - Z\underline{\beta})'(\underline{X} - Z\underline{\beta})\right] + \left[2(Z\underline{\beta} - Z\hat{\underline{\beta}}_y)'(E_{\underline{x}}(\underline{X}) - Z\underline{\beta})\right]$$

$$+ \left[(Z\underline{\beta} - Z\hat{\underline{\beta}}_y)'(Z\underline{\beta} - Z\hat{\underline{\beta}}_y)\right].$$

The middle term above vanishes because $E_{\underline{x}}(\underline{X}) = Z\underline{\beta}$. Also, the first of the three terms above is identical to $E_{\underline{x}}(\epsilon'\epsilon) = n\sigma^2$. So we have the result

$$E_{\underline{x}}\left[(\underline{X} - Z\hat{\underline{\beta}}_y)'(\underline{X} - Z\hat{\underline{\beta}}_y)\right] = n\sigma^2 + \left[(Z\underline{\beta} - Z\hat{\underline{\beta}}_y)'(Z\underline{\beta} - Z\hat{\underline{\beta}}_y)\right].$$

Using this partial result we have

$$T = E_{\hat{\theta}(\underline{y})}\left[-\frac{n}{2}\log(\hat{\sigma}_y^2)\right] - \frac{1}{2}E_{\hat{\theta}(\underline{y})}\left[\frac{n\sigma^2 + \left[(Z\underline{\beta} - Z\hat{\underline{\beta}}_y)'(Z\underline{\beta} - Z\hat{\underline{\beta}}_y)\right]}{\hat{\sigma}_y^2}\right].$$

The first term above does not need to be evaluated further because it is also the leading term in the expected log-likelihood. Also, at this point, we can drop the designation of θ as being based on sample $\underline{y}$. The designations $\underline{x}$ or $\underline{y}$ are really just dummy arguments in integrals. Consequently, in the above, the notation could be in terms of $\underline{y}$ or $\underline{x}$, or this notation can just be dropped. Thus we have

$$T = \mathrm{E}_{\hat{\underline{\theta}}}\left[-\frac{n}{2}\log(\hat{\sigma}^2)\right] - \frac{1}{2}\mathrm{E}_{\hat{\underline{\theta}}}\left[\frac{n\sigma^2 + \left[(Z\underline{\beta} - Z\hat{\underline{\beta}})'(Z\underline{\beta} - Z\hat{\underline{\beta}})\right]}{\hat{\sigma}^2}\right]. \quad (7.45)$$

Now we make use of another well-known result in theoretical statistics: Under a linear model structure with errors as *iid* normal$(0, \sigma^2)$, the MLE's $\hat{\underline{\beta}}$ and $\hat{\sigma}^2$ are independent random variables. Therefore, the second expectation term in (7.45) partitions into two multiplicative parts, as follows:

$$T = \mathrm{E}\left[-\frac{n}{2}\log(\hat{\sigma}^2)\right] - \frac{1}{2}\mathrm{E}_{\hat{\underline{\beta}}}\left[n\sigma^2 + \left[(Z\underline{\beta} - Z\hat{\underline{\beta}})'(Z\underline{\beta} - Z\hat{\underline{\beta}})\right]\right]\mathrm{E}_{\hat{\sigma}^2}\left[\frac{1}{\hat{\sigma}^2}\right].$$

As a next step, rewrite the needed expectation of the quadratic form in the above as

$$\mathrm{E}\left[(Z\underline{\beta} - Z\hat{\underline{\beta}})'(Z\underline{\beta} - Z\hat{\underline{\beta}})\right] = \mathrm{tr}\left[(Z'Z)\mathrm{E}\left[(\hat{\underline{\beta}} - \underline{\beta})(\hat{\underline{\beta}} - \underline{\beta})'\right]\right].$$

The expectation on the right-hand side above, i.e., $\mathrm{E}[(\hat{\underline{\beta}} - \underline{\beta})(\hat{\underline{\beta}} - \underline{\beta})']$, is the sampling variance–covariance matrix of $\hat{\underline{\beta}}$, which is known to be $\sigma^2(Z'Z)^{-1}$. Thus, for the $K - 1$ square identity matrix I,

$$\mathrm{E}\left[(Z\underline{\beta} - Z\hat{\underline{\beta}})'(Z\underline{\beta} - Z\hat{\underline{\beta}})\right] = \mathrm{tr}[\sigma^2 I] = \sigma^2(K - 1).$$

Putting it all together to this point in the derivation, we have

$$T = \mathrm{E}\left[-\frac{n}{2}\log(\hat{\sigma}^2)\right] - \frac{1}{2}\left[(n + K - 1)\sigma^2\right]\mathrm{E}_{\hat{\sigma}^2}\left[\frac{1}{\hat{\sigma}^2}\right]. \quad (7.46)$$

To finish the process we relate $\hat{\sigma}^2$ to a central chi-squared random variable, namely χ^2_{df} on $n - (K - 1)$ degrees of freedom, df. These results also are well known in statistical theory:

$$\frac{n\hat{\sigma}^2}{\sigma^2} \sim \chi^2_{n-K+1}.$$

So we now rearrange (7.46) to be

$$T = \mathrm{E}\left[-\frac{n}{2}\log(\hat{\sigma}^2)\right] - \frac{1}{2}\left[(n + K - 1)n\right]\mathrm{E}\left[\frac{1}{n\hat{\sigma}^2/\sigma^2}\right],$$

$$T = \mathrm{E}\left[-\frac{n}{2}\log(\hat{\sigma}^2)\right] - \frac{n}{2}(n + K - 1)\mathrm{E}\left[\frac{1}{\chi^2_{n-K+1}}\right].$$

Yet another known exact result is

$$\mathrm{E}\left[\frac{1}{\chi^2_{df}}\right] = \frac{1}{df - 2}$$

(assuming df > 2).

Using the last result above we have reduced (7.46) to

$$T = \mathrm{E}\left[-\frac{n}{2}\log(\hat{\sigma}^2)\right] - \frac{n}{2}(n + K - 1)\left[\frac{1}{n - K - 1}\right]. \tag{7.47}$$

This result is exact. No approximations were made in its derivation; however, it applies only to the particular context of its derivation, which includes the constraint $f \subseteq g$. Some more simplification of (7.47):

$$\begin{aligned}
T &= \mathrm{E}\left[-\frac{n}{2}\log(\hat{\sigma}^2)\right] - \frac{n}{2}\left[\frac{n + K - 1}{n - K - 1}\right] \\
&= \mathrm{E}\left[-\frac{n}{2}\log(\hat{\sigma}^2)\right] - \frac{n}{2}\left[1 + \frac{2K}{n - K - 1}\right] \\
&= \mathrm{E}\left[-\frac{n}{2}\log(\hat{\sigma}^2)\right] - \frac{n}{2} - \frac{nK}{n - K - 1} \\
&= \mathrm{E}\left[-\frac{n}{2}\log(\hat{\sigma}^2) - \frac{n}{2}\right] - \frac{nK}{n - K - 1}.
\end{aligned}$$

The term above within the expectation operator is the maximized log-likelihood. Thus we have

$$\begin{aligned}
T &= \mathrm{E}\left[\log(g(\underline{x} \mid \hat{\underline{\theta}}(x)))\right] - \frac{nK}{n - K - 1} \\
&= \mathrm{E}\left[\log(g(\underline{x} \mid \hat{\underline{\theta}}(x)))\right] - \frac{(n - K - 1 + K + 1)K}{n - K - 1} \\
&= \mathrm{E}\left[\log(g(\underline{x} \mid \hat{\underline{\theta}}(x)))\right] - K - \frac{K(K + 1)}{n - K - 1}.
\end{aligned}$$

If we convert this to an AIC result, we have, as an exact result in this context,

$$\begin{aligned}
-2T &= -2\mathrm{E}\left[\log(g(\underline{x} \mid \hat{\underline{\theta}}(x)))\right] + 2K + \frac{2K(K + 1)}{n - K - 1} \\
&= \mathrm{E}(\mathrm{AIC}) + \frac{2K(K + 1)}{n - K - 1} = \mathrm{E}(\mathrm{AIC}_c).
\end{aligned} \tag{7.48}$$

This result thus motivates use of the term $2K(K + 1)/(n - K - 1)$ as a small-sample-size bias-correction term added to AIC. The result assumes a fixed-effects linear model with normal errors and constant residual variances. Under different sorts of models, a different small-sample correction to AIC arises (the matter is explored some in the next subsection). However, the result given by (7.48) seems useful in other contexts, especially if n is large but K is also large relative to n. Without exception, if sample size n is small, some sort of "AIC$_c$" is required for good model selection results, and we recommend (7.48) unless a more exact small-sample correction to AIC is known.

7.4.2 Lack of Uniqueness of AIC$_c$

The result in (7.48) is not universal in that other assumed univariate (and more so for multivariate) models, with $g = f$ to facilitate a derivation, or ways of deriving a small-sample adjustment to AIC, will lead to different adjustment terms. This section is just a brief elaboration of this idea.

The simplest case to present arises for a situation analogous to one-way ANOVA, but we let the within-subgroup variance differ for each subgroup. This can be generalized to having m subsets of data, each of sample size n_i, and the full model is as used in Section 7.4.1 above, but with parameter set $\{\theta\} = \{\beta_i, \sigma_i^2, i = 1, \ldots, m\}$ (this might be a global model in some cases). Let each parameter subset be of size K_i; hence $K = K_1 + \cdots + K_m$. It should be almost obvious, after some thought, that for this situation the small-sample correction to AIC is

$$-2T = -2\mathrm{E}\left[\log(g(\underline{x} \mid \hat{\underline{\theta}}(\underline{x})))\right] + 2\sum_{i=1}^{m}\left[K_i + \frac{2K_i(K_i + 1)}{n_i - K_i - 1}\right];$$

hence,

$$\mathrm{AIC}_c = -2\log(g(\underline{x} \mid \hat{\underline{\theta}}(\underline{x}))) + 2K + \sum_{i=1}^{m}\left[\frac{2K_i(K_i + 1)}{n_i - K_i - 1}\right].$$

The reason that there are m "correction" terms is that we had to estimate m different variance parameters. One can thus envision many other models where the form of AIC$_c$ must differ from that of the simple normal-model case with only one estimated σ^2.

Another informative exact calculation of the bias term, $T - \mathrm{E}_{\underline{x}}\left[\log(g(\underline{x} \mid \hat{\theta}))\right]$, is obtained for the case of the model and truth being the one-parameter negative exponential distribution (hence $K = 1$):

$$g(x \mid \lambda) = \frac{1}{\lambda}e^{-x/\lambda}.$$

For an *iid* sample from $g(x \mid \lambda)$, let $S = x_1 + \cdots + x_n$. Then

$$\log(g(\underline{x} \mid \lambda)) = -n\log(\lambda) - S/\lambda.$$

The MLE is $\hat{\lambda} = S/n$, so

$$\log(g(\underline{x} \mid \hat{\lambda})) = -n\log(\hat{\lambda}) - n.$$

The target to be unbiasedly estimated is

$$T = \mathrm{E}_{\underline{x}}\mathrm{E}_y\left[-n\log(\hat{\lambda}) - S_{\underline{y}}/\hat{\lambda}\right],$$

where the sum S_y is based on an independent sample of size n, while $\hat{\lambda}$ is based on sample $\underline{x}$. It is easy to evaluate the above T to be

$$T = \mathrm{E}\left[\log(g(\underline{x} \mid \hat{\lambda}))\right] + n - n^2\lambda\mathrm{E}\left[\frac{1}{S}\right]. \tag{7.49}$$

The sum S is a random variable with a simple gamma distribution; hence the expectation of $1/S$ in (7.49) is known to be exactly $1/(\lambda(n-1))$. It is thus easy to derive the exact result, expressed in "AIC" form (i.e., as $-2T$):

$$-2T = -2\mathrm{E}(\log(g(\underline{x} \mid \hat{\lambda}))) + 2 + \frac{2}{n-1}.$$

Recall that here $K = 1$, so the corresponding total bias-correction term under the AIC_c form would be $2 + 4/(n-2)$. The point is that the exact form of AIC would be $2K$ plus a small-sample correction term that would vary according to the model assumed. It is reasonable to think that this small-sample correction term should be $O(1/n)$.

Theoretically, when $f \subset g$ the error in using K as the bias correction to $\hat{T} = \log(g(\underline{x} \mid \hat{\theta}))$ is always $O(1/n)$, and Hurvich and Tsai's form seems like a good general choice. There is, however, considerable need for research on improved bias terms for AIC_c-type criteria. In this regard, an area offering research opportunities is that of when the random variable is discrete (see, e.g., Sugiura 1978, Shibata 1997b), such as Poisson, binomial, or Bernoulli (hence also logistic regression), because then we can get parameter MLEs taking on the value 0. This creates a problem in evaluating the theoretical target model selection criterion because we encounter the need to compute $y \cdot \log(0)$, which is not defined (see, e.g., Burnham et al. 1994). AIC is still defined, but its small-sample properties are now more problematic, as is the small-sample bias-correction term needed to define an AIC_c. Operating characteristics of AIC-based model selection for count-type data need more study for small sample sizes.

7.5 Derivation of AIC for the Exponential Family of Distributions

A generalization of normality-based models is found in the exponential family of distributions. The realizations that (1) many common applications of statistical analyses are based on exponential family models, and (2) normality-based regression is in the exponential family and leads to exact K-L model selection results (i.e., AIC_c) motivated us to show the derivation of AIC theory under this restricted but very useful case. The canonical representation of an exponential family pdf involves sums of functions of the sample values. It is convenient to denote these sums by S_j.

A suitable canonical representation for the exponential family of probability distributions is

$$g(\underline{x} \mid \underline{\theta}) = \exp\left[\left[\sum_{j=1}^{K} S_j \theta_j\right] + H(\underline{\theta}) + G(\underline{S})\right]$$

$$= \exp\left[\underline{S}'\underline{\theta} + H(\underline{\theta}) + G(\underline{S})\right]. \tag{7.50}$$

Each S_j is a function (hence $S_j(\underline{x})$) of the full sample $\underline{x} = \mu(\theta) + \epsilon$ and any covariates Z involved in representing $\mu(\theta)$, on which we condition. The K-element vector of sufficient statistics is $\underline{S} = (S_1, \ldots, S_K)'$.

In the canonical representation of (7.50) the parameter θ is generally some 1-to-1 transformation of another K-dimensional parameter of direct interest. There is no loss of generality in allowing any such 1-to-1 transformation. We will revisit this matter and show why it is so at the end of this section.

Our goal is to evaluate

$$T = E_{\underline{x}}E_{\underline{y}}\left[\log(g(\underline{x}\mid\hat{\theta}_y))\right] = E_{\underline{x}}E_{\underline{y}}\left[\underline{S}'_x\hat{\theta}_y + H(\hat{\theta}_y) + G(\underline{S}_x)\right]. \qquad (7.51)$$

Here, $\underline{S}_x$ and $\hat{\theta}_y$ are thought of as based on independent samples $\underline{x}$ and $\underline{y}$. We also simplified the notation, now using $\hat{\theta}_y$ rather than $\hat{\theta}(y)$.

Formula (7.51) above can be rewritten as

$$\begin{aligned}
T &= E_{\underline{x}}E_{\underline{y}}\left[(\underline{S}_x - \underline{S}_y + \underline{S}_y)'\hat{\theta}_y + H(\hat{\theta}_y) + G(\underline{S}_x)\right] \\
&= E_{\underline{x}}E_{\underline{y}}\left[(\underline{S}_x - \underline{S}_y)'\hat{\theta}_y + \underline{S}'_y\hat{\theta}_y + H(\hat{\theta}_y) + G(\underline{S}_x)\right] \\
&= E_{\underline{x}}E_{\underline{y}}\left[\underline{S}'_y\hat{\theta}_y + H(\hat{\theta}_y) + G(\underline{S}_x)\right] + E_{\underline{x}}E_{\underline{y}}\left[(\underline{S}_x - \underline{S}_y)'\hat{\theta}_y\right] \\
&= \left[E_{\underline{y}}(\underline{S}'_y\hat{\theta}_y + H(\hat{\theta}_y)) + E_{\underline{x}}(G(\underline{S}_x))\right] + E_{\underline{y}}\left[(E_{\underline{x}}(\underline{S}_x) - \underline{S}_y)'\hat{\theta}_y\right].
\end{aligned}$$

The interchangeability of integration arguments now is used. This is permissible because both expectations are with respect to f; hence $E_{\underline{x}}(G(\underline{S}_x)) = E_{\underline{y}}(G(\underline{S}_y))$. Also, for simplicity we will use $E_{\underline{x}}(\underline{S}_x) = E(\underline{S})$, and now we get

$$T = E_{\underline{y}}\left[\underline{S}'_y\hat{\theta}_y + H(\hat{\theta}_y)) + G(\underline{S}_y)\right] + E_{\underline{y}}\left[(E(\underline{S}) - \underline{S}_y)'\hat{\theta}_y\right].$$

Changing the argument from y to x in the first part above, just to emphasize the result for this exponential family case, we have

$$\begin{aligned}
T &= E_{\underline{x}}(\log(g(\underline{x}\mid\hat{\theta}))) + E_{\underline{y}}\left[(E(\underline{S}) - \underline{S}_y)'\hat{\theta}_y\right] \\
&\equiv E_{\underline{x}}(\log(g(\underline{x}\mid\hat{\theta}))) - E_{\underline{y}}\left[(\underline{S}_y - E(\underline{S}))'\hat{\theta}_y\right]. \qquad (7.52)
\end{aligned}$$

Formula (7.52) is an exact result and clearly shows the bias to be subtracted from $E_{\underline{x}}(\log(g(\underline{x}\mid\hat{\theta})))$ to get T:

$$\text{Bias} = E_{\underline{y}}\left[(\underline{S}_y - E(\underline{S}))'(\hat{\theta}_y - \underline{\theta}_*)\right].$$

The notation used here is $E(\hat{\theta}_y) = \theta_*$ to denote the exact expectation of the MLE for the given sample size n and model g; $\theta_o \approx \theta_*$ with asymptotic equality.

To further simplify notation, now that only one "sample" is involved, we use

$$\text{Bias} = \mathrm{E}\left[(\underline{S} - \mathrm{E}(\underline{S}))'(\hat{\theta} - \theta_*)\right], \tag{7.53}$$

$$\text{Bias} = \mathrm{tr}\,\mathrm{E}\left[(\hat{\theta} - \theta_*)(\underline{S} - \mathrm{E}(\underline{S}))'\right] = \mathrm{tr}\left[\mathrm{COV}(\hat{\theta}, \underline{S})\right].$$

Hence for the exponential family *an exact result* is

$$T = \mathrm{E}_x(\log(g(\underline{x}\,|\,\hat{\theta}))) - \mathrm{tr}\left[\mathrm{COV}(\hat{\theta}, \underline{S})\right] \tag{7.54}$$

(something similar appears in Bonneu and Milhaud 1994). The $K \times K$ matrix of covariance elements, $\mathrm{COV}(\hat{\theta}, \underline{S})$, can be approximated by Taylor series methods. If the exact covariance matrix can be found, then we have an exact result for the needed bias term above (Hurvich and Tsai 1989, in effect, did such an exact evaluation for the normal distribution case). The result (7.54) may seem not very useful because it seems to apply only to the canonical form of the exponential family. This is not true; the matter of generality of the canonical result will be addressed below.

Before further evaluation of the bias term, we consider the MLEs and the Hessian. First,

$$\log(g(\underline{x}\,|\,\theta)) = \underline{S}'\theta + H(\theta) + G(\underline{S}),$$

so

$$\frac{\partial \log(g(\underline{x}\,|\,\theta))}{\partial \theta} = \underline{S} + \frac{\partial H(\theta)}{\partial \theta},$$

$$\frac{\partial^2 \log(g(\underline{x}\,|\,\theta))}{\partial \theta^2} = \frac{\partial^2 H(\theta)}{\partial \theta^2},$$

and thus

$$I(\theta_o) = \mathrm{E}_f\left[-\frac{\partial^2 \log(g(\underline{x}\,|\,\theta_o))}{\partial \theta^2}\right] = -\frac{\partial^2 H(\theta_o)}{\partial \theta^2}. \tag{7.55}$$

It follows that the MLE satisfies

$$\underline{S} = -\frac{\partial H(\hat{\theta})}{\partial \theta}.$$

It is worth noting here that θ_o satisfies

$$\mathrm{E}_f(\underline{S}) = -\frac{\partial H(\theta_o)}{\partial \theta}.$$

This is an exact result, whereas $\mathrm{E}(\hat{\theta}) \approx \theta_o$ is (in general) only $O(1/\sqrt{n})$.

The formula for $J(\theta_o)$, based on (7.12), becomes

$$J(\theta_o) = \mathrm{E}_f\left[\underline{S} - \mathrm{E}_f(\underline{S})\right]\left[\underline{S} - \mathrm{E}_f(\underline{S})\right]', \tag{7.56}$$

which is the true variance–covariance matrix of $\underline{S}$. The n *iid* observations produce a set of statistics $\underline{s}_i$ that sum to $\underline{S}$; hence an estimator of $J(\underline{\theta}_o)$ as given by (7.56) is

$$\hat{J}(\underline{\theta}_o) = \frac{n}{n-1} \left[\sum_{i=1}^{n} \left[\underline{s}_i - \underline{\bar{s}} \right] \left[\underline{s}_i - \underline{\bar{s}} \right]' \right]. \tag{7.57}$$

Returning now to the evaluation of the bias term, a first-order Taylor series expansion gives us

$$-\frac{\partial H(\hat{\underline{\theta}})}{\partial \underline{\theta}} \approx -\frac{\partial H(\underline{\theta}_o)}{\partial \underline{\theta}} - \frac{\partial^2 H(\underline{\theta}_o)}{\partial \underline{\theta}^2}(\hat{\underline{\theta}} - \underline{\theta}_o);$$

hence

$$\underline{S} \approx \mathrm{E}(\underline{S}) + I(\underline{\theta}_o)(\hat{\underline{\theta}} - \underline{\theta}_o), \qquad O_p(1/\sqrt{n}). \tag{7.58}$$

Inserting (7.58) into the exact result (7.53) as well as also using $\underline{\theta}_o$ to approximate $\underline{\theta}_*$ (inasmuch as we are now replacing an exact result with an approximate result anyway), we have

$$\begin{aligned}
\text{Bias} &\approx \mathrm{E}\left[\left[I(\underline{\theta}_o)(\hat{\underline{\theta}} - \underline{\theta}_o) \right]' (\hat{\underline{\theta}} - \underline{\theta}_o) \right] \\
&= \mathrm{E}\left[(\hat{\underline{\theta}} - \underline{\theta}_o)' I(\underline{\theta}_o)(\hat{\underline{\theta}} - \underline{\theta}_o) \right] \\
&= \mathrm{E}\,\mathrm{tr}\left[I(\underline{\theta}_o)(\hat{\underline{\theta}} - \underline{\theta}_o)(\hat{\underline{\theta}} - \underline{\theta}_o)' \right] \\
&= \mathrm{tr}\left[I(\underline{\theta}_o)\mathrm{E}\left[(\hat{\underline{\theta}} - \underline{\theta}_o)(\hat{\underline{\theta}} - \underline{\theta}_o)' \right] \right] = \mathrm{tr}\left[I(\underline{\theta}_o)\Sigma \right].
\end{aligned}$$

Thus we have shown that in this common case of an exponential family model,

$$T \approx \mathrm{E}_{\underline{x}}(\log(g(\underline{x} \mid \hat{\underline{\theta}}))) - \mathrm{tr}\left[I(\underline{\theta}_o)\Sigma \right] \tag{7.59}$$

(the approximation is to $O(1/n)$). Note that this derivation did not encounter any problems like those in approximation (7.28) in the general derivation of AIC in Section 7.2.

These results can be extended to any parametrized form of an exponential family model, because then we just have a 1-to-1 transformation from $\underline{\theta}$ to (say) $\underline{\beta}$ via some set of K functions, denoted here by $\underline{W}(\underline{\theta}) = \underline{\beta}$. Now $\underline{W}(\underline{\theta}_o) = \underline{\beta}_o$, and let Σ_θ and Σ_β be the variance–covariance matrices for the MLEs under the two parametrizations. An expected matrix of mixed second partial derivatives, as per (7.55) exists for the $\underline{\beta}$ parametrization; denote it by $I(\underline{\beta})$. Let the $K \times K$ Jacobian of $\underline{W}$, evaluated at $\underline{\theta}_o$, be

$$J_w = \left\{ \frac{\partial W_i(\underline{\theta}_o)}{\partial \theta_j} \right\}.$$

Then

$$J_w \Sigma_\theta J_w' = \Sigma_\beta,$$

and

$$(J_w')^{-1} I(\underline{\theta}_o)(J_w)^{-1} = I(\underline{\beta}_o).$$

Both the K–L–based target and the expected log-likelihood are invariant to 1-to-1 parameter transformations, so this must also be true for the theoretical bias correction. That is, any likelihood and MLE-based model selection criterion ought to be invariant to 1-to-1 reparametrizations of the models used. This is the case here:

$$
\begin{aligned}
\mathrm{tr}(I(\underline{\beta}_o)\Sigma_\beta) &= \mathrm{tr}\left[(J_w')^{-1} I(\underline{\theta}_o)(J_w)^{-1} J_w \Sigma_\theta J_w'\right] \\
&= \mathrm{tr}\left[(J_w')^{-1} I(\underline{\theta}_o)\Sigma_\theta J_w'\right] \\
&= \mathrm{tr}\left[I(\underline{\theta}_o)\Sigma_\theta J_w'(J_w')^{-1}\right] = \mathrm{tr}\left[I(\underline{\theta}_o)\Sigma_\theta\right].
\end{aligned}
$$

Note, however, that if we were to estimate this trace term, the estimator might perform better under some parametrizations than under others.

One last point here: It is certainly still true that

$$\mathrm{tr}\left[I(\underline{\theta}_o)\Sigma\right] = \mathrm{tr}\left[J(\underline{\theta}_o)[I(\underline{\theta}_o)]^{-1}\right].$$

So an alternative to (7.59) is

$$T \approx \mathrm{E}_{\underline{x}}(\log(g(\underline{x}\,|\,\hat{\underline{\theta}}))) - \mathrm{tr}\left[J(\underline{\theta}_o)[I(\underline{\theta}_o)]^{-1}\right].$$

This could be directly proven here, based on the simple result

$$\frac{\partial \log(g(\underline{x}\,|\,\underline{\theta}))}{\partial \underline{\theta}} = \underline{S} - \mathrm{E}(\underline{S})$$

and (7.58) to derive $[I(\underline{\theta}_o)]^{-1} J(\underline{\theta}_o)[I(\underline{\theta}_o)]^{-1} = \Sigma$.

For TIC we can use $\hat{J}(\underline{\theta}_o)$ from (7.57) and from (7.55),

$$\hat{I}(\underline{\theta}_o) = -\frac{\partial^2 H(\hat{\underline{\theta}})}{\partial \underline{\theta}^2},$$

getting an estimator of $\mathrm{tr}\left[J(\underline{\theta}_o)[I(\underline{\theta}_o)]^{-1}\right]$ that can be used (because of invariance) even if the parametrization of interest (and used for MLEs) is some $\underline{\beta}_o = \underline{W}(\underline{\theta}_o)$, not $\underline{\theta}$.

Working with exponential family cases facilitates some informative evaluation of both $\mathrm{tr}\left[J(\underline{\theta}_o)[I(\underline{\theta}_o)]^{-1}\right]$, relative to the value K, and the variability of the estimator $\mathrm{tr}\left[\hat{J}(\underline{\theta}_o)[\hat{I}(\underline{\theta}_o)]^{-1}\right]$. These topics, and others, are explored in the next section.

7.6 Evaluation of $\mathrm{tr}(J(\underline{\theta}_o)[I(\underline{\theta}_o)]^{-1})$ and Its Estimator

The general derivation of a K–L–based model selection criterion results in (7.31) and hence (7.33). By "a general derivation," we mean a derivation in which there is no assumption that the true data-generating distribution f is one of the models g in the set of models considered. Hence, the general result for K–L–based model selection does not appear to be AIC. Rather, the large-sample bias correction term subtracted from the expected maximized log-likelihood to get T is tr($J(\theta_o)[I(\theta_o)]^{-1}$) (Takeuchi 1976). In deriving this result there is not, and need not be, any assumption that any of the candidate models represent truth. However, in general we know with certainty that tr($J(\theta_o)[I(\theta_o)]^{-1}$) = K only when $J(\theta_o) = I(\theta_o)$ (this is sufficient but not necessary), and the latter equality is certain, in general, only when f is a special case of g, hence, when model g equals or is a generalization of "truth." This condition is unrealistic to expect, so how good is the approximation tr($J(\theta_o)[I(\theta_o)]^{-1}$) = K when the truth is more general than the model, but the model is a good approximation to truth? We make here some limited, but useful, progress on this issue. Extensive theory, simulation studies, and experience (e.g., Linhart and Zucchini 1986:176–182, especially results such as in their Table 10.3) are needed to give us full confidence in when we can expect reliable results from AIC, versus when we might have to use TIC. Below, we establish theory and results for some models within the exponential family of distributions.

7.6.1 Comparison of AIC Versus TIC in a Very Simple Setting

We consider two simple one-parameter distributions: negative-exponential and half-normal. For each distribution we can determine the trace tr($J(\theta_o)[I(\theta_o)]^{-1}$), assuming either that the distribution is truth, hence $f = g$, or that the other distribution is truth f and the given distribution is a model g (so $f \neq g$). We will also examine the estimators of the traces that can be used in TIC model selection and contrast TIC selection with AIC selection for these two distributions as models. This is a convenient situation to explore, partly because both distributions are in the exponential family.

For the negative-exponential distribution let $S = x_1 + \cdots + x_n = n\bar{x}$; then

$$g(\underline{x} \mid \lambda) = \frac{1}{\lambda^n} e^{-S/\lambda},$$

$$\log(g(\underline{x} \mid \lambda)) = -n \log(\lambda) - S/\lambda,$$

$$\frac{\partial}{\partial \lambda} \log(g(\underline{x} \mid \lambda)) = -n/\lambda + S/\lambda^2,$$

so $\hat{\lambda} = S/n = \bar{x}$; also here E($x$) = λ.

Direct verification yields the following results:

$$I(\lambda) = \frac{n}{\lambda^2},$$

$$\hat{I}(\hat{\lambda}) = \frac{n}{\hat{\lambda}^2},$$

i.e., the empirical Hessian here is the same as $I(\hat{\lambda})$, and

$$\hat{J}(\lambda) = \sum_{i=1}^{n} \left[\frac{-1}{\lambda} + \frac{x_i}{\lambda^2} \right]^2 = \left[\sum_{i=1}^{n} \frac{(x_i)^2}{\lambda^4} \right] - \frac{n}{\lambda^2}.$$

The true $J(\lambda) = \hat{E}_f(\hat{J}(\lambda))$, and $\hat{J}(\hat{\lambda})$ is the empirical estimator of $J(\lambda)$. If we assume that the negative-exponential model is truth, then from the above (because $E(x^2) = 2\lambda^2$),

$$J(\lambda) = \frac{n}{\lambda^2}.$$

Clearly, if $f = g$, then here $\operatorname{tr}[J(\lambda)[I(\lambda)]^{-1}] = 1$. The direct empirical estimator of this trace is

$$\operatorname{tr}[\hat{J}(\hat{\lambda})[\hat{I}(\hat{\lambda})]^{-1}] = \frac{1}{n\bar{x}^2} \left[\sum_{i=1}^{n} (x_i)^2 \right] - 1,$$

or

$$\operatorname{tr}[\hat{J}(\hat{\lambda})[\hat{I}(\hat{\lambda})]^{-1}] = \frac{\sum_{i=1}^{n}(x_i - \bar{x})^2}{n\bar{x}^2}. \tag{7.60}$$

This estimator of the trace is the same as $\frac{n-1}{n}(\hat{cv})^2$ (when we use $n-1$ in the denominator of the sample s^2 as per convention), which we can expect to be quite variable. The trace estimator of (7.60) does converge to 1 as $n \to \infty$ when $f = g$. This trace estimator is scale-invariant, so we can calculate its distributional properties by Monte Carlo methods with a single run of 100,000 samples at each n, and for each case of truth being either the negative-exponential (any value of λ can be used) or half-normal distribution (any σ can be used). For the case where the negative-exponential model is truth, the simulation motivated a revised, nearly unbiased, version of (7.60):

$$\widehat{\operatorname{tr}[J(\lambda)[I(\lambda)]^{-1}]} = \frac{n}{n-1}(\hat{cv})^2. \tag{7.61}$$

The estimated mean and standard deviation of the estimated trace function based on (7.60) and (7.61) are given below, based on 100,000 samples:

	Eq. (7.60)		Eq. (7.61)	
n	mean	st.dev.	mean	st.dev.
20	0.90	0.37	1.00	0.41
50	0.96	0.26	1.00	0.27
100	0.98	0.19	1.00	0.20
500	1.00	0.09	1.00	0.09

Notice the substantial standard deviation of either trace estimator; it is this sort of large variability in the trace estimator that has given theoretical pause to routine use of TIC. We also see that the direct estimator has bias for modest sample sizes; analytical reduction of that bias would not be practical in nontrivial applications, and yet such bias in any trace estimator might be important.

Consider now the half-normal distribution

$$f(x \mid \sigma^2) = \sqrt{\frac{2}{\pi\sigma^2}} \exp\left[-\frac{1}{2}\left(\frac{x}{\sigma}\right)^2\right].$$

Under this distribution $E(x^2) = \sigma^2$, and direct integration gives

$$E(x) = \sigma\sqrt{\frac{2}{\pi}}.$$

If the half-normal distribution is truth and the negative-exponential is a model, then λ_o as a function of σ^2 is found from

$$E_f(x) = \lambda_o = \sigma\sqrt{\frac{2}{\pi}}.$$

For example, with $\sigma^2 = 1$, $\lambda_o = 0.79788$ is the K-L best choice of λ.

Note the usage and concepts here: We denote the K-L best value of λ by λ_o to distinguish that the corresponding negative-exponential distribution (i.e., that based on λ_o) is the K-L best negative-exponential distribution to use as the model for the underlying truth. By denoting this value of λ by λ_o we are emphasizing that all we have is the K-L best negative-exponential model, but it may be a poor model; it certainly may not be truth.

The above expectations producing $J(\lambda)$ and $I(\lambda)$ were with respect to the negative-exponential as g (ignoring what f might be), but now we want to take those expectations with respect to f as half-normal. Direct verification yields $I(\lambda_o) = n/\lambda_o^2$, whereas

$$J(\lambda_o) = nE_f\left[\frac{x}{\lambda_o^2} - \frac{1}{\lambda_o}\right]^2$$

$$= I(\lambda_o)\left[\frac{\sigma^2}{\lambda_o^2} - 1\right]$$

$$= I(\lambda_o)\left[\frac{\pi}{2} - 1\right].$$

Hence, when f is half-normal and g is negative-exponential,

$$\mathrm{tr}\left[J(\lambda_o)[I(\lambda_o)]^{-1}\right] = \frac{\pi}{2} - 1 = 0.5708.$$

This trace term is not very close to 1, the number AIC would assume. This big relative difference (i.e., 0.5708 versus 1) results because the negative-exponential model is a very poor approximation to the half-normal distribution. Note that this trace term is < 1 (K is 1 here).

Next we determine the trace estimator under the half-normal as the model. We therein have

$$\hat{\sigma}^2 = \frac{\sum_{i=1}^{n}(x_i)^2}{n},$$

$$I(\sigma^2) = \frac{n}{2\sigma^4},$$

$$\hat{J}(\sigma^2) = \sum_{i=1}^{n}\left[\frac{-1}{2\sigma^2} + \frac{(x_i)^2}{2\sigma^4}\right]^2$$

$$= I(\sigma^2)\left[\frac{1}{2n}\right]\sum_{i=1}^{n}\left[\frac{(x_i)^2}{\sigma^2} - 1\right]^2.$$

The key part of the empirical estimator $\hat{J}(\hat{\sigma}^2)$ is again a squared coefficient of variation, but here it is for the variable x^2. Denote this $\widehat{cv}$ by $\widehat{cv}(x^2)$, and we can use the notation

$$\hat{J}(\hat{\sigma}^2) = \hat{I}(\hat{\sigma}^2)\left[\frac{n-1}{2n}\right]\left[\widehat{cv}(x^2)\right]^2.$$

Thus, for the half-normal distribution being the model, the TIC estimator of the trace is

$$\text{tr}[\hat{J}(\hat{\sigma}^2)[\hat{I}(\hat{\sigma}^2)]^{-1}] = \left[\frac{n-1}{2n}\right]\left[\widehat{cv}(x^2)\right]^2. \tag{7.62}$$

If truth is the half-normal model, this quantity will converge to 1; and again, (7.62) is scale-invariant. For the half-normal model as truth we computed the mean and standard deviation of (7.62) by Monte Carlo methods with 100,000 samples at each n. This led to a nearly unbiased version of (7.62),

$$\widehat{\text{tr}}[J(\sigma^2)[I(\sigma^2)]^{-1}] = \frac{1}{2}\left[\frac{n}{n-1}\right]^2\left[\widehat{cv}(x^2)\right]^2, \tag{7.63}$$

that we then also used in the simulations. The results are below:

	Eq. (7.62)		Eq. (7.63)	
n	mean	st.dev.	mean	st.dev.
20	0.87	0.39	1.01	0.45
50	0.94	0.30	1.00	0.32
100	0.97	0.23	1.00	0.23
500	0.99	0.11	1.00	0.11

The main point from the above is how variable the trace estimator is.

To complete a set of analytical results (useful for validating Monte Carlo results) we computed the value of $\text{tr}[J(\sigma_o^2)[I(\sigma_o^2)]^{-1}]$ for the half-normal model when truth is the negative-exponential distribution. First we need $\sigma_o^2 = E(x^2) = 2\lambda^2$, because the expectation of x must be taken with respect to the negative-exponential distribution. We find that for the half-normal model and the negative-exponential as truth,

$$J(\sigma_o^2) = I(\sigma_o^2)\left[\frac{1}{2}\right]E\left[\frac{x^2}{\sigma_o^2} - 1\right]^2 = 2.5I(\sigma_o^2).$$

Thus when truth is the negative-exponential, and the model is half-normal (a terrible model in this case), $\text{tr}[J(\sigma_o^2)[I(\sigma_o^2)]^{-1}] = 2.5$ (not 1 as AIC assumes). Done the other way around we had the trace < 1. It turns out that this trace function under model misspecification can be either above or below K; it varies by situation, and in some situations the trace function can equal K even with a misspecified model.

We can now compare AIC versus TIC model selection when the choices and truth are negative-exponential or half-normal.
For the negative-exponential model:

$$\text{AIC} = 2n[\log(\overline{x}) + 1] + 2,$$

$$\text{TIC} = 2n[\log(\overline{x}) + 1] + 2\frac{n-1}{n}(\widehat{\text{cv}}(x))^2,$$

$$\text{TICu} = 2n[\log(\overline{x}) + 1] + 2\frac{n}{n-1}(\widehat{\text{cv}}(x))^2.$$

For the half-normal model (using $\hat{\sigma}^2 = $ mean of the x_i^2):

$$\text{AIC} = n\left[\log(\hat{\sigma}^2) + 1 - \log(2/\pi)\right] + 2,$$

$$\text{TIC} = n\left[\log(\hat{\sigma}^2) + 1 - \log(2/\pi)\right] + \left[\frac{n-1}{n}\right]^2\left[\widehat{\text{cv}}(x^2)\right]^2,$$

$$\text{TICu} = n\left[\log(\hat{\sigma}^2) + 1 - \log(2/\pi)\right] + \left[\frac{n}{n-1}\right]^2\left[\widehat{\text{cv}}(x^2)\right]^2.$$

In both cases here TICu means just that the estimator of $\text{tr}[J(\theta_o))[I(\theta_o)]^{-1}]$ is almost unbiased, as opposed to the direct, biased, plug-in estimators of the needed coefficients of variation.

Table 7.1 shows some results. The point of this brief comparison was to learn something about AIC versus TIC in a simple setting, especially whether or not they would give greatly different results. The context here is so simple that only two models are compared. Moreover, one or the other model was used as the data-generating distribution (i.e., truth). We did not consider prediction here, so the only possible criterion to use to compare performance of AIC versus TIC is rate of selection of the true model. We did not wish to do, in this book, any serious evaluation of AIC versus TIC under full-blown realistic conditions of complex truth, and a set of approximating models, wherein the correct basis of evaluation is how well a selection procedure does at selecting the K-L best model (technically, we would be selecting the expected K-L best model).

Several inferences supported by Table 7.1, and by all other sample sizes examined for this situation, surprised us. For the case that the negative-exponential model is true, the selection results based on TIC were uniformly

TABLE 7.1. Percentage of correct selection when one of the models (negative-exponential or half-normal) is truth, based on 100,000 samples; see text for AIC, TIC, and TICu formulas; average percent correct is based on equal weighting of the two cases.

sample size, n	selection criterion	truth negative expon.	truth half-normal	average percent correct
20	AIC	64	85	75
	TIC	73	77	75
	TICu	75	75	75
50	AIC	82	92	87
	TIC	87	87	87
	TICu	87	87	87
100	AIC	93	97	95
	TIC	95	95	95
	TICu	95	95	95
500	AIC	100	100	100
	TIC	100	100	100
	TICu	100	100	100

as good or better than those under AIC. The improvement is not large except at small sample sizes, wherein an "AIC$_c$" should be used anyway. Conversely, for the case that the half-normal model is true, the selection results based on AIC were uniformly as good or better than those under TIC. In either case, bias-correction of the trace estimator makes no real difference. We would not know a priori which (if either) model was true. If we compute an average percent-correct selection based on the idea that we have no information to justify any weighting other than a 50:50 weighting of these results, we get, on average, no advantage at all for TIC over AIC. Clearly, we do not know the extent to which these results would generalize.

7.6.2 Evaluation Under Logistic Regression

Logistic regression is used often, therefore we illustrate that it is a case of an exponential family model, and we explore the above trace question for this model. Let x_i be a Bernoulli random variable with true probability μ_i of being 1 (and probability $1 - \mu_i$ of being 0). For a sample of n independent x_i we base analysis on some assumed model for the μ_i. In order to distinguish truth from model we adopt the notation for the model as $p_i \equiv p_i(\theta)$, for some structure imposed on these p_i, as a function of a K-dimensional parameter vector θ. The relevant pdf, or likelihood (the same notation continues to serve this dual role), for the model is

$$g(\underline{x} \mid \underline{\theta}) = \prod_{i=1}^{n}(p_i)^{x_i}(q_i)^{1-x_i}.$$

We assume that known covariates $\underline{z}_i$, as $K \times 1$ column vectors, are associated with each observation, x_i, and an explanatory structural model is

$$p_i = \frac{1}{1 + e^{-\underline{z}_i'\underline{\theta}}}, \quad \text{or } q_i = \frac{e^{-\underline{z}_i'\underline{\theta}}}{1 + e^{-\underline{z}_i'\underline{\theta}}},$$

which is equivalent to

$$\log{[p_i/(1 - p_i)]} = \underline{z}_i'\underline{\theta}.$$

A modest amount of algebra gives the result

$$g(\underline{x} \mid \underline{\theta}) = \exp\left[\left[\sum_{i=1}^{n}(x_i\underline{z}_i)\right]'\underline{\theta} + \left[\sum_{i=1}^{n}\left(-\log\left(1 + e^{\underline{z}_i'\underline{\theta}}\right)\right)\right]\right],$$

which is in the canonical form of the exponential family for

$$H(\underline{\theta}) = \sum_{i=1}^{n}\left(-\log\left(1 + e^{\underline{z}_i'\underline{\theta}}\right)\right) \tag{7.64}$$

and

$$\underline{S} = \sum_{i=1}^{n}(x_i\underline{z}_i) = \sum_{i=1}^{n}\underline{s}_i$$

$(G(\underline{\theta}) = 0)$. We will need the true expectation of $\underline{S}$:

$$E_f(\underline{S}) = \left[\sum_{i=1}^{n}(\mu_i\underline{z}_i)\right].$$

Also, from (7.64), $H(\underline{\theta}) = \sum_{i=1}^{n}\log(1 - p_i)$ is an equivalent form for $H(\underline{\theta})$. Two key quantities we need are

$$I(\underline{\theta}) = -\frac{\partial^2 H(\underline{\theta})}{\partial\underline{\theta}^2}$$

(see 7.55) and $-\partial H(\underline{\theta})/\partial\underline{\theta} = E_f(\underline{S})$. Some straightforward mathematics leads to the results

$$-\frac{\partial H(\underline{\theta})}{\partial\underline{\theta}} = \sum_{i=1}^{n}p_i\underline{z}_i$$

and

$$I(\underline{\theta}) = -\frac{\partial^2 H(\underline{\theta})}{\partial\underline{\theta}^2} = \sum_{i=1}^{n}p_iq_i\underline{z}_i\underline{z}_i'.$$

These formulas can be put in matrix notation. To do so we define an $n \times 1$ column vector $\underline{P}(\theta)$ with ith element $p_i(\theta)$, and an $n \times n$ diagonal matrix V_p with ith diagonal element $p_i q_i$, and an $n \times K$ matrix Z where the ith row is $\underline{z}_i'$. Then

$$E_f(\underline{S}) = Z' \mu,$$

$$-\frac{\partial H(\theta)}{\partial \theta} = Z' \underline{P}(\theta), \tag{7.65}$$

and

$$I(\theta) = Z' V_p Z. \tag{7.66}$$

The MLE $\hat{\theta}$ is found by setting (7.65) to $\underline{S}$ and solving the resultant K nonlinear equations for θ, hence solving $\underline{S} = Z' \underline{P}(\hat{\theta})$. The true parameter value θ_o that applies here, given truth $\underline{\mu}$ and the model, is found by solving the same K equations but with $\underline{S}$ replaced by its true expectation, hence solving

$$Z' \underline{\mu} = Z' \underline{P}(\theta_o),$$

or

$$Z'(\underline{\mu} - \underline{P}(\theta_o)) = \underline{0}.$$

In partly nonmatrix notation, we solve

$$\sum_{i=1}^{n} (\mu_i - p_i(\theta_o)) \underline{z}_i = \underline{0}.$$

If truth $\underline{\mu}$ is not given exactly by the assumed model evaluated at θ_o, then $\underline{\mu} = \underline{P}(\overline{\theta}_o)$ will not hold even though the above equations will have a unique solution in θ_o, just as the MLE equations will have a unique solution as $\hat{\theta}$.

To proceed we also need to know the general formula for $J(\theta_o)$. From (7.12) we have

$$J(\theta_o) = E_f \left[\left[\underline{S} - Z' \underline{P}(\theta_o) \right] \left[\underline{S} - Z' \underline{P}(\theta_o) \right]' \right].$$

In partly nonmatrix form this formula is

$$J(\theta_o) = E_f \left[\left[\sum_{i=1}^{n} (\underline{s}_i - p_i \underline{z}_i) \right] \left[\sum_{i=1}^{n} (\underline{s}_i - p_i \underline{z}_i) \right]' \right]$$

$$= \sum_{i=1}^{n} \sum_{j=1}^{n} E_f (\underline{s}_i - p_i \underline{z}_i)(\underline{s}_j - p_j \underline{z}_j)'.$$

Here, using $\underline{s}_i = x_i \underline{z}_i$ and $E_f(x_i) = \mu_i$ the above becomes

$$J(\theta_o) = \sum_{i=1}^{n} \sum_{j=1}^{n} E_f (x_i - p_i)(x_j - p_j)[\underline{z}_i \underline{z}_j']$$

$$= \sum_{i=1}^{n} E_f(x_i - p_i)^2[\underline{z}_i \underline{z}_i'] + \sum_{i \neq j}^{n} \sum^{n} E_f(x_i - p_i)(x_j - p_j)[\underline{z}_i \underline{z}_j']$$

$$= \sum_{i=1}^{n} \left[(\mu_i(1 - \mu_i)) + (\mu_i - p_i)^2 \right] [\underline{z}_i \underline{z}_i']$$

$$+ \sum_{i \neq j}^{n} \sum^{n} (\mu_i - p_i)(\mu_j - p_j)[\underline{z}_i \underline{z}_j'].$$

Completing the square in the trailing term above, we get

$$J(\underline{\theta}_o) = \sum_{i=1}^{n} \left[(\mu_i(1 - \mu_i)) + (\mu_i - p_i)^2 \right] [\underline{z}_i \underline{z}_i']$$

$$+ \left[\sum_{i=1}^{n} (\mu_i - p_i(\underline{\theta}_o))\underline{z}_i \right]\left[\sum_{i=1}^{n} (\mu_i - p_i(\underline{\theta}_o))\underline{z}_i \right]'$$

$$- \sum_{i=1}^{n} \left[(\mu_i - p_i)^2 \right] [\underline{z}_i \underline{z}_i'].$$

The middle term of the above is zero because of the equation defining $\underline{\theta}_o$, and the third term cancels with part of the first term, so we have

$$J(\underline{\theta}_o) = \sum_{i=1}^{n} \mu_i(1 - \mu_i)[\underline{z}_i \underline{z}_i'],$$

or in pure matrix terms,

$$J(\underline{\theta}_o) = Z'V_\mu Z. \tag{7.67}$$

Here, V_μ is an $n \times n$ diagonal matrix with ith diagonal element $\mu_i(1 - \mu_i)$. Contrast (7.67) to $I(\underline{\theta}_o) = Z'V_p Z$.

It is easy, but not very informative, now to write

$$\text{tr}\left[J(\underline{\theta}_o)[I(\underline{\theta}_o)]^{-1} \right] = \text{tr}\left[(Z'V_\mu Z)(Z'V_p Z)^{-1} \right] \tag{7.68}$$

$$= K + \text{tr}\left[(Z'(V_\mu - V_p)Z)(Z'V_p Z)^{-1} \right].$$

The above makes it easier to realize that the trace term is exactly K if for all i $\mu_i(1 - \mu_i) = p_i(\underline{\theta}_o)(1 - p_i(\underline{\theta}_o))$. However, these equalities may fail to hold, yet we can still get tr $\left[J(\underline{\theta}_o)[I(\underline{\theta}_o)]^{-1} \right] = K$; hence this latter equality can hold with a model that does not match truth, i.e., where $g \subset f$ with strict inequality.

The above results are totally general, so they apply to the case where, say, w replicate observations are taken at each of $j = 1, \ldots, r$ covariate values. The total sample size is then $n = r * w$, but we will have only r different values of μ_j to specify for truth and only r values of $p_j(\underline{\theta})$ to consider under any model. Hence, to gain some insights here we used the simple model $\log\left[p_j/(1 - p_j) \right] = a + bj$ for $j = 1, \ldots, r$, with w replicate observations at each j. Thus $K = 2$, $\underline{\theta} = (a, b)'$, and $\underline{z}_j = (1, j)'$. In fact, for numerical or analytical results we do not have to specify what w is (just r is needed), and we can actually proceed as if $n = r$ with just one sample at each value of j. However, the results so derived apply reasonably well only to cases where $n = r * w$ would be "large," say 100 or more (given $K = 2$). Thus to explore the trace term under this use of a simple logistic regression model we need only specify a set of $\mu_1, \ldots, \mu_r$, solve

$$\sum_{i=1}^{r}(\mu_i - (a_o + b_o i))\underline{z}_i = \underline{0}$$

for $\theta_o = (a_o, b_o)'$, and compute $J(\theta_o)$ (7.67), $I(\theta_o)$ (7.66), and then $\text{tr}\left[J(\theta_o)[I(\theta_o)]^{-1}\right]$ (also denoted by "bias"). In doing this we focused on sets of μ that were near to fitting the logistic structural model, either by generating a $\underline{p}(\theta)$ vector that fit the model, then perturbing some (or all) of the p_j, or by starting with $\mu_j = j/(r + 1)$, which is not a logistic regression structural model but is not too far from fitting such a structural model.

With at most modest deviation of truth from any actual simple logistic regression model structure we found that the trace term value stayed near $K = 2$ (between about 1.8 and 2, sometimes going a little above 2, say to 2.1). For the case of truth being the simple linear model ($\mu_j = j/(r + 1)$), the trace term varied monotonically from 1.98 at $r = 5$, to 1.91 at $r = 50$. Table 7.2 gives some results for $r = 10$, based on truth being perturbed values from the logistic model $\text{logit}(p_j) = 3.0 - 0.5j$. The first line of Table 7.2 gives the true μ_j computed from this model (scaled by 1,000).

What one can see in Table 7.2 (and other computations we did corroborate this) is that the true μ_j have to be here a very poor approximation to an exact simple logistic model before the trace term deviates much from $K = 2$. Thus, if the data seem at all well fit by a logistic model, then the use of trace $= K$ (as opposed to any attempted estimation of the trace) seems quite suitable. This is especially important here because V_μ, hence $J(\theta_o)$, cannot be estimated at all unless there is replication at each $\underline{z}_i$, and there would need to be substantial such replication; this condition rarely occurs with logistic regression.

Formula (7.68) was corroborated by direct Monte Carlo evaluation of the target bias (trace term) for a few cases in Table 7.2. The completely general, and hence most direct, way to do this is to evaluate using simulation the value of

$$\text{bias} = E_{\underline{x}}E_{\hat{\underline{\theta}}(\underline{y})}\left[\log(g(\underline{x} \mid \hat{\underline{\theta}}(\underline{y})))\right] - E_{\underline{x}}\left[\log(g(\underline{x} \mid \hat{\underline{\theta}}(\underline{x})))\right]. \qquad (7.69)$$

Hence for one Monte Carlo replicate (generating *iid* $\underline{x}$ and $\underline{y}$) we get

$$\widehat{\text{bias}} = \log(g(\underline{x} \mid \hat{\underline{\theta}}(\underline{y}))) - \log(g(\underline{x} \mid \hat{\underline{\theta}}(\underline{x}))).$$

Averaged over many samples (m), if large sample size n is used, the average $\widehat{\text{bias}}$ will equal $\text{tr}\left[J(\theta_o)[I(\theta_o)]^{-1}\right]$.

For many models the first term on the right-hand side of (7.69) will be linear in $\underline{x}$, so we can analytically take the expectation with respect to $\underline{x}$. For this

TABLE 7.2. Some values of trace $= \text{tr}\left[J(\underline{\theta}_o)[I(\underline{\theta}_o)]^{-1}\right]$ for the simple logistic model $\text{logit}(p_j) = a + bj$, $(j = 1, \ldots, 10)$ fit to μ_j as perturbed values of p_j from $\text{logit}(p_j) = 3 - 0.5j$; values of μ_j are shown, scaled by 1,000; case one (i.e., the first line) exactly fits the logistic model, but none of the other cases are a perfect fit to the assumed model form. The results are reasonably applicable if w is at least 10 or 20.

μ_1	μ_2	μ_3	μ_4	μ_5	μ_6	μ_7	μ_8	μ_9	μ_{10}	trace
924	881	818	731	622	500	378	269	182	119	2
900	900	900	900	622	500	378	269	182	119	2.063
924	881	970	757	725	530	410	231	186	134	2.047
900	944	898	773	666	522	349	264	176	127	2.046
864	993	971	821	564	514	366	236	150	135	2.023
894	990	870	826	583	457	372	252	216	127	2.00011
924	720	818	796	749	632	492	188	260	164	1.993
924	881	768	831	622	400	378	269	282	119	1.980
924	952	650	838	638	448	278	189	204	132	1.967
874	874	874	558	558	558	558	190	190	190	1.925
924	881	818	731	400	600	378	269	182	119	1.925
924	881	818	731	300	700	378	269	182	119	1.857
924	881	818	731	622	500	378	269	182	119	1.828
924	881	568	831	622	100	378	269	432	119	1.774
924	881	818	731	200	800	378	269	182	119	1.768
674	981	818	731	622	500	128	669	182	119	1.765
924	881	818	731	622	500	900	900	900	900	1.494

logistic example we thus get, expressed in basic form,

$$\widehat{\text{bias}} = \sum_{i=1}^{n} E_{\underline{y}}\left[n_i\mu_i \log(p_i(\hat{\underline{\theta}})) + n_i(1 - \mu_i)\log(q_i(\hat{\underline{\theta}}))\right]$$
$$- \sum_{i=1}^{n} E_{\underline{y}}\left[y_i \log(p_i(\hat{\underline{\theta}})) + (n_i - y_i)\log(q_i(\hat{\underline{\theta}}))\right]. \qquad (7.70)$$

Here the MLE $\hat{\underline{\theta}}$ is based on data $\underline{y}$.

For direct Monte Carlo evaluation we used replicate covariate values, as noted above with the same number of replicates w for each $j = 1, \ldots, r$. Then we generated a large number m of independent samples from the generating model, fit the model-based MLE to each sample, computed bias using (7.70), by sample, and got its average and empirical standard error.

All this is quite obvious; where we are going here is that this direct Monte Carlo evaluation is poor in the sense of needing a huge number of samples. The problem is that as w (i.e., n) increases, the number of samples needed to get a small standard error (like 0.005) on the estimate bias increases because the variance of bias, for one sample, increases with increasing sample size, and that variance can be quite large. For a large sample size n (which is required for the trace approximation to hold very well) it can take one million Monte Carlo samples to get even a moderately small standard error on estimated bias. For example, for the case in Table 7.2 where the trace is computed to be 1.774, for $w = 100$ (hence sample size $n = 1,000$ Bernoulli trials) for one representative set of 10,000 ($= m$) Monte Carlo samples we got average $\widehat{\text{bias}} = 1.528$ with an estimated standard error of 0.135. Other runs verified that it takes about one million Monte Carlo samples to get a standard error (on estimated bias) of about 0.014 with $w = 100$ in this example. But we might need a bigger w for the trace formula to apply exactly; for $w = 1,000$ and 10,000 Monte Carlo samples we got the average $\widehat{\text{bias}} = 1.982$ with $\widehat{\text{se}} = 0.424$. This phenomenon is the reverse of what we expect; i.e., we expect to get increasing precision (for the same number, m, of Monte Carlo samples) as sample size n increases. The reverse phenomenon occurs here because the expected difference in likelihoods involved in direct computation of $\widehat{\text{bias}}$ (i.e., (7.70)) is constant independent of sample size n, but the variance of each of those two likelihood sums in (7.70) is proportional to n and the two terms are not highly correlated. Thus as sample size n increases, the precision of the estimated bias, given a fixed number of Monte Carlo samples (m), actually decreases. So to evaluate well, with this brute-force approach, the adequacy of the trace term approximation at large sample sizes, it takes a huge number of Monte Carlo samples.

With models that are in the exponential family there is an alternative way to do exact Monte Carlo evaluation of the bias that must be subtracted from the maximized log-likelihood for exact K-L based model selection. Formula (7.54) is an exact result for any sample size:

$$\text{bias} = \text{tr}\left[\text{COV}(\hat{\underline{\theta}}, \underline{S})\right].$$

While $\underline{\theta}$ and $\underline{S}$ are only for the canonical form of the model, the result will apply for any parametrization of the assumed model because of the invariance of the result to 1-to-1 transformations of $\underline{\theta}$ (see end of Section 7.5). Thus the alternative Monte Carlo evaluation is simply to take for each sample the already computed MLE and minimal sufficient statistic and, from this set of records of size m, estimate the covariances $\text{cov}(\hat{\theta}_i, S_i)$, $i = 1, \ldots, K$, and then sum these K estimates. The result is $\widehat{\text{bias}}$, and this approach is much more efficient. For the same case in Table 7.2 (i.e., trace $= 1.774$), using $w = 100$ ($r = 10$, hence $n = 1,000$) and 10,000 Monte Carlo samples we got $\widehat{\text{bias}} = 1.804$ and its $\widehat{\text{se}} = 0.019$ using the covariance approach. Based on this and other runs there was a clear suggestion that $w = 100$ was not quite big enough for the trace (7.68) to apply reliably to three digits (it was then reliable to two digits). Using $w = 1,000$ and 10,000 Monte Carlo samples we got $\widehat{\text{bias}} = 1.771$, $\widehat{\text{se}} = 0.017$. This result held up on more study: (7.68) seemed to be excellent for $w = 1,000$ (which here meant $n = 10,000$).

As another example consider the last case in Table 7.2, where trace $= 1.494$ ($=$ bias). Using $w = 100$, for one run of 10,000 Monte Carlo samples (the only such run made) we got the direct result based on (7.70) as $\widehat{\text{bias}} = 1.959$, $\widehat{\text{se}} =$

0.171. In contrast, for that same simulated data set the covariance approach yielded $\widehat{\text{bias}} = 1.516$, $\widehat{\text{se}} = 0.016$. Clearly, in working with models in the exponential family, Monte Carlo or bootstrap evaluation of the needed K-L trace term should be based on (7.54).

It is worth noting a basis for the estimated standard error of $\widehat{\text{bias}} = \widehat{\text{tr}}\left[\text{COV}(\underline{\hat{\theta}}, \underline{S})\right]$. For the point estimate, use all the simulation samples to compute means; then for component i,

$$\widehat{\text{cov}}(\hat{\theta}_i, S_i) = \frac{\sum_{j=1}^{m}(\hat{\theta}_{i,j} - \bar{\hat{\theta}}_i)(S_{i,j} - \bar{S}_i)}{m-1},$$

and

$$\widehat{\text{bias}} = \sum_{i=1}^{K} \widehat{\text{cov}}(\hat{\theta}_i, S_i).$$

However, to estimate the standard error we must partition the set of m samples, say into 25 equal-sized subsets (for $m = 10{,}000$ then each subset has size 400). Compute by the above formulas $\widehat{\text{bias}}_s$ for each subset s; then estimate the standard error of $\widehat{\text{bias}}$ from these 25 independent estimates (whose mean will almost equal $\widehat{\text{bias}}$, but will not be equal due to nonlinearities).

The standard error of $\widehat{\text{bias}}$ from this covariance approach is stable as a function of data sample size n because of how the product involved behaves. It suffices to consider the product $(\hat{\theta}_i - \theta_{o,i})(S_i - E_f(S_i))$ (for any component i). This product has variance virtually independent of n because the first term converges (in n) at rate proportional to $1/\sqrt{n}$, while the second term converges at rate proportional to $\sqrt{n}$. As a result, the standard error of this covariance-based bias (hence trace) estimator is almost independent of sample size. This is much better behavior (as a function of n) than the standard error of the estimator of bias based directly on the likelihood function. The latter method (i.e., (7.69)) also requires more calculations beyond first getting $\underline{\hat{\theta}}$ and $\underline{S}$.

7.6.3 Evaluation Under Multinomially Distributed Count Data

We here assume that we have count data $n_1, \ldots, n_r$ that sum to the sample size n. Truth is the multinomial distribution $\text{mult}(n, \mu_1, \ldots, \mu_r)$ with cell probabilities μ_i summing to 1, and $0 < \mu_i < 1$. To know truth in this context we only need to know the true μ_i (assuming that the counts are multinomially distributed; they could have overdispersion, which violates this assumption). We might totally fail to know how these true probabilities arise in general in relation to any explanatory variables, or what would happen if the cells were defined in some other way. Thus, deeper truth may exist regarding the situation, but it is irrelevant to model selection purposes once we restrict ourselves to a particular multinomial setting.

For a constrained model we assume that cell probabilities $p_i(\theta)$ are known functions of a K-dimensional parameter θ with $1 \le K < r - 1$ ($K = r - 1$ is not to be considered, since then the fitted model matches truth in the sense of being a perfect match to the data). The theory in Sections 7.1 and 7.2 is now used; note that here

$$\log(g(\underline{n} \mid \underline{\theta})) = \sum_{i=1}^{r} n_i \log(p_i(\theta)).$$

First, $\underline{\theta}_o$ is determined as the solution to (7.5), which here becomes

$$\sum_{i=1}^{r} \frac{\mu_i}{p_i(\theta_o)} \frac{\partial p_i(\theta_o)}{\partial \underline{\theta}} = \underline{0}. \tag{7.71}$$

In (7.71) if we replace μ_i by n_i, we have the likelihood equations. Thus one can treat the μ_i as data and find θ_o by MLE methods. Equivalently, θ_o is the MLE when the data are replaced by their true expected values, $E_f(n_i) = n\mu_i$.

Second, applying (7.7) we directly get

$$I(\underline{\theta}_o) = n \left[\sum_{i=1}^{r} \frac{\mu_i}{[p_i(\theta_o)]^2} \left(\frac{\partial p_i(\theta_o)}{\partial \underline{\theta}} \right) \left(\frac{\partial p_i(\theta_o)}{\partial \underline{\theta}} \right)' \right]$$
$$- n \left[\sum_{i=1}^{r} \frac{\mu_i}{p_i(\theta_o)} \left(\frac{\partial^2 p_i(\theta_o)}{\partial \underline{\theta}^2} \right) \right].$$

Finally, applying the definition in (7.12), we have

$$J(\underline{\theta}_o) = E_f \left[\sum_{i=1}^{r} \frac{n_i}{p_i(\theta_o)} \left(\frac{\partial p_i(\theta_o)}{\partial \underline{\theta}} \right) \right] \left[\sum_{i=1}^{r} \frac{n_i}{p_i(\theta_o)} \left(\frac{\partial p_i(\theta_o)}{\partial \underline{\theta}} \right) \right]'.$$

The evaluation of $J(\theta_o)$ does take some algebra and knowledge of the multinomial distribution, but it is mostly a straightforward exercise, so we just give the result:

$$J(\underline{\theta}_o) = n \left[\sum_{i=1}^{r} \frac{\mu_i}{[p_i(\theta_o)]^2} \left(\frac{\partial p_i(\theta_o)}{\partial \underline{\theta}} \right) \left(\frac{\partial p_i(\theta_o)}{\partial \underline{\theta}} \right)' \right]. \tag{7.72}$$

Define the matrix A as

$$A = n \left[\sum_{i=1}^{r} \frac{\mu_i}{p_i(\theta_o)} \left(\frac{\partial^2 p_i(\theta_o)}{\partial \underline{\theta}^2} \right) \right],$$

and we have $I(\underline{\theta}_o) = J(\underline{\theta}_o) - A$. Furthermore, if the model is truth, then $\mu_i = p_i(\theta_o)$, and A reduces to the null matrix; hence then $I(\underline{\theta}_o) = J(\underline{\theta}_o)$.

Using these results we can write

$$\text{tr} \left[J(\underline{\theta}_o)[I(\underline{\theta}_o)]^{-1} \right] = \text{tr} \left[(J(\underline{\theta}_o) - A + A)[I(\underline{\theta}_o)]^{-1} \right],$$

whence

$$\mathrm{tr}\left[J(\underline{\theta}_o)[I(\underline{\theta}_o)]^{-1}\right] = K + \mathrm{tr}\left[A[I(\underline{\theta}_o)]^{-1}\right].$$

For a long time a nagging question for us was whether the trace term would always be either $> K$ or $< K$ when the model did not exactly match truth yet the model is logically known to be simpler than truth (i.e., $g \subset f$ in some general sense). Stated differently, if the Kullback–Leibler discrepancy is positive, i.e., K-L $= I(f, g) > 0$, then must $\mathrm{tr}\left[J(\underline{\theta}_o)[I(\underline{\theta}_o)]^{-1}\right] > K$ (or maybe $< K$) always occur when the model is some form of constrained truth (hence the model can be said to approximate, but not equal, truth). The answer is no, as was indicated by the logistic regression examples in Section 7.6.2); however, a more convincing answer is given here: The trace can be either $> K$ or $< K$ and there need be no consistency as to which will occur. A related question also explored below is, If $\mathrm{tr}\left[J(\underline{\theta}_o)[I(\underline{\theta}_o)]^{-1}\right] = K$, must $I(f, g) = 0$? That answer is also no.

Because the cell probabilities sum to 1, the sum of the matrices of second partials is the null matrix, O, of all zeros (the vector of first partials also sums to a null vector). Therefore, an equivalent expression for matrix A is

$$A = n\left[\sum_{i=1}^{r} \frac{\mu_i - p_i(\underline{\theta}_o)}{p_i(\underline{\theta}_o)}\left(\frac{\partial^2 p_i(\underline{\theta}_o)}{\partial \underline{\theta}^2}\right)\right].$$

The weights in this linear combination of second partial derivative matrices must be either identically zero (hence K-L is 0), or some are negative and some positive. This would suggest that A might not always have the same sign, unless the second partials are very strangely related to the model and truth. But a more detailed case is need to get an example, and it seems best to use $K = 1$ for an example, such as by using a binomial model.

Let us further assume that the data arise from n independent samples of an integer random variable, y, taking values 0 to $r - 1$. The data are then just the frequency counts n_i of times $y = i - 1$. A very simple model for the cell probabilities, μ_i, is thus to assume that this underlying random variable is a binomial random variable. This corresponds to imposing an ordering on the multinomial cells, without loss of generality, and thus the model for the cell i probability is

$$p_i(\theta) = \binom{r-1}{i-1}\theta^{i-1}(1 - \theta)^{r-i}, \qquad i = 1, \ldots, r.$$

Thus we have, as our model, an assumed underlying binomial random variable $y \sim \mathrm{bin}(r - 1, \theta)$ and a random sample of size n of this random variable. In fact, $y (= 0, 1, \ldots, r - 1)$ has the distribution given by the μ_{y+1} as its true distribution. We will need the functions below, involving first and second partial derivatives:

$$P1_i = \frac{1}{p_i(\theta)}\left(\frac{\partial p_i(\theta)}{\partial \theta}\right) = \frac{(i - 1) - \theta(r - 1)}{\theta(1 - \theta)},$$

$$P2_i = \frac{1}{p_i(\theta)}\left(\frac{\partial^2 p_i(\theta)}{\partial\theta^2}\right) = \left[\frac{(i-1)-\theta(r-1)}{\theta(1-\theta)}\right]^2$$
$$-\left[\frac{(i-1)(1-2\theta)+\theta^2(r-1)}{(\theta(1-\theta))^2}\right].$$

We solve (7.71), which is $\sum \mu_i P1_i = 0$, to find θ_o; this is exactly the same process as finding an MLE (again, the only tricky aspect is the indexing assumed here):

$$\theta_o = \frac{\sum_{i=1}^{r}\mu_i(i-1)}{r-1}.$$

This θ_o is the true expected value of $y/(r-1)$ regardless of any assumed model. We compute (7.72) as $n\sum \mu_i(P1_i)^2$:

$$J(\theta_o) = n\left[\sum_{i=1}^{r}\mu_i\left(\frac{(i-1)-\theta_o(r-1)}{\theta_o(1-\theta_o)}\right)^2\right] \equiv nE_f\left(\frac{y-E_f(y)}{\theta_o(1-\theta_o)}\right)^2.$$

We find matrix A as $n\sum \mu_i P2_i$:

$$A = J(\theta_o) - \frac{n(r-1)}{\theta_o(1-\theta_o)};$$

hence

$$I(\theta_o) = \frac{n(r-1)}{\theta_o(1-\theta_o)}.$$

It is now easy to find the trace:

$$\mathrm{tr}\left[J(\theta_o)[I(\theta_o)]^{-1}\right] = \sum_{i=1}^{r}\mu_i\frac{[(i-1)-\theta_o(r-1)]^2}{(r-1)\theta_o(1-\theta_o)}. \qquad (7.73)$$

For the case of $\mu_i = p_i(\theta_o)$, then (7.73) is 1 (this can be directly verified); hence using $\mu_i \equiv p_i(\theta_o) + (\mu_i - p_i(\theta_o))$ in (7.73) we obtain

$$\mathrm{tr}\left[J(\theta_o)[I(\theta_o)]^{-1}\right] = 1 + \sum_{i=1}^{r}(\mu_i - p_i(\theta_o))\frac{[(i-1)-\theta_o(r-1)]^2}{(r-1)\theta_o(1-\theta_o)},$$

whereupon it should be essentially obvious that the term added to 1 ($= K$) can be either positive or negative. However, we will give numerical examples, mostly for $r = 3$ because this is the smallest r we can use for our purposes here, and small r is desirable when we need to display truth.

Our model is thus bin$(2, \theta)$; hence $p_1 = (1-\theta)^2$, $p_2 = 2\theta(1-\theta)$, and $p_3 = \theta^2$. The approach is to specify the μ_i and compute $\theta_o = (\mu_2/2) + \mu_3$, and from (7.73), for $r = 3$,

$$\mathrm{tr}\left[J(\theta_o)[I(\theta_o)]^{-1}\right] = \frac{4\mu_1(\theta_o)^2 + \mu_2(1-2\theta_o)^2 + 4\mu_3(1-\theta_o)^2}{2\theta_o(1-\theta_o)}. \qquad (7.74)$$

We will also consider the values of $I(f, g)$, so we note that this K-L discrepancy is here $\sum(\mu_i) \log[\mu_i / p_i(\theta_o)]$. Numerical values are given below for three cases of truth (the μ_i) in relationship to the K-L best approximating binomial model. Case one exactly fits a binomial model. Cases two and three are fit terribly by even the K-L best approximating binomial model. In all three cases, $\theta_o = 0.5$. In what is below, "Trace" means the value computed from (7.74) for $\theta_o = 0.5$, and K-L is the Kullback–Leibler information discrepancy between truth and the best approximating binomial model ("Bias-MC" is explained below):

μ_1	μ_2	μ_3	Trace	K-L	Bias-MC
0.25	0.50	0.25	1.000	0.000	1.003
0.05	0.90	0.05	0.200	0.368	0.194
0.45	0.10	0.45	1.800	0.368	1.816

(we note that in this situation the trace term (7.73) seems to be bounded above by 2). Clearly, this bias-correction trace term can be either less than or greater than 1 when the model does not match truth. This is because the theoretical variance of y can be either larger or smaller than the theoretical binomial variance for y implied by the K-L best-fitting binomial model.

We build on this example by doing an exact Monte Carlo evaluation of the expected log-likelihood and the K-L–based target model selection criterion to verify the asymptotic derivation of the bias as being the trace term. In the above, "Bias-MC" denotes the results (accurate to two decimal places), for sample size $n = 200$, from one million Monte Carlo samples to evaluate the bias that the trace term measures based on asymptotic theory.

For a truth that cannot be well approximated here by a binomial model it is clear that the trace (equation 7.74) can be far from 1. Rather than explore this example for models that are arbitrarily poor (like cases two and three, above) we should consider models that are closer to truth, because with AIC (or TIC) the term $-2\log(\mathcal{L})$ will prevent the selection of really poor models (hence for those models a choice between the use of K or $\widehat{\mathrm{trace}}$ is irrelevant) if the set of models has some good candidates.

So we looked at one set of cases where a binomial model was not terribly wrong to use. We chose a θ, generated $p_1 = (1 - \theta)^2$, $p_2 = 2\theta(1 - \theta)$, and $p_3 = \theta^2$, then perturbed these cell probabilities to get a truth that was close to a binomial model by setting $\mu_i = p_i + \epsilon_i$, where $\epsilon_i = \delta_i - \bar{\delta}$ for $\delta_i \sim iid$ uniform$(-h, h)$. Inadmissible sets of μ_i were not generated. Given a set of μ_1, μ_2, and μ_3, (7.74) was evaluated; thus, this is not a Monte Carlo study. Rather, we use Monte Carlo methods only as a convenience in generating sets of true μ_i that are close to a binomial model.

For $\theta = 0.5$ and $h = 0.1$ (and 1,000 generated sets of truth) we got the following results for the trace given by (7.74): min $= 0.747$, max $= 1.234$, and mean $= 0.996$. These results support practical use of K rather than $\widehat{\mathrm{tr}}\left[J(\theta_o)[I(\theta_o)]^{-1}\right]$. However, it is fair to ask about estimating this trace term

TABLE 7.3. Some Monte Carlo results evaluating the (7.75) estimator of tr $\left[J(\underline{\theta}_o)[I(\underline{\theta}_o)]^{-1} \right]$ for the case of an assumed binomial$(2, \theta)$ model when truth (μ_1, μ_2, μ_3) may be more general; the true trace value is known for these cases; the mean and standard deviation of (7.75) are given based on one million samples.

μ_1	μ_2	μ_3	n	Trace	Mean	St. dev.
0.25	0.50	0.25	50	1.000	0.990	0.142
0.25	0.50	0.25	100	1.000	0.995	0.101
0.25	0.50	0.25	200	1.000	0.997	0.071
0.04	0.32	0.64	200	1.000	0.997	0.072
0.20	0.55	0.25	50	0.897	0.887	0.141
0.20	0.60	0.20	50	0.800	0.790	0.139
0.30	0.40	0.30	50	1.200	1.190	0.140
0.30	0.45	0.25	50	1.098	1.088	0.140
0.05	0.90	0.05	200	0.200	0.199	0.043
0.45	0.10	0.45	200	1.800	1.799	0.043

here (hence using TIC), as can be done by plugging $\hat{\theta}_o$ and $\hat{\mu}_i = n_i/n$ into (7.74); after simplification,

$$\widehat{\text{tr}} \left[J(\theta_o)[I(\theta_o)]^{-1} \right] = \frac{4\hat{\mu}_1(\hat{\theta}_o)^2 + \hat{\mu}_2(1 - 2\hat{\theta}_o)^2 + 4\hat{\mu}_3(1 - \hat{\theta}_o)^2}{2\hat{\theta}_o(1 - \hat{\theta}_o)}. \quad (7.75)$$

A small Monte Carlo evaluation of this estimator was done to see whether it was badly biased or highly variable. Variables in this study are the three μ_i and sample size n. Results, given in Table 7.3 based on one million samples, are the theoretical trace value, and the mean and standard deviation of (7.75) evaluated by simulation, accurate to two decimal places. In Table 7.3 if the trace is 1, then the binomial distribution is truth; otherwise, it is not truth.

From Table 7.3 it appears that the trace estimator has good properties, so it is reasonable to consider using TIC rather than AIC; at least the comparison of the two seems worth doing here. For the sets of true μ_i considered above we compared AIC to TIC for the binomial model ($K = 1$; hence a reduced model, R) and the parameter-saturated general model ($K = 2$; G). Let the corresponding maximized likelihoods be $\mathcal{L}_R$ and $\mathcal{L}_G$. Hence,

$$\text{AIC}_R = -2\log(\mathcal{L}_R) + 2,$$
$$\text{AIC}_G = -2\log(\mathcal{L}_G) + 4,$$
$$\text{TIC}_R = -2\log(\mathcal{L}_R) + 2\,\widehat{\text{tr}} \left[J(\theta_o)[I(\theta_o)]^{-1} \right],$$
$$\text{TIC}_G = -2\log(\mathcal{L}_G) + 4,$$

where $\widehat{\text{tr}} \left[J(\theta_o)[I(\theta_o)]^{-1} \right]$ is given by (7.75). Because there are only two models here, and because we want to keep matters simple, we just compared selection methods based on how often they selected the same model and how often they selected the correct data-generating model. Results are based on 10,000 Monte Carlo samples, which suffices here to get standard errors ≤ 0.005

TABLE 7.4. Some Monte Carlo results evaluating AIC versus TIC model selection for a binomial$(2, \theta)$ model $(K = 1)$ versus a saturated multinomial model $(K = 2)$; the true generating model varied (R for the binomial, G for the multinomial); each case is based on 10,000 samples; column AIC (or TIC) denotes the proportion of cases where AIC (or TIC) selected the correct data-generating model; column "Match" means that both criteria selected the same model whether or not it was the data-generating model (see text for more details).

μ_1	μ_2	μ_3	n	Truth	AIC	TIC	Match
0.25	0.50	0.25	50	R	0.83	0.84	0.93
0.25	0.50	0.25	100	R	0.85	0.84	0.96
0.25	0.50	0.25	200	R	0.84	0.84	0.97
0.04	0.32	0.64	200	R	0.84	0.83	0.99
0.20	0.55	0.25	50	G	0.69	0.76	0.91
0.20	0.60	0.20	50	G	0.57	0.53	0.89
0.30	0.40	0.30	50	G	0.46	0.53	0.92
0.30	0.45	0.25	50	G	0.24	0.28	0.93
0.05	0.90	0.05	200	G	1.00	1.00	1.00
0.45	0.10	0.45	200	G	1.00	1.00	1.00
				means	0.73	0.75	0.95

for estimated proportions. "Truth" denotes the correct generating model. The "AIC" and "TIC" columns denote the proportion of samples for which these methods selected the correct model. "Match" denotes the proportion of samples in which both methods selected the same model, regardless of which model it was.

We looked at many more results than are given above to compare AIC and TIC in this limited context; there was then no change from the above in the obvious conclusion: no meaningful difference in performance here of AIC versus TIC. More study is surely warranted; this limited look was done in the spirit that maybe something dramatic would result. It did not; as a tentative conclusion (based on all the considerations we have done on the matter, not just those of this section), it seems that simplicity strongly favors use of AIC over TIC.

We return to an interesting theoretical question posed above. It is known that if $f = g$, then tr$\left[J(\theta_o)[I(\theta_o)]^{-1}\right] = K$. However, if tr$\left[J(\theta_o)[I(\theta_o)]^{-1}\right] = K$, does this mean $f = g$?

A counterexample shows that the assertion is false; hence there are situations wherein truth is more complex than the models used for analysis and yet AIC is appropriate to use (as opposed to TIC, which would then unnecessarily be just estimating K).

For this multinomial context we have shown that $I(\theta_o) = J(\theta_o) - A$ (considerations here are for any value of r). So if $A = O$ (i.e., is all zeros), then $I(\theta_o) = J(\theta_o)$ and tr$\left[J(\theta_o)[I(\theta_o)]^{-1}\right] = K$ regardless of whether or not the model is truth (which requires $\mu_i = p_i(\underline{\theta}_o)$ for all i). Recall that

$$A = n\left[\sum_{i=1}^{r} \frac{\mu_i}{p_i(\underline{\theta}_o)}\left(\frac{\partial^2 p_i(\underline{\theta}_o)}{\partial \underline{\theta}^2}\right)\right];$$

therefore, if all second partial derivatives of the model cell probabilities are zero, we do get $A = O$. This will occur for any linear model of the cell probabilities; that is, $p_i(\underline{\theta}) = \underline{x}_i'\underline{\theta}$ for a set of known vectors, $\underline{x}_i$. Of course, such models are discouraged because they can generate fitted cell estimates out of range.

As an example we revert to the case of $r = 3$ and use the model structure $\mu_1 = \mu_3 = \theta/2$ and $\mu_2 = 1 - \theta$; so $\log(g(\underline{n}\,|\,\underline{\theta})) = (n_1 + n_3)\log(\theta/2) + n_3\log(1 - \theta)$. Here, $\theta_o = 1 - \mu_2$. Upon computing $I(\underline{\theta}_o)$ and $J(\underline{\theta}_o)$ from their basic definitions, we do in fact get $I(\underline{\theta}_o) = J(\underline{\theta}_o) = n/[\theta_o(1-\theta_o)]$ irrespective of the values of the μ_i. This means that here is a situation and a model where AIC rather than TIC is the correct selection procedure even though the model does differ from truth (i.e., $f \subseteq g$ is not true, yet this condition is sometimes cited as always required for the theoretical validity of AIC).

For the case of general r and the binomial model we can use (7.73) to investigate this trace term and AIC versus TIC. But even (7.73) is too complex to derive any insights from it directly, and numerical methods are needed. So all we really need are usable computational formulas to compute TIC, i.e., estimate $\text{tr}[J(\theta_o)[I(\theta_o)]^{-1}]$. We can get the needed formulas for any postulated model for multinomial data. First, we can find $\hat{\underline{\theta}}_o$ by solving

$$\sum_{i=1}^{r} \frac{n_i/n}{p_i(\underline{\theta}_o)}\frac{\partial p_i(\underline{\theta}_o)}{\partial \underline{\theta}} = \underline{0},$$

which we do anyway, since this is just our MLE of $\underline{\theta}$ under the assumed model. We do have to compute the set of first and second partial derivatives of the model cell structures evaluated at the MLE, but even that can be done numerically. Thus we can get, hence use and explore, TIC:

$$\hat{I}(\underline{\theta}_o) = n\left[\sum_{i=1}^{r} \frac{n_i/n}{[p_i(\hat{\underline{\theta}}_o)]^2}\left(\frac{\partial p_i(\hat{\underline{\theta}}_o)}{\partial \underline{\theta}}\right)\left(\frac{\partial p_i(\hat{\underline{\theta}}_o)}{\partial \underline{\theta}}\right)'\right]$$
$$- n\left[\sum_{i=1}^{r} \frac{n_i/n}{p_i(\hat{\underline{\theta}}_o)}\left(\frac{\partial^2 p_i(\hat{\underline{\theta}}_o)}{\partial \underline{\theta}^2}\right)\right],$$
$$\hat{J}(\underline{\theta}_o) = n\left[\sum_{i=1}^{r} \frac{n_i/n}{[p_i(\hat{\underline{\theta}}_o)]^2}\left(\frac{\partial p_i(\hat{\underline{\theta}}_o)}{\partial \underline{\theta}}\right)\left(\frac{\partial p_i(\hat{\underline{\theta}}_o)}{\partial \underline{\theta}}\right)'\right].$$

Clearly, we can also compute theoretical values of these quantities for any postulated truth and model. Such studies would be informative, but are beyond the intention of this book.

In a paper on model selection for multinomial distributions deLeeuw (1988) assumes a certain general model selection criterion and pursues it. He does so under the philosophy we espouse here: Models used for data analysis are not truth; full truth is very complex; one's analytic goal should be to find a best approximating fitted model. deLeeuw concludes that the most reasonable (essentially, compelling) explicit model selection criterion to use is AIC. In particular, he says (deLeeuw 1988:132), "This gives a justification for using the AIC, even if the model is not true." On an important related issue it also seems worth quoting deLeeuw (1988:136–137): "The independence assumption, for example, which is at the basis of most work in statistics, cannot really be falsified. As we have seen, the independence assumption merely corresponds with a particular framework of replication, for which we have to decide whether it is relevant or not."

7.6.4 Evaluation Under Poisson-Distributed Data

The purpose of this subsection is to see whether a result for multinomial count data extends to the Poisson-distributional case. We assume a sample of size r of observed Poisson counts with unknown means μ_i (= truth, *assuming* that the data are Poisson distributed). The model for these means is $\lambda_i(\underline{\theta}_o)$, $i = 1$, ..., r. Some results:

$$\log(g(\underline{n} \mid \underline{\theta})) = \sum_{i=1}^{r}\left[-\lambda_i(\underline{\theta}) + n_i \log(\lambda_i(\underline{\theta}))\right];$$

$\underline{\theta}_o$ is determined as the solution to

$$\sum_{i=1}^{r}\left[\frac{\mu_i}{\lambda_i(\underline{\theta}_o)} - 1\right]\frac{\partial\lambda_i(\underline{\theta}_o)}{\partial\underline{\theta}} = \underline{0};$$

$$I(\underline{\theta}_o) = \left[\sum_{i=1}^{r}\frac{\mu_i}{[\lambda_i(\underline{\theta}_o)]^2}\left(\frac{\partial\lambda_i(\underline{\theta}_o)}{\partial\underline{\theta}}\right)\left(\frac{\partial\lambda_i(\underline{\theta}_o)}{\partial\underline{\theta}}\right)'\right]$$
$$- \left[\sum_{i=1}^{r}\left(\frac{\mu_i}{\lambda_i(\underline{\theta}_o)} - 1\right)\left(\frac{\partial^2\lambda_i(\underline{\theta}_o)}{\partial\underline{\theta}^2}\right)\right],$$

$$J(\underline{\theta}_o) = \left[\sum_{i=1}^{r}\frac{\mu_i}{[\lambda_i(\underline{\theta}_o)]^2}\left(\frac{\partial\lambda_i(\underline{\theta}_o)}{\partial\underline{\theta}}\right)\left(\frac{\partial\lambda_i(\underline{\theta}_o)}{\partial\underline{\theta}}\right)'\right].$$

We define the $K \times K$ matrix B as

$$B = \sum_{i=1}^{r}\left(\frac{\mu_i}{\lambda_i(\underline{\theta}_o)} - 1\right)\left(\frac{\partial^2\lambda_i(\underline{\theta}_o)}{\partial\underline{\theta}^2}\right);$$

then $I(\underline{\theta}_o) = J(\underline{\theta}_o) - B$ and tr$\left[J(\underline{\theta}_o)[I(\underline{\theta}_o)]^{-1}\right] = K + $ tr$\left[B[I(\underline{\theta}_o)]^{-1}\right]$. If matrix B is zero, then regardless of how much μ_i and $\lambda_i(\underline{\theta}_o)$ differ for the r pairs of these values (hence $I(f, g) > 0$ occurs), we still have tr$\left[J(\underline{\theta}_o)[I(\underline{\theta}_o)]^{-1}\right] =$

K, so AIC is justified. This same result, which also holds for multinomial models, is achieved here by any linear model of the form $\lambda_i(\underline{\theta}) = \underline{x}_i'\underline{\theta}$.

7.6.5 Evaluation for Fixed-Effects Normality-Based Linear Models

The fixed-effects linear model based on n iid normally distributed residuals is so common that it seems almost mandatory that we consider tr $\left[J(\theta_o)[I(\theta_o)]^{-1} \right]$ under this model for some tractable "truth." The model is $\underline{Y} = X\underline{\beta} + \underline{\epsilon}, \underline{\epsilon} \sim$ multivariate-normal($\underline{0}, \sigma^2 I$), and without loss of generality the $n \times (K - 1)$ matrix X is assumed of full rank. Truth has a structural component $E(\underline{Y}) = \underline{\mu}$ (which can be estimated by $\underline{Y}$) and a stochastic component for $\underline{\epsilon} = \underline{Y} - \underline{\mu}$, distributed in some unknown way, the properties of which cannot be estimated without strong assumptions. If the model is truth, then both structural (i.e., $\underline{\mu} = X\underline{\beta}$) and distributional assumptions of the model are true.

In reality, the ϵ_i may not be independent, may not be identically distributed, and may not be normally distributed. In fact, they may not exist, in the sense that some or all ϵ_i are zero with probability 1. In this latter case truth is deterministic; that is, there is some sufficiently complex computing algorithm (perhaps a formula, with many covariates) such that if we knew that algorithm, we could predict $\underline{Y}$ with certainty (measurement error would become problematic before this level of model accuracy was reached). Hence, for unknown truth we cannot, for cases of real data, evaluate K-L–based model selection for models of continuous random variables.

We can, however, derive informative results under general models for truth that are better approximations to reality (by assumption) than the model to be used for data analysis. Therefore, we assume here that truth is $\underline{Y} = \underline{\mu} + \underline{\epsilon}, \underline{\epsilon} \sim$ multivariate-normal($\underline{0}, \tau^2 I$), where τ^2 may be zero. If fact, we can even drop the full distributional assumption, as we will demonstrate below, because the relevant evaluations require only the first four moments of the true distribution. More generally, results could be gotten under the assumption of $\underline{\epsilon} \sim$ multivariate-normal($\underline{0}, \Sigma$) for given Σ, but the more restricted framework will suffice.

We first need basic notation and results: $\underline{\theta}$ denotes the $K \times 1$ vector $(\beta', \sigma^2)'$, for β a $(K - 1) \times 1$ vector of the structural parameters. We take σ^2, not σ, as the parameter to estimate. The model pdf for the data is

$$g(\underline{y} \mid \underline{\theta}) = \frac{1}{\sqrt{2\pi\sigma^2}} \exp\left[-\frac{1}{2\sigma^2}(\underline{Y} - X\underline{\beta})'(\underline{Y} - X\underline{\beta}) \right];$$

and we take, without loss of generality,

$$\log(g(\underline{y} \mid \underline{\theta})) = -\frac{n}{2} \log(\sigma^2) - \frac{1}{2\sigma^2}(\underline{Y} - X\underline{\beta})'(\underline{Y} - X\underline{\beta}),$$

$$\frac{\partial \log(g(\underline{y} \mid \underline{\theta}))}{\partial \underline{\beta}} = \frac{1}{\sigma^2} X'(\underline{Y} - X\underline{\beta}), \tag{7.76}$$

$$\frac{\partial \log(g(\underline{y} \mid \underline{\theta}))}{\partial \sigma^2} = -\frac{n}{2\sigma^2} + \frac{1}{2(\sigma^2)^2}(\underline{Y} - X\underline{\beta})'(\underline{Y} - X\underline{\beta}). \tag{7.77}$$

As per theory, we take the expectations of (7.76) and (7.77) with respect to what is here just assumed truth, f (evaluation under absolute truth now being impossible) to get the equation

$$\frac{1}{\sigma_o^2} X'(\underline{\mu} - X\underline{\beta}_o) = \underline{0}$$

from (7.76), and then from (7.77) we derive

$$-\frac{n}{2\sigma_o^2} + \frac{1}{2(\sigma_o^2)^2} \mathrm{E}_f(\underline{Y} - \underline{\mu} + \underline{\mu} - X\underline{\beta}_o)'(\underline{Y} - \underline{\mu} + \underline{\mu} - X\underline{\beta}_o)$$

$$= -\frac{n}{2\sigma_o^2} + \frac{1}{2(\sigma_o^2)^2} \left[\mathrm{E}_f(\underline{\epsilon}'\underline{\epsilon}) + (\underline{\mu} - X\underline{\beta}_o)'(\underline{\mu} - X\underline{\beta}_o) \right]$$

$$= -\frac{n}{2\sigma_o^2} + \frac{1}{2(\sigma_o^2)^2} \left[n\tau^2 + \|\underline{\mu} - X\underline{\beta}_o\|^2 \right] = 0.$$

It is now a simple matter to find

$$\underline{\beta}_o = (X'X)^{-1} X'\underline{\mu},$$

$$\sigma_o^2 = \tau^2 + \frac{\|\underline{\mu} - X\underline{\beta}_o\|^2}{n}. \tag{7.78}$$

These parameter values define the vector $\underline{\theta}_o = (\underline{\beta}_o', \sigma_o^2)'$. Formula (7.78) shows that lack-of-fit variation, from the assumed structural model, ends up as part of residual (unexplained) variation.

To find $I(\underline{\theta}_o)$ we need the expected second mixed partials, from (7.76) and (7.77), as follows:

$$\mathrm{E}_f \left[-\frac{\partial^2 \log(g(\underline{y} \mid \underline{\theta}_o))}{\partial \underline{\beta}^2} \right] = \frac{1}{\sigma_o^2} X'X,$$

$$\mathrm{E}_f \left[-\frac{\partial^2 \log(g(\underline{y} \mid \underline{\theta}_o))}{\partial \sigma^2 \partial \sigma^2} \right] = -\frac{n}{2(\sigma_o^2)^2} + \frac{1}{(\sigma_o^2)^3} n\sigma_o^2 = \frac{n}{2(\sigma_o^2)^2},$$

and

$$\mathrm{E}_f \left[-\frac{\partial^2 \log(g(\underline{y} \mid \underline{\theta}_o))}{\partial \underline{\beta} \partial \sigma^2} \right] = \frac{1}{\sigma_o^2} X'(\underline{\mu} - X\underline{\beta}_o) = \underline{0}.$$

The last vector above is zero because of the defining equation for $\underline{\beta}_o$. Thus we have

$$I(\underline{\theta}_o) = \begin{bmatrix} \dfrac{1}{\sigma_o^2}X'X & 0 \\ \underline{0}' & \dfrac{n}{2(\sigma_o^2)^2} \end{bmatrix}.$$

The evaluation of $J(\underline{\theta}_o)$ is harder, and more dependent upon assumed f. Evaluation of $I(\underline{\theta}_o)$ required only the second moment of f, whereas evaluation of $J(\underline{\theta}_o)$ also requires third and fourth moments. Both derivations rely critically on the independence of the ϵ_i. The upper left $(K-1) \times (K-1)$ submatrix of $J(\underline{\theta}_o)$ is

$$\mathrm{E}_f\left[\frac{1}{(\sigma_o^2)^2}X'(\underline{Y} - X\underline{\beta}_o)(\underline{Y} - X\underline{\beta}_o)'X\right] = \frac{\tau^2}{(\sigma_o^2)^2}X'X.$$

The last $(K-1) \times 1$ column vector (of the first $K-1$ rows) is

$$\mathrm{E}_f\left[\frac{1}{\sigma_o^2}X'(\underline{Y} - X\underline{\beta}_o)\left[-\frac{n}{2\sigma_o^2} + \frac{1}{2(\sigma_o^2)^2}(\underline{Y} - X\underline{\beta}_o)'(\underline{Y} - X\underline{\beta}_o)\right]\right].$$

Making use of $X'(\underline{\mu} - X\underline{\beta}_o) = \underline{0}$, and some algebra, we can reduce the above to

$$\mathrm{E}_f\left[\frac{1}{2(\sigma_o^2)^3}X'(\underline{Y} - \underline{\mu})\left[(\underline{Y} - \underline{\mu})'(\underline{Y} - \underline{\mu}) + 2(\underline{Y} - \underline{\mu})'(\underline{\mu} - X\underline{\beta}_o)\right]\right],$$

and then to

$$\mathrm{E}_f\left[\frac{1}{2(\sigma_o^2)^3}\left[X'(\underline{Y} - \underline{\mu})[(\underline{Y} - \underline{\mu})'(\underline{Y} - \underline{\mu})] + 2X'(\tau^2 I)(\underline{\mu} - X\underline{\beta}_o)\right]\right]$$

$$= \mathrm{E}_f\left[\frac{1}{2(\sigma_o^2)^3}\left[X'(\underline{Y} - \underline{\mu})[(\underline{Y} - \underline{\mu})'(\underline{Y} - \underline{\mu})]\right]\right].$$

Now write the needed expectation in terms of the hypothetical residuals, which are *iid* $N(0, \tau^2)$; hence

$$\mathrm{E}_f\left[(\underline{Y} - \underline{\mu})[(\underline{Y} - \underline{\mu})'(\underline{Y} - \underline{\mu})]\right] = \mathrm{E}_f\left[\underline{\epsilon}'\left[\sum_{i=1}^n (\epsilon_i)^2\right]\right].$$

The jth element of this vector is $\mathrm{E}_f\left((\epsilon_j)^3 + \sum_{i \neq j} \epsilon_j(\epsilon_i)^2\right)$, which by virtue of the mutual independence is $\mathrm{E}_f(\epsilon_j)^3$. Because the ϵ_j are assumed to be normally distributed, their third central moment is 0. Hence, we have

$$\mathrm{E}_f\left[\underline{\epsilon}'\left[\sum_{i=1}^n (\epsilon_i)^2\right]\right] = \underline{0},$$

and the desired part of $J(\underline{\theta}_o)$ is $\underline{0}$.

The final needed element is

$$J_{KK}(\underline{\theta}_o) = E_f \left[-\frac{n}{2\sigma_o^2} + \frac{1}{2(\sigma_o^2)^2}(\underline{Y} - X\underline{\beta}_o)'(\underline{Y} - X\underline{\beta}_o) \right]^2 .$$

Several straightforward steps reduce the above to

$$J_{KK}(\underline{\theta}_o) = \frac{1}{(2\sigma_o^2)^2} \left[-n^2 + \frac{1}{(\sigma_o^2)^2} E_f \left[(\underline{Y} - X\underline{\beta}_o)'(\underline{Y} - X\underline{\beta}_o) \right]^2 \right] .$$

Define the ith row vector of X as $\underline{x}'_i$. Then

$$E_f \left[(\underline{Y} - X\underline{\beta}_o)'(\underline{Y} - X\underline{\beta}_o) \right]^2$$

$$= E_f \left[\sum_{i=1}^{n} (y_i - \underline{x}'_i\underline{\beta}_o)^2 \right]^2$$

$$= E_f \left[\sum_{i=1}^{n} \sum_{j=1}^{n} (y_i - \underline{x}'_i\underline{\beta}_o)^2 (y_j - \underline{x}'_j\underline{\beta}_o)^2 \right]$$

$$= E_f \left[\sum_{i=1}^{n} (y_i - \underline{x}'_i\underline{\beta}_o)^4 + \sum_{i\neq j} (y_i - \underline{x}'_i\underline{\beta}_o)^2 (y_j - \underline{x}'_j\underline{\beta}_o)^2 \right] .$$

By virtue of mutual independence, the expectation of the second summation above is easily found, giving

$$E_f \left[(\underline{Y} - X\underline{\beta}_o)'(\underline{Y} - X\underline{\beta}_o) \right]^2$$

$$= E_f \left[\sum_{i=1}^{n} (y_i - \underline{x}'_i\underline{\beta}_o)^4 \right] + (n\sigma_o^2)^2 - \sum_{i=1}^{n} \left[\tau^2 + (\mu_i - \underline{x}'_i\underline{\beta}_o)^2 \right]^2 .$$

For the case $\tau^2 = 0$ note that $\underline{Y} = \mu$, and so the above directly gives $E_f \left[(\underline{Y} - X\underline{\beta}_o)'(\underline{Y} - X\underline{\beta}_o) \right]^2 = (n\sigma_o^2)^2$; hence

$$J_{KK}(\underline{\theta}_o) = \frac{1}{(2\sigma_o^2)^2} \left[-n^2 + \frac{1}{(\sigma_o^2)^2}(n\sigma_o^2)^2 \right]^2 = 0 .$$

It is thus clear that if $\tau^2 = 0$, then $J(\underline{\theta}_o) = O$.

The next steps are valid only if $\tau^2 > 0$. Let

$$\sqrt{\lambda_i} = \frac{\mu_i - \underline{x}'_i\underline{\beta}_o}{\tau}$$

and

$$z_i = \frac{y_i - \mu_i}{\tau} .$$

The z_i are *iid* normal$(0, 1)$, and we have

$$E_f\left[\sum_{i=1}^{n}(y_i - x_i'\underline{\beta}_o)^4\right] = \tau^4\left[\sum_{i=1}^{n}E_f\left[z_i + \sqrt{\lambda_i}\right]^4\right].$$

The needed expectation is now easily found because it is just the fourth moment of a normal random variable with a nonzero mean; or it can be expressed as a function of the first four moments of a standard normal random variable. We find it easier to note that the needed expectation is that of the square of a noncentral chi-square random variable on 1 df and noncentrality parameter λ_i. The result is

$$E_f\left[\sum_{i=1}^{n}(y_i - x_i'\underline{\beta}_o)^4\right] = \tau^4\left[\sum_{i=1}^{n}[3 + 6\lambda_i + \lambda_i^2]\right].$$

Now, by carefully constructing the full result from all the above pieces and simplifying it, we get

$$J_{KK}(\underline{\theta}_o) = \frac{n}{2(\sigma_o^2)^2}\left[\frac{2\tau^2\sigma_o^2 - \tau^4}{(\sigma_o^2)^2}\right].$$

While derived only for $\tau > 0$, the above result can also be validly used for the case of $\tau^2 = 0$.

Finally,

$$J(\underline{\theta}_o) = \begin{bmatrix} \dfrac{\tau^2}{(\sigma_o^2)^2}X'X & 0 \\ \underline{0}' & \dfrac{n}{2(\sigma_o^2)^2}\left[\dfrac{2\tau^2\sigma_o^2 - \tau^4}{(\sigma_o^2)^2}\right] \end{bmatrix}.$$

The result we sought can now be found:

$$\text{tr}\left[J(\underline{\theta}_o)[I(\underline{\theta}_o)]^{-1}\right] = \frac{\tau^2}{\sigma_o^2}\left[K + 1 - \frac{\tau^2}{\sigma_o^2}\right]. \tag{7.79}$$

If model equals truth, we have $\mu = X\beta$, so that $\sigma_o^2 = \tau^2$ (otherwise, $\sigma_o^2 > \tau^2$), and the trace term equals K. By continuity in τ^2 we must also define this trace term as K when $\tau^2 = 0$ if the model is true. However, there are deep philosophical issues and problems associated with a truth in which $\tau^2 = 0$, so we will consider only the situation wherein even for truth there is substantial unexplainable uncertainty. In particular, if true replication is used in an experiment (or study), we suggest that it is most useful to consider that τ^2 is then the variance within true replicates (assuming variance homogeneity). This is a definition of convenience, as even truth can be at different levels, and we are mostly interested in structural truth of our models in the face of nontrivial, irreducible uncertainty inherent in data for finite sample sizes.

Surprisingly enough, we see from (7.79) that for this limited evaluation and context, $\text{tr}\left[J(\underline{\theta}_o)[I(\underline{\theta}_o)]^{-1}\right] < K$ under a misspecified model (the key limitation here was assuming truth as a normal distribution). If a good model could achieve, say, $\sigma_o^2 \leq 1.2\tau^2$, then the trace would be well within 80% of K; hence use of AIC (rather than TIC) seems acceptable, and should err, on average, on the side of parsimony. Moreover, estimation of this trace term (hence, TIC) seems very problematic, since τ^2 cannot be estimated in a study lacking true replication (cf. Linhart and Zucchini 1986:78). Even what we call true replication in an experiment provides only an estimate of τ^2, by definition, if we restrict our concept of truth to what we can predict under the design structure and independent variables used in the given experiment. Philosophically, we might be able to predict some (hence a smaller τ^2) or all (hence $\tau^2 = 0$) of the observed differences among replicate responses if we knew ultimate truth.

This example can be easily generalized; that is, we retain the assumed model and generalize truth somewhat. Whereas we assumed truth as $\epsilon \sim$ normal$(0, \tau^2)$, the only way this entered the derivations was via the first four central moments of ϵ. If we retain the *iid* assumption, we can derive generalized results; note that we retain E$(\epsilon) = 0$ with loss of generality. We could allow an asymmetric distribution for ϵ; we will not do so: We assume E$(\epsilon^3) = 0$. Thus all we need is the fourth moment of ϵ, which we will express in standardized form as

$$\gamma = E(\epsilon^4)/[E(\epsilon^2)]^2.$$

For assumed normal truth, $\gamma = 3$. For f as a logistic distribution, $\gamma = 4.2$; for a Laplace distribution, $\gamma = 6$; and for a uniform$(-h, h)$ distribution, $\gamma = 1.8$. The last two are extreme cases; one might think that γ lies approximately in the range 2 to 4.

Redoing the derivations for this more general way of representing truth is straightforward; the results are the same for $I(\underline{\theta}_o)$; but for $J(\underline{\theta}_o)$,

$$J(\underline{\theta}_o) = \begin{bmatrix} \dfrac{\tau^2}{(\sigma_o^2)^2}X'X & \underline{0} \\ \underline{0}' & \dfrac{n}{2(\sigma_o^2)^2}\left[\dfrac{2\tau^2\sigma_o^2 + \tau^4\left(\frac{\gamma-1}{2}-2\right)}{(\sigma_o^2)^2}\right] \end{bmatrix}.$$

For the trace function we get

$$\text{tr}\left[J(\underline{\theta}_o)[I(\underline{\theta}_o)]^{-1}\right] = \dfrac{\tau^2}{\sigma_o^2}\left[K + 1 + \dfrac{\tau^2}{\sigma_o^2}\left[\dfrac{\gamma-1}{2} - 2\right]\right]. \qquad (7.80)$$

Hence, for these fixed-effects linear models assuming normality, the effect of structural misspecification appears in σ_o^2 as manifest via the ratio τ^2/σ_o^2. However, the effect of error distribution misspecification is only via the fourth moment, γ (assuming symmetric errors). We have stressed a focus on complex models wherein K will not be trivially small; from (7.80) we see that as K gets large the effect of error distribution misspecification upon this trace function becomes trivial. In contrast, the effect of structural misspecification, in the sense of then having the ratio $\tau^2/\sigma_o^2 < 1$, remains equally important at any K. Once one achieves a good structural fit, then the effect of minor to modest misspecification of the error distribution becomes trivial for large K as regards use of the approximation tr $\left[J(\underline{\theta}_o)[I(\underline{\theta}_o)]^{-1} \right] = K$, hence further justifying use of AIC rather than TIC. These musings seem likely to apply also to general linear models.

One last point: How good is use of $[I(\underline{\theta}_o)]^{-1}$ for the variance–covariance matrix of $\hat{\underline{\theta}}$ under model misspecification here (ignoring selection uncertainty, that is)? As was shown (and is known in general) in Section 7.1, the correct asymptotic variance–covariance matrix is $V(\hat{\underline{\theta}}) = [I(\underline{\theta}_o)]^{-1}J(\underline{\theta}_o)[I(\underline{\theta}_o)]^{-1}$. Thus, here, $V(\hat{\underline{\beta}}) = \tau^2(X'X)^{-1}$, and

$$V(\hat{\sigma}^2) = \frac{2(\sigma_o^2)^2}{n} \left[\frac{2\tau^2\sigma_o^2 + \tau^4\left(\frac{\gamma-1}{2} - 2\right)}{(\sigma_o^2)^2} \right].$$

If a good structural fit has been achieved (so $\tau^2 \approx \sigma_o^2$), then we have

$$V(\hat{\sigma}^2) = \frac{2(\sigma_o^2)^2}{n} \left[\frac{\gamma - 1}{2} \right].$$

The drastic bias induced by a γ not near 3 (but assumed as 3) might motivate one to use an estimator of γ, and with the same estimator to then use TIC, not AIC. This approach can be recommended only weakly, at best, because the estimator of the fourth moment is so highly variable.

7.7   Additional Results and Considerations

7.7.1   Selection Simulation for Nested Models

The detailed stochastic characteristics of the model selection process have to be studied mostly by Monte Carlo simulation methods. Currently, it seems as if the only completely general approach is to specify a data-generation process and a set of models to be fit to each generated sample, and then generate samples and do all the calculations associated with model fitting and selection. This is very useful, but is not a study of properties of model selection strategies in the abstract. Rather, each application has some underlying specific models and type of truth (as generated data), and may require extensive computations that are peripheral to the heart of the model selection process.

An exception arises if we restrict ourselves to a single chain of nested models and to selection methods based on log-likelihood differences between fitted models. This scenario includes simulation of AIC, AIC_c, BIC, and likelihood ratio testing-based methods (and can be easily adapted to simulate QAIC and $QAIC_c$ model selection). All we need to generate are the independent, noncentrally distributed log-likelihood chi-square random variables between adjacent models. It is the selection process itself we then study without reference (at least for large n) to any specific models. Maximum likelihood estimation is assumed, but no parameters need actually be postulated or estimated (a disadvantage is that we cannot simultaneously study properties of parameter estimators). The important restriction here is that we can correctly generate the needed random variables only for what would be a single chain of nested models. The advantage of the method is speed and generality; this allows quick insights into properties of some model selection procedures.

At the heart of this procedure we have (conceptual) pairs of models g_i and g_{i+j}; model g_i is nested in model g_{i+j}, and the difference in number of parameters is j. The method can be developed in general, but we will only give it, and use it, with $j = 1$. That is, our conceptual set of models satisfies $g_1 \subset g_2 \subset \cdots \subset g_R$, and each incremented model (i.e., g_i versus g_{i+1}) has only 1 added parameter. We assume that large sample theory and ML parameter estimation conceptually underlie such Monte Carlo simulations. Hence, here the usual likelihood ratio test statistic has, in general, a noncentral chi-square distribution on 1 df; denote that random variable by $\chi_1^2(\lambda_i)$. The noncentrality parameter for model g_i versus g_{i+1} is λ_i.

For our set of R models we have $R - 1$ noncentrality parameters $\lambda_1, \ldots,$ λ_{R-1}. These λ_i would be functionally related to the true data-generating model, the model structures assumed $g_i(\underline{x} \mid \cdot)$, and the specific parameter values, $\underline{\theta}_{o,i}$, that specify the actual best approximating model in each family of models. However, we will be able to bypass all of those specifications in the simulation method below. We do need to be able to interpret sets of the λ_i. A $\lambda_i > 0$ would reflect the failure of at least model g_i to perfectly match truth. If we had $\lambda_{i-1} > 0$ and $\lambda_i = \cdots = \lambda_{R-1} = 0$, we would interpret this as model g_i being the true data-generating model (we will ignore pathologies that might invalidate this interpretation). Also, it is possible to have real situations where, for example, $\lambda_1 = 0$ but $\lambda_2 > 0$. Then both models g_1 and g_2 seem to be equally bad approximations to truth, because model g_3 improves as an approximation to truth compared to model g_2, but g_2 does not improve over g_1. Often, for a real situation we would have the set of λ_i monotonically decreasing, and the issue is that of which model provides the AIC best model, i.e., the expected K-L best model, when parameter estimation occurs. We continue now considering how to do simulation in this context.

Let the fitted model log-likelihoods be $\log(\mathcal{L}_i)$. From basic theory,

$$2 \log(\mathcal{L}_{i+1}) - 2 \log(\mathcal{L}_i) \sim \chi_1^2(\lambda_i).$$

Let the number of parameters in the simplest model be K_1. Then we can write the above as

$$-(\text{AIC}_{i+1} - 2K_1 - 2i) + (\text{AIC}_i - 2K_1 - 2i + 2) \sim \chi_1^2(\lambda_i),$$

or

$$\text{AIC}_i - \text{AIC}_{i+1} \sim \chi_1^2(\lambda_i) - 2. \tag{7.81}$$

Alternatively, from result (7.81) we can write a symbolic equation relating random variables:

$$\text{AIC}_i = \text{AIC}_{i+1} + \chi_1^2(\lambda_i) - 2. \tag{7.82}$$

Thus for purposes of a simulation study, if we know λ_i and AIC_{i+1}, we can generate AIC_i. We just need to be able to generate a noncentral 1 df chi-square random variable (there are routines for this, such as CINV in SAS).

Based on (7.82), we can do a backwards recursive generation of AIC_{R-1} to AIC_1 starting with $i = R - 1$, given a value for AIC_R. This idea reduces to the formula

$$\text{AIC}_i = \text{AIC}_R + \sum_{j=i}^{R-1}(\chi_1^2(\lambda_j) - 2); \tag{7.83}$$

we just need a value for AIC_R. Because everything we care about under AIC model selection depends only on the relative differences, such as $\text{AIC}_i - \text{AIC}_j$ or Δ_i or Δ_p, it suffices to set $\text{AIC}_R = R$ (any constant would suffice; this one has advantages) for every sample of AICs generated. A sample now corresponds to a realization of a set of independent $\chi_1^2(\lambda_i)$, $i = 1, \ldots, R - 1$.

We have used this approach to do simulation studies of model selection under AIC and other likelihood-based methods. To evaluate BIC we use $\text{BIC}_i = \text{AIC}_i - 2i + i \log(n)$. Because n should now vary, one must also define the noncentrality parameters on a per-unit sample size basis, hence be able to compute $\lambda_{i,n} = n\lambda_{i,1}$. The $\lambda_{i,1}$ should be very small, but otherwise their scale is arbitrary. To mimic AIC_c selection, use

$$\text{AIC}_{c,i} = \text{AIC}_i + 2\frac{K_i(K_i + 1)}{n - K_i - 1}, \tag{7.84}$$

where $K_i = K_1 + i$. Now one must specify K_1, the number of parameters envisioned in model g_1, as well as use $\lambda_{i,n} = n\lambda_{i,1}$.

To mimic QAIC model selection is a little more involved. Specify a true value of c (variance inflation factor; $c \geq 1$) and its df (in reality, the df may vary over samples). Generate $\hat{c} = \chi_{df}^2(df(c - 1))/df$ for each sample (i.e., $df(c - 1)$ is the noncentrality parameter for this chi-square random variable). Then use

$$\text{QAIC}_i = \text{QAIC}_R + \sum_{j=i}^{R-1}\left[\frac{\chi_1^2(c - 1 + c \cdot \lambda_j)}{\hat{c}} - 2\right], \qquad i = 1, \ldots, R - 1,$$

and $\text{QAIC}_R = R$. Also,

$$\text{QAIC}_{c,i} = \text{QAIC}_i + 2\frac{K_i(K_i + 1)}{n - K_i - 1}.$$

A few results are given below using this simulation approach to gain insights into model selection. For the most part, however, it is not practical to publish extensive tables of simulation results. We encourage interested persons to do their own extensive simulations, based on (7.83) and (7.84), and learn from them.

From this setup which allows simulating model selection for nested models, we can also compute theoretical expected AIC *differences*, hence determine the expected AIC best model exactly. From (7.83) we get, for models incrementing by just one parameter,

$$E(AIC_i) = E(AIC_R) + \sum_{j=i}^{R-1} (E[\chi_1^2(\lambda_j)] - 2),$$

$$E(AIC_i) = E(AIC_R) + \sum_{j=i}^{R-1} (\lambda_j - 1).$$

Let $\lambda_{i+} = \lambda_i + \cdots + \lambda_{R-1}$, $i = 1, \ldots, R-1$, and $\lambda_{R+} = 0$, and we get

$$E(AIC_i) = (E(AIC_R) - R) + (\lambda_{i+} + i).$$

Then compute the set of values V_i given by

$$V_i = \lambda_{i+} + i, \quad i = 1, \ldots, R, \tag{7.85}$$

find their minimum, V_{min}, and then compute

$$E(\Delta_i) = V_i - V_{min}. \tag{7.86}$$

As an example, if $R = 10$ and (in order) we have λ_i as 2, 6, 10, 6, 3, 1.5, 0.8, 0.4, and 0.2, then the $E(\Delta_i)$ are, in order, 22.5, 21.5, 16.5, 7.5, 2.5, 0.5, **0**, 0.2, 0.8, and 1.6. Thus the best expected AIC selected model is g_7. Some theoretical variances can also be computed, but nothing directly useful to random minima like Δ_p.

7.7.2   Simulation of the Distribution of Δ_p

The random variable $\Delta_p = AIC_{best} - AIC_{min}$ was introduced in Chapter 4 (Section 4.5). For a set of models indexed $i = 1, \ldots, R$, a given sample size n, and a conceptually well-defined repeated sampling framework (hence, a sample space), we let model g_{best} represent the best model, on average, to fit under the AIC selection criterion. Monte Carlo simulation can be used to determine this actual best model (sometimes theory suffices). In applications we are not saying that model g_{best} is truth; it is just that one of the R models must be the best model, on average, to use for all possible samples, and that is the truth that model g_{best} represents.

For each simulation-generated sample we can compute $\Delta_p = AIC_{best} - AIC_{min}$. This AIC_{min} and the value of AIC_{best} vary by sample. However, the value of *best* is fixed for all samples; for example, model g_4 might be the actual best model to use (hence *best* = 4). If model g_{best} is selected as best in the sample, then $\Delta_p = 0$; otherwise, $\Delta_p > 0$. We can compute the probability distribution, hence percentiles, of this pivotal under the simulation scenario of Section 7.7.1. However, there are too many variables to make extensive tabulations of general results feasible (at a minimum we must specify values for R and $\lambda_1, \ldots, \lambda_{R-1}$; if we use AIC_c, we also need K_1 and n).

Results about Δ_p under AIC model selection, for a few values of R with all $\lambda_i = 0$, are feasible to show. In this case model g_1 is the true data-generating model ($k = 1$ in Δ_p). This scenario is clearly at odds with what we believe applies to real data analysis (all models for data analysis are just approximations to truth). However, it can be used as a benchmark for percentiles of Δ_p. To the extent that this situation is too simple, it may serve only as a lower bound on the percentiles of the cumulative distribution function of Δ_p, at least for nested models, or for real problems where there is substantial nesting of many of the models considered.

For the case of $R = 2$ and large sample size, the qth percentile of Δ_p $(0 < q < 1)$, $\Delta_{p,q}$, is

$$\Delta_{p,q} = \max\{0, (\chi^2_{1,q} - 2)\}$$

(easily derivable from 7.81). Here, $\chi^2_{1,q}$ is the qth percentile of a central chi-square random variable on 1 df. For example, $\chi^2_{1,0.95} = 3.84$; hence $\Delta_{p,0.95} = 1.84$. We used Monte Carlo simulation to determine some percentiles of Δ_p for values of $R > 2$. One million samples were used for each value of R (as four independent runs of 250,000 samples, so we can estimate precision). Results below for $R > 2$ have a cv of about 0.5%:

	percentiles of Δ_p			
R	80%	90%	95%	99%
2	0.00	0.71	1.84	4.63
3	0.11	1.37	2.67	5.77
4	0.35	1.71	3.33	6.40
5	0.49	1.93	3.40	6.86
10	0.75	2.34	3.97	7.61
20	0.82	2.47	4.15	8.05

We have done many of these simulations to find the distribution of Δ_p for sets of noncentrality parameters wherein $\lambda_i > 0$; the percentiles are then somewhat larger as compared to the case where all $\lambda_i = 0$. For example, let $R = 10$ and $\lambda_1, \ldots, \lambda_9$ be 2, 6, 10, 6, 3, 1.5, 0.8, 0.4, 0.2. Now truth f is not in the set of models (actually, g_R could be truth; we cannot rule that out). Based on 20,000 Monte Carlo samples (two sets of 10,000), we find that model g_7 is the expected K-L (i.e., AIC) best model. The averages of the sample Δ_i values (rescaled so their minimum is 0), in order, are 22.6, 21.5, 16.5, 7.5, 2.5, 0.5, 0, 0.2, 0.8, 1.6 (reliable to $\pm 0.1 = 2$ se); compare these values to their theoretical expectations from the end of Section 7.7.1: 22.5, 21.5, 16.5, 7.5, 2.5, 0.5, 0, 0.2, 0.8, and 1.6.

Based on $\Delta_p = AIC_7 - AIC_{min}$ over these 20,000 samples, some percentiles of Δ_p are 3.3 (80%), 4.6 (90%), 6.4 (95%), 10.6 (99%) (cv's are about 1%). From these sorts of simulations, and others with explicit models (especially linear regression or capture–recapture models), we have risked saying that in real applications with at least several models ($R \geq 5$) and some nested sequences, a model g_i for which $\Delta_i \approx 4$ is implausible as the actual K-L best model structure, and $\Delta_i \approx 7$ is strong evidence against model structure g_i as being the K-L best model (and $\Delta_i \geq 10$ is very strong evidence against model g_i). The Akaike weights provide a refined interpretation of the Δ_i.

7.7.3 Does AIC Overfit?

The conceptual framework underlying valid use of AIC is one where truth has infinitely many parameters. Overfitting is often defined in a framework where there is a simple true model, with a finite number of parameters, and that true model is in the set of models considered. Then if the true model structure is nested within the selected model structure, the selected model is said to overfit: One has estimated more parameters than are in the true model.

This simplistic concept of overfitting does not apply in the K-L model selection framework. However, there is a best expected K-L model, which is the model we should use as our basis for data analysis. If that target model is nested within the selected model, might we claim that AIC has selected an overfit model? We decline to use this definition because there is natural variability in the model selected. If we miss the target model by a few parameters (or, what is the same, that the structure of the selected model is not quite the same as that of the target model), we should not say that we failed. This issue is philosophically the same as being concerned that a parameter estimator $\hat{\theta}$ may sometimes give a point estimate far away from the true parameter. We are upset only by cases where $\hat{\theta}$ is quite far from θ. But if this happens only with suitably small probability, we consider $\hat{\theta}$ as an acceptable estimator. The argument becomes somewhat circular at this point because we have mentally accustomed ourselves to being satisfied if $\hat{\theta}$ is within about $\pm 2 \, \mathrm{se}(\hat{\theta})$ of θ; hence $\hat{\theta}$ is unacceptable with a probability of only about 0.05. Similarly, practitioners of null hypothesis testing typically are willing to accept (at least de facto) a 0.05 probability of type I error (and probably a higher type II error probability in most applications).

Something similar should apply to possible overfitting for AIC model selection (underfitting is at best a minor concern with AIC model selection). We need some idea of how far from the actual K-L best model a fitted model can be before it is regarded as an overfit model. We have to allow that often we would do well to select a model within, say, 1 or 2 parameters of the actual K-L best model. In contrast, if the selected model has 10 or 20 parameters more than the target model, we think that most people will agree that the model is overfit. We can use simulation (in some cases theory exists, see Shibata 1976, Speed and Yu 1993) to find the probability distribution of the selected model index, say $\widehat{best}$, for a nested sequence of models g_1 to g_R and actual K-L best model as model g_{best}. We want to know about the tail of the probability distribution of $\widehat{best}$, hence about selection probabilities when $\widehat{best} - best$ gets large.

Here is a typical example of what can happen. For a sequence of nested models with $R = 10$ and the λ_i as 2, 6, 10, 6, 3, 1.5, 0.8, 0.4, 0.2, the K-L best model was found to be g_7. We extend this to R being 20 and then 30 with all additional $\lambda_i = 0$ (and assume that sample size will be quite large). This means that the true data-generating model is model g_{10}, so models g_{11} to g_{30} are overfit, if selected, in the sense that they do contain truly superfluous structure. The K-L best model remains model g_7 even for $R > 10$. Below we give the model selection frequencies, in model order g_1 to g_R, based on 10,000 Monte Carlo samples:

R g_7
10: 11 5 50 746 2282 2635 **1924** 1161 735 451
20: 16 3 20 405 1557 2399 **1970** 1373 816 542 274 192 125 93 67 51 39 21 18 19
30: 15 3 23 407 1689 2351 **1954** 1360 800 522 237 155 121 90 74 37 32 27 19 18
                                               11 8 10 11 9 5 2 4 2 4

The long-tailed nature of the distribution of selected models is typical of AIC when there are many "big" models (models with too many unneeded parameters) containing the K-L best model structure. For the case of $R = 30$ we have a probability of about 0.01 of selecting a model with 19 or more parameters (models g_{19} to g_{30}), hence having estimated 12 or more unneeded parameters. For both $R = 20$ and 30 there is about a 0.06 probability of selecting a model with five or more unneeded parameters (i.e., models g_{12} or higher). In general, if we say that we can accept a procedure that has about 5% of its cases a bit misleading, then we should not be upset that AIC can overfit the K-L best model by about 5 or more parameters with probability of roughly 0.06 if many such too-general models are in the set of models considered.

Note that these results are effectively for very large sample size because AIC was used, not AIC_c. For not-large n, use of AIC_c would substantially reduce the long tail at $R = 20$ and 30 (say $n = 100$). Hence, these given results are worst-case scenarios.

Here is a worst-case scenario for one linear sequence of nested models: All $\lambda_i = 0$, so model g_1 is the K-L best model. Based on one million Monte Carlo samples, the estimated model selection probabilities (good to a standard error $\leq 0.05\%$) are below:

R $\hat{\pi}_i \times 100\%$, in order $i = 1, 2, \ldots$
3   78.7   13.3 8.0
4   76.0   12.5 6.7 4.8
5   74.4   12.0 6.4 4.1 3.1
10   71.8   11.4 5.8 3.5 2.4 1.7 1.2 .9 .7 .6
20   71.2   11.2 5.7 3.4 2.3 1.6 1.1 .8 .6 .5 .4 .3 .2 .2 .1 .1 .1 .1 .1 .1

(For this case of a single nested sequence of models, these selection probabilities are known theoretically for large n and all $\lambda_i = 0$; see, e.g., Shibata 1981, 1989). The long tail is disturbing, yet with probability about 0.94 the selected model will, in this worst-case scenario, be within four parameters of the K-L best model. Based on many simulations that attempted to mimic key features of realistic AIC model selection, we claim that this is a typical result for large-sample AIC (about 0.06 probability of overfitting by five or more superfluous parameters if such general, overparametrized models are in the set considered). Given models with a large amount of unneeded structure (parameters), AIC can select overfitted models, but the probability of a seriously overfit model is arguably less than the total error probabilities (type I plus type II) in traditional hypothesis testing.

What will reduce the probability of getting a badly overfit model? Use AIC_c, which helps considerably when sample size, relative to K, is not large. Otherwise, the only recommendation we have to avoid the uncommon event of a much overfit AIC-selected model is to be very thoughtful about the a priori set of models considered. In particular, do not casually include models with a great many parameters more than you think are really needed. In regression variable selection this would mean do not simply consider every imaginable regressor variable and include it for possible selection. If you do this, you risk having large numbers of variables that have no explanatory value (they have a $\lambda \approx 0$), and that leaves you with a small but real probability of selecting a model with many worthless variables.

7.7.4 Can Selection Be Improved Based on All the Δ_i?

Given the potentially long-tailed nature of K-L–based model selection (it depends on the set of models), it seems natural to ask whether there might be information in the full set of Δ_i values that would allow us to identify those cases where we have selected a very overfit model. If so, can we change our selection to a better model based on information in the entire set of Δ_i values? We have explored this matter for a single series of nested models (as considered in the above three subsections). The idea was that perhaps the pattern in the Δ_i would be like that below in the event of selecting a badly overfit model (line one is model number i; line two is Δ_i):

1	2	3	4	5	6	7	8	9	10	11	12	13	14	15	16	17	18	19
30	10	5	0.1	1	2.5	2	1.5	2.7	3	0.8	2.3	1	0	2.4	4	5	6	7

Here, AIC has selected model g_{14}, but we might suspect that the better model to use is g_4 ($\Delta_4 = 0.1$ and none of models g_5 to g_{13} have a very big value of Δ_i). So we could change our choice to model g_4.

There is no theory to help here regarding properties of the patterns in Δ_1 to Δ_R. So we looked at a large number of simulated results for $R \geq 10$ under the worst-case scenario of all $\lambda_i = 0$. Hence, the simplest model, g_1, is both the true data-generating model and the K-L best model. Table 7.5 displays, for $R = 10$, some selected cases of $\Delta_1, \ldots, \Delta_{10}$ (a case is one realization of model selection results, for a large sample size). Cases 1, 2, and 3 are typical in that model g_1 is selected in about 72% of all samples here. Case 4 is representative

TABLE 7.5. Some sets of large-sample AIC differences, Δ_i, for $R = 10$ and all $\lambda_i = 0$. These are selected cases: T is a typical pattern obtained in 72% of samples (model g_1 selected); A is atypical but not rare (11% of cases, model g_2 selected); R is rare, about 6% of all cases, and for these, overfit models are selected.

					Δ_i					
pattern	$i = 1$	2	3	4	5	6	7	8	9	10
1 T	**0.0**	1.7	3.7	5.0	6.1	8.1	9.5	11.3	10.9	9.6
2 T	**0.0**	0.5	0.1	2.8	4.7	6.6	8.5	8.1	8.8	8.9
3 T	**0.0**	0.6	0.5	2.5	0.8	2.8	4.3	6.2	8.2	10.1
4 A	1.5	**0.0**	0.6	1.5	3.2	3.6	5.5	3.4	5.2	7.1
5 R	7.5	2.0	2.7	0.8	1.5	**0.0**	3.5	5.3	6.3	7.5
6 R	9.5	10.4	9.6	7.6	5.6	5.1	**0.0**	0.3	1.9	3.6
7 R	0.2	1.1	2.0	3.6	1.4	**0.0**	2.7	4.3	4.9	5.3
8 R	2.0	2.2	0.8	2.8	2.5	**0.0**	1.9	3.4	3.9	5.7
9 R	1.1	1.8	3.8	5.4	7.3	5.5	3.3	**0.0**	0.6	1.7
10 R	1.1	3.1	0.2	1.7	2.2	4.2	**0.0**	2.1	3.5	5.5
11 R	7.9	9.0	11.0	6.2	1.3	**0.0**	2.0	4.0	5.7	6.8
12 R	0.1	2.1	3.3	1.5	1.1	2.9	3.9	3.3	1.4	**0.0**
13 R	1.1	1.8	3.8	5.4	7.3	5.5	3.3	**0.0**	0.9	1.7
14 R	10.4	12.2	8.8	8.5	8.6	6.3	**0.0**	0.8	1.6	3.0

of samples wherein model g_2 is selected (about 11% of samples). Cases 5 to 14 are rare (models g_6 to g_{10} selected; it happens here in about 6% of all samples). In these latter cases we would say that an overfit model was selected. Only in case 12 might we feel justified in rejecting model 10 in favor of model 1 as the selected model, but even there nothing in the 10 values of Δ_i makes us think that this decision is particularly justified. In the other rare cases there is nothing in the nature of the pattern of the ten Δ_i that gives us any confidence that we are justified in selecting a model as best other than the AIC best model (for which $\Delta_k = 0$). That is, changing our selected model may lead to worse, not better, results: We cannot tell based on the data. In fact, for some cases (6, 11, 14, and perhaps 5) the "evidence" in the set of Δ_i seems clearly to support the correctness of the AIC best model.

We also simulated the performance of various ad hoc modified AIC model selection procedures to change the selection to a more parsimonious model. Such a change was done if the AIC-selected model had K much bigger than a more parsimonious model that had Δ_i very near zero (e.g., $\Delta_i < 0.5$). The exact algorithms tried varied, but none of those ad hoc methods made any meaningful difference to the overall AIC model selection relative frequencies. Also, we did not visually perceive any useful information in patterns of all the Δ_i. While more work could be done along these lines, we are pessimistic that it would be fruitful. Basically, if the data "lie" to you (i.e., a poor model is selected because the sample is atypical), there are no diagnostics computable from that sample to that tell you that it has "lied."

7.7.5 Linear Regression, AIC, and Mean Square Error

We present here some theoretical formulas relevant to model selection, from which informative results can be computed, for a certain case of linear regression under constant error variance. Specifically, we assume that truth is a linear regression based on orthogonal regressors, but our models include only a subset of the regressors (the global model may include all regressors). Let $z_1, \ldots, z_m$ be *iid* normal(0,1). Independently, let ϵ be normal(0, σ^2), and the response variable x, based on the regressors, is given by truth as

$$x = \beta_o + \sum_{j=1}^{m} \beta_j z_j + \epsilon,$$

so

$$E(x \mid \underline{z}) = \beta_o + \sum_{j=1}^{m} \beta_j z_j.$$

Results below are scale-invariant in terms of the regressors because β_j and z_j occur in the models only as a product. To keep notation consistent with Section 7.7.1, we define here the base model, g_1, to be $x = \beta_o + \beta_1 z_1 + \delta$, with δ assumed as normal(0, σ_1^2). The number of parameters in this model is $K_1 = 3$. The normal assumption is true, but $\sigma_1^2 \neq \sigma^2$. In general, for $r \geq 1$, model g_r is

$$x = \beta_o + \sum_{j=1}^{r} \beta_j z_j + \delta,$$

with δ assumed as normal(0, σ_r^2); the number of parameters is $K_r = r + 2$.

We can consider the sequence of nested models for $r = 1, \ldots, R \leq m$. The ordering of regressors is arbitrary, but is used in the formulas below when nested models are considered sequentially (as per the theory in Section 7.7.1). It is then convenient, but not required, to specify the regression coefficients to satisfy $|\beta_j| > |\beta_{j+1}|$. Doing so yields insights into AIC model selection more easily (such a structured situation is also considered in Speed and Yu 1993).

From the point of view of insights to be gained, the assumption of orthogonal regressors is not restrictive if $R = m$, because any regression problem can be transformed into the case of orthogonal regressors, for example by resorting to regression on principal components. However, if we consider cases for $R < m$, the orthogonality assumption is restrictive, because we cannot make our observed regressors orthogonal to the regressors not observed (this is more realistic of real data). For this reason there is no advantage in considering cases with $R < m$.

The regressors are random variables. Therefore, to get analytical results we take an additional expectation over certain matrices that are based on the random (row) vector $\underline{z}_r = (1, z_1, \ldots, z_r)$. For a sample of size n, the model in matrix notation is $\underline{x} = Z\beta + \underline{\delta}$, so matrices such as $Z'Z$ and $(Z'Z)^{-1}$ arise. The rows of matrix Z (an $n \times r + 1$ matrix) are the n observed sample vectors $\underline{z}_r$. For large n it is acceptable (to get theoretical results) to replace these matrices by their expectations with respect to the random $\underline{z}_r$ (for any n, $E(Z'Z) = I$; for large n, $E(Z'Z)^{-1} \approx I$). However, the nonlinearities involved mean that some theoretical formulas below are only large-sample approximations.

Under this scenario of regressor independence we determined the sequential noncentrality parameters of Section 7.7.1 that apply, for large sample sizes, to the nested sequence of models defined here (g_1 to g_R):

$$\lambda_i = n \log \left[1 + \frac{(\beta_{i+1})^2}{(\beta_{i+2})^2 + \cdots + (\beta_m)^2 + \sigma^2} \right], \qquad i = 1, \ldots, m-2,$$

$$\lambda_{m-1} = n \log \left[1 + \frac{(\beta_m)^2}{\sigma^2} \right].$$

Given this context and the theory in Section 7.7.1 we can compute (approximate) expected Δ_i values for AIC and AIC_c, hence determine the (approximate) theoretical expected K-L optimal model. We can also simulate actual sets of Δ_i values. The needed partial sums of noncentrality parameters are

$$\lambda_{i+} = n \log \left[1 + \frac{\sum_{j=i+1}^{m} (\beta_j)^2}{\sigma^2} \right], \qquad i = 1, \ldots, m-1. \qquad (7.87)$$

The value of exploring this situation is that we can also determine here other theoretical quantities that can be related, or compared, to AIC model selection. Under any of the models the regression coefficient estimators are unbiased for the true parameters (because all regressors are orthogonal, a condition not expected in general). Under model g_r the value of σ_r^2 is

$$\sigma_r^2 = (\beta_{r+1})^2 + \cdots + (\beta_m)^2 + \sigma^2.$$

In notation used elsewhere in Chapter 7, the above σ_r^2 is σ_o^2 under model g_r. The usual (conditional on the model) parameter cv's under model g_r are

$$\text{cv}(\hat{\beta}_r) = \frac{\beta_r}{\text{se}(\hat{\beta}_r \mid g_r)} = \frac{\sqrt{n}\beta_r}{\sigma_r}, \qquad r = 1, \ldots, R.$$

However, rather than compute the above cv's it is informative just to compute the cv's under the global model, $\sqrt{n}\beta_r/\sigma$, and observe the magnitudes of these cv's versus what parameters are included or excluded from the theoretically optimal model.

We can also determine the overall mean square error (MSE) of a fitted model. Minimum MSE is generally accepted as a good theoretical basis for model selection; here we have

$$\text{MSE}(Z) = \sum_{j=1}^{n} E_f \left[\hat{E}(x_j \mid \underline{z}_r) - E(x_j \mid \underline{z}_m) \right]^2.$$

This MSE(Z) is for fitted model g_r conditional on the regressors. Expectation is over ϵ, hence truth. Again, another level of expectation needs to be taken here over Z to get the unconditional result

$$\text{MSE} = \sigma_r^2(r+1) + \sum_{j=r+1}^{m} n(\beta_j)^2. \tag{7.88}$$

Sometimes model selection is based on minimum MSE of prediction of the response variable for a single additional (independent of sample) vector of regressors. Then this average mean square error of prediction (MSEP) for model g_r is MSEP = MSE $+n\sigma^2$. Thus, it suffices to consider only MSE. Mallows's C_p implements minimum MSE model selection for regression (Mallows 1973, 1995).

The K-L–based target criterion T (7.20) can be determined exactly here. We express that result as $-2T$ for direct comparison to expected AIC$_c$- and MSE-based results, and we will label it here as KL. Hence, KL ($= -2T$) for model g_r is exactly

$$\text{KL} = n\text{E}[\log(\chi_{n-r-1}^2(0))] + n \log\left[\frac{\sigma_r^2}{n}\right]$$

$$+ \frac{n}{n-r-3}\left[(r+1) + \left(\frac{n\sigma^2}{\sigma_r^2}\right) + \sum_{j=r+1}^{m} n\left(\frac{\beta_j}{\sigma_r}\right)^2\right]. \tag{7.89}$$

In MSE (7.88) the term $\sum_{j=r+1}^{m} n(\beta_j)^2$ is bias squared due to excluded regressors. Note that in MSE these components of bias are absolute, not relative to theoretical precision of the excluded $\hat{\beta}_j$, whereas in KL, the biasing effect of excluded regressors is "judged" relative to the theoretical precision these $\hat{\beta}_j$ have. That is, the comparable term reflecting bias is in terms of the ratios $(\beta_j/\sigma_r)^2$, not just $(\beta_j)^2$. This feature of the KL criterion seems more desirable to us than just optimizing on pure bias versus variance as MSE does.

Most insights based on all of these results will need to come from numerical examples and simulation. Because of the possible volume of such results (considering all the variables here), we leave such computing up to the interested reader. We have done a lot of computing and simulation of results based on these formulas and Section 7.7.1 results. One result (known in the literature about C_p versus AIC) is that K-L–based model selection for regression is just about the same as selection based on minimum theoretical MSE. We can determine this by computing the theoretical criteria KL and MSE. In so doing it is convenient to rescale the R values of those criteria to have their minimum at zero. Table 7.6 gives such results for one case: $R = m = 10$, $n = 30$, $\beta_i = (0.6)^{i-1}$, and σ taking several values in the range 0.025 to 1.

We have not undertaken a detailed analysis of these criteria for linear regression, let alone for this case of orthogonal regressors. However, the above results are representative of cases we have looked at in that the theoretical K-L best model has either the same number of parameters, or is actually more

TABLE 7.6. Comparison of model selection theoretical criterion KL, (7.89), and MSE, (7.88), for models g_1 to g_R, for $n = 30$, $R = m = 10$, and $\beta_r = (0.6)^{r-1}$, for several values of σ; results for both KL and MSE have been rescaled so that their minima are zero, hence clearly indicating the theoretical optimal model under these criteria.

	$\sigma = 1.0$		$\sigma = 0.5$		$\sigma = 0.25$		$\sigma = 0.05$		$\sigma = 0.025$	
r	KL	MSE	KL	MSE	KL	MSE	KL	MSE	KL	MSE
1	8.01	14.08	27.17	16.66	57.72	17.58	142.67	17.99	180.83	17.99
2	1.70	3.76	11.16	5.59	33.52	6.32	113.80	6.66	151.78	6.68
3	**0.00**	0.56	2.75	1.64	15.09	2.18	85.50	2.46	123.00	2.47
4	0.59	**0.00**	**0.00**	0.33	4.34	0.68	58.46	0.90	94.69	0.91
5	2.26	0.42	0.28	**0.00**	0.20	0.17	34.20	0.33	67.31	0.33
6	4.52	1.21	2.01	0.04	**0.00**	0.02	15.33	0.11	41.98	0.12
7	7.20	2.13	4.50	0.21	1.74	**0.00**	4.10	0.04	20.87	0.04
8	10.29	3.10	7.52	0.43	4.48	0.03	**0.00**	0.01	6.74	0.01
9	13.83	4.09	11.03	0.67	7.90	0.09	0.46	0.00	0.45	0.00
10	17.90	5.08	15.09	0.91	11.92	0.14	3.33	**0.00**	**0.00**	**0.00**

parsimonious than the theoretically best model under minimum MSE. Actual expected models selected under operational criteria such as AIC_c or C_p can differ slightly from these results (but less so as n gets large).

7.7.6 AIC_c and Models for Multivariate Data

The derivation of large-sample AIC in Section 7.2 *does* apply to the case of n independent multivariate observations, each with p nonindependent components. The small-sample improvement of AIC that applies for univariate (i.e., $p = 1$) linear models with homogeneous normal residuals, AIC_c, *does not* apply in the corresponding multivariate case. This problem has been studied by Fujikoshi and Satoh (1997). They focused on selection of model structure, that is, inclusion or exclusion of the same set of possible regressor variables in each of the p regressions. They assume that a general $p \times p$ variance–covariance matrix Σ applies for the residual vector of each observation. Thus for a model with k regressors (this may include an intercept) there are $k \times p$ structural parameters. Each model also includes $p(p + 1)/2$ unknown parameters in Σ. Thus, $K = (k \cdot p) + p(p + 1)/2$.

For their data analysis context Fujikoshi and Satoh (1997) derived an exact AIC, CAIC in their notation, analogous to the univariate case AIC_c. Their result (their formula 7) can be expressed as follows:

$$CAIC = AIC + 2\frac{K(k + 1 + p)}{n - k - 1 - p}. \tag{7.90}$$

The univariate case corresponds to $p = 1$, and then $K = k + 1$. Hence, the general result in (7.90) reduces to the univariate AIC_c. Fujikoshi and Satoh

(1997) do not consider other multivariate problems, nor do they consider the general case of models with reduced numbers of parameters in Σ. Thus, for other multivariate modeling problems we do not know whether (7.90) applies. Our key point here is that the univariate result for AIC_c does not apply to the multivariate setting.

The form of (7.90), by virtue of including variable k, is unique to the restricted context considered by Fujikoshi and Satoh (1997). By eliminating k from (7.90) we hypothesize a generalization of univariate AIC_c to corresponding multivariate applications:

$$\text{AIC}_c = \text{AIC} + 2\frac{K(K+v)}{np - K - v}. \tag{7.91}$$

In (7.91) v is the number of distinct parameters used in, and estimated for, Σ; $1 \le v \le p(p+1)/2$. Note that the count K includes v. Formula (7.91) is correct for the univariate case wherein $v = p = 1$. Interim use of (7.91) seems reasonable until a derivation is published for the needed generalization of AIC_c to multivariate applications.

The emphasis of this section is the generalization of AIC_c to multivariate applications. Many multivariate analysis methods, such as multivariate regression, analysis of variance or covariance, are done in a least squares framework while assuming a multivariate normal model such as $\text{MVN}(X\beta, \Sigma)$ (Σ must be full rank). Given that least squares is used, the software may not provide the value of the maximized log-likelihood. However, the residual sum of squares and cross products matrix (SSCP) is nearly always provided by commercial software packages and the MLE of Σ is $\hat{\Sigma} = \text{SSCP}/n$. Furthermore, the maximized log-likelihood is proportional to $-(n/2)\log(|\hat{\Sigma}|)$, where $|\hat{\Sigma}|$ denotes the determinant of $\hat{\Sigma}$. Hence, as long as all models considered assume multivariate normal residuals, we may use

$$\text{AIC} = n \times \log(|\hat{\Sigma}|) + 2K$$

and

$$\text{AIC}_c = n \times \log(|\hat{\Sigma}|) + 2K + \frac{2K(K+1)}{n - K - 1}.$$

In the univariate case SSCP is just the residual sum of squares RSS ($\equiv |\text{RSS}|$), and the MLE of σ^2 is RSS/n. Thus the multivariate case reduces to the univariate case. Theory for the multivariate case is summarized by Seber (1984:61).

If the determinant $|\text{SSCP}|$ is directly available, one could make use of $|\hat{\Sigma}| = n^{-p}|\text{SSCP}|$. However, because $\log(|\hat{\Sigma}|) = -p \times \log(n) + \log(|\text{SSCP}|)$ and p and n are constants we can just as well take, for example,

$$\text{AIC} = n \times \log(|\text{SSCP}|) + 2K.$$

As long as all models considered assume multivariate normal residuals the difference between the two expressions for AIC is just an additive constant that drops out of all inferential uses of AIC.

7.7.7   There Is No True TIC_c

Fujikoshi and Satoh (1997) also consider a small sample version of TIC; in essence they want to extend AIC_c to TIC_c. However, AIC_c (Hurvich and Tsai 1989) arises by computing the exact value of the target model selection criterion (7.20) for a linear model with constant normally distributed residuals under the condition (assumption) that this model is the true data-generating model. TIC is derived without any assumption that truth, f, is the same as the model g. That derivation can be justified only for large sample sizes. To compute a small sample, an exact version of TIC would require us to specify the exact form of the distribution f (i.e., specify truth). Even if we could do this computation in general, or at all, the result would depend upon assumed, but unknown, truth. Thus no defensible, general small-sample analytical version of TIC (i.e., a TIC_c) seems possible.

The issue did not escape the attention of Fujikoshi and Satoh (1997). What they did (and they knew it) was to assume that the linear, with normal errors, global model defined by using all available regressors contained the true model as an unknown submodel. Thus the true model is, by assumption, in the set of models considered and is a special case of the global model. Under these assumptions Fujikoshi and Satoh (1997) derived an analytical formula for any sample size, for the target criterion of (7.20). Their formula has a component, beyond $-2\log(\mathcal{L})$, that must be estimated from the data, as opposed to components that are simple functions of known n and K, and that extra component is estimable only by virtue of the strong assumptions made. We can elect to use the same small-sample-size adjustments with TIC as we use for AIC, and doing so may be a good idea; but we cannot find truly general small-sample-size adjustments for TIC.

7.7.8   Kullback–Leibler Information Relationship to the Fisher Information Matrix

The Fisher information matrix is defined by (7.6) for any $\underline{\theta} \in \Theta$:

$$\mathcal{I}(\underline{\theta}) = E_g\left[-\frac{\partial^2 \log(g(\underline{x} \mid \theta))}{\partial\theta_i \partial\theta_j}\right].$$

In taking this expectation it is assumed that the true data-generating model is $g(\underline{x} \mid \underline{\theta})$ (hence the underlying integration is with respect to g). We use $\underline{\theta} \equiv \underline{\theta}_o$ when this one particular member (i.e., g at $\underline{\theta}_o$) of the set of models defined for fixed structure, and any $\underline{\theta} \in \Theta$, is the generating model for the data. From

(7.7), we have

$$I(\underline{\theta}_o) = \mathrm{E}_f\left[-\frac{\partial^2 \log(g(\underline{x}\,|\,\underline{\theta}_o))}{\partial \theta_i \partial \theta_j}\right];$$

in general, $I(\underline{\theta}_o) \neq \mathcal{I}(\underline{\theta}_o)$. Moreover, $\mathcal{I}(\underline{\theta}_o)^{-1}$ is guaranteed to be the large-sample variance–covariance matrix of the MLE $\hat{\underline{\theta}}$ only when $g = f$ is true. For a value of $\underline{\theta} \in \Theta$ that is near the K-L minimizing value of $\underline{\theta}_o$, a valid quadratic approximation to the K-L difference is

$$I(f, g(\cdot\,|\,\underline{\theta})) - I(f, g(\cdot\,|\,\underline{\theta}_o)) \approx \frac{1}{2}(\underline{\theta} - \underline{\theta}_o)' I(\underline{\theta}_o)(\underline{\theta} - \underline{\theta}_o)'.$$

For the case of $f = g$ we get the result

$$I(g(\cdot\,|\,\underline{\theta}_o), g(\cdot\,|\,\underline{\theta})) \approx \frac{1}{2}(\underline{\theta} - \underline{\theta}_o)' \mathcal{I}(\underline{\theta}_o)(\underline{\theta} - \underline{\theta}_o)'.$$

Thus if one member of the set of models $g(\underline{x}\,|\,\theta)$, $\underline{\theta} \in \Theta$, is the data-generating distribution, then the approximate K-L information loss that results from using a nearby distribution as the approximating model is the above quadratic form in the Fisher information matrix. This is not a profound result given the definitions of both $\mathcal{I}(\underline{\theta})$ and $I(g(\cdot\,|\,\underline{\theta}_o), g(\cdot\,|\,\underline{\theta}))$, but it does serve to show that the two underlying concepts of "information" are related, albeit quite different, concepts. It was in the 1920s that Fisher chose to name this expected matrix of second mixed partials of a probability distribution "information." There is no relationship to information theory, which is a subject developed mostly since Shannon's pioneering work in the late 1940s that fundamentally deals with logs of probabilities. The Fisher information matrix fundamentally relates to the precision of ML estimators. The Kullback and Leibler paper of 1951 was a result of their attempt to understand and explain what Fisher meant by "information" in relation to sufficiency (personal communication, R. A. Leibler).

7.7.9 Entropy and Jaynes Maxent Principle

In Section 2.9 we noted that the Akaike weights w_i can be motivated by a (semi) Bayesian approach based on prior probabilities τ_i. To choose these prior probabilities in a manner philosophically consistent with the rest of this book we suggest resorting to the use of the Jaynes maximal entropy (maxent) principle (Jaynes 1957, 1982, Jessop 1995). This principle arises from information theory. The maxent principle says that if we must completely specify a probability distribution with only partial knowledge about moments, or other features, of that distribution, then we should choose the distribution that is maximally uninformative with regard to missing information. This means that we choose the distribution that has maximal entropy subject to any informative constraints we can justify, such as constraints based on data. Mathematically, we find the set of positive numbers $\tau_1, \ldots, \tau_R$ that maximize entropy $-\sum \tau_i \log(\tau_i)$ subject to the constraint $\sum \tau_i = 1$ and any other functional constraints we can justify (for example, constraints about the mean and/or variance of the distribution). The result is a distribution that conveys no information other than what we explicitly build into it. If the only constraint we impose is that the prior probabilities sum to 1, then the maxent distribution is given by $\tau_i = 1/R$.

That we can justify this uninformative prior for the models based on information theory is yet another example of how deeply information and entropy theory underlie statistical model selection. We will not divert from our objective of exploring information-theoretic data-based model selection. However, we recommend that interested readers pursue some of the references given here on the subject. An introductory reference that ties together some aspects of statistics and information theory, including the Jaynes maxent principle, is Jessop (1995). A nontechnical reference is Lucky (1991). For a biological perspective see Yockey (1992), while Cover and Thomas (1991) give a very thorough overview of information theory. Short, highly mathematical treatments are given by Wehrl (1978) and Ullah (1996).

As a general comment we emphasize the extensive foundations and extent of information and entropy theory, and how these basic ideas occur in many scientific and technical areas (from Boltzmann to Einstein to Shannon to Kullback–Leibler, for example). There is thus a deep foundation to Kullback–Leibler information measure and a firm basis for its use in model selection and other aspects of statistics. K-L is not just another (of many) possible measures of discrepancies between probability distributions; it is unique as a basis for data-based model selection in science when truth is very complex, data are "noisy," and models can be only approximations to truth.

In saying that this theoretical foundation for use of K-L information is deep, we would liken it to the theoretical basis for the importance of the constant e (≈ 2.7183) in mathematics. It is not at all obvious why such a strange, irrational number should universally be the basis for logarithms and exponentials in most of mathematics and science. But just as with K-L information, there is a compelling, deep reason, not easily perceived, for the importance of "e."

7.7.10 Akaike Weights w_i Versus Selection Probabilities π_i

The model selection probabilities can be expressed as expectations of indicator random variables that are a function of the sample data:

$$M_i(\underline{x}) = \begin{cases} 1 \text{ if model } i \text{ is selected by AIC,} \\ 0 \text{ otherwise.} \end{cases}$$

By definition, $\mathrm{E}(M_i(\underline{x})) = \pi_i$. We assume no ties for the best model.

The Akaike weights (see Section 2.9) defined by

$$w_i = \frac{\exp(-\frac{1}{2}\Delta_i)}{\sum_{r=1}^{R} \exp(-\frac{1}{2}\Delta_r)}$$

are also random variables and can be related to the above $M_i(\underline{x})$. Let k index the selected model. Because $\Delta_k = 0$ and $\Delta_i > 0$ for $i \neq k$, we have, for any $0 < \gamma < \infty$,

$$w_k(\gamma) = \frac{1}{1 + \sum_{r \neq k} \exp(-\gamma \Delta_r)}$$

and

$$w_i(\gamma) = \frac{\exp(-\gamma \Delta_i)}{1 + \sum_{r \neq k} \exp(-\gamma \Delta_r)}, \quad i \neq k.$$

In the limit as γ goes to infinity we have the result

$$\lim_{\gamma \to \infty} w_i(\gamma) = M_i(\underline{x}),$$

whence

$$\lim_{\gamma \to \infty} E(w_i(\gamma)) = \pi_i$$

(the implied interchange of limits will be valid here). Therefore, it must generally be the case that $E(w_i(0.5)) \equiv E(w_i) \neq \pi_i$; also, $E(w_i)$ and π_i are not unrelated.

This result does not rule out $E(w_i) \approx \pi_i$, which simulation supports as sometimes a useful approximation. Moreover, use of the *set* of Akaike weights as an estimator for the *set* of selection probabilities seems useful in formulas where such $\hat{\pi}_i$ are needed. (Research could be done to find improved $\hat{\pi}_i$ based on the Akaike weights).

7.8 Kullback–Leibler Information Is Always ≥ 0

It is not obvious that the Kullback–Leibler discrepancy,

$$I(f, g) = \int f(x) \log \left(\frac{f(x)}{g(x)} \right) dx,$$

is strictly nonnegative for any possible $g(x)$. Here we reduce the notation for the model g to just $g(x)$ rather than $g(x \mid \theta)$. Also, the possible multidimensional nature of f and g is not emphasized in the proofs in this section.

Rigorous proofs exist that $I(f, g) \geq 0$ and that $I(f, g) = 0$ if and only if $g(x) \equiv f(x)$ for all x. Here we give a valid, but not rigorous, proof that $I(f, g) \geq 0$. We do so for both the case of continuous distributions and the case of discrete distributions such as the Poisson, binomial, or multinomial, wherein

$$I(f, g) = \sum_{i=1}^{k} p_i \log \left(\frac{p_i}{q_i} \right).$$

In the discrete case there are k possible outcomes of the underlying random variable. Then the true probability of the ith outcome is given by p_i, while the $q_1, \ldots, q_k$ constitute the approximating probability distribution (i.e., the model). Hence, here f and g correspond to the p_i and q_i, respectively.

In the first case, both $f(x)$ and $g(x)$ must be valid probability distributions, hence satisfy $f(x) \geq 0$, $g(x) \geq 0$ and both integrate to 1:

$$\int f(x)dx = 1, \qquad \int g(x)dx = 1.$$

The exact limits of integration need not be specified here, but must be the same for both f and g. Moreover, without loss of generality we can assume $f(x) > 0$, $g(x) > 0$; hence the ratio $f(x)/g(x)$ is never 0 or $1/0$, which is undefined (but may be taken as ∞). For the discrete case we have $0 < p_i < 1$, $0 < q_i < 1$, and

$$\sum_{i=1}^{k} p_i = 1, \qquad \sum_{i=1}^{k} q_i = 1.$$

We consider first the case of continuous probability distributions. The key to one line of proof is to define a new function

$$h(x) = \frac{g(x) - f(x)}{f(x)};$$

thus,

$$\frac{g(x)}{f(x)} = 1 + h(x).$$

The lower bound on $h(x)$ is -1, because for any x over which integration is performed, $g(x)$ can be arbitrarily close to 0. The upper bound on $h(x)$ is ∞, thus $-1 < h(x) < \infty$. Note also that $\log(a) = -\log(1/a)$. Hence,

$$
\begin{aligned}
I(f, g) &= \int f(x) \log \left(\frac{f(x)}{g(x)} \right) dx \\
&= -\int f(x) \log \left(\frac{g(x)}{f(x)} \right) dx \\
&= 0 - \int f(x) \log \left(\frac{g(x)}{f(x)} \right) dx \\
&= \int f(x)h(x)dx - \int f(x) \log \left(\frac{g(x)}{f(x)} \right) dx.
\end{aligned}
$$

The last step above uses the fact that

$$
\begin{aligned}
0 &= \int f(x)h(x)dx = \int f(x) \frac{g(x) - f(x)}{f(x)} dx \\
&= \int (g(x) - f(x))dx
\end{aligned}
$$

$$= \int g(x)dx - \int f(x)dx = 1 - 1 = 0.$$

Returning now to the main proof, we have

$$I(f, g) = \int f(x)h(x)dx - \int f(x)\log\left(\frac{g(x)}{f(x)}\right)dx$$

$$= \int f(x)h(x)dx - \int f(x)\log(1 + h(x))dx$$

$$= \int f(x)\big[h(x) - \log(1 + h(x))\big]dx$$

$$= \int f(x)t(h(x))dx,$$

where $t(h(x)) = h(x) - \log(1 + h(x))$. We do not need to care about the actual values of $t(h(x))$. Nor do we need to consider $t(\cdot)$ as a function x; hence, also, x may be univariate or multivariate. It suffices to consider the function $h - \log(1 + h)$, hence $t(h)$, over the full range of h, $-1 < h < \infty$, that is possible by varying x. All we care about is some basic aspects of this *function*, namely that is it strictly nonnegative. It is.

Calculus can be used to show that $t(h) \geq 0$, and that $t(0) = 0$ is the unique minimum, and for any $h \neq 0$, then $t(h) > 0$. A simple heuristic "proof" is just to plot $t(h)$ over, say, $-1 < h \leq 5$, and check $t(h)$ at a few bigger values of h (Figure 7.1, and $t(10) = 7.6021$, $t(20) = 16.9555$, $t(100) = 95.3849$, $t(1000) = 993.0913$). Given that $t(x) \geq 0$ for all x, then $f(x)t(x) \geq 0$

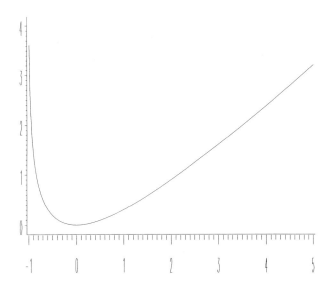

FIGURE 7.1. Plot of the function $t(h)$ near 0.

for all x; hence

$$I(f, g) = \int f(x) \log \left(\frac{f(x)}{g(x)} \right) dx \equiv \int f(x) t(x) dx \geq 0.$$

The calculus proof that $t(x) \geq 0$ makes use of the first and second derivatives of the function $t(h) = h - \log(1 + h)$:

$$t'(h) = \frac{h}{1 + h},$$

$$t''(h) = \frac{1}{(1 + h)^2}.$$

The set of critical points (which includes minima, maxima, and inflection points) of $t(h)$ consists of the solutions to $t'(h) = 0$ plus the limiting endpoints (-1 and infinity). In this case the unique solution is $h = 0$ (it does not matter that $h(x) = 0$ could occur for more than one value of x). The nature of this extremum is deduced from $t''(0) = 1$, which, because it is positive, proves that $h = 0$ is a minimum of the function $t(h)$ (and by uniqueness of the solution, it is the only minimum). Therefore, for all h (and hence all x), $t(h(x)) \geq 0$. Also, from these results, $t(h(x))$ is a convex function.

Deeper mathematical theory is required to prove that $I(f, g) = 0$ only if $f(x) = g(x)$ for all x (in the relevant range of integration). It is obvious that if $f(x) = g(x)$, then $I(f, g) = 0$. Part of the "deeper" mathematics referred to says that when $f(x)$ is a continuous probability density function and if $t(x) \geq 0$, then

$$I(f, g) = \int f(x) t(x) dx = 0$$

if and only if $t(x) = 0$ for all x in the range of integration. This statement seems reasonably intuitive, so we will not belabor the point. Thus we have

$$h(x) - \log(1 + h(x)) \equiv 0, \quad \text{for all } x,$$

or

$$h(x) = \log(1 + h(x)),$$

and finally,

$$e^{h(x)} = 1 + h(x).$$

The standard series expansion for e^h can be used here, whence

$$1 + h(x) + \sum_{i=2}^{\infty} \frac{1}{i!} [h(x)]^i = 1 + h(x),$$

or

$$\sum_{i=2}^{\infty} \frac{1}{i!} [h(x)]^i = 0.$$

If $h(x) > 0$, then the above could not be true; therefore, it has to be that if $I(f, g) = 0$, then $h(x) \leq 0$ at all x. However, if we allow $h(x) < 0$ over any set of x values $\mathcal{N}$ for which

$$\int_{\mathcal{N}} f(x)dx > 0,$$

then we would have

$$\int g(x)dx < \int f(x)\,dx,$$

which cannot be true. Thus, because both $f(x)$ and $g(x)$ are probability density functions, $h(x) \leq 0$ for all x, which implies that we must then in fact have $h(x) = 0$ for all x.

Now we consider (in less detail) the discrete case

$$I(f, g) = \sum_{i=1}^{k} p_i \log\left(\frac{p_i}{q_i}\right),$$

such that $0 < p_i < 1, 0 < q_i < 1$ for all i, and $\sum_{i=1}^{k} p_i = 1, \sum_{i=1}^{k} q_i = 1$. For fixed k, this $I(f, g)$ is a function of $k - 1$ variables, which can be taken as $q_1, q_2, \ldots, q_{k-1}$ (the p_i distribution is considered fixed here). Let

$$h_i = \frac{q_i - p_i}{p_i}, \qquad i = 1, \ldots, k,$$

whence

$$\frac{q_i}{p_i} = 1 + h_i, \qquad -1 < h_i < \infty, \qquad i = 1, \ldots, k.$$

As in the continuous case,

$$\sum_{i=1}^{k} p_i h_i = \sum_{i=1}^{k} (q_i - p_i) = 0,$$

so we can derive

$$I(f, g) = \sum_{i=1}^{k} p_i \left(h_i - \log(1 + h_i)\right) = \sum_{i=1}^{k} p_i t(h_i).$$

It was proved above that $t(h) \geq 0$; thus it must be that even in the discrete case $I(f, g) \geq 0$.

It is clear that if $p_i = q_i$ for all i, then $I(f, g) = 0$. Assume $I(f, g) = 0$. Then it must be that $t(h_i) = 0$ for all i (otherwise, $I(f, g)$ will be > 0). Therefore, we must have

$$e^{h_i} = 1 + h_i, \qquad i = 1, \ldots, k.$$

The set of indices $\{1, \ldots, k\}$ can be partitioned into two sets $\mathcal{N}$ and $\mathcal{P}$ wherein $h_i < 0$ for i in $\mathcal{N}$ and $h_i \geq 0$ for i in $\mathcal{P}$. For i in $\mathcal{P}$, $h_i > 0$ leads to a contradiction because then we would have to have $e^{h_i} > 1 + h_i$. Thus we conclude that for i in $\mathcal{P}$, $h_i = 0$, which is equivalent to $p_i = q_i$ for i in $\mathcal{P}$. Next, note that

$$
\begin{aligned}
0 &= \sum_{i=1}^{k} p_i h_i \\
&= \sum_{i=1}^{k} (q_i - p_i) \\
&= \sum_{i \,\mathrm{in}\, \mathcal{N}} (q_i - p_i) + \sum_{i \,\mathrm{in}\, \mathcal{P}} (q_i - p_i) \\
&= \sum_{i \,\mathrm{in}\, \mathcal{N}} (q_i - p_i).
\end{aligned}
$$

But i in $\mathcal{N}$ means $h_i < 0$, or $q_i - p_i < 0$, which would mean that the above sum would be strictly < 0, which is a contradiction. This contradiction means that the set $\mathcal{N}$ is empty: There cannot be any $h_i < 0$ if $I(f, g) = 0$. Thus if $I(f, g) = 0$, then $f \equiv g$ (i.e., $p_i = q_i$, for all i in the discrete case).

7.9 Summary

Most of this chapter is quite technical; we will try to provide a high-level summary of key points or results. Sections 7.1 through 7.6 provide foundational mathematical theory for K-L information-theoretic model selection. The general theory is given (Sections 7.1–7.3) along with several important special cases (Sections 7.4–7.5) and some specific exploration of AIC versus TIC (Section 7.6). In particular, a very detailed derivation of TIC is given in Section 7.2, along with the relationship to AIC. Then Section 7.6 is a detailed (but not exhaustive) examination of the issue of whether we can use AIC (actually, AIC_c) rather than TIC; the results strongly support use of AIC_c as not only acceptable, relative to TIC, but actually preferable. Section 7.7 (in 7.7.1–7.7.5) provides simple (though not general) methods to explore key properties of model selection that are operationally based on the log-likelihood; some theoretical results are also given. Sections 7.7.6–7.7.10 give a few miscellaneous results and considerations that do not fit elsewhere. Section 7.7 is overall much less technical than Sections 7.1–7.6, and we urge you to read Section 7.7 for the general insights therein. Section 7.8 gives a proof of $I(f, g) \geq 0$.

There are several rigorous derivations from Kullback–Leibler information, leading to various information-theoretic criteria: The most relevant general derivation leads to Takeuchi's (1976) information criterion (TIC). The exact derivation of AIC_c is given in detail in Section 7.4. It is also noted that there is no unique small-sample version of AIC, but AIC_c is recommended for general use (that could change in the future, especially for discrete distributions). Section 7.5 gives a derivation of AIC for the exponential family of models; this family of models is used in most parametric data analysis. When done just for the exponential family of models, the derivation of information-theoretic criteria from K-L information is much more exact, relies on fewer assumptions, and is easier to understand.

The fact that such derivations exist is important to know. The derivations and explanations are very detailed because the theory underlying model selection based on K-L information is important to have clearly stated to allow understanding. Such understanding of the theory puts one in a much better position to accept use of the information-theoretic criteria and understand its strengths and weaknesses.

While Kullback–Leibler information is the logical basis for likelihood-based model selection, it turns out we must use *expected* (over $\hat{\theta}$) Kullback–Leibler information as the quantity of interest when model parameters must be estimated. This, of course, is the reality of actual data analysis. The bias versus variance tradeoff and the associated model parsimony achieved by K-L–based model selection is an important byproduct of the approach. That is, the derivations make it clear that K-L–based model selection does not start with the explicit objective of meeting the principle of parsimony in model selection. Rather, it is a natural consequence of data-based K-L model selection that this bias–variance tradeoff happens. In fact, it is because the criterion, with estimated parameters, must minimize *expected* K-L information that we get the cross-validation property of AIC.

The detailed derivations make it clear that use of information-theoretic criteria in the analysis of real data is not based on the existence of a "true model" or the notion that such a true model is in the set of candidates being considered. Literature contrary to this point is mistaken.

Model selection attempts to establish some rigorous basis to achieve proper parsimony in the model(s) used for inference. The relationship of TIC to AIC is made clear, and investigations were undertaken to show that often AIC is a good proxy for TIC. It seems poetic that AIC can be thought of as a parsimonious implementation of the more general TIC. The trace term $\mathrm{tr}[J(\underline{\theta}_o)I(\underline{\theta}_o)^{-1}]$ is about equal to K for "good" models and does not depend on sample size, once sample size is large. Some insights are provided to help in understanding the relationship between $\mathrm{tr}[J(\underline{\theta}_o)I(\underline{\theta}_o)^{-1}]$ and K in a variety of practical situations. Evaluations were conducted for logistic regression, multinomial, and Poisson count data, and normal regression models. In all cases we examined, the trace term of TIC is very close to being K as long as the model structure and assumed error distribution are not drastically different from truth. When the model is not the true data-generating model, the trace term was not systematically $> K$ or $< K$. Rather, the matter is unpredictable; the model can be misspecified, and still the trace term can be any of $= K$, $> K$, or $< K$. For all the cases examined, however, if the model was less general than truth (the real-world case), we predominantly found $\mathrm{tr}[J(\underline{\theta}_o)I(\underline{\theta}_o)^{-1}] < K$. Thus, use of AIC should then often lead to slightly more parsimonious models than use of TIC (to the extent that there is any appreciable difference in results from these two procedures).

If the set of models contains one or more good models, then poor models will tend not to be selected (or even ranked high), because $-2\log(\mathcal{L})$ will be relatively large for a poor model and this term will dominate the criterion value for that model, hence rendering the issue of use of TIC versus AIC largely moot. As noted above, for good models use of AIC is acceptable. Use AIC$_c$ for small samples and even for large samples if values of K become large relative to sample size. More research on such second-order improvements is needed, especially for discrete random variables.

Monte Carlo methods seem to be the only tool to assess general stochastic aspects of model selection and methods to incorporate model selection uncertainty. In some cases, asymptotic results can be obtained, but these seem to be of little interest or practical use. We present some quick ways to explore AIC model selection using Monte Carlo simulation in the case of nested sequences of models. For that same context, the theoretical expected values of the Δ_i can be easily found, and this is explored in some detail for linear regression models with normal errors. The issue of AIC overfitting is clarified and explored. Extreme overfitting can occur, but the probability of this event is low, and one way to minimize the problem is to keep the set of models considered small. Doing searches over "all possible models" (e.g., all-subsets selection) increases the risk of overfitting. In linear regression it seems that AIC selection is very similar to model selection based on minimum theoretical MSE (of course, in the analysis of real data we cannot do selection based on minimum theoretical MSE).

8
Summary

This book covers some philosophy about data analysis, some theory at the interface between mathematical statistics and information theory, and some practical statistical methodology useful in the applied sciences. In particular, we present a general strategy for modeling and data analysis. We provide some challenging examples from our fields of interest, provide our ideas as to what not to do, and suggest some areas needing further theoretical development. We side with the fast-growing ranks that see limited utility in statistical null hypothesis testing. Finally, we provide references from the diverse literature on these subjects for those wishing to study further.

Conceptually, there is information in the observed data, and we want to express this information in a compact form via a "model." Such a model represents a scientific hypothesis and is then a basis for making inferences about the process or system that generated the data. One can view modeling of information in data as a change in "coding" like a change in language. A concept or emotion expressed in one language (e.g., French) loses some exactness when expressed in another language (e.g., Russian). A given set of data has only a finite, fixed amount of information. The (unachievable) goal of model selection is to attain a perfect 1-to-1 translation such that no information is lost in going from the data to a model of the information in the data. Models are only approximations, and we cannot hope to perfectly achieve this idealized goal. However, we can attempt to find a model of the data that is best in the sense that the model loses as little information as possible. This thinking leads directly to Kullback–Leibler information $I(f, g)$: the information lost when model g is used to approximate full reality f. We wish then to select a model that minimizes K-L information loss. Because we must estimate model parameters from the data, the best we can do is to minimize (estimated) expected K-L information loss. However, this can easily be done using one of the information-theoretic criteria (e.g., AIC, AIC_c, QAIC, or TIC). Then a *good* model allows the efficient and objective separation or filtration of *information* from *noise*. In an important sense, we are not really trying to model the *data*; instead, we are trying to model the *information* in the data.

While we use the notation f to represent truth or full reality, we deny the existence of a "true model" in the life sciences. Conceptually, let f be the process (truth) that generates the sample data we collect. We want to make inferences about truth, while realizing that full reality will always be beyond us when we have only sample data. Data analysis should not be thought of as an attempt to identify f; instead, we must seek models that are good approximations to truth and from which therefore we can make valid inferences concerning truth. We do not want merely to describe the data using a model that has a very large number of parameters; instead, we want to use the data to aid in the selection of a parsimonious model that allows valid inferences to be made about the system or process under study. A parsimonious model, representing a well-defended scientific hypothesis, aids in our understanding of the system of interest.

Relatively few statistics books provide a summary of the key points made and yet fewer provide an effective, unified strategy for data analysis and inference where there is substantial complexity. The breadth of the technical subjects covered here makes a summary difficult to write. Undergraduate students occasionally ask the professor, "What is important for me to know for the final examination?" The professor is typically irritated by such a question. Surely, the student should realize that it is *all* important! Indeed, our interpretation of Akaike's pioneering work is that it *is* all important. The information-theoretic paradigm is a *package*; each of the package's contents is important in itself, but it is the integration of the contents that makes for an effective philosophy, a consistent strategy, and a practical and powerful methodology. The part of this package that has been so frequently left out is the critical thinking, hypothesis generation, and modeling *before* examination of the data; ideally, much of this thinking should occur prior even to data collection. This is the point where the science of the issue formally enters the overall "analysis" (Anderson and Burnham 1999a).

The information-theoretic methods we present can be used to select a single best model that can be used in making inferences from empirical data. AIC is often portrayed in the literature in this simple manner. The general approach is much richer than this simplistic portrayal of model selection might suggest. **In fact, an emphasis of this second edition is multimodel inference (MMI). MMI has several advantages; all relate to the broad subject of model selection uncertainty.** One can easily rank alternative models (hypotheses) from best to worst using the convenient differences Δ_i. The likelihood for each model, given the data [i.e., $\mathcal{L}(g_i \mid data)$], can be easily computed, and these can be normalized to obtain Akaike weights (w_i) which can be interpreted as probabilities. Confidence sets of models can be defined to aid in identifying a subset of good models. Evidence ratios are useful for comparing relative support of one model versus another, given the data; such ratios are useful, irrespective of other models in the set.

Model selection uncertainty can be easily quantified using Akaike weights (the bootstrap is an alternative). Estimates of this component of uncertainty can be incorporated into unconditional estimates of precision using several methods. For many problems (e.g., prediction) model-averaging has advantages, and we treat this important issue in Chapters 4–5. Thus, we often recommend formal inference from all models in the set.

For those who have scanned through the pages of this book there might be surprise at the general lack of mathematics and formulas (Chapters 6 and 7 being the exceptions). That has been our intent. The *application* of the information-theoretic methods is relatively simple. They are easy to understand and use ("low tech"), while the underlying theory is quite deep (e.g., Chapter 7). As we wrote the book and tried to understand Akaike's various papers (see Parzen et al. 1998) we found the need to delve into various issues that are generally philosophical. The science of the problem has to be brought into modeling *before* one begins to rummage through the data (data dredging). In some critical respects, applied statistics courses are failing to teach statistics as an integral part of scientific discovery, with little about modeling and model selection methods or their importance, while succeeding (perhaps) in teaching null hypothesis testing methods and data analysis methods based on the assumption that the model is both true and given. Sellke et al. (2001:71) note, "The standard approach in teaching—stressing the formal definition of a *p* value while warning against its misinterpretation—has simply been an abysmal failure." It seems necessary to greatly reduce the reporting of *P*-values (Anderson et al. 2001b and d).

8.1   The Scientific Question and the Collection of Data

The formulation of the research question is crucial in investigations into complex systems and processes in the life sciences. A good answer to a poor question is a mistake all too often seen in the published literature and is little better than a poor answer to a poor question. Investigators need to continually readdress the importance and quality of the question to be investigated. Good scientific hypotheses, represented by models, must have a place at the head of the table.

A careful program of data collection must follow from the hypotheses posed. Particular attention should be placed on the variables to be measured and interesting covariates. Observational studies, done well, can show patterns, associations, and relationships and are confirmatory in the sense that certain issues stem from a priori considerations. More causal inference must usually come from more formal experimentation (i.e., important confounding factors are controlled or balanced, experimental units are randomly assigned to treatment and control groups with adequate replication), but see Anderson et al. (1980), Gail (1996), Beyers (1998), and Glymour (1998) for alternative philosophies. Valid inference must assume that these basic important issues have been carefully planned and conducted. Before one should proceed, two general questions must be answered in the affirmative:

Are the study objectives sound, relevant, and achievable?
Has there been proper attention to study design and laboratory or field protocol?

8.2   Actual Thinking and A Priori Modeling

Fitting models, each representing a scientific hypothesis, to data has been important in many biological, ecological, and medical investigations. Then statistical inferences about the system of interest are made from an interpretable parsimonious model of the observational or experimental data. We expect to see this activity increase as more complicated scientific and management issues are addressed. In particular, a priori modeling becomes increasingly important as several data sets are collected on the same issue by different laboratories or at widely differing field sites over several years.

We recommend much more emphasis on thinking! Leave the computer idle for a while, giving time to think hard about the overall problem. What useful information is contained in the published literature, even on issues only somewhat related to the issue at hand? What nonlinearities and threshold effects might be predicted? What interactions are hypothesized to be important? Can two or more variables be combined to give a more meaningful variable for analysis? Should some variables be dropped from consideration? Discussions should be encouraged with the people in the field or laboratory that were close to the data collection. What parameters might be similar across groups (i.e., data sets)? Model building should be driven by the underlying science of the issue combined with a good understanding of mathematical models. Ideally, this important conceptual phase might take several days or even weeks of effort; this seems far more time than is often spent under current practice.

Biologists generally subscribe to the philosophy of "multiple working hypotheses" (Chamberlain 1890, Platt 1964, Mayr 1997), and these should form the basis for the set of candidate models to be considered formally. Model building can begin during the time that the a priori considerations are being sorted out. Modeling must carefully quantify the science hypotheses of interest. Often it is effective to begin with the global model and work toward some lower-dimensional models. Others may favor a bottom-up approach. The critical matter here is that one arrives, eventually, at a small set of good candidate models, prior to examination of the empirical data. We advise the inclusion of all models that are reasonably justified prior to data analysis; however, every attempt should be made to keep the number of candidate models small.

Critical Thinking

Our science culture does not do enough to regularly expect and enforce critical thinking. This failure has slowed the scientific discovery process.

We fail to fault the trivial content of the typical ecological hypothesis.

There is a need for more *careful thinking* (than is usually evident) and a *better balance* between scientific hypotheses, data, and analysis theory.

Chamberlin's concept of *multiple working hypotheses*, suggested well over 100 years ago, has a deep level of support among science philosophers. He thought the method led to "certain distinctive habits of mind and had prime value in education." Why has this principle not become the standard, rather than the rare exception, in so many fields of applied science?

Platt (1964) noted that years and decades can be wasted on experiments, unless one thinks carefully in advance about what the most important and conclusive experiments would be.

With the information-theoretic approach, there is no concept of a "null" hypothesis, or a statistical hypothesis test, or an arbitrary α-level, or questionable power, or the multiple testing problem, or the fact that the so-called null hypothesis is nearly always *obviously* false in the first place. Much of the application of statistical hypothesis testing arbitrarily classifies differences into meaningless categories of "significant" and "nonsignificant," and this practice has little to contribute to the advancement of science (Anderson et al. 2000). We recommend that researchers stop using the term "significant," since it is so overused, uninformative, and misleading. The results of model selection based on estimates of expected (relative) Kullback–Leibler information can be very different from the results of some form of statistical hypothesis testing (e.g., the simulated starling data, Section 3.4, or the sage grouse data, Section 3.5).

So, investigators may proceed with inferential or confirmatory data analysis if they feel satisfied that they can objectively address two questions:

Was the set of candidate models derived a priori?
What justifies this set?

The justification should include a rationale for models both included and excluded from the set. A carefully defined set of models is crucial whether information-theoretic methods are used to select the single best model, or the entire set of models is used to reach defensible inferences. If so little is known about the system under study that a large number of models must be included in the candidate set, then the analysis should probably be considered only exploratory (if models are developed as data analysis progresses, it is both exploratory and risky). **One should check the fit or adequacy of the global model using standard methods. If the global model is inadequate (after,**

perhaps, adjusting for overdispersed count data), then more thought should be put into model building and thinking harder about the system under study and the data collected. There is no substitute for good, hard thinking at this point (Platt 1964).

8.3   The Basis for Objective Model Selection

Statistical inference from a data set, *given a model*, is well advanced and supported by a very large amount of theory. Theorists and practitioners are routinely employing this theory, either likelihood or least squares, in the solution of problems in the applied sciences. The most compelling question is, *"what model to use?"* Valid inference must usually be based on a good approximating model, but which one?

Akaike chose the celebrated Kullback–Leibler discrimination information as a basis for model selection. This is a fundamental quantity in the sciences and has earlier roots in Boltzmann's concept of *entropy*, a crowning achievement of nineteenth-century science. The K-L distance between conceptual truth f and model g is defined for continuous functions as the integral

$$I(f, g) = \int f(x) \log \left(\frac{f(x)}{g(x \mid \theta)} \right) dx,$$

where log denotes the natural logarithm and f and g are n-dimensional probability distributions. Kullback and Leibler (1951) developed this quantity from "information theory," thus the notation $I(f, g)$ as it relates to the "information" lost when model g is used to approximate truth f. Of course, we seek an approximating model that loses as little information as possible; this is equivalent to minimizing $I(f, g)$ over the models in the set. Full reality is considered to be fixed. An interpretation equivalent to minimizing $I(f, g)$ is that we seek an approximating model that is the "shortest distance" from truth. Both interpretations seem useful and compelling.

The K-L distance can be written equivalently as

$$I(f, g) = \int f(x) \log(f(x)) \, dx - \int f(x) \log(g(x \mid \theta)) \, dx.$$

The two terms on the right in the above expression are statistical expectations with respect to f (truth). Thus, the K-L distance (above) can be expressed as a difference between two expectations,

$$I(f, g) = E_f[\log(f(x))] - E_f[\log(g(x \mid \theta))],$$

each with respect to the true distribution f. The first expectation, $E_f[\log(f(x))]$, is a constant that depends only on the unknown true distribution. Therefore, treating this unknown term as a constant, only a measure of *relative* distance is possible. Then

$$I(f, g) = \text{constant} - \mathrm{E}_f[\log(g(x \mid \theta))],$$

or

$$I(f, g) - \text{constant} = -\mathrm{E}_f[\log(g(x \mid \theta))].$$

Thus, the term $\big(I(f, g) - \text{constant}\big)$ is a *relative* distance between truth f and model g. This provides a deep theoretical basis for model selection if one can compute or estimate $\mathrm{E}_f[\log(g(x \mid \theta))]$.

Akaike (1973, 1974, 1985, 1994) showed that the critical quantity for estimating relative K-L information was

$$\mathbf{E}_y \mathbf{E}_x[\mathbf{log}(g(x|\hat{\theta}(y)))],$$

where y and x are independent random samples from the same distribution and both statistical expectations are taken with respect to truth (f). This double expectation, both with respect to truth f, is the target of model selection approaches based on K-L information.

8.4 The Principle of Parsimony

Parsimony is the concept that a model should be as simple as possible with respect to the included variables, model structure, and number of parameters. Parsimony is a desired characteristic of a model used for inference, and it is usually visualized as a suitable tradeoff between squared bias and variance of parameter estimators (Figure 1.3). Parsimony lies between the evils of underfitting and overfitting (Forster and Sober 1994, Forster 1999). Expected K-L information is a fundamental basis for achieving proper parsimony in modeling.

The concept of parsimony has a long history in the sciences. Often this is expressed as "Occam's razor": shave away all that is unnecessary. The quest is to make things "as simple or small as possible." Parsimony in statistics represents a tradeoff between bias and variance as a function of the dimension of the model (K). A good model is a proper balance between underfitting and overfitting, given a particular sample size (n). Most model selection methods are based on the concept of a squared bias versus variance tradeoff. Selection of a model from a set of approximating models must employ the concept of parsimony. These philosophical issues are stressed in this book, but it takes some experience and reconsideration to reach a full understanding of their importance.

8.5   Information Criteria as Estimates of Expected Relative Kullback–Leibler Information

---

Roots of Theory

As deLeeuw (1992) noted, Akaike found a formal relationship between Boltzmann's entropy and Kullback–Leibler information (dominant paradigms in information and coding theory) and maximum likelihood (the dominant paradigm in statistics).

This finding makes it possible to combine estimation (point and interval estimation) and model selection under a single theoretical framework: optimization.

Akaike's (1973) breakthrough was the finding of an estimator of the expected relative K-L information, based on a bias-corrected maximized log-likelihood value. His estimator was an approximation and, under certain conditions, asymptotically unbiased. He found that

estimated expected (relative) K-L information $\approx \log(\mathcal{L}(\hat{\theta})) - K,$

where $\log(\mathcal{L}(\hat{\theta}))$ is the maximized log-likelihood value and K is the number of estimable parameters in the approximating model (this is the bias-correction term). Akaike multiplied through by -2 and provided Akaike's information criterion (AIC)

$$\text{AIC} = -2\log(\mathcal{L}(\hat{\theta})) + 2K.$$

Akaike considered his information-theoretic criterion an extension of Fisher's likelihood theory. Conceptually, the principle of parsimony is enforced by the added "penalty" (i.e., $2K$) while minimizing AIC.

Assuming that a set of a priori candidate models has been carefully defined, then AIC is computed for each of the approximating models in the set, and the model where AIC is minimized is selected as best for the empirical data at hand. This is a simple, compelling concept, based on deep theoretical foundations (i.e., K-L information). Given a focus on a priori issues, modeling the relevant scientific hypotheses, and model selection, *the inference is the selected model.* In a sense, parameter estimates are almost byproducts of the selected model. This inference relates to the estimated best approximation to truth and what information seems to be contained in the data.

Important refinements followed shortly after the pioneering work by Akaike. Most relevant was Takeuchi's (1976) information criterion (termed TIC), which provided an asymptotically unbiased estimate of relative expected K-L information. TIC is little used, since it requires the estimation of $K \times K$ matrices of first and second partial derivatives of the log-likelihood function, and its practical use hinges on the availability of a relatively large sample size. In a sense, AIC can be viewed as a parsimonious version of TIC. A second refinement was motivated by Sugiura's (1978) work, and resulted in a series of papers by Hurvich and Tsai (1989, 1990b, 1991, 1994, 1995a and 1995b, 1996). They provided a second order approximation, termed AIC_c, to estimated, expected relative K-L information,

$$AIC_c = -2\log(\mathcal{L}(\hat{\theta})) + 2K + \frac{2K(K+1)}{(n-K-1)},$$

where n is sample size The final bias-correction term vanishes as n gets large with respect to K (and AIC_c becomes AIC), but the additional term is important if n is not large relative to K (we suggest using AIC_c if $n/K < 40$ or, alternatively, always using AIC_c).

A third extension was a simple modification to AIC and AIC_c for overdispersed count data (Lebreton et al. 1992). A variance inflation factor $\hat{c}$ is computed from the goodness-of-fit statistic, divided by its degrees of freedom, $\hat{c} = \chi^2/$ df. The value of the maximized log-likelihood function is divided by the estimate of overdispersion to provide a proper estimate of the log-likelihood. These criteria are denoted by QAIC and $QAIC_c$ as they are derived from quasi-likelihood theory (Wedderburn 1974),

$$QAIC = -[2\log(\mathcal{L}(\hat{\theta}))/\hat{c}] + 2K,$$

and

$$QAIC_c = -[2\log(\mathcal{L}(\hat{\theta}))/\hat{c}] + 2K + \frac{2K(K+1)}{n-K-1}$$
$$= QAIC + \frac{2K(K+1)}{n-K-1}.$$

When no overdispersion exists, $c = 1$, and the formulas for QAIC and $QAIC_c$ reduce to AIC and AIC_c, respectively. There are other, more sophisticated, ways to account for overdispersion in count data, but this simple method is often quite satisfactory. Methods are given in Chapter 6 to allow different partitions of the data to have partition-specific estimates of overdispersion. Note that the number of estimable parameters (K) must include the number of estimates of c. Thus, if males and females have different degrees of overdispersion and these are to be estimated from the data, then K must include 2 parameters for these estimates.

AIC is often presented in the scientific literature in an ad hoc manner, as if the bias-correction term K (the so-called penalty term) was arbitrary. Worse yet, perhaps, is that AIC is often given without reference to its fundamental link with Kullback–Leibler information. Such shallow presentations miss the point, have had very negative effects, and have misled many into thinking that there is a whole class of selection criteria that are "information-theoretic" (Chapter 6). Criteria such as AIC, AIC_c, QAIC, and TIC are estimates of expected (relative) Kullback–Leibler distance and are useful in the analysis of real data in the "noisy" sciences.

8.6   Ranking Alternative Models

Because only *relative* K-L information can be estimated using one of the information criteria, it is convenient to rescale these values such that the model with the minimum AIC (or AIC$_c$ or TIC) has a value of 0. Thus, information-criterion values can be rescaled as simple differences,

$$\Delta_i = \text{AIC}_i - \text{AIC}_{min}$$
$$= \hat{E}_{\hat{\theta}}[\hat{I}(f, g_i)] - \min \hat{E}_{\hat{\theta}}[\hat{I}(f, g_i)].$$

While the value of minimum $\hat{E}_{\hat{\theta}}[\hat{I}(f, g_i)]$ is not known (only the relative value), we have an estimate of the size of the increments of information loss for the various models compared to the estimated best model (the model with the minimum $E_{\hat{\theta}}[\hat{I}(f, g_i)]$). The Δ_i values are easy to interpret and allow a quick comparison and ranking of candidate models and are also useful in computing Akaike weights. As a rough rule of thumb, models having Δ_i within 1–2 of the best model have substantial support and should receive consideration in making inferences. Models having Δ_i within about 4–7 of the best model have considerably less support, while models with $\Delta_i > 10$ have either essentially no support and might be omitted from further consideration or at least fail to explain some substantial structural variation in the data. If the observations are not independent (but are treated as such) or if the sample size is quite small, or it there is a very large number of models, then the simple guidelines above cannot be expected to hold.

There are cases where a model with $\Delta_i > 10$ might still be useful, particularly if the sample size is very large (e.g., see Section 6.8.2). For example, let model A, with year-specific structure on one of the parameters, be the best model in the set ($\Delta_A = 0$) and model B, with less structure on the subset of year-specific parameters, have $\Delta_B = 11.4$. Assume that all models in the candidate set were derived prior to data analysis (i.e., no data dredging). Clearly, model A is able to identify important variation in a parameter across years; this is important. However, in terms of understanding and generality of inference based on the data, it might sometimes be justified to use the simpler model B, because it may seem to "capture" the important fixed effects. Models A and B should both be detailed in any resulting publication, but understanding and interpretation might be enhanced using model B, even though some information in the data would be (intentionally) lost. Such lost information could be partially recovered by, for example, using a random effects approach (see Section 3.5.5) to estimate the mean of the time-effects parameter and the variance of its distribution.

The principle of parsimony provides a philosophical basis for model selection; Kullback–Leibler information provides an objective target based on deep, fundamental theory; and the information criteria (particularly AIC and AIC$_c$) provide a practical, general methodology for use in data analysis. Objective model selection and model weighting can be rigorously based on these principles. In practice, one need not assume that any "true model" is contained in the set of candidates (although this is sometimes stated, erroneously, in the technical literature). [We note that several "dimension-consistent criteria" have been published that attempt to provide asymptotically unbiased (i.e., "consistent") estimates of the dimension (K) of the "true model." Such criteria are only estimates of K-L information in a strained way, are based on unrealistic assumption sets, and often perform poorly (even toward their stated objective) unless a very large sample size is available (or where σ^2 is negligibly small, such as in many problems in the physical sciences). We do not recommend these dimension-consistent criteria for the analysis of real data in the life sciences.]

8.7 Scaling Alternative Models

The information-theoretic approach does more than merely estimate which model is best for making inference, given the set of a priori candidate models and the data. The Δ_i allow a ranking of the models from an estimated best to the worst; the larger the Δ_i, the less plausible is model i. **In many cases it is not reasonable to expect to be able to make inferences from a single (best) model; biology is not simple; why should we hope for a simple inference from a single model?** The information-theoretic paradigm provides a basis for examination of alternative models and, where appropriate, making formal inference from more than one model (MMI).

The simple transformation $\exp(-\frac{1}{2}\Delta_i)$ results in the (discrete) likelihood of model i, given the data $\mathcal{L}(g_i|x)$. These are functions in the same sense that $\mathcal{L}(\theta|x, g_i)$ is the likelihood of the parameters θ, given the data (x) and the model (g_i). These likelihoods are very useful; for example, the evidence ratio for model i versus model j is merely

$$\mathcal{L}(g_i|x)/\mathcal{L}(g_j|x).$$

It is convenient to normalize these likelihoods such that they sum to 1, as

$$w_i = \frac{\exp(-\frac{1}{2}\Delta_i)}{\sum_{r=1}^{R} \exp(-\frac{1}{2}\Delta_r)},$$

and interpret these as a weight of evidence. Akaike (e.g., Akaike 1978b, 1979, 1980, and 1981b; also see Kishino 1991 and Buckland et al. 1997) suggested these values, and we have found them to be simple and very useful. The evidence ratio of model i versus model j is then just w_i/w_j; this is identical to the ratio of the likelihood $\mathcal{L}(g_i|x)/\mathcal{L}(g_j|x)$. Drawing on Bayesian ideas we can interpret w_i as the estimated probability that model i is the K-L best model for the data at hand, given the set of models considered (see Section 6.4.5).

An interesting and recent finding is that AIC can be derived under a formal Bayesian framework, and this fact has led to some deeper insights. The breakthrough here was to consider priors on models that are a function of both n and K (we call this class of model priors "savvy," i.e., shrewdly informative); then AIC and AIC_c fall out as a strictly Bayesian result. Indeed, as AIC has a Bayesian derivative, it is compelling to interpret the Akaike weights as posterior model probablilities. While many (objective) Bayesians are comfortable with the use of a defuse or noninformative prior on model parameters (e.g., a uniform prior on a model parameter), use of such defuse priors on models (such as $1/R$) may have poor properties or unintended consequences. That is, some priors on models may be uninformative, but not innocent. In the end, the Bayesian derivation of AIC (or AIC_c) and BIC differ only in their priors on models. However, these criteria are fundamentally different in a variety of substantive ways. In this book we place an emphasis on the derivation of AIC and AIC_c as bias-corrected estimates of Kullback–Leibler information because this seems so much more objective and fundamental.

The w_i are useful as the "weight of evidence" in favor of model i as being the actual K-L best model in the set. The bigger the Δ_i, the smaller the weight and the less plausible is model i as being the best approximating model. Inference is conditional on both the data and the set of a priori models considered.

Alternatively, one could draw B bootstrap samples (B should often be 10,000 rather than 1,000), use the appropriate information criterion to select a best model for each of the B samples, and tally the proportion of samples whereby the ith model was selected. Denote such bootstrap-selection frequencies by $\hat{\pi}_i$. While w_i and $\hat{\pi}_i$ are not estimates of exactly the same entity, they are often closely related and provide information concerning the uncertainty in the best model for use. The Akaike weights are simple to compute, while the bootstrap weights are computer-intensive and not practical to compute in some cases (e.g., the simulated starling experiment, Section 3.4), because thousands of bootstrap repetitions must be drawn and analyzed.

Under the hypothesis-testing approach, nothing can generally be said about ranking or scaling models, particularly if the models were not nested. In linear least squares problems one could turn to adjusted R^2 values for a rough ranking of models, but other kinds of models cannot be scaled using this (relatively very poor) approach (see the analogy in Section 2.5).

8.8   MMI: Inference Based on Model Averaging

Rather than base inferences on a single selected best model from an a priori set of models, we can base our inferences on the entire set by using model-averaging. The key to this inference methodology is the Akaike weights. Thus, if a parameter θ is in common over all models (as θ_i in model g_i), or our goal is prediction, by using the weighted average we are basing point inference on the entire set of models,

$$\hat{\bar{\theta}} = \sum_{i=1}^{R} w_i \hat{\theta}_i,$$

or

$$\hat{\bar{\theta}} = \sum_{i=1}^{R} \hat{\pi}_i \hat{\theta}_i.$$

This approach has both practical and philosophical advantages. Where a model-averaged estimator can be used, it appears to have better precision and reduced bias compared to $\hat{\theta}$ from the selected best model.

If one has a large number of closely related models, such as in regression-based variable selection (all-subsets selection), designation of a single best model is unsatisfactory, because that estimated "best" model is highly variable from data set to data set. In this situation model-averaging provides a relatively much more stabilized inference. The concept of inference being tied to all the models can be used to reduce model selection bias effects on regression-coefficient estimates in all-subsets selection. For the regression coefficient associated with predictor x_j we use the estimate $\hat{\bar{\beta}}_j$, which is the estimated regression coefficient β_j averaged over all models in which x_j appears:

$$\hat{\bar{\beta}}_j = \frac{\sum_{i=1}^{R} w_i I_j(g_i) \hat{\beta}_{j,i}}{w_+(j)},$$

$$w_+(j) = \sum_{i=1}^{R} w_i I_j(g_i),$$

where i is for model $i = 1, \ldots, R$, j is for predictor variable j, and

$$I_j(g_i) = \begin{cases} 1 \text{ if predictor } x_j \text{ is in model } g_i, \\ 0 \text{ otherwise.} \end{cases}$$

Conditional on model g_i being selected, model selection has the effect of biasing $\hat{\beta}_{j,i}$ away from zero. Thus a new estimator, denoted by $\tilde{\beta}_i$, is suggested:

$$\tilde{\beta}_i = w_+(i) \hat{\bar{\beta}}_i.$$

Investigation of this idea, and extensions of it, is an open research area. The point here is that while $\hat{\bar{\beta}}_j$ can be computed ignoring models other than the ones x_j appears in, $\tilde{\beta}_i$ does require fitting all R of the a priori models.

8.9 MMI: Model Selection Uncertainty

At first, one might think that one could use an information critrion to select an approximating model that was "close" to truth (remembering the bias versus variance tradeoff and the principle of parsimony) or that "lost the least information" and then proceed to use this selected model for inference as if it had been specified a priori as the only model considered. Actually, this approach would not be terrible, since at least one would have a reasonable model, selected objectively, based on a valid theory and a priori considerations. This approach would often be superior to much of current practice. Except in the case where the best model has an Akaike weight > 0.9, the problem with considering only this model, and the usual measures of precision *conditional on this selected model*, is that this tends to overestimate precision. Breiman (1992) calls the failure to acknowledge model selection uncertainty a "quiet scandal." [We might suggest that the widespread use of statistical hypothesis testing and blatant data dredging in model selection represent "loud scandals."] In fact, there is a variance component due to model selection uncertainty that should be incorporated into estimates of precision such that these are unconditional (on the selected model). While this is a research area needing further development, several useful methods are suggested in this book, and others will surely appear in the technical literature in the next few years, including additional Bayesian approaches.

The Akaike (w_i) or bootstrap (π_i) weights that are used to rank and scale models can also be used to estimate unconditional precision where interest is in the parameter θ over R models (model g_i, for $i = 1, \ldots, R$),

$$\widehat{\text{var}}(\hat{\bar{\theta}}_i) = \left[\sum_{i=1}^{R} w_i \sqrt{\widehat{\text{var}}(\hat{\theta}_i \mid g_i) + (\hat{\theta}_i - \hat{\bar{\theta}})^2} \right]^2,$$

$$\widehat{\text{var}}(\hat{\bar{\theta}}_i) = \left[\sum_{i=1}^{R} \pi_i \sqrt{\widehat{\text{var}}(\hat{\theta}_i \mid g_i) + (\hat{\theta}_i - \hat{\bar{\theta}})^2} \right]^2.$$

These estimators, from Buckland et al. (1997), include a term for the conditional sampling variance, given model g_i (denoted by $\widehat{\text{var}}(\hat{\theta}_i \mid g_i)$ here) and incorporate a variance component for model selection uncertainty $(\hat{\theta}_i - \hat{\bar{\theta}})^2$. These estimators of unconditional variance are also appropriate in cases where one wants a model-averaged estimate of the parameter when θ appears in all models.

Chapter 4 gives some procedures for setting confidence intervals that include model selection uncertainty, and it is noted that achieved confidence-interval coverage is then a useful measure of the utility of methods that integrate model selection uncertainty into inference. Only a limited aspect of model uncertainty can be currently handled. *Given* a set of candidate models and an objective selection method, we can assess selection uncertainty. The uncertainty in defining the set of models cannot be addressed; we lack a theory for this issue. In fact, we lack good, general guidelines for defining the a priori set of models. We expect papers to appear on these scientific and philosophical issues in the future.

8.10   MMI: Relative Importance of Predictor Variables

Inference on the importance of a variable is similarly improved by being based on all the models. If one selects the best model and says that the variables in it are the important ones and the other variables are not important, this is a very naive, unreliable inference. We suggest that the relative importance of variable x_j be measured by the sum of the Akaike weights over all models in which that variable appears:

$$w_+(j) = \sum_{i=1}^{R} w_i I_j(g_i).$$

Thus again, proper inference requires fitting all the models and then using a type of model-averaging. A certain balance in the number of models each with model j, must be achieved. When possible, one should use inference based on all the models, via model-averaging and selection bias adjustments, rather than risk making inference based only on the model estimated to be the best and, often, ignoring other models that are also quite good.

8.11   More on Inferences

Information-theoretic methods do not offer a mechanical, unthinking approach to science. While these methods can certainly be misused, they elicit careful thinking as models are developed to represent the multiple scientific hypotheses that must be the focus of the entire study. A central theme of this book is to call attention to the need to ask better scientific questions in the applied sciences (Platt 1964). Rather than test trivial null hypotheses, it is better to ask deeper questions relating to well-defined alternative hypotheses. For this goal to be achieved, a great deal more hard thinking will be required.

There needs to be increased attention to separating those inferences that rest on a priori considerations from those resulting from some degree of data dredging. White (2000:1097) comments, "Data snooping is a dangerous practice to be avoided, but in fact is endemic."

Essentially no justifiable theory exists to estimate precision (or test hypotheses, for those still so inclined) when data dredging has taken place (the theory (mis)used is for a priori analyses, assuming that the model was the only one fit to the data). A major concern here is the finding of effects and relationships that are actually spurious where inferences are made post hoc (see Lindsey 1999b, Anderson et al. 2001b). This glaring fact is either not understood by practitioners and journal editors or is simply ignored. Two types of data dredging include (1) an iterative approach, in which patterns and differences observed after initial analysis are "chased" by repeatedly building new models with these effects included and (2) analysis of "all possible models." Data dredging is a poor approach to making inferences about the sampled population, and both types of data dredging are best reserved for more exploratory investigations and are not the subject of this book.

The information-theoretic paradigm avoids statistical null hypothesis testing concepts and focuses on relationships of variables (via selection) and on the estimation of effect size and measures of its precision. This paradigm is primarily in the context of making inferences from a single selected model or making robust inference from many models (e.g., using model-averaging based on Akaike weights). Data analysis is a process of learning what effects are supported by the data and the degree of complexity of the best models in the set. Often, models other than just the estimated best model contain valuable information. Evidence ratios and confidence sets on models help in making inferences on all, or several of the best, models in the set. Information-theoretic approaches should not be used unthinkingly; a good set of candidate models is essential, and this involves professional judgment and representation of the scientific hypotheses into the model set.

When the analysis of data has been completed under an information-theoretic approach, one should gather and report on the totality of the evidence at hand. The primary evidence might be the selected model and its parameter estimates and appropriate measures of precision (including a variance component for model selection uncertainty.) The ranks of each of the R models and the Akaike weights should be reported and interpreted. Model-averaged parameter estimates are often important, particularly for prediction. Evidence ratios, confidence sets on the K-L best model, and a ranking of the relative importance of predictor variables are often useful evidence. When appropriate, quantities such as adjusted R^2 and $\hat{\theta}^2$ should be reported for, at least, the best model. The results from an analysis of residuals for the selected model might also be important to report and interpret. Every effort should be made to fully and objectively report on all the evidence available. If some evidence arose during post hoc activities, this should be clearly stated in published results. Figure 8.1 provides a simplistic graphical representation of the information-theoretic approach. The point of Figure 8.1 is to reinforce some foundational issues (bottom building blocks) and the practical tools and methods (middle row of blocks) that rest on these foundations. If these are used carefully and objectively, one can hope to provide compelling evidence allowing valid inferences. The weakest link seems often to be the left block on the bottom—thinking deeply about the science problem and the alternative hypotheses!

It seems worth noting that K-L information and MMI can be used in certain types of conflict resolution where data exist that are central to the possible resolution of the conflict (Anderson et al. 1999, 2001c). Details here would take us too far afield; however, as Hoeting et al. (1999) noted (in a Bayesian context), "Model averaging also allows users to incorporate several competing models in the estimation process; thus model averaging may offer a committee of scientists a better estimation method than the traditional approach of trying to get the committee to agree on a best model."

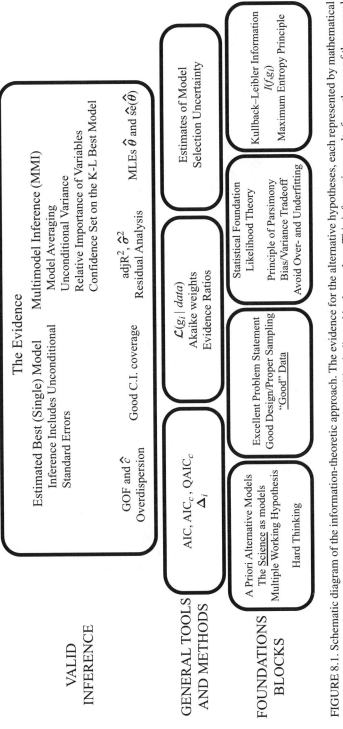

FIGURE 8.1. Schematic diagram of the information-theoretic approach. The evidence for the alternative hypotheses, each represented by mathematical models, and the analysis results are provided by the methods and quantities indicated in the top box. This information results from the use of the general methods in three linked, general tool boxes, which rest on the concepts and deep theory in four basic foundation blocks.

8.12   Final Thoughts

At a conceptual level, reasonable data and a *good* model allow a separation of "information" from "noise." Here, information relates to the structure of relationships, estimates of model parameters, and components of variance. Noise then refers to the residuals; variation left unexplained. We can use the information extracted from the data to make proper inferences.

Summary

We want an approximating model that minimizes information loss $I(f, g)$ and properly separates noise (noninformation, or entropy) from structural information. The philosophy for this separation is the principle of parsimony; the conceptual target for such partitioning is Kullback–Leibler information; and the tactic for selection of a best model is an information criterion (e.g., AIC, AIC_c, $QAIC_c$, or TIC). The notion of data-based model selection and resulting inference is a very difficult subject, but we do know that substantial uncertainty about the selected model can often be expected and should be incorporated into estimates of precision.

Still, model selection (in the sense of parsimony) is the critical issue in data analysis. In using the more advanced methods presented here, model selection can be thought of as a way to compute Akaike weights. Then one uses one or more models in the set as a way to make robust inferences from the data (MMI). More research is needed on the quantification of model uncertainty, measures of the plausibility of alternative models, ways to reduce model selection bias, and ways to provide effective measures of precision (without being conditional on a given model). Confidence intervals with good achieved levels should be a goal of inference following data-based model selection.

Information-theoretic methods are relatively simple to understand and practical to employ across a very wide class of empirical situations and scientific disciplines. The information-theoretic approach unifies parameter estimation and model selection under an optimization framework, based on Kullback–Leibler information and likelihood theory. With the exception of the bootstrap, the methods are easy to compute by hand if necessary (assuming that one has the MLEs, maximized log-likelihood values, and $\widehat{\text{var}}(\hat{\theta}_i \mid g_i)$ for each of the R models). Researchers can easily understand the information-theoretic methods presented here; we believe that it is *very* important that researchers understand the methods they employ.

References

Agresti, A. (1990). *Categorical data analysis*. John Wiley and Sons, New York, NY.

Aitkin, M. (1991). Posterior Bayes factors (with discussion). *Journal of the Royal Statistical Society*, Series B, **53**, 111–143.

Akaike, H. (1973). Information theory as an extension of the maximum likelihood principle. Pages 267–281 *in* B. N. Petrov, and F. Csaki, (eds.) *Second International Symposium on Information Theory*. Akademiai Kiado, Budapest.

Akaike, H. (1974). A new look at the statistical model identification. *IEEE Transactions on Automatic Control AC* **19**, 716–723.

Akaike, H. (1976). Canonical correlation analysis of time series and the use of an information criterion. Pages 27–96 *in* R.K. Mehra, and D.G. Lainiotis (eds.) *System Identification: Advances and Case Studies*. Academic Press, New York, NY.

Akaike, H. (1977). On entropy maximization principle. Pages 27–41 *in* P.R. Krishnaiah (ed.) *Applications of statistics*. North-Holland, Amsterdam, The Netherlands.

Akaike, H. (1978a). A new look at the Bayes procedure. *Biometrika* **65**, 53–59.

Akaike, H. (1978b). A Bayesian analysis of the minimum AIC procedure. *Annals of the Institute of Statistical Mathematics* **30**, 9–14.

Akaike, H. (1978c). On the likelihood of a time series model. *The Statistician* **27**, 217–235.

Akaike, H. (1979). A Bayesian extension of the minimum AIC procedure of autoregressive model fitting. *Biometrika* **66**, 237–242.

Akaike, H. (1980). Likelihood and the Bayes procedure (with discussion). Pages 143–203 *in* J.M. Bernardo, M.H. De Groot, D.V. Lindley, and A.F.M. Smith (eds.) *Bayesian statistics*. University Press, Valencia, Spain.

Akaike, H. (1981a). Likelihood of a model and information criteria. *Journal of Econometrics* **16**, 3–14.

Akaike, H. (1981b). Modern development of statistical methods. Pages 169–184 *in* P. Eykhoff (ed.) *Trends and progress in system identification*. Pergamon Press, Paris.

Akaike, H. (1983a). Statistical inference and measurement of entropy. Pages 165–189 *in* G.E.P. Box, T. Leonard, and C-F. Wu (eds.) *Scientific inference, data analysis, and robustness*. Academic Press, London.

Akaike, H. (1983b). Information measures and model selection. *International Statistical Institute* **44**, 277–291.

Akaike, H. (1983c). On minimum information prior distributions. *Annals of the Institute of Statistical Mathematics* **35A**, 139–149.

Akaike, H. (1985). Prediction and entropy. Pages 1–24 *in* A.C. Atkinson, and S.E. Fienberg (eds.) *A celebration of Statistics*. Springer, New York, NY.

Akaike, H. (1987). Factor analysis and AIC. *Psychometrika* **52**, 317–332.

Akaike, H. (1992). Information theory and an extension of the maximum likelihood principle. Pages 610–624 *in* S. Kotz, and N.L. Johnson (eds.) *Breakthroughs in statistics*, Vol. 1. Springer-Verlag, London.

Akaike, H. (1994). Implications of the informational point of view on the development of statistical science. Pages 27–38 *in* H. Bozdogan (ed.) *Engineering and Scientific Applications*. Vol. 3, Proceedings of the First US/Japan Conference on the Frontiers of Statistical Modeling: An Informational Approach. Kluwer Academic Publishers, Dordrecht, the Netherlands.

Akaike, H., and Nakagawa, T. (1988). *Statistical analysis and control of dynamic systems*. KTK Scientific Publishers, Tokyo. (English translation by H. Akaike and M. A. Momma).

Allen, D.M. (1970). Mean square error of prediction as a criterion for selecting variables. *Technometrics* **13**, 469–475.

Amari, S. (1993). Mathematic methods of neurocomputing. Pages 1–39 *in* O.E. Barndorff-Nielson, J.L. Jensen, and W.S. Kendall (eds.) *Networks and chaos—statistical and probabilistic aspects*. Chapman and Hall, New York, NY.

Anonymous. (1997). *The Kullback Memorial Research Conference*. Statistics Department, The George Washington University, Washington, D.C.

Anderson, D. R. (2001). The need to get the basics right in wildlife field studies. *Wildlife Society Bulletin* **29**, 1294–1297.

Anderson, S., Auquier, A., Hauck, W.W., Oakes, D., Vandaele, W., and Weisberg, H.I. (1980). *Statistical methods for comparative studies*. John Wiley and Sons, New York.

Anderson, D.R., and Burnham, K.P. (1976). *Population ecology of the mallard: VI. the effect of exploitation on survival*. U.S. Fish and Wildlife Service Resource Publication No. 128.

Anderson, D.R., and Burnham, K.P. (1999a). General strategies for the collection and analysis of ringing data. *Bird Study* **46** (suppl.), S261–270.

Anderson, D.R., and Burnham, K.P. (1999b). Understanding information criteria for selection among capture–recapture or ring recovery models. *Bird Study* **46** (suppl.), S514–21.

Anderson, D.R., Burnham, K.P., and White, G.C. (1994). AIC model selection in overdispersed capture–recapture data. *Ecology* **75**, 1780–1793.

Anderson, D.R., Burnham, K.P., and White, G.C. (1998). Comparison of AIC and CAIC for model selection and statistical inference from capture–recapture studies. *Journal of Applied Statistics* **25**, 263–282.

Anderson, D.R., Burnham, K.P., Franklin, A.B., Gutierrez, R.J., Forsman, E.D., Anthony, R.G., White, G.C., and Shenk, T.M. (1999). A protocol for conflict resolution in analyzing empirical data related to natural resource controversies. *Wildlife Society Bulletin* **27**, 1050–1058.

Anderson, D.R., Burnham, K.P., and Thompson, W.L. (2000). Null hypothesis testing: problems, prevalence, and an alternative. *Journal of Wildlife Management* **64**, 912–923.

Anderson, D.R., and K.P. Burnham. 2001a. Commentary on models in ecology. *Bulletin of the Ecological Society of America* **82**, 160–161.

Anderson, D.R., Burnham, K.P., Gould, W.R., and Cherry, S. (2001b). Concerns about finding effects that are actually spurious. *Wildlife Society Bulletin* **29**, 311–316.

Anderson, D.R., Burnham, K.P., and White, G.C. (2001c). Kullback-Leibler information in resolving natural resource conflicts when definitive data exist. *Wildlife Society Bulletin* **29**, 1260–1270.

Anderson, D.R., Link, W.A., Johnson, D.H. and Burnham, K.P. (2001d). Suggestions for presenting results of data analyses. *Journal of Wildlife Management* **65**, 2001d. 373–378.

Apostol, T.M. (1957). *Mathematical analysis: a modern approach to advanced calculus.* Addison-Wesley Publishing Co., Inc. Reading, MA.

Armitage, P. (1957). Studies in the variability of pock counts. *Journal of Hygiene* **55**, 564–581.

Atilgan, T. (1996). Selection of dimension and basis for density estimation and selection of dimension, basis and error distribution for regression. *Communications in Statistics— Theory and Methods* **25**, 1–28.

Atkinson, A.C. (1978). Posterior probabilities for choosing a regression model. *Biometrika* **65**, 39–48.

Atkinson, A. C. (1980). A note on the generalized information criterion for choice of a model. *Biometrika* **67**, 413–18.

Atmar, W. (2001). A profoundly repeated pattern. *Bulletin of the Ecological Society of America* **82**, 208–211.

Augustin, N.H., Mugglestone, M.A., and Buckland S.T. (1996). An autologistic model for the spatial distribution of wildlife. *Journal of Applied Ecology* **33**, 339–347.

Azzalini, A. (1996). *Statistical inference based on the likelihood.* Chapman and Hall, London.

Bancroft, T.A., and Han, C-P. (1977). Inference based on conditional specification: a note and a bibliography. *International Statistical Review* **45**, 117–127.

Barron, A., Rissanen, J., and Yu, B. (1998). The minimum description length principle in coding and modeling. *IEEE Transactions on Information Theory* **44**, 2743–2760.

Bartlett, M.S. (1936). Some notes on insecticide tests in the laboratory and in the field. *Journal of the Royal Statistical Society*, Supplement **1**, 185–194.

Bedrick, E.J., and Tsai, C-L. (1994). Model selection for multivariate regression in small samples. *Biometrics* **50**, 226–231.

Berger, J.O., and Pericchi, L.R. (1996). The intrinsic Bayes factor for model selection and prediction. *American Statistical Association* **91**, 109–122.

Berger, J.O., and Wolpert, R.L. (1984). The likelihood principle. *Institute of Mathematical Statistics* Monograph 6.

Berk, R.H. (1966). The limiting behavior of posterior distributions when the model is incorrect. *Annals of Mathematical Statistics* **37**, 51–58.

Bernardo, J.M., and Smith, A.F.M. (1994). *Bayesian theory*. John Wiley and Sons, Chichester, UK.

Berry, D.A. (1988). Multiple comparisons, multiple tests and data dredging: a Bayesian perspective. *Bayesian Statistics* 3. Clarendon, Oxford, UK.

Berryman, A.A., Gutierrez, A.P., and Arditi, R. (1995). Credible, parsimonious and useful predator–prey models—a reply to Abrams, Gleeson, and Sarnelle. *Ecology* **76**, 1980–1985.

Beyers, D.W. (1998). Causal inference in environmental impact studies. *Journal of the North American Benthological Society* **17**, 367–373.

Bliss, C.I. (1935). The calculation of the dosage-mortality curve. *Annals of Applied Biology* **22**, 134–167.

Bickel, P., and Zhang, P. (1992). Variable selection in nonparametric regression with categorical covariates. *Journal of the American Statistical Association* **87**, 90–97.

Bishop, Y.M.M., Fienberg, S.E., and Holland, P.W. (1975). *Discrete multivariate analysis: theory and practice*. The MIT Press, Cambridge, MA.

Blansali, R.J. (1993). Order selection for linear time series models: a review. Pages 50–66 *in* T. S. Rao (ed.) *Developments in time series analysis*, Chapman and Hall, London.

Blau, G.E., and Neely, W.B. (1975). Mathematical model building with an application to determine the distribution of DURSBAN® insecticide added to a simulated ecosystem. Pages 133–163 *in* A. Macfadyen (ed.) *Advances in Ecological Research*, Academic Press, London.

Bollen, K.A., and Long, J.S. (1993). *Testing structural equations*. Sage Publ., London.

Boltzmann, L. (1877). Über die Beziehung zwischen dem Hauptsatze derzwe:Ten mechanischen Wärmetheorie und der Wahrscheinlichkeitsrechnung respective den Sätzen über das Wärmegleichgewicht. *Wiener Berichte* **76**, 373–435.

Bonneu, M., and Milhaud, X. (1994). A modified Akaike criterion for model choice in generalized linear models. *Statistics* **25**, 225–238.

Box, G.E.P. (1967). Discrimination among mechanistic models. *Technometrics* **9**, 57–71.

Box, G.E.P. (1976). Science and statistics. *Journal of the American Statistical Association* **71**, 791–799.

Box, G.E.P., and Jenkins, G.M. (1970). *Time series analysis: forecasting and control.* Holden-Day, London.

Box, G.E.P., Hunter, W.G., and Hunter, J.S. (1978). *Statistics for experimenters.* John Wiley and Sons, New York, NY, USA.

Box, G.E.P., Leonard, T., and Wu, C-F. (eds.) (1981). *Scientific inference, data analysis, and robustness.* Academic Press, London.

Box, J.F. (1978). *R. A. Fisher: the life of a scientist.* John Wiley and Sons, New York, NY.

Boyce, M.S. (1992). Population viability analysis. *Annual Review of Ecology and Systematics* **23**, 481–506.

Bozdogan, H. (1987). Model selection and Akaike's information criterion (AIC): the general theory and its analytical extensions. *Psychometrika* **52**, 345–370.

Bozdogan, H. (1988). A new model-selection criterion. Pages 599–608 *in* H.H. Bock (ed.) *Classification and related methods of data analysis.* North-Holland Publishing Company, Amsterdam, The Neterlands.

Bozdogan, H. (1994). Editor's general preface. Pages ix–xii *in* H. Bozdogan (ed.) *Engineering and Scientific Applications.* Vol. 3, Proceedings of the First US/Japan Conference on the Frontiers of Statistical Modeling: An Informational Approach. Kluwer Academic Publishers, Dordrecht, the Netherlands.

Breiman L. (1992). The little bootstrap and other methods for dimensionality selection in regression: X-fixed prediction error. *Journal of the American Statistical Association* **87**, 738–754.

Breiman, L. (1995). Better subset regression using the nonnegative garrote. *Technometrics* **37**, 373–384.

Breiman L. (1996). Heuristics of instability and stabilization in model selection. *The Annals of Statistics* **24**, 2350–2383.

Breiman, L. (2001). Statistical modeling: the two cultures (with discussion) *Statistical Science* **16**, 199–231.

Breiman, L., and Freedman, D.F. (1983). How many variables should be entered in a regression equation? *Journal of the American Statistical Association* **78**, 131–136.

Brisbin I.L., Jr., Collins, C.T., White, G.C., and McCallum, D.A. (1987). A new paradigm for the analysis and interpretation of growth data: the shape of things to come. *Auk* **104**, 552–554.

Brockwell, P.J., and Davis, R.A. (1987). *Time series: theory and methods.* Springer-Verlag, New York, NY.

Brockwell, P.J., and Davis, R.A. (1991). *Time series: theory and methods*. 2nd ed. Springer-Verlag, New York, NY.

Broda, E. (1983). *Ludwig Boltzmann: man, physicist, philosopher* (translated with L. Gay). Ox Bow Press, Woodbridge, Connecticut.

Brown, D. (1992). A graphical analysis of deviance. *Applied Statistics* **41**, 55–62.

Brown, D., and Rothery, P. (1993). *Models in biology: mathematics, statistics and computing*. John Wiley and Sons. New York, NY.

Brown, L.D., Cai, T.T., and DasGupta, A. (2001). Interval estimation for a binomial proportion. *Statistical Science* **16**, 101–135.

Brown, P.J. (1993). *Measurement, regression, and calibration*. Clarendon Press, Oxford, UK.

Brownie, C., Anderson, D.R., Burnham, K.P., and Robson, D.S. (1985). *Statistical inference from band recovery data–a handbook*. 2nd ed. U.S. Fish and Wildlife Service Resource Publication 156.

Brownie, C., Hines, J.E., Nichols, J.D., Pollock, K.H., and Hestbeck, J.B. (1993). Capture–recapture studies for multiple strata including non-Markovian transition probabilities. *Biometrics* **49**, 1173–1187.

Brush, S.G. (1965). *Kinetic theory*. Vol. 1 Pergamon Press, Oxford, UK.

Brush, S.G. (1966). *Kinetic theory*. Vol. 2 Pergamon Press, Oxford, UK.

Bryant, P.G., and Cordero-Braña, O.I. (2000). Model selection using the minimum description length principle. *The American Statistician* **54**, 257–268.

Buckland, S.T. (1982). A note on the Fourier series model for analyzing line transect data. *Biometrics* **38**, 469–477.

Buckland, S.T. (1984). Monte Carlo confidence intervals. *Biometrics* **40**, 811–817.

Buckland, S.T., Anderson, D.R., Burnham, K.P., and Laake, J.L. (1993). *Distance sampling: estimating abundance of biological populations*. Chapman and Hall, London.

Buckland, S.T., Anderson, D.R., Burnham, K.P., Laake, J.L., Borchers, D.L., and Thomas, L. (2001). *An introduction to distance sampling*. Oxford University Press, Oxford, UK.

Buckland, S.T., Burnham, K.P., and Augustin, N.H. (1997). Model selection: an integral part of inference. *Biometrics* **53**, 603–618.

Buckland, S.T., and Elston, D.A. (1993). Empirical models for the spatial distribution of wildlife. *Journal of Applied Ecology* **30**, 478–495.

Burman, P. (1989). A comparative study of ordinary cross-validation, v-hold cross-validation and repeated learning-testing methods. *Biometrika* **76**, 503–514.

Burman, P., and Nolan, D. (1995). A general Akaike-type criterion for model selection in robust regression. *Biometrika* **82**, 877–886.

Burnman, K.P. (1988). A comment on maximum likelihood estimation for finite mixtures of distributions. *Biometrical Journal* **30**, 379–384.

Burnham, K.P. (1989). Numerical survival rate estimation for capture–recapture models using SAS PROC NLIN. Pages 416–435 *in* L. McDonald, B. Manly, J. Lockwood and J. Logan (eds.) *Estimation and analysis of insect populations*. Springer-Verlag, New York, NY.

Burnham, K.P. (in review). Basic random effects models in ringing and capture–recapture data. *Ecological and Environmental Statistics*.

Burnham, K.P., and Anderson, D.R. (1992). Data-based selection of an appropriate biological model: the key to modern data analysis. Pages 16–30 *in* D.R. McCullough, and R.H. Barrett (eds.) *Wildlife* 2001: *Populations*. Elsevier Scientific Publications, Ltd., London.

Burnham, K.P., and Anderson, D.R. (2001). Kullback-Leibler information as a basis for strong inference in ecological studies. *Wildlife Research* **28**, 111–119.

Burnham, K.P., Anderson, D.R., White, G.C., Brownie, C., and Pollock, K.H. (1987). *Design and analysis methods for fish survival experiments based on release-recapture*. American Fisheries Society, Monograph **5**.

Burnham, K.P., Anderson, D.R., and White, G.C. (1994). Evaluation of the Kullback–Leibler discrepancy for model selection in open population capture–recapture models. *Biometrical Journal* **36**, 299–315.

Burnham, K.P., White, G.C., and Anderson, D.R. (1995a). Model selection in the analysis of capture–recapture data. *Biometrics* **51**, 888–898.

Burnham, K.P., Anderson, D.R., and White, G.C. (1995b). Selection among open population capture–recapture models when capture probabilities are heterogeneous. *Journal of Applied Statistics* **22**, 611–624.

Burnham, K.P., Anderson, D.R., and White, G.C. (1996). Meta-analysis of vital rates of the Northern Spotted Owl. *Studies in Avian Biology* **17**, 92–101.

Burnham, K.P., and White, G.C. (2002). Evaluation of some random effects methodology applicable to bird ringing data. *Journal of Applied Statistics* **29**, 245–264.

Bystrak, D. (1981). The North American breeding bird survey, Pages 522–532. *in* C.J. Ralph, and J.M. Scott (eds.) *Estimating numbers of terrestrial birds. Studies in Avian Biology* **6**.

Carlin, B., and Louis, T. (2000). *Bayes and empirical Bayes methods for data analysis*. (2nd ed.) Chapman and Hall, London.

Carlin, B.P., and Chib, S. (1995). Bayesian model choice via Markov chain Monte Carlo methods. *Journal of the Royal Statistical Society*, Series B **57**, 473–484.

Carlin, B.P., and Louis, T.A. (1996). *Bayes and empirical Bayes methods for data analysis*. Chapman and Hall, London.

Carpenter, S.R. (1990). Large-scale perturbations: opportunities for innovation. *Ecology* **71**, 2038–2043.

Carrol, R.J., and Ruppert, D. (1988). *Transformation and weighting in regression*. Chapman and Hall, New York, NY.

Carrol, R., Ruppert, D., and Stefanski, L. (1995). *Measurement error in nonlinear models*. Chapman and Hall, London.

Casella, G. (1995). An introduction to empirical Bayes data analysis. *The American Statistician* **39**, 83–87.

Caswell, H. (2001). *Matrix population models: construction, analysis and interpretation.* Sinauer Associates, Inc., Publishers, Sunderland, MA.

Cavanaugh, J.E., and Neath, A.A. (1999). Generalizing the derivation of the Schwarz information criterion. *Communication in Statistics—Theory and Methods* **28**, 49–66.

Cavanaugh, J.E., and Shumway, R.H. (1997). A bootstrap variant of AIC for state-space model selection. *Statistica Sinica* **7**, 473–496.

Chamberlain, T.C. (1890). The method of multiple working hypotheses. *Science* **15**, 93 (Reprinted 1965, Science **148**, 754–759.

Chamberlin, T. C. (1965). (1890) The method of multiple working hypotheses. *Science* **148**, 754-759. (reprint of 1890 paper in *Science*).

Chatfield, C. (1991). Avoiding statistical pitfalls (with discussion). *Statistical Science* **6**, 240–268.

Chatfield, C. (1995a). *Problem solving: a statistician's guide.* Second edition. Chapman and Hall, London.

Chatfield, C. (1995b). Model uncertainty, data mining and statistical inference. *Journal of the Royal Statistical Society*, Series A **158**, 419–466.

Chatfield, C. (1996). Model uncertainty and forecast accuracy. *Journal of Forecasting* **15**, 495–508.

Chen, M.-H., Shao, Q.-M., and Ibrahim, J.G. (2000). *Monte Carlo methods in Bayesian computation.* Springer, New York, NY.

Cherry, S. (1998). Statistical tests in publications of The Wildlife Society. *Wildlife Society Bulletin* **26**, 947–953.

Chib, S., and Jeliazkov, I. (2001). Marginal likelihood from the Metropolis-Hasting output. *Journal of the American Statistical Association* **96**, 270–281.

Chow, G.C. (1981). A comparison of the information and posterior probability criteria for model selection. *Journal of Econometrics* **16**, 21–33.

Chung, H-Y., Lee, K-W., and Koo, J-A. (1996). A note on bootstrap model selection criterion. *Statistics and Probability Letters* **26**, 35–41.

Clayton, D., and Hills, M. (1993). *Statistical models in epidemiology.* Oxford University Press, Oxford, UK.

Clayton, M.K., Geisser, S., and Jennings, D. (1986). A comparison of several model selection procedures. Pages 425–439 *in* P. Goel, and A. Zellner (eds.) *Bayesian inference and decision.* Elsevier, New York, NY.

Cochran, W.G. (1963). *Sampling techniques.* 2nd ed., John Wiley and Sons, Inc. New York, NY.

Cohen, E.G.D., and Thirring, W. (eds.) (1973). *The Boltzmann equation: theory and applications.* Springer-Verlag, New York, NY.

Collett, D. (1994). *Modeling survival data in medical research*. Chapman and Hall, London, UK.

Collopy, F., Adya, M., and Armstrong, J.S. (1994). Principles for examining predictive validity: the case of information systems spending forecasts. *Information Systems Research* **5**, 170–179.

Conner, M.M., McCarty, C.W., and Miller, M.W. (2000). Detection of bias in harvest-based estimates of chronic wasting disease prevalence in mule deer. *Journal of Wildlife Diseases* **36**, 691–699.

Conner, M.M., White, G.C., and Freddy, D.J. (2001). Elk movement in response to early-season hunting in northwest Colorado. *Journal of Wildlife Management* **65**, 926–940.

Cook, T.D., and Campbell, D.T. (1979). *Quasi-experimentation: design and analysis issues for field settings*. Houghton Mifflin Company, Boston, MA.

Cook, R., Cook, J.G., Murray, D.L., Zager, P., Johnson, B.K., and Gratson, M.W. (2001). Development of predictive models of nutritional condition for Rocky Mountain elk. *Journal of Wildlife Management* **65**, 973–987.

Copas, J.B. (1983). Regression, prediction and shrinkage (with discussion). *Journal of the Royal Statistical Society*, Series B, **45**, 311–354.

Cover, T.M., and Thomas, J.A. (1991). *Elements of information theory*. John Wiley and Sons, New York, NY.

Cox, D.R. (1990). Role of models in statistical analysis. *Statistical Science* **5**, 169–174.

Cox, D.R. (1995). The relation between theory and application in statistics (with discussion). *Test* **4**, 207–261.

Cox, D.R., and Reid, N. (2000). *The theory of the design of experiments*. Chapman and Hall/CRC, Boca Raton, FL.

Cox, D.R., and Snell, E.J. (1989). *Analysis of binary data*. 2nd ed., Chapman and Hall, New York, NY.

Craven, P., and Wahba, G. (1979). Smoothing noisy data with spline functions: estimating the correct degree of smoothing by the method of generalized cross-validation. *Numerical Mathematics* **31**, 377–403.

Cressie, N.A.C. (1991). *Statistics for spatial data*. John Wiley and Sons, New York, NY.

Cutler, A., and Windham, M.P. (1994). Information-based validity functionals for mixture analysis. Pages 149–170 *in* H. Bozdogan (ed.) *Engineering and Scientific Applications*. vol. 2, Proceedings of the First US/Japan Conference on the Frontiers of Statistical Modeling: An Informational Approach. Kluwer Academic Publishers, Dordrecht, the Netherlands.

Daniel, C., and Wood, F.S. (1971). *Fitting equations to data*. Wiley-Interscience, New York, NY.

de Gooijer, J.G., Abraham, B., Gould, A., and Robinson, L. (1985). Methods for determining the order of an autoregressive-moving average process: a survey. *International Statistical Review* **53**, 301–329.

deLeeuw, J. (1988). Model selection in multinomial experiments. Pages 118–138 *in* T. K. Dijkstra (ed.) *On model uncertainty and its statistical implications*. Lecture Notes in Economics and Mathematical Systems, Springer-Verlag, New York, NY.

deLeeuw, J. (1992). Introduction to Akaike (1973) information theory and an extension of the maximum likelihood principle. Pages 599–609 *in* S. Kotz, and N.L. Johnson (eds.) *Breakthroughs in statistics*. Vol. 1. Springer-Verlag, London.

Dempster, A.P. (1971). Model searching and estimation in the logic of inference. *Symposium on the Foundations of Statistical Inference*. University of Waterloo, Waterloo, Ontario, Canada.

Dempster, A.P. (1997). The direct use of likelihood for significance testings. *Statistics and Computing* **7**, 247–252.

Desu, M.M., and Roghavarao, D. (1991). *Sample size methodology*. Academic Press, Inc., New York, NY.

Dijkstra, T.K. (ed). (1988). *On model uncertainty and its statistical implications*. Lecture Notes in Economics and Mathematical Systems, Springer-Verlag, New York, NY.

Dijkstra, T.K., and Veldkamp, J.H. (1988). Data-driven selection of regressors and the bootstrap. Pages 17–38 *in* T.K. Dijkstra (ed.) *On model uncertainty and its statistical implications*. Lecture Notes in Economics and Mathematical Systems, Springer-Verlag, New York, NY.

Draper, D. (1995). Assessment and propagation of model uncertainty (with discussion). *Journal of the Royal Statistical Society*, Series B **57**, 45–97.

Draper, N.R., and Smith, H. (1981). *Applied regression analysis*. Second edition. John Wiley and Sons, New York, NY.

Eberhardt, L.L. (1978). Appraising variability in population studies. *Journal of Wildlife Management* **42**, 207–238.

Eberhardt, L.L., and Thomas, J.M. (1991). Designing environmental field studies. *Ecological Monographs* **61**, 53–73.

Edwards, A.W.F. (1976). *Likelihood: an account of the statistical concept of likelihood and its application to scientific inference*. Cambridge University Press, Cambridge, UK.

Edwards, A.W.F. (1992). *Likelihood: expanded edition*. The Johns Hopkins University Press, Baltimore, MD.

Edwards, A.W.F. (2001). Occam's bonus. *In* A. Zellner, H.A. Keuzenkamp, and M. McAleer (eds.), *Simplicity, inference and modelling*, pgs. 128–132. Cambridge University Press, Cambridge, UK.

Edwards, D. (1998). Issues and themes for natural resources trend and change detection. *Ecological Applications* **8**, 323–325.

Efron, B. (1979) Bootstrap methods: another look at the jackknife. *American Statistician* **7**, 1–26.

Efron, B. (1983). Estimating the error rate of a prediction rule: improvements on cross-validation. *Journal of the American Statistical Association* **78**, 316–331.

Efron, B. (1984). Comparing non-nested linear models. *Journal of the American Statistical Association* **79**, 791–803.

Efron, B. (1986). How biased is the apparent error rate of a prediction rule? *Journal of the American Statistical Association* **81**, 461–470.

Efron, B., and Morris, C. (1975). Data analysis using Stein's estimator and its generalizations. *Journal of the American Statistical Association* **70**, 311–319.

Efron, R., and Gong, G. (1983). A leisurely look at the bootstrap, the jackknife, and cross-validation. *The American Statistician* **37**, 36–48.

Efron, B., and Tibshirani, R.J. (1993). *An introduction to the bootstrap*. Chapman and Hall, London.

Ellison, A.M. (1996). An introduction of Bayesian inference for ecological research and environmental decision-making. *Ecological Applications* **6**, 1036–1046.

Feder, M., Merhav, N., and Gutman, M. (1992). Universal prediction of individual sequences. *IEEE Transactions on Information Theory* **38**, 1258–1270.

Fienberg, S.E. (1970). The analysis of multidimensional contingency tables. *Ecology* **51**, 419–433.

Fildes, R., and Makridakis, S. (1995). The impact of empirical accuracy studies on time series analysis and forecasting. *International Statistics Review* **63**, 289–308.

Findley, D.F. (1985). On the unbiasedness property of AIC for exact or approximating linear stochastic time series models. *Journal of Time Series Analysis* **6,** 229–252.

Findley, D.F. (1991). Counterexamples to parsimony and BIC. *Annals of the Institute of Statistical Mathematics* **43,** 505–514.

Findley, D.F., and Parzen, E. (1995). A conversation with Hirotugu Akaike. *Statistical Science* **10**, 104–117.

Finney, D.J. (1971). *Probit analysis*. 3rd. ed. Cambridge University Press, London.

Fisher, R.A. (1922). On the mathematical foundations of theoretical statistics. Royal Society of London. *Philosophical Transactions* (Series A) **222**, 309–368.

Fisher, R.A. (1936). Uncertain inference. *Proceedings of the American Academy of Arts and Sciences* **71,** 245–58.

Fisher, R.A. (1949). A biological assay of tuberculins. *Biometrics* **5,** 300–316.

Flack, V.F., and Chang, P.C. (1987). Frequency of selecting noise variables in subset regression analysis: a simulation study. *The American Statistician* **41**, 84–86.

Flather, C.H. (1992). Patterns of avian species-accumulation rates among eastern forested landscapes. Ph.D. dissertation. Colorado State University. Fort Collins, CO.

Flather, C.H. (1996). Fitting species-accumulation functions and assessing regional land use impacts on avian diversity. *Journal of Biogeography* **23**, 155–168.

Ford, E.D. (2000). *Scientific method for ecological research*. Cambridge University Press, Cambridge, UK.

Forster, M.R. (1995). Bayes or bust: the problem of simplicity for a probabilistic approach to confirmation. *British Journal of the Philosophy of Science* **46**, 1–35.

Forster, M.R. (2000). Key concepts in model selection: performance and generalizability. *Journal of Mathematical Psychology* **44**, 205–231.

Forster, M.R. (2001). The new science of simplicity. *In* A. Zellner, H.A. Keuzenkamp, and M. McAleer (eds.), *Simplicity, inference and modelling*, pgs. 83–119. Cambridge University Press, Cambridge, UK.

Forster, M.R., and Sober, E. (1994). How to tell when simpler, more unified, or less *ad hoc* theories will provide more accurate predictions. *British Journal of the Philosophy of Science* **45**, 399–424.

Franklin, A.B., Anderson, D.R., and Burnham, K.P. (2002). Estimation of long-term trends and variation in avian survival probabilities using random effects models. *Journal of Applied Statistics* **29**, 267–287.

Franklin, A.B., Shenk, T.M., Anderson, D.R., and Burnham, K.P. (2001). Statistical model selection: the alternative to null hypothesis testing. Pages 75–90 *in* T.M. Shenk and A.B. Franklin (eds.) *Modeling in Natural Resource Management*. Island Press, Washington, D. C.

Freedman, D.A. (1983). A note on screening regression equations. *The American Statistician* **37**, 152–155.

Freedman, D. (1999). From association to causation: some remarks on the history of statistics. *Statistical Science* **14**, 243–258.

Freedman, D.A., Navidi, W., and Peters, S.C. (1988). On the impact of variable selection in fitting regression equations. Pages 1–16 *in* T.K. Dijkstra (ed.) *On model uncertainty and its statistical implications*. Lecture Notes in Economics and Mathematical Systems, Springer-Verlag, New York, NY.

Fujikoshi, Y, and Satoh, K. (1997). Modified AIC and C_p in multivariate linear regression. *Biometrika* **84**, 707–716.

Gail, M.H. (1996). Statistics in action. *Journal of the American Statistical Association* **91**, 1–13.

Gallant, A.R. (1987). *Nonlinear statistical models*. John Wiley and Sons, New York, NY.

Gammerman, D. (1997). *Markov Chain Monte Carlo*. Chapman and Hall, London.

Garthwaite, P.H., Jolliffe, I.T., and Jones, B. (1995). *Statistical inference*. Prentice Hall, London.

Gause, G.F. (1934). *The struggle for existence*. Williams and Wilkins, Baltimore, MD.

Geisser, S. (1975). The predictive sample reuse method with applications. *Journal of the American Statistical Association* **70**, 320–328.

Gelfand, A.E., and Smith, A.F.M. (1990). Sampling-based approaches to calculating marginal densities. *Journal of the American Statistical Association* **85**, 398–409.

Gelfand, A., and Dey, D.K. (1994). Bayesian model choice: asymptotics and exact calculations. *Journal of the Royal Statistical Society*, Series B, **56**, 501–514.

Gelfand, A., and Ghosh, S. (1998). Model choice: a minimum posterior predictive loss approach. *Biometrika* **85**, 1–11.

Gelman, A., Carlin, J.B., Stern, H.S. and Rubin, D.B. (1995). *Bayesian Data Analysis*. Chapman and Hall, London.

George, E.I., and Foster, D.P. (2000). Calibration and empirical Bayes variable selection. *Biometrika* **87**, 731–748.

George, E.I., and McCulloch, R.E. (1993). Variable selection via Gibbs sampling. *Journal of the American Statistical Association* **88**, 881–889.

Gerard, P.D., Smith, D.R., and Weerakkody, G. (1998). Limits of retrospective power analysis. *Journal of Wildlife Management* **62**, 801–807.

Gilchrist, W. (1984). *Statistical modelling*. Chichester, Wiley and Sons, New York, NY.

Gilks, W. R., Richardson, S., and Spiegelhalter, D.J. (1996). *Markov chain Monte Carlo in practice*. Chapman and Hall, London.

Glymour, C. (1998). Causation. Pages 97–109 *in* S. Kotz (ed.), *Encyclopedia of statistical sceinces*. John Wiley and Sons, New York, NY.

Gochfeld, M. (1987). On paradigm vs. methods in the study of growth. *Auk* **104**, 554–555.

Gokhale, D.V., and Kullback, S. (1978). *The information in contingency tables*. Marcel Dekker, New York, NY.

Golub, G.H., Health, M., and Wahba, G. (1979). Generalized cross-validation as a method for choosing a good ridge parameter. *Technometrics* **21**, 215–223.

Goodman, S.N. (1993). *p* values, hypothesis tests, and likelihood: implications for epidemiology of a neglected historical debate (with discussion). *American Journal of Epidemiology* **137**, 485–501.

Goodman, S.N., and Berlin, J.A. (1994). The use of predicted confidence intervals when planning experiments and the misuse of power when interpreting results. *Annals of Internal Medicine* **121**, 200–206.

Goutis, C., and Casella, G. (1995). Frequentist post-data inference. *International Statistical Review* **63**, 325–344.

Granger, C.W.J., King, M.L., and White, H. (1995). Comments on testing economic theories and the use of model selection criteria. *Journal of Econometrics* **67**, 173–187.

Graybill, F.A., and Iyer, H.K. (1994). *Regression analysis: concepts and applications*. Duxbury Press, Belmont, CA.

Greenhouse, S.W. (1994). Solomon Kullback: 1907–1994. *Institute of Mathematical Statistics Bulletin* **23**, 640–642.

Guiasu, S. (1977). *Information theory with applications*. McGraw-Hill, New York, NY.

Guisan, A., and Zimmermann, N.E. (2000). Predictive habitat distribution models in ecology. *Ecological Modelling* **135**, 147–186.

Ha, I.D., Lee, Y., and Song, J-K. (2001). Hierarchical likelihood approach for frailty models. *Biometrika* **88**, 233–243.

Hairston, N.G. (1989). *Ecological experiments: purpose, design and execution*. Cambridge University Press, Cambridge, UK.

Hald, A. (1952). *Statistical theory with engineering applications*. John Wiley and Sons, New York, NY.

Hald, A. (1998). *A history of mathematical statistics*. John Wiley, New York, NY.

Hand, D.J. (1994). Statistical strategy: step 1. Pages 1–9 *in* P. Cheeseman, and R.W. Oldford (eds.) *Selecting models from data*. Springer-Verlag, New York, NY.

Hand, D.J. (1998). Data mining: statistics and more? *The American Statistician* **52**, 112–118.

Hand, D.J., Blunt, G., Kelly, M.G., and Adams, N.M. (2000). Data mining for fun and profit. *Statistical Science* **15**, 111–131.

Hannan, E.J., and Quinn, B.G. (1979). The determination of the order of an autoregression. *Journal of the Royal Statistical Society*, Series B **41**, 190–195.

Hansen, M.H., and Yu, B. (2001). Model selection and the principle of minimum description length. *Journal of the American Statistical Association* **96**, 746–774.

Harlow, L.L., Mulaik, S.A., and Steiger, J.H. (eds.) (1997). *What if there were no significancs tests?* Lawrence Erlbaum Associates, Publishers, Mahwah, NJ.

Hasenöhrl, F. (ed.) (1909). *Wissenschaftliche Abhandlungen*. 3 Vols. Leipzig.

Hastie, T.J., and Tibshirani, R.J. (1990). *Generalized additive models*. Chapman and Hall, London.

Haughton, D. (1989). Size of the error in the choice of a model to fit data from an exponential family. *Sankhya*, Series A **51**, 45–58.

Hayne, D. (1978). Experimental designs and statistical analyses. Pages 3–13 *in* D.P. Synder (ed). *Populations of small mammals under natural conditions*. Pymatuning Symposium in Ecology, University of Pittsburgh, Vol 5.

Henderson, H., and Velleman, P. (1981). Building multiple regression models interactively. *Biometrics* **37**, 391–411.

Heyde, C.C. (1997). *Quasi-likelihood and its application: a general approach to optimal parameter estimation*. Springer-Verlag, New York, NY.

Hilborn, R., and Mangel, M. (1997). *The ecological detective: confronting models with data*. Princeton University Press, Princeton, NJ.

Hjorth, J.S.U. (1994). *Computer intensive statistical methods: validation, model selection and bootstrap*. Chapman and Hall, London.

Hobson, A., and Cheng, B-K. (1973). A comparison of the Shannon and Kullback information measures. *Journal of Statistical Physics* **7**, 301–310.

Hocking, R.R. (1976). The analysis and selection of variables in linear regression. *Biometrics* **32**, 1–49.

Hoenig, J.M., and Heisey, D.M. (2001). The abuse of power: the pervasive fallacy of power calculations for data analysis. *The American Statistician* **55**, 19–24.

Hoeting, J.A., and Ibrahim, J.G. (1996). Bayesian predictive simultaneous variable and transformation selection in the linear model. Department of Statistics, Colorado State University, Fort Collins. Technical Report No. 96/39.

Hoeting, J.A., Madigan, D., Raftery, A.E., and Volinsky, C.T. (1999). Bayesian model averaging: a tutorial (with discussion). *Statistical Science* **14**, 382–417.

Hosmer, D.W., and Lemeshow, S. (1989). *Applied logistic regression analysis.* John Wiley and Sons, New York, NY.

Howard, R.A. (1971). *Dynamic probabilistic systems.* John Wiley and Sons. New York, NY.

Hurvich, C.M., Simonoff, J.S., and Tsai, C-L. (1998). Smoothing parameter selection in nonparametric regression using an improved Akaike information criterion. *Journal of the Royal Statistical Society*, Series B, **60**, 271–293.

Hurvich, C.M., and Tsai, C-L. (1989). Regression and time series model selection in small samples. *Biometrika* **76**, 297–307.

Hurvich, C.M., and Tsai, C-L. (1990a). Model selection for least absolute deviations regression in small samples. *Statistics and Probability Letters* **9**, 259–265.

Hurvich, C.M., and Tsai, C-L. (1990b). The impact of model selection on inference in linear regression. *The American Statistician* **44**, 214–217.

Hurvich, C.M., and Tsai, C-L. (1991). Bias of the corrected AIC criterion for underfitted regression and time series models. *Biometrika* **78**, 499–509.

Hurvich, C.M., and Tsai, C-L. (1994). Autoregressive model selection in small samples using a bias-corrected version of AIC. Pages 137–157 *in* H. Bozdogan (ed.) *Engineering and Scientific Applications.* Vol. 1, Proceedings of the First US/Japan Conference on the Frontiers of Statistical Modeling: An Informational Approach. Kluwer Academic Publishers, Dordrecht, the Netherlands.

Hurvich, C.M., and Tsai, C-L. (1995a). Relative rates of convergence for efficient model selection criteria in linear regression. *Biometrika* **82**, 418–425.

Hurvich, C.M., and Tsai, C-L. (1995b). Model selection for extended quasi-likelihood models in small samples. *Biometrics* **51**, 1077–1084.

Hurvich, C.M., and Tsai, C-L. (1996). The impact of unsuspected serial correlations on model selection in linear regression. *Statistics and Probability Letters* **27**, 115–126.

Hurvich, C.M., Shumway, R., and Tsai, C-L. (1990). Improved estimators of Kullback–Leibler information for autoregressive model selection in small samples. *Biometrika* **77**, 709–719.

Ibrahim, J.G., and Chen, M-H. (1997). Predictive variable selection for the multivariate linear model. *Biometrics* **53**, 465–478.

Inman, H.F. (1994). Karl Pearson and R. A. Fisher on statistical tests: a 1935 exchange from *Nature. The American Statistician* **48**, 2–11.

Irizarry, R.A. (2001). Information and posterior probability criteria for model selection in local likelihood estimation. *Journal of the American Statistical Association* **96**, 303–315.

Ishiguro, M., Sakamoto, Y., and Kitagawa, G. (1997). Bootstrapping log likelihood and EIC, an extension of AIC. *Annals of the Institute of Statistical Mathematics* **49**:411–434.

Jaffe, A.J., and Spirer, H.F. (1987). *Misused statistics: straight talk for twisted numbers*. Marcel Dekker, Inc., New York, NY.

James, F.C., and McCulloch, C.E. (1990). Multivariate analysis in ecology and systematics: panacea or Pandora's box? *Annual Reviews of Ecology and Systematics* **21**, 129–166.

Jaynes, E.T. (1957). Information theory and statistical mechanics. *Physics Review* **106**, 620–630.

Jaynes, E.T. (1982). On the rationale of maximum-entropy methods. *Proceedings of the IEEE* **70**, 939–952.

Jaynes, E.T. (in prep.). *Probability theory: the logic of science*. Cambridge University Press, Cambridge, UK.

Jeffreys, H. (1948). *Theory of probability*. Oxford University Press, Oxford, UK.

Jeffreys, H. (1973). *Scientific inference*. 3rd ed. Cambridge University Press, Cambridge, UK.

Jessop, A. (1995). *Informed assessments: an introduction to information, entropy and statistics*. Ellis Horwood, London.

Jevons, W. S. (1874, 1877). The principles of science. MacMillan, London, United Kingdom.

Jiménez, J.A., Hughes, K.A., Alaks, G., Graham, L., and Lacy, R.C. (1994). An experimental study of inbreeding depression in a natural habitat. *Science* **266**, 271–273.

Johnson, D.H. (1995). Statistical sirens: the allure of nonparametrics. *Ecology* **76**, 1998–2000.

Johnson, D.H. (1999). The insignificance of statistical significance testing. *Journal of Wildlife Management* **63**, 763–772.

Johnson, J.W. (1996). Fitting percentage of body fat to simple body measurements. *Journal of Statistics Education* **4** (e-journal).

Johnson, N.L., and Kotz, S. (1970). *Continuous univariate distributions*. Houghton Mifflin Company, New York, NY.

Johnson, N.L., and Kotz, S. (1992). *Univariate discrete distributions*. (2nd ed.) Wiley-Interscience Publication, New York, NY.

Jones, D., and Matloff, N. (1986). Statistical hypothesis testing in biology: a contradiction in terms. *Journal of Economic Entomology* **79**, 1156–1160.

Judge, G.C., and Yancey, T. (1986). *Improved methods of inference in econometrics*. North Holland, Amsterdam, the Netherlands.

Kabaila, P. (1995). The effect of model selection on confidence regions and prediction regions. *Econometric Theory* **11**, 537–549.

Kapur, J.N., and Kesavan, H.K. (1992). *Entropy optimization principles with applications*. Academic Press, London.

Kareiva, P. (1994). Special feature: higher order interactions as a foil to reductionist ecology. *Ecology* **75**, 1527–1559.

Kass, R.E., and Raftery, A.E. (1995). Bayes factors. *Journal of the American Statistical Association* **90**, 773–795.

Kass, R.E., and Raftery, A.E. (1995). "Bayes factors." *Journal of the American Statistical Association* **90**, 773–795.

Kass, R.E., and Wasserman, L. (1995). A reference Bayesian test for nested hypotheses and its relationship to the Schwarz criterion. *Journal of the American Statistical Association* **90**, 928–934.

Kishino, H., Kato, H., Kasamatsu, F., and Fujise, Y. (1991). Detection of heterogeneity and estimation of population characteristics from field survey data: 1987/88 Japanese feasibility study of the Southern Hemisphere minke whales. *Annals of the Institute of Statistical Mathematics* **43**, 435–453.

Kiso, K., Akamine, T., Ohnishi, S., and Matsumiya, Y. (1992). Mathematical examinations of the growth of sea-run and fluviatile forms of the female masu salmon *Oncorhynchus masou* in rivers of the southern Sanriku district, Honshu, Japan. *Nippon Suisan Gakkaishi* **58**, 1779–1784.

Kittrell, J.R. (1970). Mathematical modelling of chemical reactors. *Advances in Chemical Engineering* **8**, 97–183.

Knopf, F.L., Sedgwick, J.A., and Cannon, R.W. (1988). Guild structure of a riparian avifauna relative to seasonal cattle grazing. *Journal of Wildlife Management* **52**, 280–290.

Konishi, S., and Kitagawa, G. (1996). Generalized information criteria in model selection. *Biometrika* **83**, 875–890.

Kooperberg, C., Bose, S., and Stone, C.J. (1997). Polychotomous regression. *Journal of the American Statistical Association* **92**, 117–127.

Kreft, I., and deLeeuw, J. (1998). *Introducing multilevel modeling.* Sage Publications, Thousand Oaks, CA, USA.

Kuhn, T.S. (1970). *The structure of scientific revolutions.* 2nd ed. University of Chicago Press, Chicago, IL

Kullback, S. (1959). *Information theory and statistics.* John Wiley and Sons, New York, NY.

Kullback, S. (1987). The Kullback-Leibler distance. *The American Statistician* **41**, 340–341.

Kullback, S., and Leibler, R.A. (1951). On information and sufficiency. *Annals of Mathematical Statistics* **22**, 79–86.

Laake, J.L., Buckland, S.T., Anderson, D.R., and Burnham, K.P. (1994). *DISTANCE user's guide.* Version 2.1. Colorado Cooperative Fish and Wildlife Research Unit, Colorado State University, Fort Collins, CO.

Lahiri, P. (ed.) (2001). Model selection. Institute of Mathematical Statistics, Lecture Notes, No. 38.

Larimore, W.E. (1983). Predictive inference, sufficiency, entropy and an asymptotic likelihood principle. *Biometrika* **70**, 175–181.

Larimore, W.E., and Mehra, R.K. (1985). The problem of overfitting data. *Byte* **10**, 167–180.

Laud, P.W., and Ibrahim, J.G. (1995). Predictive model selection. *Journal of the Royal Statistical Society*, Series B **57**, 247–262.

Laud, P.W., and Ibrahim, J.G. (1996). Predictive specification of prior model probabilities in variable selection. *Biometrika* **83**, 267–274.

Leamer, E.E. (1978). *Specification searches: ad hoc inference with nonexperimental data.* John Wiley and Sons, New York, NY.

Lebreton, J-D., Burnham, K.P., Clobert, J., and Anderson, D.R. (1992). Modeling survival and testing biological hypotheses using marked animals: a unified approach with case studies. *Ecological Monograph* **62**, 67–118.

Lee, Y., and Nelder, J.A. (1996). Hierarchical generalized linear models. *Journal of the Royal Statistical Society*, Series B **58**, 619–678.

Lehmann, E.L. (1983). *Theory of point estimation.* John Wiley and Sons, New York, NY.

Lehmann, E.L. (1990). Model specification: the views of Fisher and Neyman, and later developments. *Statistical Science* **5**, 160–168.

Leirs, H., Stenseth, N.C., Nichols, J.D., Hines, J.E., Verhagen, R., and Verheyen, W. (1997). Stochastic seasonality and nonlinear density-dependent factors regulate population size in an African rodent. *Nature* **389**, 176–180.

Leonard, T., and Hsu, J.S.J. (1999). *Bayesian methods: an analysis for statisticians and interdisciplinary researchers.* Cambridge University Press, Cambridge, UK.

Leroux, B.G. (1992). Consistent estimation of a mixing distribution. *The Annals of Statistics* **20**, 1350–1360.

Lewis, P. (1998). Maximum likelihood as an alternative to parsimony for inferring phylogeny using nucleotide sequence data. Pages 132–163 *in* D. Soltis, P. Soltis, and J. Doyle (eds.) *Molecular systematics of plants II.* Kluwer Publishing, Boston, MA, USA.

Liang, K-Y, and McCullagh, P. (1993). Case studies in binary dispersion. *Biometrics* **49**, 623–630.

Lindley, D.V. (1986). The relationship between the number of factors and size of an experiment. Pages 459–470 *in* P.K. Goel, and A. Zellner (eds.) *Bayesian inference and decision techniques.* Elsevier Science Publishers, New York, NY.

Lindsey, J.K. (1995). *Modeling frequency and count data.* Oxford University Press, Oxford, UK.

Lindsey, J.K. (1996). *Parametric statistical inference.* Oxford Science Publishing Company, Oxford, UK.

Lindsey, J.K. (1999a). On the use of corrections for overdispersion. *Applied Statistics* **48**, 553–561.

Lindsey, J.K. (1999b). Some statistical heresies. *The Statistician* **48**, 1–40.

Lindsey, J.K. (1999c). *Revealing statistical principles*. Oxford University Press, New York, NY.

Lindsey, J.K., and Jones, B. (1998). Choosing among generalized linear models applied to medical data. *Statistics in Medicine* **17**, 59–68.

Linhart, H. (1988). A test whether two AIC's differ significantly. *South African Statistical Journal* **22**, 153–161.

Linhart, H., and Zucchini, W. (1986). *Model selection*. John Wiley and Sons, New York, NY.

Longford, N.T. (1993). *Random coefficient models*. Oxford University Press, Inc., New York, NY.

Longford, N.T. and Nelder, J.A. (1999). Statistics versus statistical science in the regulatory process. *Statistics in Medicine* **18**, 2311–2320.

Lucky, R.W. (1991). *Silicon dreams: information, man, and machine*. St. Martin's Press, New York, NY.

Ludwig, D. (1989). Small models are beautiful: efficient estimators are even more beautiful. Pages 274–284 *in* C. Castillo-Chavez, S.A. Levin, and C.A. Shoemaker (eds.) *Mathematical approaches to problems in resource management and epidemiology*. Springer-Verlag, London.

Lunneborg, C.E. (1994). *Modeling experimental and observational data*. Duxbury Press, Belmont, CA.

Lytle, D.A. (2002). Flash floods and aquatic insect life-history evolution: evaluation of multiple models. *Ecology* **83**, 370–385.

Madigan, D., and Raftery, A.E. (1994). Model selection and accounting for model uncertainty in graphical models using Occam's window. *Journal of the American Statistical Association* **89**, 1535–1546.

Madigan, D.M., Raftery, A.E., York, J.C., Bradshaw, J.M., and Almond, R.G. (1994). Strategies for graphical model selection. Pages 91–100 *in* P. Chesseman, and R. W. Oldford (eds.) *Selecting models from data: AI and statistics IV*. Springer-Verlag, Lecture Notes in Statistics **89**.

Mallows, C.L. (1973). Some comments on C_p. *Technometrics* **12**, 591–612.

Mallows, C.L. (1995). More comments on C_p. *Technometrics* **37**, 362–372.

Manly, B.F.J. (1991). *Randomization and Monte Carlo methods in biology*. Chapman and Hall, New York, NY.

Manly, B.F.J., McDonald, L.L., and Thomas, D.L. (1993). *Resource selection by animals: statistical design and analysis for field studies*. Chapman and Hall, New York, NY.

Manly, B.F.J. (1992). *The design and analysis of research studies*. Cambridge University Press, Cambridge, UK.

Manly, B.F.J. (2001). *Statistics for environmental science and management*. Chapman and Hall, London.

Marshall, J.R. (1990). Data dredging and noteworthiness. *Epidemiology* **1**, 5–7.

Matis, J.H., and Kiffe, T.R. (2000). *Stochastic population models: a compartmental perspective*. Springer, New York, NY.

MATLAB® (1994) *High-performance numerical computations and visualization software*. The MathWorks, Inc., Natick, MA.

Maurer, B.A. (1998). Ecological science and statistical paradigms: at the threshold. *Science* **279**, 502–503.

Mayr, E. (1997). *This is biology: the science of the living world*. The Belknap Press of Harvard University Press. Cambridge, MA.

McBride, G.B., Loftis, J.C., and Adkins, N.C. (1993). What do significance tests really tell us about the environment? *Environmental Management* **17**, 423–432.

McCullagh, P., and Nelder, J.A. (1989). *Generalized linear models*. 2nd. ed. Chapman and Hall, New York, NY.

McCullagh, P., and Pregibon, D. (1985). Discussion comments on the paper by Diaconis and Efron. *Annals of Statistics* **13**, 898–900.

McLachlan, G.J., and Peel, D. (2000). *Finite Mixture Models*. John Wiley and Sons, Inc., New York, NY.

McQuarrie, A.D. (1999). A small-sample correction of the Schwarz SIC model selection criterion. *Statistics and Probability Letters* **44**, 79–86.

McQuarrie, A.D.R., and Tsai, C-L. (1998). *Regression and time series model selection*. World Scientific Publishing Company, Singapore.

Mead, R. (1988). *The design of experiments: statistical principles for practical applications*. Cambridge University Press, New York, NY.

Miller, A.J. (1990). *Subset selection in regression*. Chapman and Hall, London.

Mooney, C.Z., and Duval, R.D. (1993). *Bootstrapping: a nonparametric approach to statistical inference*. Sage Publications, London.

Moore, D.F. (1987). Modelling the extraneous variance in the presence of extra-binomial variation. *Journal of the Royal Statistical Society* **36**, 8–14.

Morgan, B.J.T. (1992). *Analysis of quantal response data*. Chapman and Hall, London.

Morgan, B.J.T. (2000). *Applied stochastic modelling*. Arnold Press, London, UK.

Morris, C.N. (1983). Parametric empirical Bayes inference: theory and applications. *Journal of the American Statistical Association* **78**, 47–65.

Mosteller, F., and Tukey, J.W. (1968). Data analysis, including statistics. *in* G. Lindzey, and E. Aronson (eds.) *Handbook of Social Psychology*, Vol. 2. Addison-Wesley, Reading, MA.

Myers, R.A., Barrowman N.J., Hutchings, J.A., and Rosenberg, A.A. (1995). Population dynamics of exploited fish stocks at low populations levels. *Science* **269**, 1106–1108.

Naik, P.A., and Tsai, C-L. (2001). Single-index model selections. *Biometrika* **88**, 821–832.

Nester, M. (1996). An applied statistician's creed. *Applied Statistics* **45**, 401–410.

Newman, K. (1997). Bayesian averaging of generalized linear models for passive integrated transponder tag recoveries from salmonids in the Snake River. *North American Journal of Fisheries Management* **17**, 362–377.

Nichols, J.D., and Kendall, W.L. (1995). The use of multi-strata capture–recapture models to address questions in evolutionary ecology. *Journal of Applied Statistics* **22**, 835–846.

Nishii, R. (1988). Maximum likelihood principle and model selection when the true model is unspecified. *Journal of Multivariate Analysis* **27**, 392–403.

Noda, K., Miyaoka, E., and Itoh, M. (1996). On bias correction of the Akaike information criterion in linear models. *Communications in Statistics—Theory and Methods* **25**, 1845–1857.

Norris, J.L., and Pollock, K.H. (1995). A capture–recapture model with heterogeneity and behavioural response. *Environmental and Ecological Statistics* **2**, 305–313.

Norris, J.L., and Pollock, K.H. (1997). Including model uncertainty in estimating variances in multiple capture studies. *Environmental and Ecological Statistics* **3**, 235–244.

O'Connor, M.P., and Spotila, J.R. (1992). Consider a spherical lizard: animals, models, and approximations. *American Zoologist* **32**, 179–193.

O'Connor, R.J. (2000). Why ecology lags behind biology. *The Scientist* **14**, 35.

O'Hagan, A. (1995). Fractional Bayes factors for model comparison (with discussion). *Journal of the Royal Statistical Society*, Series B **57**, 99–138.

Olden, J.D., and Jackson, D.A. (2000). Torturing data for the sake of generality: how valid are our regression models? *Ecoscience* **7**, 501–510.

Otis, D.L., Burnham, K.P., White, G.C., and Anderson, D.R. (1978). Statistical inference from capture data on closed animal populations. *Wildlife Monographs* **62**, 1–135.

Pan, W. (1999). Bootstrapping likelihood for model selection with small samples. *Journal of Computational and Graphical Statistics* **8**, 687–698.

Pan, W. (2001*a*). Akaike's information criterion in generalized estimating equations. *Biometrics* **57**, 120–125.

Pan, W. (2001*b*). Model selection in estimating equations. *Biometrics* **57**, 529–534.

Parzen, E. (1994). Hirotugu Akaike, statistical scientist. Pages 25–32 *in* H. Bozdogan (ed.) *Engineering and Scientific Applications*. Vol. 1, Proceedings of the First US/Japan Conference on the Frontiers of Statistical Modeling: An Informational Approach. Kluwer Academic Publishers, Dordrecht, the Netherlands.

Parzen, E., Tanabe, K., and Kitagawa, G. (eds.) (1998). *Selected papers of Hirotugu Akaike*. Springer-Verlag Inc., New York, NY.

Pascual, M.A., and Kareiva, P. (1996). Predicting the outcome of competition using experimental data: maximum likelihood and Bayesian approaches. *Ecology* **77**, 337–349.

Peirce, C.S. (1955). Abduction and induction. Pages 150–156 *in* J. Buchler (ed.), *Philosophical writings of Peirce*. Dover, New York, NY.

Penrose, K., Nelsom, A., and Fisher, A. (1985). Generalized body composition prediction equation for men using simple measurement techniques (abstract). *Medicine and Science in Sports and Exercise* **17**, 189.

Peters, R.H. (1991). *A critique for ecology.* Cambridge University Press, Cambridge, NY, USA.

Peterson, T.S. (1960). *Elements of calculus*(2nd ed.). Harper Brothers, New York, NY.

Platt, J.R. (1964). Strong inference. *Science* **146**, 347–353.

Pollock, K.H., Nichols, J.D., Brownie, C., and Hines, J.E. (1990). *Statistical inference for capture–recapture experiments. Wildlife Monographs.* **107**, 1–97.

Pope, S.E., Fahrig, L., and Merriam, H.G. (2000). Landscape complementation and metapopulation effects on leopard frog populations. *Ecology* **81**, 2498–2508.

Posada, D., and Crandall, K. (1998). MODELTEST: testing the model of DNA substitution. *Bioinformatics* **14**, 817–818.

Posada, D., and Crandall, K. (2001). Selecting models of nucleotide substitution: an application of human immunodeficiency virus 1 (HIV-1). *Molecular Biology and Evolution* **18**, 897–906.

Poskitt, D.S., and Tremayne A.R. (1987). Determining a portfolio of linear time series models. *Biometrika* **74**, 125–137.

Pötscher, B.M. (1991). Effects of model selection on inference. *Econometric Theory* **7**, 163–185.

Qian, G., Gabor, G., and Gupta, R.P. (1996). Generalized linear model selection by the predictive least quasi-deviance criterion. *Biometrika* **83**, 41–54.

Qin, J., and Lawless, G. (1994). Empirical likelihood and general estimating equations. *Annals of Statistics* **22**, 300–325.

Quinn, J.F., and Dunham, A.E. (1983). On hypothesis testing in ecology and evolution. *American Naturalist* **122**, 22–37.

Raftery, A.E. (1995). Bayesian model selection in social research (with discussion). *Sociological Methodology* **25**, 111–195.

Raftery, A. (1996a). Approximate Bayes factors and accounting for model uncertainty in generalized linear regression models. *Biometrika* **83**, 251–266.

Raftery, A. (1996b). Hypothesis testing and model selection. Pages 163–187 *in* W.R. Gilks, S. Richardson, and D.J. Spiegelhalter (eds.), *Markov chain Monte Carlo in practice.* Chapman and Hall, London.

Raftery, A., Madigan, D.M., and Hoeting, J. (1993). *Model selection and accounting for model uncertainty in linear regression models.* Technical Report No. 262, Department of Statistics, University of Washington, Seattle, WA.

Raftery, A.E., Madigan, D., and Hoeting, J.A. (1997). Bayesian model averaging for linear regression models. *Journal of the American Statistical Association* **92**, 179–191.

Rawlings, J.O. (1988). *Applied regression analysis: a research tool.* Wadsworth, Inc., Belmont, CA.

Rencher, A.C., and Pun, F.C. (1980). Inflation of R^2 in best subset regression. *Technometrics* **22**, 49–53.

Renshaw, E. (1991). *Modelling biological populations in space and time.* Cambridge University Press, Cambridge, UK.

Reschenhofer, E. (1996). Prediction with vague prior knowledge. *Communications in Statistics—Theory and Methods* **25**, 601–608.

Reschenhoffer, E. (1999). Improved estimation of the expected Kullback-Leibler discrepancy in case of misspecification. *Econometric Theory* **15**, 377–387.

Rexstad, E. (2001). Back cover of T.M. Shenk and A.B. Franklin, (eds.) (2001). *Modeling in natural resource management.* Island Press, Washington, D.C.

Rexstad, E.A., Miller, D.D., Flather, C.H., Anderson, E.M., Hupp, J.W., and Anderson, D.R. (1988). Questionable multivariate statistical inference in wildlife habitat and community studies. *Journal of Wildlife Management* **52**, 794–798.

Rexstad, E.A., Miller, D.D., Flather, C.H., Anderson, E.M., Hupp, J.W., and Anderson, D.R. (1990). Questionable multivariate statistical inference in wildlife habitat and community studies: a reply. *Journal of Wildlife Management* **54**, 189–193.

Ripley, B.D. (1996). *Pattern recognition and neural networks.* Cambridge University Press, Cambridge, UK.

Rissanen, J. (1989). *Stochastic complexity in statistical inquiry.* World Scientific, Series in Computer Science, Vol 15. Singapore.

Rissanen, J. (1996). Fisher information and stochastic complexity. *IEEE Transactions on Information Theory* **42**, 40–47.

Robert, C.P., and Casella, G. (1999). Monte Carlo statistical methods. Springer-Verlag, New York, NY.

Roecker, E.B. (1991). Prediction error and its estimation for subset-selected models. *Technometrics* **33**, 459–468.

Ronchetti, E., and Staudte, R.G. (1994). A robust version of Mallows' C_p. *Journal of the American Statistical Association* **89**, 550–559.

Rosenblum, E.P. (1994). A simulation study of information theoretic techniques and classical hypothesis tests in one factor ANOVA. Pages 319–346 *in* H. Bozdogan (ed.) *Engineering and Scientific Applications.* Vol. 2, Proceedings of the First US/Japan Conference on the Frontiers of Statistical Modeling: An Informational Approach. Kluwer Academic Publishers, Dordrecht, the Netherlands.

Roughgarden, J. (1979). *Theory of population genetics and evolutionary ecology: an introduction.* Macmillan Publishing Company, New York, NY.

Royall, R.M. (1997). *Statistical evidence: a likelihood paradigm.* Chapman and Hall, London.

Royle, J.A., and Link, W.A. (2002). Random effects and shrinkage estimation in capture–recapture methods. *Journal of Applied Statistics* **29**, 329–351.

Sakamoto, Y. (1982). Efficient use of Akaike's information criterion for model selection in a contingency table analysis. *Metron* **40**, 257–275.

Sakamoto, Y. (1991). *Categorical data analysis by AIC*. KTK Scientific Publishers, Tokyo, Japan.

Sakamoto, Y., and Akaike, H. (1978). Analysis of cross classified data by AIC. *Annals of the Institute of Statistical Mathematics Part B* **30**, 185–197.

Sakamoto, Y., Ishiguro, M., and Kitagawa, G. (1986). *Akaike information criterion statistics*. KTK Scientific Publishers, Tokyo, Japan.

Santer, T.J., and Duffy, D.E. (1989). *The statistical analysis of discrete data*. Springer-Verlag, New York, NY.

SAS Institute Inc. (1985). SAS® language guide for personal computers, Version 6 Edition. SAS Institute Inc., Cary, NC.

SAS Institute. (1988). SAS/STAT®user's guide. Edition 6.03. SAS Institute, Cary, NC.

Sauerbrei, W., and Schumacher, M. (1992). A bootstrap resampling procedure for model building: application to the Cox regression model. *Statistics in Medicine* **11**, 2093–2109.

Sawa, T. (1978). Information criteria for discriminating among alternative regression models. *Econometrica* **46**, 1273–1291.

Scheiner, S.M., and Gurevitch, J. (eds.) (1993). *Design and analysis of ecological experiments*. Chapman and Hall, London.

Schmidt, B.R., and Anholt, B.R. (1999). Analysis of survival probabilities of female common toads. *Amphibia-Reptilia* **20**, 97–108.

Schoener, T.W. (1970). Nonsynchronous spatial overlap of lizards in patchy habitats. *Ecology* **51**, 408–418.

Schreuder, H.T., Gregoire, T.G., and Wood, G.B. (1993). *Sampling methods for multiresource forest inventory*. John Wiley and Sons, New York, NY.

Schwarz, G. (1978). Estimating the dimension of a model. *Annals of Statistics* **6**, 461–464.

Sclove, S.L. (1987). Application of some model-selection criteria to some problems in multivariate analysis. *Psychometrika* **52**, 333–343.

Sclove, S.L. (1994a). Small-sample and large-sample statistical model selection criteria. Pages 31–39 *in* P. Cheeseman, and R.W. Oldford (eds.) *Selecting models from data*. Springer-Verlag, New York, NY.

Sclove, S.L. (1994b). Some aspects of model-selection criteria. Pages 37–67 *in* H. Bozdogan (ed.) *Engineering and Scientific Applications*. Vol. 2. Proceedings of the First US/Japan Conference on the Frontiers of Statistical Modeling: An Informational Approach. Kluwer Academic Publishers, Dordrecht, the Netherlands.

Seber, G.A.F. (1977). *Linear regression analysis*. John Wiley and Sons, New York, NY.

Seber, G.A.F. (1984). *Multivariate observations*. John Wiley and Sons, New York, NY.

Seber, G.A.F., and Wild, C.J. (1989). *Nonlinear regression*. John Wiley and Sons, New York, NY.

Sellke, M.J., Bayarri, J., and Berger, J.O. (2001). Calibration of *p* values for testing precise null hypotheses. *The American Statistician* **55**, 62–71.

Severini, T.A. (2000). *Likelihood methods in statistics*. Oxford University Press, Oxford, UK.

Shannon, C.E. (1948). A mathematical theory of communication. *Bell System Technical Journal* **27**, 379–423 and 623–656.

Shao, J. (1993). Linear model selection by cross-validation. *Journal American Statistician Association* **88**, 486–494.

Shao, J. (1996). Bootstrap model selection. *Journal of the American Statistical Association* **91**, 655–665.

Shao, J. (1997). An asymptotic theory for linear model selection. *Statistica Sinica* **7**, 221-264.

Shao, J, and Tu, D. (1995). *The jackknife and bootstrap*. Springer-Verlag, New York, NY.

Shefferson, R.P., Sandercock, B.K., Proper, J., and Beissinger, S.R. (2000). Estimating dormancy and survival of a rare herbaceous perennial using mark-recapture models. *Ecology* **82**, 145–156.

Shenk, T.M., and Franklin, A.B., (eds.) (2001). *Modeling in natural resource management*. Island Press, Washington, D. C.

Shi, P., and Tsai, C-L. (1998). A note on the unification of the Akaike information criterion. *Journal of the Royal Statistical Society*, Series B, **60**, 551–558.

Shi, P., and Tsai, C-L. (1999). Semiparametric regression and model selection. *Journal of Statistical Planning and Inference* **77**, 341–349.

Shibata, R. (1976). Selection of the order of an autoregressive model by Akaike's information criterion. *Biometrika* **63**, 117–26.

Shibata, R. (1980). Asymptotically efficient selection of the order of the model for estimating parameters of a linear process. *Annals of Statistics* **8**, 147–164.

Shibata, R. (1981). An optimal selection of regression variables. *Biometrika* **68**, 45–54. Correction (1982). **69**, 492.

Shibata, R. (1983). A theoretical view of the use of AIC. Pages 237–244 *in* O.D. Anderson (ed.) *Time series analysis: theory and practice*. Elsevier Science Publication, North-Holland, the Netherlands.

Shibata, R. (1986). Consistency of model selection and parameter estimation. Pages 127–141 *in* J. Gani, and M.B. Priestly (eds.) *Essays in time series and allied processes. Journal of Applied Probability*, Special Volume 23A.

Shibata, R. (1989). Statistical aspects of model selection. Pages 215–240 *in* J.C. Willems (ed.) *From data to model*. Springer-Verlag, London.

Shibata, R. (1997a). Bootstrap estimate of Kullback–Leibler information for model selection. *Statistica Sinica* **7**, 375–394.

Shibata, R. (1997b). Discrete models selection. Pages D.20–D.29 *in* K.T. Fang, and F.J. Hickernell (eds.) *Contemporary multivariate analysis and its applications.* Hong Kong Baptist University, Hong Kong, China.

Shimizu, R. (1978). Entropy maximization principle and selection of the order of an autoregressive Gaussian process. *Annals of the Institute of Statistical Mathematics* **30**, 263–270.

Shono, H. (2000). Efficiency of the finite correction of Akaike's information criteria. *Fisheries Science* **66**, 608–610.

Silverman, B.W. (1982). Algorithm AS 176: kernel density estimation using the fast Fourier transform. *Applied Statistics* **31**, 93–99.

Silvey, S.D. (1975). *Statistical inference.* Chapman and Hall, London.

Simonoff, J.S., and Tsai, C-L. (1999). Semiparametric and additive model selection using an improved AIC criterion. *Journal of Computational and Graphical Statistics* **8**, 22-40.

Skalski, J.R., Hoffman, A., and Smith, S.G. (1993). Testing the significance of individual- and cohort-level covariates in animal survival studies. Pages 9–28 *in* J.-D. Lebreton and P.M. North (eds.) *Marked individuals in the study of bird population.* Birkhäuser-Verlag, Basel, Switzerland.

Skalski, J.R., and Robson, D.S. (1992). *Techniques for wildlife investigations: design and analysis of capture data.* Academic Press, New York, NY.

Smith, G.N. (1966). Basic studies on DURSBAN® insecticide. *Down to Earth* **22**, 3–7.

Smith, S.C., Skalski, J.R., Schlechte, J.W., Hoffman, A., and Cassen, V. (1994). *SURPH.1 Statistical Survival Analysis of Fish and Wildlife Tagging Studies.* Centre for Quantitative Sciences, University of Washington, Seattle, WA.

Sober, E. (1999). Instrumentalism revisited. *Crítica* **31**, 3–39.

Sober, E. (2001). Instrumentalism, parsimony and the Akaike framework. *Proceedings of the Philosophy of Science Association* (in press).

Sommer, S., and Huggins, R.M. (1996). Variables selection using the Wald test and a robust C_p. *Applied Statistics* **45**, 15–29.

Soofi, E.S. (1994). Capturing the intangible concept of information. *Journal of the American Statistical Association* **89**, 1243–1254.

Southwell, C. (1994). Evaluation of walked line transect counts for estimating macropod density. *Journal of Wildlife Management* **58**, 348–356.

Speed, T.P., and Yu, B. (1993). Model selection and prediction: normal regression. *Annals of the Institute of Statistical Mathematics* **1**, 35–54.

Spiegelhalter, D.J., Best, N.G., and Carlin, B.P., and van der Linde, A. (2002). Bayesian measures of model complexity and fit. *Journal of the Royal Statistical Society*, Series B **64**, 1–34.

Sprott, D.A. (2000). *Statistical inference in science.* Springer Series in Statistics, London.

Steel, M., and Penny, D. (2001). Parsimony, likelihood and the role of models in molecular phylogenetics. *Molecular Biology and Evolution* **17**, 839–850.

Stein, A., and Corsten, L.C.A. (1991). Universal kriging and cokriging as a regression procedure. *Biometrics* **47**, 575–587.

Sterling, T.D., Rosenbaum, W.L., and Weinkam, J.J. (1995). Publication decisions revisited: the effect of the outcome of statistical tests on the decision to publish and vice versa. *The American Statistician* **49**, 108–112.

Stewart-Oaten, A. (1995). Rules and judgments in statistics: three examples. *Ecology* **76**, 2001–2009.

Stigler, S.M. (1986), *The history of statistics*. Harvard University Press, Cambridge, MA.

Stoica, P., Eykhoff, P., Janssen, P, and Söderström, T. (1986). Model-structure selection by cross-validation. *International Journal of Control* **43**, 1841–1878.

Stone, M. (1974). Cross-validatory choice and assessment of statistical predictions (with discussion). *Journal of the Royal Statistical Society*, Series B **39**, 111–147.

Stone, M. (1977). An asymptotic equivalence of choice of model by cross-validation and Akaike's criterion. *Journal of the Royal Statistical Society*, Series B **39**, 44–47.

Stone, C.J. (1982). Local asymptotic admissibility of a generalization of Akaike's model selection rule. *Annals of the Institute of Statistical Mathematics* Part A **34**, 123–133.

Stone, M., and Brooks, R.J. (1990). Continuum regression: cross-validated sequentially constructed prediction embracing ordinary least squares, partial least squares and principal components regression (with discussion). *Journal of the Royal Statistical Society*, Series B **52**, 237–269.

Stromborg, K.L., Grue, C.E., Nichols, J.D., Hepp, G.R., Hines, J.E., and Bourne, H.C. (1988). Postfledging survival of European starlings exposed to an organophosphorus insecticide. *Ecology* **69**, 590–601.

Sugiura, N. (1978). Further analysis of the data by Akaike's information criterion and the finite corrections. *Communications in Statistics, Theory and Methods* **A7**, 13–26.

Takeuchi, K. (1976). Distribution of informational statistics and a criterion of model fitting. *Suri-Kagaku* (Mathematic Sciences) **153**, 12–18. (In Japanese).

Taub, F.B. (1993). Book review: Estimating ecological risks. *Ecology* **74**, 1290–1291.

Taubes, G. (1995). Epidemiology faces its limits. *Science* **269**, 164–169.

Thabane, L., and Haq, M.S. (1999). On Bayesian selection of the best normal population using the Kullback–Leibler divergence measure. *Statistica Neerlandica* **53**, 342–360.

Thompson, S.K. (1992). *Sampling*. Wiley, New York, NY.

Thompson, M.E. (1997). *Theory of sample surveys*. Chapman and Hall, London.

Thompson, W.L., and Lee, D.C. (2000). Modeling relationships between landscape-level attributes and snorkel counts of chinook salmon and steelhead parr in Idaho. *Canadian Journal of Fisheries and Aquatic Sciences* **57**, 1834–1842.

Tibshirani, R. (1996). Regression shrinkage and selection via the lasso. *Journal of the Royal Statistical Society*, Series B **58**, 267–288.

Tjur, T. (1998). Nonlinear regression, quasi likelihood, and overdispersion in generalized linear models. *The American Statistician* **52**, 222–227.

Tong, H. (1994). Akaike's approach can yield consistent order determination. Pages 93–103 *in* H. Bozdogan (ed.) *Engineering and Scientific Applications*. Vol. 1, Proceedings of the First US/Japan Conference on the Frontiers of Statistical Modeling: An Informational Approach. Kluwer Academic Publishers, Dordrecht, the Netherlands.

Tukey, J.W. (1980). We need both exploratory and confirmatory. *The American Statistician* **34**, 23–25.

Turchin, P., and Batzli, G.O. (2001). Availability of food and the population dynamics of arvicoline rodents. *Ecology* **82**, 1521–1534.

Ullah, A. (1996). Entropy, divergence and distance measures with econometric applications. *Journal of Statistical Planning and Inference* **49**, 137–162.

Umbach, D.M., and Wilcox, A.J. (1996). A technique for measuring epidemiologically useful features of birthweight distributions. *Statistics in Medicine* **15**, 1333–1348.

Venter, J.H., and Snyman, J.L.J. (1995). A note on the generalized cross-validation criterion in linear model selection. *Biometrika* **82**, 215–219.

Ver Hoef, J.M. (1996). Parametric empirical Bayes methods for ecological applications. *Ecological Applications* **6**, 1047–1055.

Wade, P.R. (2000). Bayesian methods in conservation biology. *Conservation Biology* **14**, 1308–1316.

Walters, C.J. (1996). Computers and the future of fisheries. Pages 223–238 *in* B. A. Megrey, and E. Mcksness (eds.) *Computers in fisheries research*. Chapman and Hall, London.

Wang, C. (1993). *Sense and nonsense of statistical inference: controversy, misuse, and subtlety*. Marcel Dekker, Inc., New York, NY.

Wang, P., Puterman, M. L., Cockburn, I., and Le, N. (1996). Mixed Poisson regression models with covariate dependent rates. *Biometrics* **52**, 381–400.

Wasserman, L. (2000). Bayesian model selection and model averaging. *Journal of Mathematical Psychology* **44**, 92–107.

Wedderburn, R.W.M. (1974). Quasi-likelihood functions, generalized linear models, and the Gauss–Newton method. *Biometrika* **61**, 439–447.

Wehrl, A. (1978). General properties of entropy. *Reviews of Modern Physics* **50**, 221–260.

Weiner, J. (1995). On the practice of ecology. *Journal of Ecology* **83**, 153–158.

Weisberg, S. (1985). *Applied linear regression*. 2nd ed. Wiley, New York, NY.

Wel, J. (1975). Least squares fitting of an elephant. *Chemtech* Feb. 128–129.

Westfall, P.H., and Young, S.S. (1993). *Resampling-based multiple testing: examples and methods for p-value adjustment*. John Wiley and Sons, New York, NY.

White, G.C. (1983). Numerical estimation of survival rates from band-recovery and biotelemetry data. *Journal of Wildlife Management* **47**, 716–728.

White, G.C. (2000). Population viability analysis: data requirements and essential analysis. Pages 288-331 *in* L. Boitani and T. K. Fuller (eds.), *Research Techniques in Animal Ecology: Controversies and Consequences.* Columbia University Press, New York, NY.

White, G.C., Anderson, D.R., Burnham, K.P., and Otis, D.L. (1982). *Capture-recapture and removal methods for sampling closed populations.* Los Alamos National Laboratory, LA-8787-NERP, Los Alamos, NM.

White, G.C., and Burnham, K.P. (1999). Program MARK-survival estimation from populations of marked animals. *Bird Study* **46**.

White, G.C., Burnham, K.P., and Anderson, D.R. (2001). Advanced features of program MARK. *in* R. Fields (ed.) *Integrating People and Wildlife for a Sustainable Future*, Proceedings of the Second International Wildlife Management Congress. The Wildlife Society, Bethesda, MD.

White, H. (1994). *Estimation, inference and specification analysis.* Cambridge University Press, Cambridge, UK.

White, H. (2000). A reality check for data snooping. *Econometrica* **68**, 1097–1126.

Williams, B.K., Nichols, J.D., and Conroy, M.J. (2002). *Analysis and management of animal populations: modeling, estimation, and decision making.* Academic Press, San Diego, CA.

Williams, D.A. (1982). Extra-binomial variation in logistic linear models. *Applied Statistics* **31**, 144–148.

Wood, S.N. and Thomas, M.B. (1999). Super-sensitivity to structure in biological models. *Proceedings of the Royal Society* **266**, 565–570.

Woods, H., Steinour, H.H., and Starke, H.R. (1932). Effect of composition of Portland cement on heat evolved during hardening. *Industrial and Engineering Chemistry* **24**, 1207–1214.

Ye, J. (1998). On measuring and correcting the effects of data mining and model selection. *Journal of the American Statistical Association* **93**, 120–131.

Yockey, H.P. (1992). *Information theory and molecular biology.* Cambridge University Press.

Yoccoz, N.G. (1991). Use, overuse, and misuse of significance tests in evolutionary biology and ecology. *Bulletin of the Ecological Society of America* **72**, 106–111.

Young, L.J., and Young, J.H. (1998). *Statistical ecology.* Kluwer Academic Publishers, London, UK.

Yu, B. (1996). Minimum description length principle: a review. In *Proceedings of the 30th Conference on Information Sciences and Systems*, Princeton University, NJ.

Yu, B. (1999). Coding and model selection: a brief tutorial on the principle of minimum description length. *Statistical Computing and Graphics* **9**, 1, 27–32.

Zablan, M.A. (1993). Evaluation of sage grouse banding program in North Park, Colorado. M.S. thesis, Colorado State University, Fort Collins, CO.

Zeger, S.L., and Karim, M.R. (1991). Generalized linear models with random effects: a Gibbs sampling approach. *Journal of the American Statistical Association* **86**, 79–86.

Zellner, A., Keuzenkamp, H. A., and McAleer, M., (eds.) (2001). *Simplicity, inference and modelling: keeping it sophisticatedly simple*. Cambridge University Press, Cambridge, UK.

Zhang, P. (1992a). Inferences after variable selection in linear regression models. *Biometrika* **79**, 741–746.

Zhang, P. (1992b). On the distributional properties of model selection criteria. *Journal of the American Statistical Association* **87**, 732–737.

Zhang, P. (1993a). Model selection via multifold cross-validation. *Annals of Statistics* **20**, 299–313.

Zhang, P. (1993b). On the convergence rate of model selection criteria. *Communications in Statistics*, Part A—Theory and Methods **22**, 2765–2775.

Zhang, P. (1994). On the choice of penalty term in generalized FPE criterion. Pages 41–49 *in* P. Cheeseman, and R.W. Oldford (eds.) *Selecting models from data*. Springer-Verlag, New York, NY.

Zucchini, W. (2000). An introduction to model selection. *Journal of Mathematical Psychology* **44**, 41–61.

Index

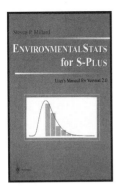

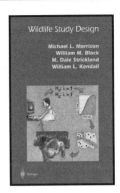